PENGUIN REFERENCE

THE PENGUIN DICTIONARY OF BUILDING

Born in 1915, John S. Scott was a chartered structural and mining engineer, a certificated colliery manager and a qualified linguist. He worked in Austria, France, Germany, Saudi Arabia and the UK. He spent ten years in civil engineering and then went on to write and translate from French, German and Russian. He is also the author of *The Penguin Dictionary of Civil Engineering*. John S. Scott died in 1997.

James H. MacLean was born in 1942 in Melbourne and has ten years' site experience, lastly as a site manager for Costain Construction in London. He worked in Paris for many years and wrote two dictionaries for translating building terms between French and English, which led him to meet John Scott. His dissertation for a masters degree in building, taken in Melbourne, was based on these two dictionaries. He now lives on Flinders Island in Tasmania.

THE PENGUIN DICTIONARY OF
BUILDING

JAMES H. MACLEAN AND
JOHN S. SCOTT

FOURTH EDITION

Cross-referenced to
The Penguin Dictionary of Civil Engineering

PENGUIN BOOKS

PENGUIN BOOKS

Published by the Penguin Group
Penguin Books Ltd, 27 Wrights Lane, London w8 5tz, England
Penguin Putnam Inc., 375 Hudson Street, New York, New York 10014, USA
Penguin Books Australia Ltd, Ringwood, Victoria, Australia
Penguin Books Canada Ltd, 10 Alcorn Avenue, Toronto, Ontario, Canada m4v 3b2
Penguin Books (NZ) Ltd, Private Bag 102902, NSMC, Auckland, New Zealand

Penguin Books Ltd, Registered Offices: Harmondsworth, Middlesex, England

First published 1964
Second edition 1974
Reprinted with revisions 1979
Third edition 1984
Fourth edition 1993
Reprinted with minor revisions 1995

8

Copyright © John S. Scott, 1964, 1974, 1979, 1984
Copyright © James H. MacLean and John S. Scott, 1993, 1995
All rights reserved

The moral right of the authors has been asserted

Set in 8½/9½ pt Monophoto Imprint
Typeset by Datix International Limited, Bungay, Suffolk
Printed in England by Clays Ltd, St Ives plc

CONTENTS

PREFACE TO THE FOURTH EDITION

Because this book cannot conveniently exceed its present size and because there are so many building terms, we have two volumes, this one and *The Penguin Dictionary of Civil Engineering*, referred to in cross-references by the letter *C*. We are sorry for this unavoidable annoyance. The reader may also find *The Penguin Dictionary of Architecture* and *The Penguin Dictionary of Science* useful.

The term building covers the construction of houses, offices, shops, factories, etc., their use, and the operation of their mechanical and electrical services. House building differs from commercial and industrial building, although many methods are used in both. Innovation in building materials brings new terms, but there are also terms with such a long tradition that they cannot be left out, particularly as renovation is very much part of the building industry today.

Different names are often used for the same thing and, in common speech, part of a term may be omitted for economy of language or where the meaning is clear from the context. The explanation of each term is given only once, with its alternative terms, or at least some of them. At the alphabetical location of the less-used term the entry simply gives a cross-reference (indicated by the use of italic type) to the most commonly used one.

The Penguin Dictionary of Building gives references to British Standards (BS), European Standards (EN), and Building Research Establishment publications (BRE) for further information on particular subjects. Lists of other equivalent standards are usually available from any national standard body.

Useful annual publications include building price books, *Specification* and, one of the most essential, the British Standards Institution Catalogue, the key to the encyclopaedia of information found in the ten thousand or so British Standards. The BSI Catalogue has over a thousand pages and lists all BSI publications, including European Standards. Database services include NormImage (full text) and Perinorm (bibliographical records), both on CD–ROM compact disc, and BSI's Standard Line, accessed through the Pergamon host computer. A few local reference libraries still hold BSs in hard copy, but most have gone to microfiche. BSI Handbook No. 3 summarizes one thousand five hundred British Standards concerned with building and is published in four loose-leaf binders. BS 6100 is a vast document that covers all standardized terms in building and civil engineering. It is organized into separate chapters, with an easy-to-use numbering system. British Standards describe only accepted practice: they cannot describe ultra-modern techniques because that might inhibit the development of these techniques.

Her Majesty's Stationery Office's free, well-indexed, sectional list (Building) gives the titles of Building Research Establishment publications, including the monthly BRE Digests.

The Construction Industry Research and Information Association (CIRIA) green book gives outlines of hundreds of information organizations and sources, with addresses and phone/fax numbers.

The Health and Safety Executive provides information and guidance on regulations relating to construction work.

ACKNOWLEDGEMENTS

John Scott wrote the first three editions of *The Penguin Dictionary of Building*. This fourth edition is a joint work with James MacLean, who wishes to acknowledge help given by the following people and organizations: Md Najib Ibrahim and Dr Tadj Oreszczyn of the Bartlett School of Architecture, Tony Stowe of Blockley's Bricks, Christopher Powell of the Brick Development Association, the staff of Bromley reference library, the Building Bookshop, the Chartered Institute of Building, the Consumers' Association, John Weir of the Glass and Glazing Federation, Patrick Hanks, the Health and Safety Executive headquarters library, the Heating and Ventilation Contractors' Association, Paul Grimwood of the London Fire Brigade, M & H plant hire, Alan Maddison, Philip Maton, Rosamund Moon, Pilkington Glass advisory service, Prof. Juan Sager of UMIST on languages for special purposes, Dr Matthew Smith and Alan Williams of *Specification*, M. T. Sherratt of Taunton, and the Zinc Development Association.

In Australia thanks go to Harry Marrayatt of Wangaratta on fire protection, Dr Miles Lewis, Julian Nance and Peter Williams of Melbourne, and Colleen Patterson of Flinders Island, Tasmania.

Information from British Standards Institution publications is reproduced by permission of the BSI, 2 Park Road, London W1A 2BS.

IMPERIAL-TO-METRIC (MAINLY SI)
CONVERSION FACTORS AND
ABBREVIATIONS OF UNITS

The building industry in Britain and Europe (and many other places) uses the metric system. To make the text easier to read, almost all measurements of length in *The Penguin Dictionary of Building* are given in millimetres only, without conversion to feet and inches.

The five hundred pages of BS 350 (*Conversion Factors and Tables*) are the best source of conversion factors, but those given below are the few that builders really need. Others are printed in *The Penguin Dictionary of Civil Engineering*.

Exact conversions are given in bold type.

Length   **25.4** millimetres (mm) = 1 in.
101.6 mm = 4 in.
304.8 mm = 1 ft.
SWG to millimetres, see the entry **Standard wire gauge**.

Area   1 in^2 = **6.4516** cm^2 = **645.16** mm^2.
1 yd^2 = 0.8361 m^2.
10.764 ft^2 = 1 m^2.

Volume   1 US wet gallon (US gal) = 3.785 litres.
1 m^3 = 1.308 cubic yards or 35.315 cubic feet.

Temperature   Freezing point is 0 ° Celsius (C) or 32° Fahrenheit (F) and
each 5 °C = 9 °F.

Weight, mass   1 kg = 2.2046 lb.
1 tonne (1000 kg) = 2204.6 lb.
1 US ton (2000 lb) = 907.2 kg.

Force   1 pound force (lb f) = 4.448 newton (N)
1 N = 0.2248 lb f.
1 kilogram force (kg f) = 2.2046 lb f = 9.807 N.

Pressure, stress   1 bar (100 kPa) = 14.5036 lb per in^2 (psi).
1 standard atmosphere = 1.01325 bar = 1013.25 hPa
(millibars) = 101.325 kPa.
1 psi = 6895 N/m^2 = 6.895 kPa.
1000 psi = 6.895 N/m^2.

Heat	1 British thermal unit (Btu) = 1.055 kilojoule (kJ).
	1 Btu/hour = 0.293 J/second = 0.293 watt (W).
	1 ton of refrigeration = 3024 calories.
	The French unit frigorie = − 1 calorie.

K-value	1 Btu/in.ft².hour.degree F = 0.144 watt/metre.degree C.

U-value	1 Btu/ft².hour.degree F = 5.678 watt/m².degree C.

Metric multipliers	tera	T	1 000 000 000 000	pico	p	1/1000 000 000 000
	giga	G	1 000 000 000	nano	n	1/1000 000 000
	mega	M	1 000 000	micro	μ	1/1000 000
	kilo	k	1 000	milli	m	1/1000
	hecto	h	100	centi	c	1/100
	deca	da	10	deci	d	1/10

Cross-references are indicated by the use of italic or heavy type. The letter (*C*) shows that the entry referred to is in *The Penguin Dictionary of Civil Engineering*.

Authorities

The Penguin Dictionary of Building could not have been compiled without the official guidance of the following bodies:

ASHVE   American Society of Heating and Ventilation Engineers
ASTM   American Society for Testing and Materials
BRE   Building Research Establishment
BSI   British Standards Institution
CEN   European Committee for Standardization

A

ablution fitting A large *sanitary fitting* in which several people at the same time can wash their hands, arms, or faces. Ablution fittings are made for rough use and have water sprayers, often with *touchless controls* or foot controls. Trough types can be set in ranges, while washing fountains have a circular pedestal bowl.

above ground level (USA **a. grade**) Higher than ground level, particularly with reference to the *superstructure*, and to work after a building is *out of the ground*.

abrasion resistance The ability of a *finish* to stand up to wear from rubbing, e.g. the wear of paint by fingers.

abrasives Rough, hard materials (usually powders, grits, or stones) in various forms such as *sandpaper, grinding wheels*, or *grinding discs*, used to smooth or clean surfaces or to sharpen and hone *edge tools*. From hard to soft they include: diamond, garnet, corundum (emery), carborundum, powdered glass, and silica sand. The hardest abrasives cause others to wear the most while lasting the longest themselves. Coarse abrasives remove the most material, fine abrasives give the smoothest surface.

ABS *Acrylonitrile butadiene styrene.*

absorption (1) The taking-up of a liquid or gas by a porous body. In Britain the absorption of *bricks* is tested by boiling in water for five hours. It was formerly used as an indication of *frost resistance*. The total absorption is not the *absorption rate*.

(2) The loss of energy into a surface in the form of sound, heat, or light. The total sound absorption (measured in *sabins*) within a room is increased by soft finishings, open windows, the presence of people, or *acoustic finishings*, all of which have a high coefficient of absorption. This reduces echo and reverberation, although it does not affect *transmission* into another room, except in *discontinuous construction*. Heat absorption depends on *absorptivity* and raises the temperature of a body. Light absorption makes a surface appear darker.

absorption rate, initial rate of a. (USA) The speed at which water is taken up by a brick when partly immersed for one minute, used to measure initial *suction*.

absorption refrigerator A refrigerator without a compressor which uses heat to evaporate lithium bromide from water. It is silent and not very efficient but can be run on waste heat.

absorptive form (USA) *Permanent formwork*.

absorptivity, absorptive power The relative rate of *absorption* of heat by a body compared with a similarly shaped black body in the same conditions. Compare **emissivity** and see **solar collector**.

abstracting Assembling data from *taking-off* so that similar types of work are brought together under one *item* ready for *billing*. It is the first part of the *working-up* of *bills of quantities*.

ABT (Association of Building Technicians) A specialist section of UCATT (Union of Construction,

Allied Trades and Technicians) that includes building design and supervisory staff, *architects, civil engineers* (*C*), *clerks of works* and *foremen*.

abutment An intersection between a roof slope and a wall that rises above it.

abutment flashing A *flashing* over an *abutment*. At the head of the slope it is level, at the side of a slope it is *raking* or *stepped*.

abutment piece (USA) A *sill* or sole plate.

accelerated weathering, artificial ageing The testing of materials by exposure to cycles of sunlight, heat, frost, and wetness or dryness that are more severe than in nature.

accelerator (1) accelerating admixture An *admixture* that hastens setting and increases the early strength of concrete, usually causing an increase in temperature. This allows earlier removal of formwork and may give *frost protection*. Accelerators containing *calcium chloride* are virtually banned in the UK for use in reinforced or prestressed concrete.
(2) Any catalyst, such as *driers* in paint, a curing agent, or a hardener.

acceptance (1) Taking over. Acceptance of materials at delivery, or of a building, or its installation at *handover*, usually involves a transfer of *responsibility*. It can be implied when somebody takes over and continues the work of another person, or one *trade* follows another. Acceptance tests are often made to check for compliance with the *specifications*.
(2) Consent. It may be the act of reaching agreement with somebody who has made an *offer*, forming a binding *contract* with rights and obligations.

access A way in or out, including any *circulation area* for walking (such as a stair, balcony, or ramp), vehicle access to a site (such as a road or *wayleave*), or even hand access for *inspection*. Special requirements apply to access for fire fighting and for the *disabled*.

access chamber A space big enough for a person to get into.

access cover, a. eye, inspection fitting (USA **cleanout**) A *pipe fitting* with a removable plate for inspection, testing, or *rodding*. Access covers should be reachable and surrounded by *working space*, in an area above flood level, and where water and filth will not do damage.

access floor, raised f. A floor above the structural *floor*, creating a space between them, for cables going to office work stations, to allow easy *wire management* for data, telephone, power and lighting. Access floors can be deep demountable *platform floors* or shallow access floors, laid on battens, with a cavity less than 100 mm high, and a grid of holes. Access floors are a building *element*.

access hole An opening large enough for a person to get through for work on installations, or for a tool to be passed through for making adjustments. Access holes normally have a *cover*, such as a panel or trap.

accessories Components used to fix, join, or connect building *elements* or *services*, including *hardware, fixings*, and roof *flashings*.

accessory box, a. enclosure A small housing in a wall to protect electrical accessories, usually of PVC, with preformed *popout* holes for *conduit*.

accident Accidents on construction sites in the UK in which anybody dies or is seriously injured, or is off work for more than three days, must be reported as soon as possible, usually by phone to the local office of the

Health and Safety Executive. Records must be kept on Form F2508 from HMSO Books. See **safety**.

accordion door A *folding door*.

accreditation An approval of one official body by a higher authority, e.g. for building products *certification*.

acetal resin Strong, low-priced synthetic *resin* made by polymerizing formaldehyde. Acetal plastics are used in plumbing fittings, particularly for threaded valve components and intricate shapes.

acoustic, acoustical Concerned with sound. Sound or noise in buildings is controlled either by reducing *transmission* between rooms by the method of construction, or by increasing *absorption* within a room, with *acoustic finishings*. In sound measurements the acoustical transmission factor is the amount of sound that passes through a type of construction, while the acoustical reduction factor is its reciprocal, implying the effectiveness of *sound insulation*.

acoustic board Wall and ceiling *acoustic finishings* made with a porous core of *insulating board* or *mineral-fibre tiles* and a surface that has many tiny round holes, or is fissured or rilled. It can be fixed as full boards or made into *acoustic tiles*. Painting can block the holes and reduce sound absorbency; careful spraying with many thin coats of *emulsion paint* reduces mainly the high-frequency sound absorption.

acoustic clip A *floor clip* with a flexible rubber pad to reduce *transmission* of impact sound through a timber *floating floor*.

acoustic construction The improvement of *sound insulation* by using thick walls, heavy materials of low stiffness, or discontinuous construction such as a *floating floor*, which reduce noise *transmission*.

acoustic finishings Materials inside which sound energy is lost by *absorption*. However, they usually have high sound *transmission*. See next entry.

acoustic plaster Gypsum *plaster* with lightweight aggregate, used as an *acoustic finishing*.

acoustic tile A square of *acoustic board*.

acrylic adhesive A fast-curing *emulsion* adhesive, usually white in colour and fairly non-staining. It is used wet, but goes tacky quickly, and can hold down curling edges of vinyl floor tiles.

acrylic paints *Emulsion paints* which hold their colour well, are very durable and resistant to oils, fats and grease, and have sufficient resilience to avoid cracking. They are used for fast-drying *floor paint* and as *organic coatings* which are *stoved* on to metals.

acrylic resin (polymethyl methacrylate) The synthetic *resin* in acrylic paints, sealants, adhesives, and sheet. It has poor resistance to alkalis.

acrylic sealants *One-part* sealants that set to a tough rubbery material by *solvent release*. They have good adhesion to most *substrates* without the need for a primer, but should not be used in *wet areas* or in contact with cement-based products, which are alkaline.

acrylic sheet Fairly strong, lightweight, easily moulded *plastics*, mainly used for clear or tinted *glazing*, e.g. Perspex or Plexiglas, and coloured *sanitary fittings*, such as baths or basins, which usually have a backing layer of *glassfibre-reinforced polyester*. Acrylic sheet melts and burns easily, so that it is a *fire hazard* as well as having low *fire resistance*, although it can be broken to allow escape and, unlike glass, does not splinter into sharp pieces. Its surface is attacked by the caustic soda in some cleaning compounds.

acrylonitrile butadiene styrene (ABS) A tough copolymer, used to make *plastics pipes* for cold-water supply, or soil and waste drains. It has good impact resistance down to − 40°C, better than PVC. Pipes can have *solvent welded joints*, made with special ABS cement, or *push-fit joint* rings. Its softening point is higher than that of polypropylene, so that it can handle hot waste water up to 80°C. However, petrol, oil, and *linseed oil* can damage it. ABS pipe should be bent at about 125°C; no *bending spring* is needed. Like all plastics, ABS is *combustible*.

active fire protection Mechanical and electrical devices to warn of or extinguish fires, such as *sprinklers*, *alarms*, *fire detectors*, and fans or dampers for a *pressurized escape route*. They can be very effective but require regular inspection and maintenance, and may need standby power. See **passive fire protection**.

active leaf The leaf of a *double door* which is most used, held closed by latching to the *inactive leaf*.

active solar heating Heating systems that use *solar collectors* and mechanical devices such as pumps or fans. Compare **passive solar heating**.

activity Work shown on a *programme*, such as an *arrow diagram*, that has been worked out by the *critical path method*. Activities are described by *trade* and *element*, e.g. 'column casing', 'brickwork to external walls', 'lifts', 'bathroom tiling', etc. They take up time and resources, and have start and finish *events*. They can be critical, non-critical, splittable, non-splittable, or dummy.

act of God *Force majeure.*

actuator A mechanical linkage (such as a plunger or rod) to create or transmit movement such as pushing, pulling, or slight rotation, which enable *automatic controls* to control a flow of fluid (water or air) or the current in an electrical system.

adaptor A device which matches different objects, e.g. the connections between pipes or ducts of different sizes or sections, or between British electrical plugs and Continental ones.

additional work An increase in the amount of work in a contract becomes a *variation* to the contract and usually results in an *extra*.

addressable system, intelligent fire alarm A system with *fire detectors* that reply to a computer digital code, sending back a message giving their location and status. Analysis of the data gives the precise location of a fire.

adds Overall quantities from *taking off* that bring together all similar work before *deducts*, such as window openings, are subtracted.

adhesion The sticking together of components by making a chemical *bond*, using an *adhesive*, cement, or bonding compound. Timber parts are stuck with glue, roofing feltwork is bonded with heated asphalt, and Portland cement bonds strongly to steel. The *binder* in paints makes the dried *film* adhere to the *substrate* on which it is applied.

adhesive, glue A liquid that in hardening sticks things together by *adhesion*. It allows joints or fixings to be made without spoiling a surface with nails or screws. Careful selection of adhesives is vital to success. There are many different types, each suited to a particular job and to a variety of materials and service conditions. Problems of *compatibility* can also occur. Surface *prepara-*

tion may be needed, then *application* within the *open assembly time*. Liquid adhesives come in tins, squeeze bottles, *gun-grade* cartridges, or twin plunger units for *two-part* adhesives, which are usually the strongest but have a short *working life*. Solid types come as *hot-melt* sticks and as *film glue*. There are many methods of *application*: some adhesives cover the full contact area, others are combed on or applied in *dabs*. While an adhesive is setting and *curing* the joint may need *pressure*, except for *contact adhesives*. The *binder* in adhesives is usually a synthetic *resin* and they are basically similar to glues, cements, and bonding agents. For guidance on adhesives for wall and floor tiles see BS 5385, for ceramic tiles, BS 5980, for flooring, wall coverings, and wood, BS 5442 and BRE Digest 340. A helpful checklist for site use of adhesives is given in BRE Digest 212. European classification of non-structural wood adhesives is given in BS EN 204. BS 1203 classifies synthetic resin adhesives for plywood and timber, according to *durability* and resistance to *delamination*, in decreasing order as *weatherproof*, *boil-resistant*, *moisture-resistant*, and *interior*.

adjudication Adding a *margin* to the net price of a builder's *estimate* to give the *tender* amount. Adjudication is done by the manager after discussion with the chief estimator.

administrative charges *Overheads*.

admixture A product or agent added in small quantities to the basic constituents (aggregate, cement, and water) of concrete or mortar to alter a particular combination of properties of the mix while it is fresh or after it has hardened. For fresh concrete, admixtures such as *plasticizers* improve *workability* and reduce water requirement, while *accelerators* or *retarders* change the set-

ting rate or adapt to weather conditions (frost, heat). In the set concrete, admixtures improve a secondary quality, usually with only a slight improvement on the main properties of strength and durability. The commonest types are *air-entraining agents*, mainly used for frost resistance. Pigments are added for colour. *Integral waterproofers* and *polymer* modifiers also exist. See BS 5075; see also *C*.

adobe, mud brick A moulded and sun-dried *brick* or block of clay (or the clay itself) usually containing chopped straw reinforcement, as in the Bible, used for the walls of *earth buildings*. Adobe construction is common in Central and South America (Lima cathedral is built of adobe blocks) and is occasionally used in semi-arid regions of the USA, Australia, and elsewhere.

advance (1) A payment made before it is due, such as for unfixed materials on site or for the cost of *preliminaries*. A *bond* may be required.
(2) *Ahead of schedule*.

advertisement for bids (USA) See **tendering**.

aerated concrete Lightweight and highly insulating *cellular* material made from a mix of fine sand, pulverized fuel ash, and chemical admixtures, cast into moulds and *autoclaved* to make *lightweight concrete blocks*. It is easy to saw and drive nails into, although metal fixings do not hold well and may corrode, leading to the use of *resin anchors* (*C*). Aerated concrete has high *moisture movement*.

AFNOR (Association française de normalisation) The French *standards* body. Normes françaises are identified by the letters 'NF' plus a letter and numbers arranged by subject area, group and topic, e.g. NF C 15.100.

A-frame building A building with beams straight from the ground to the roof *ridge*. The lower part of the roof slope usually takes the place of walls.

African mahogany (Khaya ivorensis) Timber from West Africa, generally Nigeria or Ghana, resembling *Brazilian mahogany*, but of a different species, less uniform and sometimes spongy or with *shake*. African mahogany varies in colour but is mainly reddish-brown. It is easy to work and fairly strong and durable. Makore and gaboon are other West African mahoganies.

after-diversity maximum demand, connected load The power actually required by an electrical installation: *installed load* multiplied by the *diversity factor*.

after-flush The small quantity of water remaining in the *cistern* after *flushing* a *water closet*. It trickles slowly down and remakes the *seal*.

after-tack The defect of a paint film which has been *tack-free* and then becomes *tacky*.

ageing (USA **aging**) (1) The storing of a material to improve its qualities. Varnish that has been aged has improved *gloss* and reduced *pinholing* and *crawling*. Ageing is done before use, unlike *maturing* or *curing*. See also *C*.
(2) The decay and breaking down of a material by natural action, usually meaning the atmosphere, sunlight, and frost, but also by *accelerated weathering*.

agent (1) A *site manager*.
(2) An *admixture*.

aggregate Any granular material used as the main constituent of concrete, mortar, or plaster. Aggregate is described by its size, as *coarse* (*C*), *fine* (*C*), or *all-in*, its source (natural or artificial) and its shape, e.g. rounded or angular. It should not be elongated. Concrete *mixes* contain varying proportions of aggregate sizes *graded* (*C*) according to use. Natural aggregate in structural concrete should be hard, dense, inert, and sound. Artificial aggregates are mostly *lightweight*. Decorative aggregate, often chosen for colour, is used as a *finish* to plaster or render, as a facing medium for *bitumen felt* (roofing felt), or as an *exposed aggregate finish* to concrete.

aggressive conditions Environmental conditions that cause damage or *corrosion*, usually from chemicals in the air or ground or from heat and cold. Aggressive atmospheres include corrosive gases, laboratory and chemical fumes, smoke from boilers, acid rain, salt air, or industrial pollution. Aggressive *sulphates* in soils can attack concrete. Aggressive heat and cold conditons include extremes of temperature and *freeze–thaw cycles*.

agreed levels A survey of *natural ground* levels before the commencement of groundworks, marked on to a drawing and signed by the *site engineer* and the *clerk of works*.

agreement A brief statement in writing which says (witnesses) that a building contract has been formed between the *employer* and the *contractor*, who sign it. The name and type of works are given, as well as the *contract sum* and *time for completion*. The other *contract documents*, such as *drawings*, are mentioned, and come below the agreement in *precedence*.

Agrément Certificate A *certificate* issued by the British Board of Agrément, which gives an independent opinion of the fitness for purpose of a building product or system, details of which can be obtained in an Assessment Report. The Board was set up in

1966 as the Agrément Board, based on the French Agrément (approval) system started in 1958, which was itself considerably modified and renamed Avis Technique (assessment) in 1969. Early certificates concerned non-traditional methods, but this scope was later widened to include traditional products with an export potential. They have been linked to the *Building Regulations* since 1985. Certificates issued in one country can be used in another, being based on shared *Methods of Assessment and Testing*, working through the European Union of Agrément (UEAtc), of which the Board is a member. This is to lead to European Technical Approvals (ETAs), allowing manufacturers to use the *CE mark*.

ahead of schedule, advance A description of an *event* that occurs or an *activity* that has been completed before the date on the *programme*.

air-admittance valve A device to let air into *sanitary pipework*, used in addition to the *soil vent*, to relieve minor differences in air pressure.

air balancing Adjusting air-conditioning *dampers* so that air is evenly distributed to rooms. It is necessary at *commissioning* or when partitions are moved.

air-blast cleaning The flushing of rubbish away from a work surface using a *blow gun* supplied with compressed air.

airborne noise Sound or noise that passes through the air only, often *flanking transmission*, as compared with *direct transmission* through walls or floors.

air brick, ventilating b. A *special* brick with perforations through it so that it can be built into a wall to ventilate a room or an *underfloor space*. Air bricks can be of normal size, two-course high (both perforated between stretcher faces), or made to be laid as a *brick-on-edge*.

air brush A small *spray gun*.

air change A quantity of fresh air equal to the volume of the room being ventilated. The ventilation rate is the number of air changes per hour. Offices need about 30 changes, boiler houses and laundries 10 to 20, classrooms 6, reading rooms 2 and storerooms 1.5. Kitchens often have high air-change rates, although if heat and smoke are withdrawn through hoods, excessive rates can be avoided.

air conditioner A small- to medium-capacity packaged *air-treatment* unit or cabinet, usually without ductwork and often mounted in a window.

air conditioning The supply of cool or warm, dry, and filtered air using *mechanical services* to maintain conditions of temperature and humidity within the *comfort zone*, with the indirect benefit of a quiet environment, as windows are kept closed against traffic noise. *Automatic controls* operate the *air-treatment* plant and determine the settings of *air-handling* equipment.

air curtain A strong flow of warm air directed towards a doorway from the outside, to stop cold air entering in winter when the door is left open. Small air curtains are blown from the sides, but larger units have a high-velocity air current from overhead and a return duct in the floor.

air diffuser An *air terminal unit* that controls the direction of *supply air* for distribution into a room or occupied area.

air distribution The supply of treated air from a central *air-conditioning* plant by blowing it through ductwork to *air terminal units* and into a space.

air-dried timber, air-seasoned t.

Sawn timber that has had natural *seasoning*, by stacking cut planks in the open air, usually with *stickers* between them. See **kiln-dried timber**.

air-dry A description of *timber* which is at *equilibrium moisture content* averaged over a year. Some *moisture movement* can therefore occur between seasons, but cracks from *drying shrinkage* are avoided. Moisture content can be up to 17% for exterior work, or as low as 8% if close to a heat source.

air eliminator A plumbing fitting for automatic release of gases, as from a cold-water supply pipe, to prevent an *air lock*.

air-entraining agent An *admixture* for concrete. See *C*.

air gap The height of a water draw-off *tap* above the rim of a sanitary fitting (or *cistern* ballvalve outlet above an overflow), to prevent siphonage *backflow* from contaminating the supply pipe.

air grille, air grating A simple type of *air terminal unit*: a perforated metal plate that allows air to pass, while acting as a sight screen.

air handling equipment Equipment in an *air-conditioning* system for moving air into and out of rooms, usually by blowing with fans through ducts. The *supply air* from the *air-treatment* plant goes to *air terminal units* discharging into the space where air is distributed, with recirculation of *return air* back to the air-treatment plant, unless the system is *all-air*.

air-handling luminaire A *luminaire* through which *extract air* is drawn, to reduce the cooling load on the *air-conditioning* system as well to improve light output and to extend lamp life.

air-handling unit (AHU), air handler A packaged *air conditioner* – either a large individual unit for a

special environment, such as a kitchen or computer room, or a simple *fan coil unit*.

air heater A *unit heater*.

air house, pneumatic structure A 'balloon' structure, either *air-supported* or *air-inflated*. Air houses pay for themselves quickly when used for temporary shelter, because they can span large areas, are inexpensive and easy to transport, and allow fast erection or demounting. In case of fire, the burnt hole can provide both a means of escape and ventilation.

air-inflated structure A type of *air house* with a double membrane kept taut by low-pressure air. The membrane is carried by a frame of steel or aluminium, or by high-pressure air-filled 'tubes'.

airless spraying Painting by a *spray gun* with a miniature high-pressure pump (140–210 bars) forcing paint through a fine nozzle. With this method there is almost no 'spray mist' and less *overspray* than with compressed-air spraying.

air lock (1) A bubble of air, trapped in a high point of a pipe, obstructing the flow of water. Air locks are not likely to occur in well-laid-out cold-water pipes, but for other types of systems, particularly heating or hot water, some way of releasing or *bleeding* air is often provided. Sealed systems of central heating and mains pressure *unvented* hot-water systems are less prone to air locks than older low-pressure *open-vented systems*. See diagram.
(2) Any device for preventing the flow of air, often merely a *lobby* with two doors far enough apart so that the first closes behind a person before the second is reached. Air-lock doors should open in the same direction.

air receiver, air vessel A *pressure*

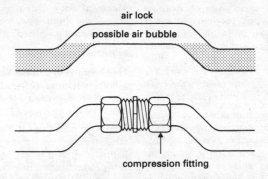

air lock

possible air bubble

compression fitting

An air lock. A compression joint is fitted at a high point in a pipe circuit, so as to leak out possible air locks. When the flow ceases (evidence of a bubble), the fitting is unscrewed until the air is heard to escape. When the noise ceases, drops of water flow out and the joint is screwed up again. The flow should then restart.

vessel (*C*) connected to the outlet from an air compressor for storage of *compressed air* until it is used. One may be connected to the delivery side of a pump to extract air from the water.

air-release valve, pet cock A small valve for *bleeding* air from pipework, a pump casing, pressure vessel, etc.

air right The right of the owner of a low city building with fewer floors than allowed by the *plot ratio* to sell the space above for another building to be stacked on top. However, this may mean also allowing foundations and columns to be built under and through the property to provide support.

air-seasoned timber *Air-dried timber*.

air set *Warehouse set*.

air shaft A *light well*.

air space A *cavity*.

air-supported structure A type of *air house* with a single membrane carried by low-pressure air. The edges need *anchorage* from concrete beams in the ground or sandbags, and the entrance needs an *air lock*.

air terminal unit Equipment on the end of *air-conditioning* ductwork to distribute *supply air* into a space or collect *return air*. There are many types, with different acoustic and aerodynamic performance, from a simple *grille* or *diffuser* to *registers* or more complex *reheat units*, *fan coil units*, *induction units*, and *air-handling luminaires*. The *mixing box* or air-handling equipment connected to the air terminal unit may be regarded as part of it, and often gives the whole system its name, as do *variable air-volume boxes* or *constant-volume boxes*.

air termination network The parts of a *lightning protection system* on a roof. Flat roofs have *roof conductors*. Ridges or high points have a sharp *tip* to emit ions and neutralize the high static voltages which cause lightning strikes. Both lead any discharge that does occur to the *down conductors*.

air test A *drain test* using air pressure. The pipe run is blocked at the top and bottom with *screw plugs*, then air is pumped in until 100 mm of *water gauge* is shown on a glass *U-gauge*. The test is satisfactory if the pressure does not drop below 75 mm within five minutes. Air tests are used for *gravity pipes*, e.g. the *sanitary pipework* above ground. See **hydraulic test** (*C*).

airtight inspection cover A cast-iron plate over an *inspection chamber* (*C*). Covers are removable, non-ventilating, and bolted down to a frame, which has a groove filled with grease. They may be required over a *soil drain* indoors, and are made to resist flooding.

air-to-air heat-transmission coefficient The *U-value*.

air treatment Altering the air temperature and humidity, and removing dust and impurities with *filters*, to make it suitable for use in an *air-conditioning* system. In winter air treatment involves normal *heating*, which may be shared with *radiators* in a *split system*, and the use of *humidifiers*. In summer or damp weather a *chiller* runs the *cold coils* and any *dehumidifiers*.

air valve A valve for *bleeding* air from water pipes.

air vessel An *air receiver*.

air void A *blowhole*.

air washer, wet air filter A chamber in which air is mixed with water, removing contaminants such as dust and gases. Suited to industrial exhausts, it is no longer used for *air-conditioning* supply air owing to costly maintenance and because it is a possible source of undesirable bacteria.

alarm An alarm system is an automatic *communications installation* to indicate fire, intruders, etc. A fire alarm may be set off by a manual *call point* or an automatic *fire detector* and usually operates a warning bell or other *sounder* as a signal to evacuate.

alarmed door A door with an alarm set off by built-in detectors, which show when it is opened or forced.

alburnum *Sapwood*.

alkali-resistant glassfibre Special glassfibre with up to 20% by weight of zirconium dioxide, for *glass-reinforced concrete*.

alkali-resistant paint Few paints are alkali-resistant and there is a risk of failure if applied to young concrete or any product containing cement in the first year or two. *Oil paints* have little alkali resistance and suffer *saponification*, but *emulsion paints* can resist mild alkali attack. The main alkali-resistant paints are *cement paints*, *masonry paints*, bitumens, *epoxy paints*, and *chlorinated rubber paint*, but *pigments* also need to be alkali-resistant. They need normal surface *preparation* and usually a dry *substrate*, although not always a primer.

alkali-resistant primer A *primer* used under *oil paint* on concrete.

alkyd paint Durable exterior *oil paint* or varnish, with varying percentages of *alkyd resins*, usually in *gloss*, which are easy to brush (but are easily over-brushed too thin, so require care to give good *film* thickness). They are fast-drying and have good weather and abrasion resistance, as well as low permeability to water *vapour*. Since alkyd paint, widely used for site decoration of timber, tends to become brittle with age, external joinery should be re-painted every three or four years.

alkyd resin A synthetic polyester *resin* made from an alcohol combined with an acid, used in *alkyd paints*.

all-air, 100% air *Air conditioning* in which the air is used only once, then rejected to the outside.

all-glass door A frameless door with

a leaf of solid *toughened glass* and patch fittings.

all-in contract A *design–build* contract.

all-in material *Aggregate* with all sizes up to a stated maximum (e.g. minus 60 mm), but without accurate *grading* (*C*), mainly used for bedding drain pipes and to make non-structural concrete.

alternate bay construction *Ground slabs* cast in two stages, often like the squares of a chequerboard, to form *contraction joints* (*C*) in the concrete.

alternate lengths work *Underpinning* under old foundations, placed in two stages so that support is continuously available. Each length (or stool) is about a metre long, with 'odds' and 'evens' excavated, concreted, and *pinned up* separately.

alternate proposal (USA) A proposal different from what is described in the *tender* or *bid* documents.

aluminium (USA **aluminum**) A lightweight and fairly strong metal, normally used as an alloy, in the form of castings, sheet or *extrusions* (*C*). Aluminium has good corrosion resistance unless in contact with *dissimilar metals*. It is soft and easily worked with woodworking or *carbide-tipped* tools, or joined by welding. When polished it has high *reflectance* of light and heat. Although non-combustible, it is not a reliable structural material if used above 100 to 150°C. When heated it has a *critical temperature* of 225°C and melts at 650°C, thus is unsuitable as a *fire-stop* material. After heating it is soft: later it 'age-hardens'. It is nearly three times as flexible as mild steel, which also has about three times its density, and has quite high *thermal movement* even for a metal. Aluminium can be painted, but

requires surface *preparation* including *degreasing* as well as a suitable *metal primer*, usually based on *zinc chromate*. Factory-made aluminium windows or metalwork are usually *self-finished* by *anodizing* or with an *organic coating*.

aluminium cable Electrical cable with aluminium *conductors*. It is cheaper and lighter than copper cable, but because special electrical terminals are needed which do not crush soft aluminium, it is not used in house wiring.

aluminium foil Aluminium *sheet* which is thinner than 150 *microns*, often only 20 microns. It reflects heat, light, and *ultraviolet*, and has *low emissivity* and very low *permeability*. The main uses in building are for surfacing materials, such as *insulation*, roofing felt, or gypsum plasterboard, and/or as a *vapour barrier*. See **poultice attack**.

aluminium paint A finishing paint made by *leafing* aluminium powder or foil and blending with a suitable *medium*. It is both bright and reflective (at least when new) and a good *sealer* with high heat resistance. Its main use is as *solar protection* over asphalt roofing felt.

aluminium primer (1) A wood *priming paint* noted for its high water resistance. Its uses are as *knotting* on patches of resin or more generally on very resinous timbers, and as a sealer over timber *preservatives*.

(2) A *metal primer* for aluminium, e.g. *zinc chromate primer*.

aluminium roofing Durable sheet *roofing*. The commonest is *self-supporting*, about 0.8 mm thick, *profiled* for stiffness, and *organic coated*, but it is also used for *supported sheetmetal roofing*. It is more easily damaged than steel roofing – people walking on it may deform its shape or tread on hard debris such as nails or chippings. Its softening temperature is low enough for it to act as a *fire vent*. Fixing nails or screws for aluminium roofing should also be in aluminium, and all gutters and downpipes, *flashings*, and lightning *roof conductors* either in aluminium or *plastics*, so that only one metal is present. Copper corrodes aluminium, even without direct contact. The first aluminium roof, on the church of San Giorgio in Rome, was made of material less pure than that used today but is reported to be still in good condition.

aluminium windows External windows, mostly factory-made with *gasket glazing*. Frames and sashes are slender, crisply shaped sections, usually *anodized* or *organic coated* but sometimes clad with stainless steel. *Mill finish* is no longer specified. See **curtain wall**.

aluminium–zinc coatings Corrosion-protective coatings for sheet-steel roofing developed in the USA, similar to *galvanizing* but longer-lasting. However, they are easily damaged by contact with steel or other *dissimilar metals*.

ambient conditions The quality of the surrounding air, lighting, and acoustics.

ambient counter A *servery* at room temperature, not heated or chilled.

amendment A revision or modification to a *contract document* made to correct an error or to show a *variation* in the works.

American National Standards Institute (ANSI) The umbrella body for standardization in the USA, publishers of the *standards* of each of the 100 or so sponsor organizations.

American Society for Testing and Materials (ASTM) A source of voluntary consensus *standards* on materials, products, systems and services, republished every year.

analogue detector A *fire detector* which gives electronic signals to represent what it senses, e.g. a data code. The central computer decides if there is a fire.

anchor (1) A *primary fixing* permanently built into a structure, to hold components, either directly or through *secondary fixings*. Anchor types include *expansion bolts* (C) and *resin anchors* (C), or they may be grouted.

(2) A *secondary fixing* made from a short length of *angle*, or from a steel plate, such as an L-shaped *clip* or metal cleat. Anchors have bolt holes for connections to the *primary fixing* and the component to be secured.

anchorage A system for securing a component against a force, such as the uplift from a pressurized *air house* or the tension in a tendon for *prestressed concrete* (C).

anchor block A heavy block of concrete cast round a pipeline with *push-fit joints*, one near each joint, to prevent sideways movement. Compare **thrust block**.

anchor plate, floor p. A steel floor paving, about 305 × 305 mm and 3.2 mm thick, with downward-projecting lugs or cross flanges. It is fully *bedded* in mortar 40 mm thick and is used for very heavy traffic and impact loads.

ancient lights Windows in an external wall that have enjoyed light for at least twenty years and cannot be built over without the owner's consent under the Prescription Act 1832. A sign 'ancient lights' is put up to avoid having to prove an *easement*, which may have to date back to 1189.

ancient monument A building listed

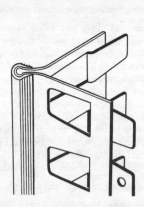

Galvanized-steel angle beads to protect plaster corners. (Courtesy Expamet.)

under the UK Ancient Monuments Acts. Owners must notify the Department of the Environment, Directory of Ancient Monuments and Historic Buildings, at the planning stage, if they intend to repair or even clean them. Local planning officers can usually advise.

angiosperms A large group of flowering plants that includes deciduous trees, therefore all *hardwoods*, but not pine trees.

angle (1) **corner** A change in direction, or the space between two lines or surfaces, such as an internal or external angle.
(2) **right angle** L-shaped.
(3) A section or shape in steel, aluminium, or plastics with two flanges at right angles. The flanges may be of equal or unequal length.

angle bead, plaster b. A thin line of *metal trim* for *protection* of the outer *arris* of a plaster corner against knocks. Large holes in the metal provide bond and are used for nailing. They are plastered over, using the bead

edge as a permanent *screed rail*. See diagram.

angle block, glue b., blocking A small wood block, usually shaped like a right-angled triangle, glued and screwed into an angle as a stiffener.

angle brace, a. tie A bar fixed across an angle in a frame to stiffen it.

angle cleat, clip anchor An L-shaped *anchor* to hold purlins to a roof truss.

angle closer A brick cut specially to complete, i.e. close, the *bond* at the corner of a wall.

angled, angling At an angle, usually not a right angle.

angled tee A pipe fitting, such as a *junction*, with the inlet leg at a shallow angle, used at drain connections.

angle fillet A triangular strip or moulding in the corner between two surfaces, often in *cement* or *asphalt*. It makes two sharp 45° angles.

angle float A plasterer's tool for shaping an internal corner.

angle gauge A *template* made for setting out or checking (or trying) angles.

angle grinder, disc g. A *power tool* with a right-angle drive, used with rigid abrasive discs, for the hand grinding and cutting of metal or masonry, for which different abrasive wheels are available. Angle grinders have less tendency to buck than *disc sanders* and can be large and powerful. See diagram, p. 346.

angle joint A carpentry *joint* at a corner, not a *lengthening joint*.

angle rafter A *hip rafter*.

angle section, a. iron A steel *angle*.

angle staff An *angle bead*.

angle tie An *angle brace*.

angle tile, angular t., arris t. A roofing *accessory* made to cover the angle at a *hip* or *ridge*, or for *tiling* at the corner of a building.

angle trowel, twitcher A plasterer's trowel, either with upturned edges for working internal angles, or vee-shaped for working external angles. See diagram, p. 332.

angle valve A *screwdown valve* with its outlet at a right angle to the pipeline.

angling See **angled**.

anhydrous Containing no water, usually from being *calcined* (*C*) by heat, for example *quicklime*. In the UK anhydrous gypsum is no longer produced.

animal black Paint *pigment* made by heating animal products, such as bone.

annealed glass Ordinary sheet *glass* that has not been *toughened*.

annealing Softening a metal or glass by heating and slow cooling. Annealed copper pipes are used underground or for concealed *panel heating*. Annealed steel *binding wire* is used for *steelfixing*.

annual ring, growth r., year r. One ring of *springwood* and *summerwood* added to a growing tree each year. Since ring thickness is affected by seasonal variations it is possible to date old timber by comparing the sequences of its rings with other timbers grown in the same district, which (if of the same date) will have rings in a similar sequence. Some timbers grown near the equator show almost no sign of variations in their *rate of growth*.

annular Shaped like a ring.

annunciator An *indicator panel*.

anodizing A hard and attractive *coating* to aluminium. It has good *durability* except for attack by alkalis, as from splashing by fresh cement or from water off walls containing cement. It is used on *self-finished* items such as windows, or as a reliable base for *organic coatings*. Anodizing on outdoor surfaces should generally be cleaned every six months, though less in a dry climate and more often in towns or near the sea.

ANSI *American National Standards Institute*.

anti-bandit glass *Security glazing* that delays access through a window for a short time, such as 10 mm thick *laminated glass*. It is not *bullet-resistant*.

anti-climb paint Paint that stays soft and slippery, applied above normal reaching height to drain pipes on the outside of walls.

anti-condensation paint A coating containing cork dust or *vermiculite*, put on pipes as *insulation* and to absorb moisture. It has little effect on really heavy *condensation*, which needs *lagging* as well as a *vapour barrier* and good ventilation. Anti-condensation paint is also expensive and gets dirty easily.

anti-corrosive paints Paints which contain *inhibiting pigments* to delay corrosion of metal surfaces. *Metal primer* for steel is anti-corrosive, but needs to be sealed by the later coats of the complete *paint system*.

anti-dust product A *dust-proofer*.

anti-frost agent A frost-proofing *admixture*, or a cold-weather concreting aid.

anti-graffiti treatment A *textured finish* with enough small stones to discourage markings with felt pens, lipstick, etc. Other treatments exist, e.g. the use of *cement-rubber latex*.

anti-intruder chain-link fencing A security fence made of woven wire.

anti-siphon pipe A *branch vent*.

anti-siphon trap A waste *trap* which resists *unsealing*, usually by having a *deep seal* of water in it, achieved by having a lower *dip*. In the *single-stack system* of internal pipework these traps may save the expense of *branch vents*.

anti-slip paint A paint containing hard sand, cork dust or similar material used for finishing wood floors or decks.

antistatic flooring An electrically conducting *floor finish*, such as *flexible PVC* or carpet, that can earth *static*, used in *computer* rooms and operating theatres. Antistatic carpets may have a conducting backing, chemicals factory-mixed with the dyes, or metal and other fibres woven into them. Antistatic floors usually need special laying and cleaning techniques.

anti-sun glass *Solar-control glazing*.

anti-vibration mounting, flexible m., resilient m. A flexible pad (often rubber) for mechanical equipment, to prevent *transmission* of *equipment noise* to the fabric of the building.

apartment (mainly USA) A *flat, maisonette*, or similar dwelling.

appliance ventilation duct A *ducted flue*.

application (1) Putting a coating (such as paint), adhesive, or sealant on to a surface. The surface may be called the *substrate* and usually requires suitable *preparation*, and for the first coats *pretreatment* or a *primer*. Hand application includes brushing, rolling, and trowelling. Mechanical application is by spray gun, power roller, or projection. Paintwork can suffer from many *defects* of application, such as *beads, brush marks, cracking, curtaining, orange peeling, runs, skips*, and wrinkling.

(2) A request in writing, e.g. an application to the local authority for a *building permit* or a *sewer connection*.

applicator (1) A tool used for placing adhesive, plaster, etc.

(2) Somebody trained to place special coatings, sealants, etc.

appraisal The examination and testing of building materials or processes and the assessment of their suitability for a particular use, often as a step towards *approval*. Appraisals are carried out by *certification* bodies such as the British Board of *Agrément*, and the *Timber Research and Development Association*. BRE Digests 166 and 167 describe appraisal procedures in Denmark, Germany, France, the Netherlands, and Sweden.

apprentice A young person who agrees to work under a skilled master on small pay for a given number of years in order to learn a *craft*. Today this is combined with theory classes at a polytechnic and independent testing of skills.

apprenticeship The time served by an *apprentice*, according to a written agreement called the *indenture*. Apprenticeship is encouraged as a way of gaining *National Vocational Qualifications* in many of the *building trades*.

approval Approval to build, alter, or demolish is obtained from the local planning authority, often within the district council in the U K. Shop drawings have to be approved and stamped before work is commenced. Samples of materials, or a change to an equal product, may need written approval from the specifier. Like *appraisal*, the usual aim of approval is *quality assurance* but it may be limited to the visual appearance of the work, owing to the risk of *transferred responsibility*.

Approved Document Documents associated with the U K *Building Regulations*, which give guidance on detailed matters: structure; fire spread; site preparation and resistance to moisture; drains; noise; conservation of fuel and power, etc.

approximate quantities Preliminary *bills of quantities* used where the extent of the work is difficult to measure because it is hidden, time is short, or the aim is only to see if the project is feasible; also for particular types of contract, such as *two-stage tendering* or *schedules of prices*. Quantities are often worked out by indirect methods such as *cubing*.

appurtenant works (USA) All labour and materials required for the satisfactory completion of a job, which are included even if not described in detail in an *item*.

apron A horizontal or vertical panel, e.g. the inside wall behind a window *spandrel*, a *fascia*, or the *flat* in front of a *dormer*.

apron flashing A one-piece L-shaped *flashing* usually on the lower side of a chimney.

apron lining (Scotland **breastplate**) Joinery *casing* over the vertical face of a stair *well*.

apron wall A *spandrel* wall.

arbitration When a *dispute* over a building *contract* has reached deadlock, both parties may agree to its settlement by an arbitrator named in the *conditions of contract*, to avoid costly and lengthy court proceedings. Arbitrators may call expert witnesses, but evidence is not taken under oath.

arcade An arched passage, often lined with shops. Compare **colonnade**.

arch (1) A curved beam of stones or bricks that works in compression, forcing the stones together and thrusting against their *springing*. Arches (and *vaults*) can span greater distances than a solid stone beam. The arch stones or bricks are called *voussoirs*. Types include barrel, jack, and relieving arches. (2) A *lintel* over a window, usually a flat arch of *bricks-on-edge*.

arch bar A flat steel bar or *angle* to carry the bricks of a window *arch*.

architect A person who designs and supervises the construction of buildings. Starting with the brief given by the client, the services of an architect include feasibility studies at the start, development of a suitable scheme, then preparation of *contract documents* and the calling for *tenders*. After the contract is signed, supervision includes site inspections before issuing each *certificate*, and chairing the *site meeting*, often on the same day. In Britain the title 'registered architect' should be used only by those who have studied at university degree level, mainly in design and building construction, but

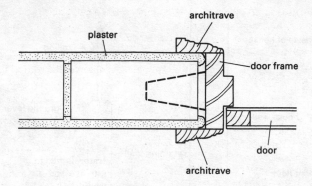

plaster

architrave

door frame

door

architrave

An architrave.

also building law, mechanical and electrical services, fine arts, professional practice, etc. and who meet the requirements of the Architect's Registration Board of the United Kingdom (ARCUK). Other countries have similar requirements.

architectural concrete Concrete with a textured surface, either *direct-finish* or the result of surface treatment after forms are stripped.

architectural drawings *Contract documents* prepared by the *architect*, showing the layout and details of work for a new building. On large projects they form a separate set of drawings from those for structural work and building services.

architectural metalwork Decorative or ornamental *metalwork*, such as gates, railings, staircases, balustrades, screens and ducts, or access covers.

architectural sections Drawn, extruded, or folded shapes made of aluminium or stainless steel, used as decorative *trim* around the outside of windows, the porches of industrial or commercial buildings, and so on.

architrave Joinery *trim* which is *planted* to cover the small gap between

a door frame (or *jamb lining*) and the wall finishing. It has *mitred joints* and may be fixed with nails or architrave adhesive. See diagram

architrave bead *Metal trim* used as a *stop bead* for plaster, fixed to the wall beside a door or window opening and covered by the *architrave*.

architrave block, plinth b., skirting b., foot b. A block at the foot of a door *architrave* against which the skirting board also fits.

architrave trunking A hollow *architrave* to carry wiring in *skirting trunking* across a door opening.

area, dry a. (USA **areaway**) An open space below ground level round a *basement* to separate it from the surrounding ground and keep it dry. See diagram, p. 18.

arm (1) The *outreach arm* of a lighting column or a bracket for carrying lighting fittings.
(2) The *dipper* arm of a *backhoe* or similar excavator.

armour *Mechanical protection* such as steel sheeting on a *fire door*.

armoured cable (USA **armored c.**) An electrical cable with stainless-

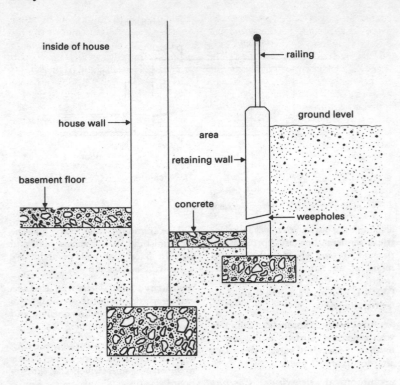

An area with weepholes. In a house with no damp course, the excavation of the earth and the building of the wall on the right help to dry out the house wall. Where the height of earth to be held up is 0.6 m or less a much weaker wall (e.g. half-brick) will suffice, with an area only 250 mm wide, costing about a quarter of that shown.

steel strip or galvanized wire wound over the conductors and insulation, often with an outer plastics sheath (or serving), for main distribution supply and buried *feeders* (*C*). The insulation is a *thermoset* such as *cross-linked polyethylene* or hard ethylene-propylene rubber.

array A repetitive series of similar components.

arris The edge at a corner or angle, particularly of finishings such as joinery or plasterwork. A sharp arris in hardwood can cut a person's skin and is prone to damage, as is any paint applied over it, which becomes thin from surface tension. Arrises are usually *eased* or *pencil*-rounded.

arris gutter A V-shaped wooden *Yankee gutter*.

arris-wise, arris-ways Diagonal, with reference to laying bricks, slates, or tiles or sawing timber.

arrow diagram, network A *programme* on which *activities* are represented by arrows, joined to show their sequence and logical relationships, usually worked out by the *critical-path method*, manually or by computer. The length of each arrow can be *time-scaled* to show activity *duration*, with the head and tail of each arrow being at an *event*. The name, duration and *float* of each activity is printed. Arrow diagrams are used on site for detailed planning or as the basis for simpler *bar charts*. See diagram, p. 352.

articulation (1) A hinge or pin joint in a framework to allow angular movement, instead of *continuity* (*C*) and transfer of stress.

(2) An *event* on an *arrow diagram* where several *activities* meet.

artificial (1) Created by man, not occurring naturally. Synthetic fibres and reconstituted materials (e.g. *reconstituted marble*) manufactured from waste are examples of artificial materials.

(2) Of finishings, an imitation of an expensive material using a cheap one.

artificial ageing, a. weathering *Accelerated weathering*.

artificial aggregate Nearly all *lightweight aggregates* except pumice, but sometimes also blastfurnace *slag* and *clinker*.

artificial lighting, illumination The use of electric *luminaires* for *lighting*. See **daylight**.

artificial stone *Cast stone*.

asbestos A mineral crystal consisting of thin, tough fibres like textile, which can withstand high temperatures when pure. Asbestos was much used in building in the past for high-temperature *insulation*, often *sprayed* for the encasement of steelwork, or as reinforcement in asbestos cement building board, corrugated roofing, and wallboard (which became brittle with age). However,

since the risk to health from breathing in asbestos fibres (and *dust* or any fine fibres) has been realized, *asbestos-free materials* have taken their place. If in doubt, reference should be made in the UK to the Health and Safety Executive's publications or the Control of Asbestos at Work Regulations 1987, and in the USA to Environmental Protection Agency guidelines. See **asbestos removal**.

asbestos cement sheet A product now superseded by *fibre cement boards*.

asbestos encapsulation In-place treatment of *sprayed asbestos* with a durable and impact-resistant coating to seal in the asbestos fibres, used in cases where *asbestos removal* is impracticable. The coating should have *flamespread* rating of Class 1 to BS 476, which is difficult to achieve over high-temperature insulation materials such as asbestos that keep in the heat.

asbestos-free board, fibre cement b. *Building board* with fibre reinforcement and a binder such as *calcium silicate* or *cement*. It has low *movement* from changes in temperature and humidity. Low-density boards (about 450 kg/m^3) containing *vermiculite* are used for *encasement*, while medium and dense boards (up to 1200 kg/m^3) are used for general purposes. Although it is possible to cut them with a *carbide-tipped* circular-saw blade, it is more usual to employ special shears or a masonry disc in a *circular saw*.

asbestos-free materials Non-combustible materials made as replacements for *asbestos*, e.g. *sprayed mineral insulation* and *asbestos-free board*.

asbestos removal Specialist work for removing *asbestos* placed in buildings. It involves wearing protective clothing, breathing apparatus to avoid inhaling the asbestos fibres, loading the old material into sealed bags, and special methods of disposal. If removal is not practicable, *asbestos encapsulation* may be required. In Britain, the local area office of the Health and Safety Executive should be contacted for guidance.

as-built drawing A record drawing made by the builder after works have been completed and showing the effects of all drawing *revisions*.

as-drawn wire Steel wire that has been hardened by being drawn through a die, without further treatment. It is used for reinforcement *fabric*.

as-dug aggregate Quarry material that has not been crushed or screened, but loaded directly for delivery.

ashlar Walls or facings of *stonework* laid in *courses* of *dimension stone* with thin joints about 3 mm thick, which should be *raked out* 20 mm and pointed during cleaning down. The faces may be plain, *rusticated*, or vermiculated, or have a *drafted margin*. It was used by the Egyptians in 3000 BC.

ashlering A low wall, usually about 1 m high, in the side of an *attic*, from the floor to part way up the sloping ceiling, often in *blockwork*.

ASHVE American Society of Heating and Ventilation Engineers.

asiatic closet A *squatting closet*.

asphalt A mixture of *bitumen* and sand, clay, or other inert mineral filler, such as limestone. It is hard at normal temperatures, but flows stiffly when heated. Types include: compressed rock asphalt, *lake asphalt*, natural rock asphalt and *mastic asphalt*.

asphalt mixer A machine for heating and mixing *asphalt* before laying. It is usually mounted on a trailer, with a motor to drive the mixer, and is fired on *bottled gas*. Precautions in use are similar to those needed for a *bitumen boiler*.

asphalt roofing *Membrane roofing* made with two or three coats of *asphalt work*, sometimes taken to include *bitumen-felt* roofing.

asphalt work A specialist *trade* that uses *mastic asphalt* for *tanking*, *asphalt roofing*, and *waterproofing*. Blocks are melted in an *asphalt mixer*; the hot mix (180 to 210°C) is carried in steel buckets by the potman to the layer, who spreads it with a wooden float. Horizontal and vertical work are *measured separately*, as are the *labours* for working into dishings.

assembly Putting together several components or elements, or the resulting *set* or unit.

assembly gluing Constructional gluing on site, particularly of timberwork or joinery units that were factory-fabricated by *primary gluing*.

assessment of tender responses, vetting (USA **bid tabulation**) Detailed appraisal by the *architect* of *tenders* received from builders, to enable him or her to advise the client on which builder to *award* the job to. The lowest tender is usually accepted, but there is often a need for detailed comparison between prices, balanced against each builder's financial stability and technical qualifications.

assignment of contract The transfer of a *contract* to another party, usually when one contractor is unable to carry out the work and so finds another contractor who can. The consent of the *employer* has to be obtained.

Association of Building Technicians See **ABT**.

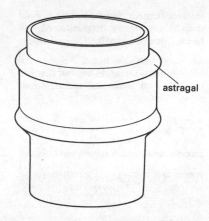

An astragal.

ASTM American Society for Testing and Materials.

astragal (1) A *half-round* moulding that stands up above a surface, for example the raised U-shape round the outside of a cast-iron pipe socket. See diagram.
(2) (USA, Scotland) A *glazing bar*, especially overhead.
(3) A strip of timber or metal fixed to the edge of a door leaf to cover the gap between it and the door frame and to improve security or reduce noise and draughts.

atomization The very fine pulverization of a liquid, as in *spray painting* or a boiler *rotary-cup burner*.

atrium A tall internal courtyard with a glazed roof that lets in daylight, often used in multi-storey hotels with *balcony* access. An atrium can be dangerous in a fire, as it makes a hole through the *compartment* and needs special *smoke control*, escapes, and *sprinklers*.

attached column A structural *column* partly projecting from a wall, as does a *pilaster*. It is therefore also partly *engaged*.

attendance Plant and equipment provided by the *main contractor*, or *builder's work in connection* done for other trades working on site, usually *sub-contractors*. It can be *general attendance* or *special attendance*.

attenuation A reduction in sound level during *transmission* from one place to another, often through a wall or floor or along ductwork.

attenuator A silencing unit in air-conditioning *ductwork* to reduce transfer of *equipment noise*, mainly from *fans*, into an occupied space or to the exterior.

attic, garret A habitable room in the *roof space*, with a sloping ceiling that follows the roof slope and often with *ashlering*.

auger A drilling tool shaped like a cork-screw, for boring holes in wood. It can have an eye forged in the top end for turning with a wooden handle, or be a *bit* turned by a *wood borer* or carpenter's *brace*.

authority Any organization with the power to make official decisions, e.g. on the suitability of materials and methods. *Local authorities* have a duty to control construction, conversion or demolition of buildings. See **certification, standard**.

autoclave A pressure vessel for treatments using steam at high temperature and pressure, e.g. for curing *calcium silicate bricks* or *aerated concrete*.

autoclaved aerated concrete The most usual type of *aerated concrete*.

automatic controls Devices that pick up the message from a *sensor* (or detector) and send power to an *actuator*. They are used for *mechanical and electrical* services, such as lifts, air conditioning, fire alarms, *door openers*, *touchless* units, etc. Automatic controls that use micro-processors can usually take the right action well before anyone would notice that action was needed. See **building management system**.

auto-suppression system An automatic *fire-extinguishing system* which is activated by *fire detectors*.

award of tender Notification by the *client* of acceptance of one *tender*, thus forming a binding *contract*. The decision is usually based on the *assessment of tender responses* and leads to the signing of a form of *agreement*.

awl A small tool with a plain-pointed steel shaft and a straight handle, mainly used as a *scriber* for marking hard surfaces. Compare **gimlet**.

awning, terrace blind An external blind of fabric, such as canvas, that can be put up for protection against sun or rain.

axe (USA **axhammer**) A *brick-layer's hammer*.

axed work A *bush-hammered* finish.

Axminster carpet Patterned *carpet* with cut wool *pile* woven into the backing.

B

back (1) A location opposite to *face*; usually hidden.

(2) To move in an opposite direction, or to return to a start position.

backacter A *backhoe*.

back boiler A small *boiler* fitted at the back of the hearth of an open fireplace or in a *room-heater* to provide hot water and sometimes to heat *radiators*.

backdraught (USA **backdraft**) The inrush of air (e.g. when a door is opened) into a closed, airtight room containing an oxygen-starved, smouldering fire, as the hot gases inside are cooled and contract, drawing smoke back towards the fire. Wood can smoulder when the oxygen content of the air is down to 4 or 5%. As fresh air (21% oxygen) raises the oxygen content to 15% the glowing fire bursts into flame and ignites the air/smoke mixture, which explodes violently. Cold smoke from a smothered fire is still explosive. In a fire, care should be taken to feel if the door is hot before opening it. It should be left closed if hot. Compare **flashover**.

backdrop, backdrops connection, drop c. (USA **sewer chimney**) A vertical or steeply sloping pipe on an inlet to a *drop manhole* in a *foul water* drain to prevent solids accumulating. An *access cover* is provided for rodding. See diagram.

backer A *backup strip*.

backerboard *Gypsum baseboard*.

backfall A slope in a gutter (or any gravity drain) that is opposite to the intended direction of flow, resulting in *ponding* or *backflow*.

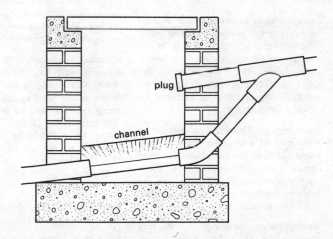

plug

channel

A backdrop.

backfill, backfilling (1) Material returned into an excavation, round foundations or over buried services and their *surround*. It is placed in layers or *lifts* that are individually compacted. Backfill over drains or cables goes on top of any gravel surround and may contain *tracer* or *breaker slabs*. Suitable *spoil* may be used, or gravel, or *dry lean concrete*.

(2) (USA) Backing brickwork or *bricknogging*.

(3) A *background*.

back-flap hinge A hinge with wide flaps, screw-fixed to the *face* of a door and frame, on a door too thin to be carried on *butt hinges*. See diagram, p. 229.

backflow Movement of a fluid in the opposite direction to the natural or intended direction of flow. In a *gravity pipe* such as a drain, backflow can be caused by a *backfall* or flooding of the outlet. In a pressure pipe, backflow can be caused by *back siphonage*.

backflow valve A *back-siphonage preventer*.

background, backing, base A surface on which *plasterwork* is applied. The surface can be *metal lathing*, brickwork, concrete, *insulating board*, *gypsum lath*, etc., some of which require suitable *preparation*. For paintwork, however, the background is called a *substrate*.

background drawing A simplified floor plan of a building, used to help *coordination* in the preparation of *shop drawings* for building services (such as air conditioning). As all trades use the same background drawing, any *clashing* can be detected.

background heating A heating system run at a low temperature, supplemented by gas or electric fires etc. during periods of occupancy and cold weather.

back gutter A roof gutter on the 'uphill' side of a chimney (or similar *penetration*), usually made from *supported sheetmetal roofing*. An alternative is to have a *saddle*.

backhoe, backacter, digger A versatile *excavator* with a bucket on the end of an arm, rather like a giant human arm. The operator controls hydraulic rams to swing, raise, or lower the whole arm, extend or draw in the forearm or *dipper*, or dig with the bucket (crowding). Apart from digging, the bucket is also used for *demolition* and for tamping *fill*. A bucket can be changed, or replaced with a *hydraulic breaker*. Small (or 180°) backhoes have a frame with *stabilizers* and are carried on a rubber-tyred road-going tractor. A 'combination backhoe' also has a front-end loader. An 'offset backhoe' has a rail so that it can be slid to each side, to dig against walls. Big (or 360°) backhoes are usually mounted on crawler tracks and need a level surface to work on, but can be 'walked' across an open trench, aided by the bucket. Mini-diggers usually have tracks and a bulldozer blade.

backing (1) Support material behind a *facing*, such as concrete behind *cast stone*, dry lean *haunching*, wooden *firring* pieces used as grounds for a *dry lining*, or a *backup strip* behind sealant.

(2) **carpet b.** The stiff underside of carpet which holds the *pile* in place. It is made of woven cotton, synthetic fibre, rubber, etc., and often coated with latex.

(3) A *background*.

backing coat Any coat of plaster other than a *finishing coat*.

backings *Firring* strips on joists, or any *backing*.

back lintel The lintel supporting the *backing* of a wall and not seen on the face.

backnut A nut used to make a tight and locked joint, as on the long thread of a *connector* or on the *pipe tail* of a pillar tap or waste. See diagram, p. 177.

back painting Painting the unseen face of *building board* to give *balanced construction* and prevent buckling or cracking from an impervious paint system applied to the face alone. The back may be painted with a cheaper paint such as bitumen.

back propping, reshoring The replacement of *props* under a slab after stripping of *formwork*, to provide support while it gains strength and to reduce *creep* (*C*).

back putty, bed p. The *putty* behind a pane of glass or other *glazing*, as *bedding* between it and the *rebate*. See **face putty**.

back saw A hand saw stiffened with a heavy fold of steel or brass along its back, such as a *tenon saw*. See diagram, p. 388.

back sawing The *breaking-down* of timber by *flat sawing*.

backset, set back The distance from the spindle of a door lock to the edge of the door leaf, providing at least 25 mm clearance for fingers or hands between a door handle or knob and the *door stop* on the frame.

back shore, jack s. An outer support in *raking shores*, under the *rider shore*. See diagram, p. 406.

back shutter, top form *Formwork* to the top face of a steeply sloping concrete slab, to hold the fresh concrete during placing and vibration.

back siphonage, s. The sucking-back of dirty water from within a building leading to *backflow* into a water main if its pressure drops. All water authorities have regulations to prevent this pollution of the main, e.g. by having an *air gap* under the draw-off taps of sinks or drinking fountains, or *ballvalves* of cisterns. A ballvalve with a *silencing pipe* must have an anti-siphon hole.

back-siphonage preventer, backflow valve A valve in a water main with two non-return mechanisms separated by a self-draining chamber.

backsplash A *splashback*.

backup strip, joint backing, backer A strip of compressible foam plastic inserted into a *movement joint* to limit the depth of *sealant* to half the joint width (with a minimum of 10 mm to give enough bonding area). This 2 : 1 ratio of width to depth allows the sealant to stretch without tearing away from the joint sides, acting as *pointing* rather than a *filled joint*. The front face of the backup strip may have a *bond breaker*. See diagram, p. 295.

backward-curved blade An efficient aerofoil-shaped blade of a *centrifugal fan*, which does not overload the motor if the discharge is blocked.

backward pass A calculation for a *programme* using the *critical path method* working from the *end event* to the *beginning event*, to find the *latest dates* for each *activity* that will still give the earliest completion date calculated from the *forward pass*.

baffle A device for breaking up a flow of air, water, or light, often used as a *diffuser*. Baffles reduce glare from *luminaires* or sound passing through an *attenuator*, and break the force of rain driving into an *open-drained joint*.

bagasse board *Cane-fibre board.*

bag filter An air *filter* used in air

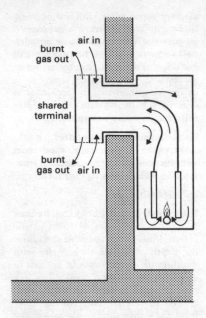

burnt
gas out

air in

shared
terminal

burnt
gas out air in

A balanced-flue gas fire. This one is hung on a wall.

conditioning in which the air passes through bags.

bagged joint A *flush joint* made by rubbing *brickwork* with sacking.

bagging, bag rendering Rubbing cement mortar over the face of *brickwork* using sacking, to fill in small holes and leave a rough-textured surface suitable for painting.

bag plug An inflatable *drain plug* used for a drain test.

baked enamel See **enamel paints, stoving**.

bakelite One of the earliest *plastics*, produced by Dr Baekeland in 1916 from *phenolic resin*, used for electrical insulation, door handles, etc.

balanced construction A method of construction of wood-based *building*

board with equal thickness of wood in both directions of the grain. For balanced *three-ply*, the face and back *veneers* are of equal thickness and the core twice as thick as either of them. Impermeable coatings on balanced construction should be applied to both faces, as in *melamine-surfaced chipboard* or painting plus *back painting*.

balanced flue A double flue made so that the *draught* is balanced by running its two halves (usually concentric) to a shared terminal. Wind gusts affect the two halves equally. One half flue supplies combustion air for a gas heater or oil-fired boiler and the other lets out the burnt gas or smoke. Balanced flues can be much shorter than a normal open flue, e.g. 600 mm high. The boiler or heater is not connected to the indoor air and small units are therefore *room-sealed* or can be mounted outside a wall. A grille may be needed to protect people from a hot low-level terminal, and its discharge should not pollute air drawn indoors. Large balanced flues and *U-ducts* have a rooftop terminal. See diagram.

balanced sash A *sash window* with counterweights or springs.

balanced step, dancing s., French flier A *tapered tread* with an eccentric *springing point*, used in the corner of a sweeping type of *geometric stair*, rarely seen outside France. Balanced steps have constant *going* on the *walking line* but the faces of several steps are swung on plan to increase the inside going of *winders* by reducing that of some *fliers*. (From the French 'balancé', in this sense usually translated as 'swung, tilted, see-saw'.)

balancing The adjustment of the flow of water or air through *radiators* or *air terminal units* (air balancing) so that each receives enough. Balancing is usually done only once, at *commissioning*.

balancing valve A *lockshield valve* for *balancing* the flow to radiators. Both it and the control valve are turned off to isolate a radiator for maintenance, thus avoiding laborious draining of the *circuit*.

balcony A platform projecting from the outside wall of an upper floor at a door or window, usually with a balustrade or parapet. It may allow *access*, e.g. from a stair. If carried only on beams (not on posts below), it has to continue inwards so that the floor weight counterbalances the balcony weight and loads.

balk (USA) See **baulk**.

ballast (1) Unscreened gravel containing sand, grit, and stones of less than boulder size, sometimes used to make concrete. See also *C*.

(2) (USA **gravel**) Material used for *loading* something against uplift from the wind or water pressure. On flat roofs, white *spar* gravel is often used as heavy *solar protection* over *built-up roofing*; such roofs need a *ballast grating*. For exposed sites or foot traffic, ballast is usually concrete slabs or a mortar screed, as gravel can be lifted in a storm, damaging nearby windows.

(3) An electrical choke in the *control gear* of a fluorescent tube.

ballast grating, gravel guard A screen in front of a *rainwater outlet* to stop gravel ballast entering.

ball catch, bullet c. A cupboard door fastener with a spring-loaded ball projecting slightly from a *mortice* in the door. The ball engages with a hole (socket) in a *striking plate* and holds by friction.

ball cock A *ballvalve*.

balloon framing (USA) House *timber-frame construction* with *studs* two full storeys high from the *bottom plate* to the roof *plate*, past the floor *joists* which are nailed to them. It can be erected more quickly than a *platform frame*.

balloon grating, b. wire A large strainer over a *rainwater outlet*.

ballottini Small glass beads, used in reflective paint.

ball test A *drain test* using a ball that fits into a pipe with about 12 mm clearance. The ball is flushed down the pipe to a manhole to detect any partial blockage, e.g. from building debris.

ballvalve, floatvalve A float-operated valve used to control the flow of water into a *cistern*. It may have a delayed but rapid shut-off to stop dribbling. Ballvalves often prevent *water hammer* by releasing over-pressure. They must be reliable and resist corrosion and furring, so that they are often all-plastics, with rubber seatings and a *silencing pipe*. The float need not be a ball. See diagram, p. 28.

baluster, banister A post in a balustrade of a bridge or a flight of *stairs*. Wooden turned balusters were first used in England in Elizabethan times and were then about 80 mm in diameter. Their square end is a *die*.

balustrade A protective guard rail to prevent people falling, at the edge of a stair, landing, or platform, with closely spaced infill such as *balusters* from the *handrail* down to the floor.

band course A *string course*.

banding A *lipping*.

band saw A power saw with a continuous steel belt driven round two or three pulleys. Used mostly for cutting timber, band saws range in size from large sawmill units with tungsten-tipped teeth for the initial *breaking*

banister

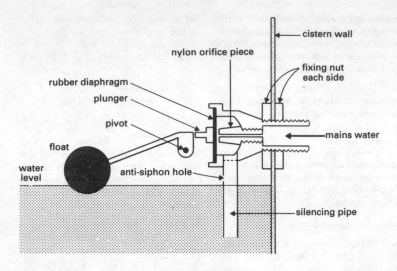

A ballvalve.

down of logs to small benchtop units for fine joinery work.

banister A *baluster*.

bank cube Undisturbed volume of ground, before it is excavated.

banker (1) **gauging board** A platform for batching and hand mixing *mortar* or *plaster*, or small batches of *concrete*, usually by shovel. It measures about 1.8 × 1.3 m, with boards 230 mm high on three sides. After mixing, the mortar or plaster is taken to a bricklayers' or plasterers' working platform or *spot board*.

(2) A mason's workbench, of stone or heavy timber.

bank guarantee A surety put up by a bank to act as a *performance bond* for a builder, a common requirement in *conditions of contract*.

bank of lifts Several *lifts* at a common location, usually with shared controls.

bar (1) A long, narrow shape, such as a metal or wooden *glazing bar* or a steel *reinforcement bar*.

(2) A practical unit of pressure widely used in Europe, close to one atmosphere. It is an exact *SI unit* (1 bar = 100 kN/m² = 100 kPa = about 10 m head of water).

bar bender A *bending tool* or a power bender.

bar chair A support for a top layer of *reinforcement bars* in a concrete slab, in the shape of a chair, either bent on site from small-diameter bar or factory-produced from wire.

bar chart A *programme* showing the work dates of each *trade* represented by bars. It is easy to understand, unlike the master *arrow diagram*. See diagram, p. 353.

barge board, verge b., gable b. Sloping roof *trim* of wood, plastics, or metal, fixed in pairs along the edge of a *gable*

28

to cover the roof timbers and protect them from rain. Old barge boards were often beautifully carved, and in Alpine areas still are. See diagram, p. 200.

barge course, verge c. A brick *coping* to a *gable* wall, or the tiles next to the gable, slightly overhanging.

barge flashing A *flashing* over a *barge board*.

barium plaster, X-ray p. A radiation-protection plaster containing *barytes* aggregate and *gypsum plaster* or Portland *cement* as a binder, used on the walls of X-ray rooms to reduce the amount of radiation penetrating them. The thickness depends on the type of X-ray equipment and the absorbing capacity of the plaster.

barrel That part of a pipe throughout which the bore and wall thickness remain uniform. Gas pipes were, at first, made from musket barrels, surplus after the battle of Waterloo.

barrel bolt A door *fastener* with a metal rod or bar that runs in a case, entering a hole in the *jamb*. Sizes range from thumb slides to *tower bolts*.

barrel light A *rooflight* with curved *glazing*, often in *polycarbonate*, with similarly curved *glazing bars*. It resembles a *barrel vault* (*C*).

barrel nipple (USA also **shoulder fitting**) A short *tubular* with a *taper thread* outside at each end, unthreaded in the middle.

barrier effect The formation of a thin film on a surface, protecting a material from further corrosion.

barrow run A temporary path for loaded *wheelbarrows* over soft ground or across floor joists, usually made by laying a series of *scaffold boards*.

bar schedule A *bending schedule*.

bar setting (USA) *Steelfixing*.

bar spacer A small plastics accessory used to hold reinforcement away from *formwork*, at the correct *cover* distance.

barytes (barium sulphate) A heavy *aggregate* used in *barium plaster*.

base (1) The lowest part of a wall or column, which may be widened into a *moulding* or *plinth* on top of the *footings*.

(2) A *pad footing* or a *machine base*.

(3) **background** The surface over which a *finishing* is applied, such as *granolithic* on a stair tread as a base for vinyl tiles or a *screed* as a base for a *topping*. Paints are applied to a *substrate*, not a 'base'.

(4) The main ingredient of a product; for *paint*, either its main *pigment* (lead base, zinc base) or the main part of the *medium* (water-based emulsion). For *two-part products* it is the main part, to which the *hardener* is added.

(5) The substance from which a product originated; *white spirit* is petroleum-based and *chipboard* is wood-based.

(6) (USA **sub-base**) A *skirting* board.

(7) Also used to mean: basic, initial, end, cap, or holder.

base bid (USA) A *bid* to which an *alternate proposal* is offered.

baseboard (1) (USA) A *skirting*.

(2) Either *gypsum baseboard* or plastering-grade *woodwool* or *strawboard*.

base course The lowest course of a brickwork or blockwork wall, laid on top of the *foundation*. The top may be *weathered*. See also *C*.

base exchange A *water-softening* process in which water passes through a mains pressure tank containing *zeolite*, a mineral reagent, to absorb the salts which harden the water. Periodically the tank must be flushed with a salt solution to regenerate the zeolite. Common in small houses, the process can excessively soften water, and so some raw water is mixed into the supply. Also it adds sodium, which may affect heart sufferers. See also *C*.

base gusset A stiffening rib between a *baseplate* and a steel column.

basement A *storey* which, in the USA, is less than half below ground level. In Britain it can be wholly below ground but is often living space. See also **area, cellar, subbasement, sub-level**.

basement wall Either the wall round a basement or a *sleeper wall* supporting a floor.

base moulding A moulding at the top of a *base*, where it narrows into a wall or column, often acting as a *weathering*.

baseplate (1) **stanchion base** A thick steel plate on the bottom of a steel column, with holes for *holding-down bolts*. It may have *base gussets*.
(2) A bearing plate on the end of a prop or jack, e.g. for an *access floor*.

base sheet, first-layer felt A *bitumen felt* fixed down to a substrate, such as *decking*, and bonded to the *intermediate sheets* of *built-up roofing*.

base shoe, s. mold (USA) A quarter-round *bead*, *planted* along the bottom of a *skirting* at the junction with the floor boards to cover the joint.

base slab A concrete slab under a structure, or a *raft foundation* (*C*).

base tie A small folded steel channel, bar, etc., across the bottom of a *pressed-metal* door frame to hold the jambs at the correct spacing while the wall is built. It may be removed or covered by the floor *screed*.

basin, washbasin A *sanitary fitting* for washing or rinsing the hands, usually installed in a bathroom. Basins are mostly made of vitreous china or fireclay. The *bowl* has a rounded rim and an outlet or *waste* with a plug hole and grating. Basins which are made to be *wall-hung* or to go on a *pedestal* have a ledge at the back with one or more *tapholes* and a *skirting* behind. Bowls to be recessed into a *vanity unit* have a flat rim.

basin mixer A *mixer* for a *basin*, either a *monobloc* type for a single taphole basin or a combination type for a three-taphole basin.

basketweave pattern A wood finishing with the grain in neighbouring squares alternately in directions at right angles to each other.

bat (1) **brickbat** A brick cut across and shorter than full length, used to complete a *bond*. The most usual is a half bat, or *snapheader*, but other sizes are possible.
(2) A rectangular panel of *insulation* covered with paper, made to fit between the *studs* of an external wall or between the *joists* of a ceiling.

batch One mixing of concrete, mortar, plaster, etc.

batch box, measuring frame, gauge b. A bottomless box filled with each constituent of a *mix*, then lifted away to leave the mix on the mixing platform or *banker*. Batch boxes were formerly used for accurate batching by volume of plaster mixes, which are now usually *premixed*. If the sand is wet, extra should be added to allow for *bulking* (*C*). See diagram.

bath A *sanitary fitting*, usually with a

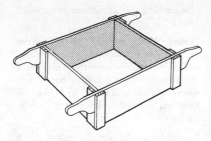

A batch box.

full-length tub and an edge roll round the rim. Domestic baths are made from cast acrylic sheet, cast iron, or steel sheet. Fittings include a mixer or *headworks* and a *waste* with overflow. Bath bottoms have a slight fall towards the outlet, when the rim is set level.

bath mixer A *water fitting* to blend hot and cold water for a *bath*.

bath panel A rectangle of *building board*, slab, etc. to conceal the underside of a *bath*, from the rim down to the floor, often with a *toe recess*.

bathroom A bathroom, being a *wet area*, commonly has *tiling* finishes. In the UK, *socket outlets* are not allowed in bathrooms, except for *shaver sockets*, and switches should be *pull-cord* type.

bathroom pod A factory-finished bathroom/toilet *module*, with its own roof and cladding. It can be placed on a prepared slab by crane and connected to water, electricity, and drainage services within a few hours.

bath/shower mixer A *water fitting* with a diverter enabling blended hot and cold water to be sent either to the *bath* or the shower.

batten A small section, normally of timber, to which sheet materials, *slates*, and roof *tiles* are fixed (BS 6100). They are usually 50 × 25 mm or less,

and spaced at batten *gauge*. See **grounds**.

battenboard A *building board* with a core of wood battens and wood *veneer* facings, bonded together with resin adhesive. Battenboard is strong and stable.

batten door A *matchboarded door*.

battening *Common grounds* fixed to a wall as a base for a *dry lining*.

batten roll A *wood roll*.

batter An artificial, uniform, steep slope, or its inclination, expressed as one horizontal to so many vertical units.

batter board (USA) A *profile*.

batter peg A peg driven into the ground to show the limits of an earth slope.

battery (1) A heating or cooling *coil* for air treatment in air conditioning.
(2) An electric accumulator battery, used as a power supply for *self-illuminating exit signs*, *uninterruptible* power supplies, etc.

batting Surfacing soft stone with a broad chisel in parallel strokes, giving a regular pattern of fluted cuts.

baulk A piece of square-sawn or hewn *timber* of equal or nearly equal cross-section dimensions (BS 6100).

bay (1) One of several uniform divisions of a building, such as the space enclosed between four columns or two beams.
(2) **pour**, **section** An area of concrete or floor *screed* cast in one operation. The maximum area, or length of structural concrete between *construction joints*, is limited to lessen *drying shrinkage*.
(3) The area of plaster, asphalt, roofing, etc., laid at one time.

bayonet cap The cap of an *incandescent lamp* for a *lamp holder*. It has two J-shaped holes and spring-loaded contacts.

bay window A window formed in a projection of a wall beyond its general line and carried on its own foundations, unlike an *oriel*. If curved it is a *bow window*.

bead (1) **beading** Usually a semicircular *moulding* that masks a joint, sometimes alongside a *quirk*. But see **glazing fillet**.

(2) A metal *angle bead* or similar *stop bead* for plaster.

(3) In *supported sheetmetal roofing* an edge bent round to a tube shape, or to 180° for stiffening, for fixing the edge of the sheet or as a *drip*.

(4) The defect of an accumulation of *paint* or varnish at a lower edge due to excessive flow during application.

beading A *bead*.

beading router, beader A *woodworking machine* for moulding *beads* on wood or for cutting grooves in which to insert beads.

beam A structural member that resists loads which bend it. Beams can be described by purpose (bond, transfer, wind), location (perimeter, ring, ground), form (arch, downstand, deep, continuous, simply supported), or material (steel, concrete, timber). Beams include *joists* and *girders*. See also *C*.

beam box (1) A *beam form*.
(2) A metal *joist hanger*.

beam casing Joinery *casing* round any type of beam, or *beam encasing*.

beam clamp Bars with screws or slots and wedges that are tightened round beam *formwork* to prevent movement or leakage during concreting.

beam encasing Concrete *encasement*

surrounding a structural steelwork beam.

beam filling, wind f. *Brick nogging* or masonry between the ends of floor or ceiling *joists* where they rest on a wall, both to give sideways support to the joist ends and to act as a *fire stop* between the wall and the floor or ceiling space.

beam form, b. box *Formwork* for a reinforced-concrete beam, or concrete *encasement* to a *structural steelwork* beam.

beam hanger A U-shaped rod placed over a structural steel beam to carry the *soffit* form for concrete *encasement*.

bearer A horizontal timber beam that carries the *joists* of a floor. It needs to be stiff and well supported itself or the floor will bounce when walked on.

bearing (1) A surface that supports a load. See also *C*.

(2) **loadbearing** Of a member or material, carrying weight.

bearing capacity The pressure that a material can withstand, particularly the *foundation* ground (or soil) for *footings*. See **good ground**.

bearing length The contact length between a beam and its support (either an end bearing or an intermediate bearing).

bearing pad A block of concrete or stone used as a *bearing plate*.

bearing plate A plate built into a wall to support the end of a beam or the base of a column, spreading the load over an area large enough to prevent the wall failing.

bearing pressure The load on a *bearing* surface divided by its area.

bearing wall A *loadbearing wall*.

bed (1) The under-surface of a *brick*,

building stone, slate, or tile laid on horizontal *bedding* to make a joint.

(2) *Bedding*.

bedding (1) A thick layer of material laid over a surface to fill the irregularities between it and a component placed on top of it. Mortar bedding is used for brickwork, and either mortar or adhesive bedding for tiling. Floor tiling needs to be fully (or solidly) bedded by *buttering* and tapping into place, but wall tiling is often laid in thin adhesive bedding applied with a *notched trowel* or with *dabs*. Buried drains are laid on granular bedding, such as sand or *pea-gravel*, or concrete bedding. Glazing for windows with traditional timber frames uses *back putty*.

(2) Laying an element on bedding and bringing it to its final position by steady pressure, moving it to and fro, or tapping it with a mallet, trowel, or *beetle*.

bedding and jointing The main operations of *bricklaying*, laying bricks in *courses* on mortar bedding and shaping the joint faces.

bedding planes Layers along which sedimentary rocks were originally deposited horizontally. Some rocks, but not *slate*, tend to split along these planes. See **face bedding**, **freestone**, **natural bed**.

bedhead panel A stainless-steel plaque on the wall above a hospital bed, with outlets or connectors for services: medical gases, telephone, call bell.

bed joint A horizontal mortar *joint* in brickwork or blockwork, usually 10 mm thick. It should be well filled during *bricklaying*. Both bed joints and *perpends* show as a *face joint*.

bedpan sink A *sanitary fitting* for washing and scalding hospital bedpans by hand. Its discharge is run to a *soil drain*.

bedpan washer A machine that washes and scalds bedpans automatically.

bed putty *Back putty*.

beech (Fagus sylvatica) A *hardwood* of Europe, Asia Minor, and Japan with typical spindle-shaped markings on *flat-sawn* boards. Reddish-brown in colour, it is a little easier to work than English oak.

beef up To strengthen or improve something, e.g. by providing extra bracing for possible earthquake loads.

beetle (1) **maul** A heavy mallet with a wooden head which does less damage than using a steel-headed sledge hammer. It is used on wooden pegs, on precast concrete paving slabs when being bedded on sand, etc.

(2) See **borer**.

beginning event The start of the first activity on a programme in the *critical path method*, the *commencement* of building.

behind schedule See **slippage of dates**.

bell-and-spigot joint (USA) A *spigot-and-socket joint*.

bellcast eaves (Scotland) Eaves with *cocking pieces*.

belled-out bored in-situ concrete pile A *deep foundation* (C) bored down to *good ground* and widened at the bottom to reduce the *bearing pressure*.

below ground level (USA **b. grade**) Said of work on the *substructure* and foundations, and for buried services.

belt sander A portable *sander* with a *sandpaper* belt, used for the fine smoothing of joinery and occasionally of metal or masonry. See diagram, p. 346.

bench A strong, firm work top for jobs requiring tools (bench tools) worked by hand or power. Benches are found in joinery shops, kitchens, offices, and test laboratories, and may have cupboards under.

bench grinder A *power tool* attached to a bench, usually with two abrasive wheels, one each side of its electric motor. Bench grinders are used for sharpening *edge tools* and removing metal from surfaces. The wheels have guards and the user should wear *goggles*.

benching Concrete in the base of an *inspection chamber* (*C*), cast to form a deep channel sloping gently up to the walls each side. It ensures that no solids are left after flooding, and is stood on when *rodding*.

bench plane A *plane* used at a bench, for dressing flat timber, e.g. a *jack plane, jointer, try plane*.

bench sander A fixed *sander* used in the workshop. The wood is brought up to it. Three types exist: *belt sanders, disc sanders*, and *orbital sanders*. All of them use strong *sandpaper*, unlike *grinders*.

bend A short length of pipe or duct for turning a corner, which also allows *thermal movement* in the pipework. For pipes a 90° bend is called a quarter bend, a 45° bend is called a one-eighth bend. The radii of pipe bends can be *quick*, normal, or *slow* and they are less abrupt than an *elbow*. For low-velocity *ducts* bends are made of sheetmetal with seamed edges and may have vanes to turn the air flow. See **lobster-back bend, duckfoot bend**. See also diagram, p. 177.

bender A *bending tool* or a machine for bending *reinforcement bars* or pipes.

bending Making a curve in a straight material (e.g. a pipe or bar) by forcing

it to change shape. By comparison, sheet materials are *folded*. See also *C*.

bending schedule, bar s. A list of *reinforcement bars* giving their location and details of *cutting and bending*, usually worked out by the steel supplier from the *structural drawings*. Bars are delivered to site in bundles, each with a label enabling the *steelfixers* to see which bending schedule to use.

bending spring A helical spring of circular cross-section inserted into a copper or plastics tube to keep it circular during hand bending, often after heating by a *torch*. The spring is kept oiled so that it can be easily withdrawn. Small-diameter tubes are inserted inside the spring and are not heated. Using a bending spring is more convenient than *loading* a pipe with sand.

bending tool, bar bender (USA **hickey**) A hand tool with two lugs and a long handle for bending steel or plastics tubes, or *reinforcement bars*.

bent A piece of timber curved by lamination or steaming. See also *C*.

bespoke system High-quality purpose-made building elements, or complete buildings, e.g. curtain walls, farm buildings.

best reed (Arundo phragmites) The true *thatching* reed.

bevel (1) The meeting of one surface with another at an angle not a right angle. A *chamfer* is a special bevel.
(2) **tee-b., b. square, sliding b.** A tool with a movable blade for *setting out* angles. The blade passes through a stock (handle) to which it can be clamped by a screw and wing nut at any desired angle. It is used by joiners, tilers, and bricklayers.

bevel halving, bevelled h. A *scarf joint* between timbers.

bevelled closer A brick with a vertical *bevel* cut running from the middle of one side to a far corner. About a quarter of the brick is cut off.

bevel siding (USA) Another name for clapboard (in Britain, *weatherboarding*).

beverages area A counter with a sink, water, and power supply for preparing hot drinks.

bib, bibcock A water draw-off *tap* fed by a horizontal supply pipe, not by a pipe from below, as is usual for a washbasin with a *pillar tap*.

bid (USA) A *tender*.

bidet A *sanitary fitting* for washing the nether parts of the body (personal ablutions). Water is supplied from an over-rim mixer or *upward spray*. The discharge is *waste water*, not *soil water*.

bi-fold door A compact cupboard door with two pairs of folding leaves.

billing Writing the *description* and *quantities* of each *item* in *bills of quantities*, using processed data from *taking-off*. The final part of *working-up*.

bills of quantities (BQ, BOQ) (USA **bill of materials**) A set of *descriptions* of the materials and labour needed for construction work, with their *quantity* and *unit price* (m, m², m³, kg etc.), listed as numbered *items*. Each building *trade* has an individual bill, numbered 'Bill No. 1', etc., hence use of the plural 'bills' for a complete document. Bills of quantities are prepared by a *quantity surveyor*, for use in *tendering*, to provide a common basis for contractors when *estimating* prices, as well as in the settlement of the *final account*. The rules for preparing bills of quantities are given in the *Standard Method of Measurement*, which also gives the *sequence of trades*.

In a *contract* 'with quantities' the bills become a *contract document* and *extended prices* have to be filled in, making a *priced bill*, which is used for pricing *variations*. Other types are bills of *approximate quantities* and *elemental bills*.

bimetallic corrosion Corrosion from *dissimilar metal contact*.

binder (1) A material that hardens to bond things together. Binders include: the *film-forming agent* in paint; the *resin* in an adhesive; the *gypsum* or *lime* in plaster; the *bitumen* between felts in *built-up roofing*; and *hydraulic binders* such as cement.

(2) Any *tie* securely fixed to a number of main framing members to stop them moving in relationship to each other, i.e. to make them act in common.

binder bar, binding Secondary *reinforcement*, e.g. *stirrups* in beams or columns, that forms a *cage* to resist shear forces and hold main bars in position.

binding wire, tying w., tie wire Soft *annealed* iron wire used for *steel-fixing*.

biodegradation The attack on materials by micro-organisms, e.g. on some sealants in wet areas, round *water closets*, or in swimming pools. All dead organic material biodegrades in the ground, providing food for growing plants or other life.

bi-parting door A *centre-opening door*.

birch (Betula) The silver birch and other birches are the hardest European *hardwoods*. Its wood is white to pale brown, usually without *figure* and more workable but stiffer than oak. It is used for making furniture and plywood.

bird screen A grille of wire netting to stop birds entering *fresh-air inlets*, *exhaust outlets*, and other ventilation openings.

birdseye, peacock's eye A *figure* in timber formed by small pointed depressions in the *annual rings* of many consecutive years, one of the few figures which show up well in *rotary cutting*, particularly with maple.

birdsmouth joint (USA **bird's mouth j.**) A V-shaped cut on the underside of a timber where it crosses another. In *roof cutting* it allows the *rafter* to fit over a *wall plate*.

bit (1) An interchangeable working tool that fits into the *chuck* of a *drill*, a *router*, or a carpenter's *brace*. The most common bit is the *drill bit*, but there are also *countersinks* and *screwdriver bits*.
(2) The working head of a *soldering iron*, usually made of copper.

bit gauge A small metal piece that can be fixed to a drill *bit* to stop drilling at the right depth.

bitty A bitty paint or varnish contains small *nibs* of paint skin, etc., which stick up above the paint surface after a coat is applied.

Bitumastic A *bituminous paint*.

bitumen A smooth, black, heavy material which becomes liquid when heated, made from natural or distilled petroleum and similar to *pitch* and *tar*, which are made from coal. Bitumen with *fillers* is *asphalt*. Bitumen is used as a waterproof *binder* and in hot *bonding compound*. It may be boiled or *blown bitumen*, or *polymer-modified*. Bitumen is degraded by sunlight and air, which cause oxidation and brittleness, unless it has some form of *solar protection*. Old bitumen feltwork can be treated with a resaturant.

bitumen boiler, kettle, cauldron A heated vessel for melting *bitumen* before it is spread. It should stand on a firm, level surface, at least 3m from its gas cylinders. Hoses must be properly connected and in good condition. The lid should be kept on as much as possible, and overheating, splashing, contact with water, and naked flames avoided. In case of fire, the lid is put on and the gas turned off. It should not be left unattended or moved while heating.

bitumen felt, bituminous f., roofing f. Sheets of fibres matted into felt and bonded by saturating with *bitumen* or *bitumen-polymer*. The fibres are usually a non-woven *polyester* fleece, or glass wool, which both have better resistance to deformation, damage or tearing than traditional felts with rag, asbestos, flax, or jute fibres. Particular bitumen felts are made for each of the layers in *built-up roofing*, and for similar work such as *flashings* and *box gutters*. Other uses include *vapour barriers*, roofing *underlays*, and *sheathing felts*. They are mostly applied by *pour and roll* or *torching-on* and may have *bonding tape* at joints. They come in rolls and the many different types are grouped according to codes given in BS 747.

bitumen-polymer sheet High-performance *bitumen felt*, made with bitumen that contains a *polymer* modifier, such as ethylene/propylene or styrene/butadiene. It is reinforced with polyester or glassfibres and mostly used as a single layer in *membrane roofing*.

bituminized fibre pipe (USA) A *pitch-fibre pipe*.

bituminous paint, bitumen p. Paint made from *asphalt* or petroleum *bitumen* and a *solvent* or a bitumen emulsion. It is inexpensive and mainly used for waterproofing or to protect metals. Only non-tainting types may be used inside water storage *cisterns*.

black steel Steel with a surface layer of dark-coloured iron oxides used for *low-pressure hot-water* heating pipes. See also **black** (*C*).

blade (1) The vane of a *fan* impeller that moves air by spinning. In a *centrifugal fan*, the blades are either *backward-curved*, forward-curved, or radial. In a *propeller fan* they are aerofoil shaped. Fan blades are usually of metal or plastics.

(2) The part of a tool (trowel, plane) or earth-moving machine (bulldozer, grader) that touches the work.

(3) A *slat* of wood or glass as infill to a *louvre*.

blade seal A thin strip of rubber projecting from the edge of a door leaf, as a *wiping seal* to *weatherstrip* the frame or for *smoke control*.

blanc fixe Very fine barium sulphate (BaSO₄) powder used as a paint *extender*, made by chemical precipitation not, as are barytes, by crushing.

blanket, quilt Insulation made of *mineral wool*, such as *glassfibre*, that comes in rolls, usually with a paper lining to each face. Some quilts are backed with *aluminium foil* to reduce heat losses by radiation. It is installed as ceiling insulation by rolling out to form an overlay on top of the ceiling joists. This creates a *cold roof*.

blank panel Opaque infill to a frame, instead of *glazing*.

blank wall, blind w. A wall with no openings for doors or windows.

blank window A window that has been walled up.

blast cleaning Driving a high-speed stream of abrasive, air or water on to a surface by forcing it out of a pipe with *compressed air*. Many different abrasives are used, such as sand (*sand blasting*), cast-iron shot, *grit*, and *corundum* (aluminium oxide) or any combination, as in *wet blasting*, depending on the surface and what is to be cleaned off. All dry blast cleaning needs *dust* con-

trol to prevent lung diseases. Blast cleaning of steel is used before *shop priming*, to remove *mill scale* (*C*) and roughen the surface as part of *preparation* for painting, and is followed immediately by a *blast primer*. Site blast cleaning can remove old paint or renovate external stonework. Skill and care are needed to avoid damaging soft stones.

blastfurnace cement An economical *blended cement* made by mixing finely ground blastfurnace *slag* with Portland cement clinker.

blastfurnace slag See **slag**.

blasting (1) *Blast cleaning*.

(2) The use of *explosives* (*C*).

blast primer A quick-drying *pretreatment primer* applied immediately after the *blast cleaning* of steel. It prevents re-rusting for a short time, and allows the steel to be handled. Welding primers are similar.

bleaching The removal of colour by chemical action, often an oxidation caused by sunlight and air. See **weathering**.

bled timber Timber from *yellow pine* trees that have been tapped for resin, usually inferior to other timber.

bleed To blow off or purge gases from a pipe carrying a liquid. See **blow down** (*C*).

bleeder pipe, b. tile (USA) A pipe through a retaining wall as a *weephole* for water.

bleeding (1) **bleed-through** The penetration of a coat of paint by substances from the *substrate* that dissolve in the paint's *medium*, usually causing discoloration.

(2) The passing of adhesives through cracks in wood veneer.

(3) The elimination of gases from a

pipe by release from an automatic *air eliminator* or a *bleed valve*.

(4) The exudation of water from concrete or of bitumen from *tarmacadam* (*C*).

bleed valve A valve for *bleeding* gases from a pipe.

blemish Anything which mars the appearance of a *finish*, without affecting strength. It is therefore not a *defect*.

blended cement Portland *cement* mixed with other materials that have some chemical reaction with it. This can be either from *pozzolanic activity* (*C*) with *free lime* liberated by cement as it sets, or from the presence of similar oxides but in different proportions, as has blastfurnace *slag*. Blended cements are economical, develop low heat in setting, have a lower water content for equal *workability*, and are more resistant to *sulphate attack* than *ordinary Portland cement*. However, they set more slowly and need longer *curing*, particularly at low temperatures.

blended water Hot and cold water that have been combined by a *mixer*, usually to give a comfortable temperature for washing or showering.

blind fixing Concealed *fasteners* or fixing systems.

blind hole A hole that does not pass right through a material, as a *bottomless hole* does.

blinding concrete (USA **slope c.**) A layer of *dry lean concrete* about 50 mm thick covering the bottom of an excavation, either over *hardcore* (*C*) or directly on the *formation*, to seal the underlying material and prevent mud from dirtying *reinforcement bars* or intruding into the concrete of *footings* or a *ground slab*. See **oversite concrete**.

blind mortice A *mortice* that does not pass through the timber.

blind nailing *Secret nailing*.

blinds Panels of textile which can be drawn down inside or outside a window to control *daylight* or heat from the sun; a specialist *trade*.

blind wall A *blank wall*.

blister figure A *quilted figure*.

blistering (1) Bubbles in a paint surface caused by moisture or resin vaporizing under the surface.

(2) (USA **blub**) Local swelling in finished plasterwork, a *defect* which may cause the finishing coat to fall away from the background. See **blowing**.

bloated clay *Expanded clay*.

blob A *manifestation* of new glazing.

block (1) **building b.** A rectangular masonry unit that is larger than a *brick* and so lifted with both hands. Blocks are also rectangular on face, no more than 650 mm long and thicker than a *slab*. The most common face size is 440 × 215 mm to course with the metric *brick format*, or 400 × 100 mm, for the 10 cm *module*. They are manufactured in many different thicknesses, surface colours, and textures and have largely replaced *common bricks* for walls to be plastered. Precast blocks are mostly made from dense or *lightweight concrete* by the *immediate demould* process, which gives accurate sizes: thus *blockwork* can be built as *facework* (or 'fairfaced') both sides. Block types may be *solid*, *hollow* or cavity, or *cellular*. Building Regulations for thermal *insulation* have led to lightweight blocks which still have acceptable loadbearing strength, often made from cellular material such as *autoclaved aerated concrete*. See diagram.

(2) A regular-shaped building compo-

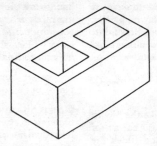

A hollow concrete block. 440 × 215 × 215 mm and many smaller sizes.

nent: *angle blocks* in joinery, *architrave blocks* for skirtings, *floor blocks* in a concrete floor slab, *glass blocks*, plastic *spacer* blocks for glazing, *wood-block flooring*, etc.

(3) A rough or irregular mass of material, e.g. *asphalt* blocks for melting or concrete *anchor blocks* for pressure pipes.

(4) A building that is rectangular in shape, usually *high-rise*, but also a low block or *podium*; or a *plantroom* that is *rectangular on plan* and has a flat roof.

(5) A *pulley* in a *sheave* (*C*), used with ropes for hoisting or pulling.

blockboard (USA **staved lumber core**) A *building board* made of wooden *core* strips up to 30 mm wide, glued between *veneers* with the grain at right angles, forming a *composite* board. Lightweight blockboards can use a low-density timber in the core. Blockboard resembles *laminboard* but is cheaper.

block bonding Connecting several *courses* of brickwork of one wall into the courses of another, often in order to *bond* shallow facing bricks into thicker common bricks in the *backing*, and to bond new work into old, in *toothing*.

block bridging (USA) *Solid bridging*.

block flooring *Wood-block flooring*.

blocking (1) Stopping up a hole, e.g. with a *fire barrier* in services ducts.
(2) A *spacer* for setting to level, an *angle block*, or *firring*.

blocklayer A *bricklayer* who builds *blockwork*.

blockmaker, blockmaking machine A machine for making precast concrete building *blocks*. Factory blockmakers are mostly fully automated and use intense *vibration* to compact a dry *mix* in a mould, followed by *immediate demould* on to a tray, transfer to pallets, and storage until hard. Some processes recycle excess water containing cement, in which case the water must be protected from pollution. Mobile blockmakers have a hand-operated hydraulic press used both for compaction and to eject the fresh blocks directly on to the ground, where they are left until strong enough to use.

blockout (USA) A *pocket*.

block plan A small-scale plan showing the broad outlines of existing buildings or of proposed new buildings.

block saw A mason's *handsaw* for cutting *lightweight concrete blocks*. It can have *carbide tips*.

blockwork Masonry of precast con-

bloom

crete building *blocks* laid in cement mortar; also the *trade* of laying blocks. Blockwork is similar to *brickwork*, but quicker and usually cheaper unless plastered. As walling or *partitions* it is more thermally insulating, but more subject to moisture and other *movement*.

bloom (1) A thin film, like the bloom on fruit, which forms on the surface of *gloss* paint or varnish and hides the colour or reduces the gloss. This defect can be due to the paint composition or to improper *application* in humid air conditions. It can sometimes be removed with a cloth. See **blushing**.
(2) *Efflorescence*.

blower (1) A *fan* for air conditioning, or a compressor.
(2) A *blow gun* for cleaning *formwork*.

blow gun, lance A length of steel pipe with a flattened tip, connected to a compressed-air hose, used for *blowing* out *formwork*.

blowhole, superficial void A bubble of air trapped against *formwork* during the *placing of concrete* and not driven out during *vibration*. It is usually filled by *patching*.

blowing (1) **pitting, popping** Small pits in the surface of plasterwork formed by the expansion of material behind the surface. The cause of this *defect* may be lime that was incompletely slaked or the slow oxidation of coal in the lime.
(2) See **blown screed**.
(3) **blowing out** *Air-blast cleaning* of *formwork* with a *blow gun* supplied with compressed air, to remove rubbish from steelfixing and other trades before the *placing of concrete*.

blowlamp, blowtorch A compact burner with a powerful flame and a built-in fuel supply, used for heating materials, mostly to soften them for hand work such as soldering and bending pipes or conduits. Blowlamps now consume *bottled gas* in throw-away canisters. They should be changed outdoors. See **heat gun**, **hot work**.

blown air *Supply air* distributed by a fan through ductwork.

blown bitumen *Bitumen* oxidized by blowing a stream of air through the heated material, which makes it more rubbery and raises its softening temperature, for use as *bonding compound*. Synthetic *polymers* (such as styrene/butadiene) are often added to improve flexibility, give better fatigue resistance, and overcome age hardening.

blown oil *Linseed oil* that has had air forced through it while also being boiled. It dries more quickly and is used in *oil paints* and *linoleum*.

blown screed A *screed* that has partially lifted off its *background* because of *bond* failure.

blowpipe A torch used for *oxy-cutting* steel.

blub (1) *Blistering*.
(2) A hole in a plaster cast formed by an air bubble.

blue brick See **Staffordshire blue brick**.

blueing Increasing the apparent whiteness of a white *pigment* or paint by adding a trace of blue.

blue stain, b. fungus, sap s. A blue fungal discoloration in the *sapwood* of timber which does not reduce its strength. It goes grey with age.

blushing Milky opalescence in *lacquer*, a defect caused by application in cold or wet weather, or a lack of *compatibility* with other materials.

board *Building board* or *panel products*, often called sheet or slab; formerly *square-sawn timbers*. See also **switch-board, distribution board, fuse-board**.

board-finish plaster, board p. Usually a low-expansion *retarded hemihydrate plaster*, suitable for single-coat finishing work on true surfaces such as *gypsum plasterboard*. It is one of the few plasters to which lime should never be added.

board foot The North American unit of *board measure* of sawn timber, formerly used in the UK: the amount of *lumber* in a piece which before sawing, planing, and *shrinkage* measured 12 × 12 × 1 in. All three of its *dressed* dimensions may be short.

boarding, sheathing boards Boards laid side-by-side; *tongued and grooved* boarding is used for *flooring*, *close boarding* for roofing, sheathing, fences, etc., *shiplap boards* for *weatherboards*, *hit and miss* boarding for fences, and so on.

boarding joists *Common joists* for floorboarding.

board measure The superficial measurement of a quantity of timber, in the UK by the *board metre* and in America by the *board foot*.

board metre The quantity of timber in a piece measuring 1m² in area and 25 mm thick.

boasted ashlar, b. surface A rough finish to stonework made by *boasting*.

boaster (Scotland **drove**) A mason's *boasting* chisel, 40 to 80 mm wide, struck by a mallet in dressing stone.

boasting The hand *wasting* or rough dressing of the surface of a stone with oblique or vertical strokes, which are usually not uniform, from a *boaster*.

boasting for carving Leaving stone over-sized and roughly dressed for sculpture or similar decoration.

boat scaffold A *flying scaffold*.

boat spike A *ship spike*.

bob A *plumb bob*.

body (1) The main part of a valve or similar plumbing *fitting*.
(2) Open-textured *fireclay* glazed to make *faience* tiles.
(3) The stiffness (consistency) of a *paint* or the solidity of the dried *film*. Non-drip paints have *false body*.

body carpet Carpet in rolls 27 in., 36 in. or 1 m wide, or to special widths.

bodying-in, b.-up The early stages of *French polishing*, including staining, filling, and the first polishing before *spiriting-off*.

boiled oil *Linseed oil* heated briefly at about 260°C (not boiled) with soluble *driers*, used to make *oil paint*.

boiler (USA **furnace**) A water heater in which the water should not boil. A boiler is used for the production of *domestic hot water* and hot water for *central heating* (or both in *combination boilers*). It is heated by a *furnace*, or can be an *electrode boiler*. Domestic *small-bore units* or commercial cast-iron sectional and *shell boilers*, fired on oil or gas, need a *flue*. Gas can be used for *condensing boilers*. Large oil-fired boilers have *rotary cup burners*, *forced draught*, automatic *purges*, soot-blowing equipment, and usually an *economizer* (*C*).

boiler house, b. room A *plantroom* for housing *boilers*.

boil-resistant adhesive (BR) An *adhesive* with good water resistance, able to withstand cold water for many years and attack by insects and fungus.

bollard A *lighting bollard*. See also *C*.

bolster (1) A broad-faced steel bricklayer's chisel with a blade about 110 mm wide.

(2) **corbel piece, crown plate, headtree, saddle** A hard timber cap over a post to help carry a beam by increasing the bearing area under a beam.

bolt (1) A *fastener* with a head and screw thread, for a nut. Examples are: *coach bolt, high-strength friction-grip bolt* (C), *holding-down bolt, hook bolt, toggle bolt,* and *U-bolt.*

(2) **draw b.** The tongue of a *lock* that prevents the door opening when it is out (thrown) and allows it to open when *withdrawn.* See **claw-bolt lock, dead bolt.**

(3) *Hardware* to hold a door or shutter closed, usually with direct or simple action. Types include: *barrel bolt, Cremona bolt, espagnolette bolt, extension bolt,* and *panic bolt.*

bolt box, b. sleeve *Formwork* round foundation *holding-down bolts* for structural steelwork, to make a *pocket* in the concrete. During concreting the bolts may move out of position by up to 20 mm (very accurate setting-out is wasted). After the concrete hardens steelwork is erected, with the bolts passing through holes in each *baseplate.* The pocket formed by the bolt box allows the baseplate to be moved into its right place. Later the bolt boxes are *grouted.* See diagram, p. 231.

bolt croppers A pair of hand shears used for cutting bolts or steel rods.

bond (1) **bond strength** A strong connection resulting either from *key,* which is mechanical, or from *adhesion,* which may be chemical, or from a combination of both, as in *high-bond bars* (C).

(2) The layout of *courses* of bricks or blocks in a wall, devised when mortars were weaker and had less *adhesion* than now. It was important then to keep every vertical joint at least a quarter length of a brick or stone from the next vertical joint above or below, using *bats* and *closers.* Brickwork today is mostly *facework* laid to *stretcher bond.* Bonds for thick walls are rare, except for: *English bond, Flemish bond, gardenwall bond, heading bond, Quetta bond,* and *rat-trap bond.*

(3) The arrangement of *roof tiles* or slates to exclude rain, by placing the joint between adjoining tiles over or near the centre of the tile of the course below.

(4) A surety put up by a bank which promises to pay the *client* if the builder does not stand by a *tender* price, or does not do the work for *performance* of a contract, or if the builder has been paid an *advance.* Bonds are not usually payable on demand – the client must have proof of *default* for it to be called.

(5) An electrical connection for earth *bonding.*

bond beam A reinforced concrete beam made inside a course of hollow blockwork from which the block ends and internal walls have been removed. The beam ties the wall together.

bond breaker A material that prevents adhesion, e.g. of a *sealant* to a *backup strip* or of the moving half of a *dowel bar* in a *ground slab.*

bond course A course of *headers* in a brickwork or blockwork wall.

bonded See **bonding.**

bonded screed A *separate screed.*

bonder, bonding brick A *header* in a *bond course,* or a bond stone that passes completely or partially through a wall.

Bonderization A common type of *passivation.*

bonding (1) The use of high-strength *adhesives,* usually thermosetting resins, elastomers, or thermoplastics, to make permanent structural joints, composite metal members, prefabricated building panels, etc. Careful pre-treatment and degreasing is usually required. See **bond**.

(2) The putting-up of *surety* by a bank, e.g. for a *performance bond*.

(3) **earth b., equi-potential b.** The interconnection of metal parts in a building, such as pipes, ducts, baths, handrails, reinforcement, etc., with heavy cables, which are then also connected to an *earth electrode* to protect against electrical shock by preventing any dangerous rise of voltage above earth potential, and to reduce electrical interference between equipment. Compare **earthing**.

bonding agent, b. primer Many *bonding treatments* exist, often of secret compositon. They are put on a *substrate* as a coat, as part of the *preparation* before the application of asphalt, paint, plaster, timber lamination, or veneer, to improve adhesion.

bonding brick A *header* in a *bond course*.

bonding capacity (USA) A rating given by a bank that indicates the financial stability of a contractor, and therefore the amount of work that can be taken on.

bonding compound, hot b.c. Hot molten *blown bitumen* put on to a roof to stick layers of *built-up roofing* together. *Sealing compound* is similar but applied cold.

bonding conductor A heavy cable for earth *bonding*.

bonding plaster A pre-mixed lightweight *gypsum plaster* containing *exfoliated vermiculite*, used as the *undercoat* in *two-coat work* on surfaces difficult to bond, such as smooth-formed concrete walls.

bonding tape Tape glued on both faces for joining sheet materials in *damp-proof courses, vapour barriers,* etc., whether of *bitumen felt, pitch-polymer,* or polyethylene.

bonding treatment Any treatment of a surface before the application of a finishing (paint, plaster) so that it will stick. This can include *keying, spatter-dash,* the use of a *bonding agent* or *pretreatment primer,* etc.

bond length The length of concrete reinforcement *bar* required at a *lap* or *curtailment* to develop full bond strength.

bond stone A *bonder*.

boning The use of T-shaped rods for *setting out* a flat surface, for excavations, *ground slabs*, drain-laying, etc., now less common than *lasers* (*C*).

boning pegs Small hardwood pegs that were placed at the edges of large stones to be dressed flat, used with *winding strips* as guides.

bonnet (1) A cover or hood, e.g. the roof over a *bay window*.

(2) A cover over the *body* of a valve or tap.

(3) A wire-netting sphere put into the top of a vent stack or chimney as a *bird screen* or to prevent sparks going out.

bonnet tile, cone t., hip b. A *hip tile* with a rounded top. It is bedded in mortar on the next bonnet tile.

bonus system A wages incentive system in which an additional payment is made on top of the normal daily payment if a set target of work is achieved.

bookmatching The cutting, sawing, or slicing of wood veneer or marble so that the two faces from each side of the cut are opened out to form a mirror image of each other.

boom The *jib* of a crane, the *dipper* arm of a *backhoe*, etc. See also *C*.

booster (1) A pump for increasing the pressure in a water supply pipe or *fire riser*, often fitted with a diaphragm tank pressure maintenance vessel.

(2) A heating element in an electric *storage water heater* which is switched on during the day if the draw-off temperature drops too low.

boot A step down formed in the edge of a concrete floor slab (or a beam or lintel) to carry the outer skin of a *cavity wall*. This allows a drop in the *damp-proof course* to drain rainwater outwards through *weepholes* in the facing brickwork. The *movement joint* under the boot is weatherproofed with a *sealant* and the boot outside face may be concealed with *brick slips*.

booth A small room or cubicle, for privacy or isolation, often with *sound insulation*.

boot lintel A lintel across a door or window opening, with a *boot*.

bore (1) To drill a hole, or the hole after it is drilled.

(2) The internal diameter of a hole or a pipe. Copper pipe for water supply is mainly *microbore*.

bore lock, key-in-knob set, tubular mortice l. A door lock with a tubular T-shaped case. It fits into a bored *mortice* made with a *hole saw* from the face and with a drill from the *stile*, which is easier than cutting a rectangular mortice.

borer, woodworm Wood-boring beetles in the larval (grub) phase, which burrow into timber for food and shelter, leaving worm holes when they emerge as adults. Borers found in buildings include the common *furniture beetle*, *powderpost beetle*, *death watch beetle*, house *longhorn beetle*, and *wood-boring weevils*, each with different habits. Treatment against attack using *preservatives* usually does not penetrate

deeply enough to reach all borers, but they are all killed by high temperatures, as in *kiln-dried timbers*. See BRE Digest 327.

borosilicate glass Heat- and chemical-resistant clear glass made from borax as well as the usual silica, giving complex compounds of sodium borosilicate. It is used in factory-shaped and tempered *fire-resisting glass*, as *glassfibre*, to make vitreous *enamel*, and in glass drain pipes.

borrowed light A window in an internal wall or partition (BS 6100).

boss (1) An area raised above a general surface, for example an upstanding flat area on the curved shell of a boiler or hot-water *cylinder*, usually with a hole for attaching a *fitting*, e.g. a pipe union for an inlet or outlet, or for inserting an *immersion heater*.

(2) The outside of a groove for a *bossed connection* in a plastics pipe.

(3) A rounded projection at the intersection between ribs in a roof vault, made by the inside end of a keystone, which is often carved.

bossed connection, b. end A groove in the *socket* end of a plastics pipe to receive an *O-ring joint*. The pipe has constant wall thickness so the groove appears on the outside as a boss.

bossing The shaping of sheet metal with a former and a mallet.

bottled gas *Propane* or *butane*, or a mixture of them, compressed and sold liquid in steel cylinders as a portable supply of gas under pressure. As both gases are heavier than air and dangerous, bottles must be kept outdoors or in a cupboard ventilated directly to the open air.

bottle trap A trap on the *waste* from a kitchen sink or bathroom basin which has a bottle-shaped cap that can be removed to clean out any rubbish causing a blockage, such as matches, hairpins, and hair. Bottle traps are usually less self-cleansing than a tubular *S-trap*.

bottom chord The lowest member of a *truss* or *girder*.

bottom-hung window A window with its opening *sash* hinged at the bottom.

bottoming Hand trimming *excavations* with a shovel, working behind an excavator such as a *backhoe*, to clean out lumps of dirt and leave a tidy *formation*.

bottomless hole, through h. A hole which passes through a material. It is therefore difficult to make a fixing in it except with a *toggle bolt*, *anchor*, etc. Compare **blind hole**.

bottom plate A member along the bottom of a *framed partition* that spreads the loads from the studs to the supporting floor.

bottom rail The horizontal bottom member of a *framed door*, *casement* or lower *sash*.

bottom tie A *base tie* of a pressed-metal door frame.

bottom-up construction The usual way of building, the opposite of *top-down construction*.

bottom ventilation An air inlet for natural ventilation, particularly to a *plantroom* or machinery enclosure.

boundary The edge of a building *site*, often enclosed by a *hoarding*.

bowing, camber Timber *warp* at right angles to a face, commonly due to uneven *drying shrinkage*. Steel bows from uneven heating (thermal bowing).

bowl (1) The part of a *sanitary fitting* made to contain water, usually with a *waste* or outlet. Kitchen *sinks* may have one or two bowls. A bowl of a WC is also called the *pan*.

(2) A light *diffuser* over a *luminaire*, often made from *acrylic sheet*.

bowl urinal, pod u. A rounded *vitreous china* urinal, usually rimless and wall-hung at varying heights to suit users, to BS 5520.

bow saw An ancient type of saw with a thin blade held at both ends in a loose H-frame and tightened with twisted wire or string. See diagram, p. 388.

bow window A projecting *bay window* which is curved on plan.

box (1) A wooden case, or a *casing* of building board on grounds or studs.

(2) *Formwork* for in-situ concrete, either for the outside of a *column* or to make a hole or *pocket*, e.g. a *bolt box*.

(3) A small electrical enclosure, unit, or case, often for *fuses*.

box cornice The edge of a roof built out to conceal the gutter, with a *fascia* in front and a *soffit board* underneath.

boxed heart The *heart centre*, usually within a 100 mm square, cut out during the *conversion* of timber and discarded. This is done with some *eucalyptus* hardwoods, which have poor heart.

box frame A sash window *cased frame*.

box gutter, trough g. A roof gutter below the general level of the roofing, with a lining in *supported sheetmetal roofing*, or bitumen roofing felt over a wooden box. Box gutters should always have a safety *overflow*, such as a *scupper* through a parapet, so that if the outlet becomes blocked with leaves or snow, the rainwater will not flood through the ceiling or overload the roof.

boxing A formwork *box*, or a *casing* of building board on grounds or studs.

boxing shutters, folding s. Internal *shutters* that fold away into a recess at the side, called the boxing.

box-out A *pocket*.

box staple, b. strike A metal box on the side of a door frame into which the latch of a *rim lock* enters when the door is closed.

boxwood A hard *hardwood* used for making chisel handles.

BQ, BOQ *Bills of quantities*.

brace (1) A strut or tie that stabilizes other members, usually placed so that a triangle is formed, e.g. a *knee brace*, the braces on a *matchboarded door*, temporary braces during construction etc. See also *C*.
  (2) **carpenter's b., bit stock** A hand tool used for turning a drilling *bit* to make holes in wood. The bit is held in the *chuck*.

braced frame A *structural timber* building frame with widely spaced, heavy posts and beams or girders which carry the main loads. It has *infill walls* which are *non-loadbearing*. Bracing for structural stability comes partly from the joints between the posts and beams, although additional *cross-bracing* can be concealed in the walls.

bracket (1) A projecting support usually fixed to a wall or column. Stair *handrails* are often on metalwork brackets. Pipework can be supported by brackets with *two-piece cleats*. See **corbel** and **cantilever** (*C*).
  (2) A short vertical board fixed to the *carriage* of a timber stair to support the *tread* directly.

bracket arm An *outreach arm*.

bracket lamp A *wall-hung lamp*.

bracket scaffold A scaffold carried on brackets bolted to a wall.

bracketed stairs Stairs with treads carried on a *cut string*, usually with overhanging *nosings*.

brad (1) A *wire nail* usually 50 to 65 mm long, with a rounded bullet head.
  (2) A *cut nail* of constant thickness but tapering width, with a square head projecting from one edge only, or an *oval-wire brad*.

bradder A small *nail gun* that drives *brads* up to about 65 mm long.

braided cable Flexible flat cable made from several strands of bare copper wire plaited together, used for the earth *bonding* of moving parts, e.g. switchboard doors.

branch A secondary pipe or cable connected to a *main* in a distribution or collection system, such as water supply, drainage, sewerage or electrical circuits. Also called a lateral or spur.

branch circuit, final sub-circuit The electrical wiring to power or lighting use points, from the fuses in the *distribution board*, usually run as either a *ring main* or a *radial circuit*.

branch manhole A *manhole* (*C*) in which a branch connection is made.

branch pipe, b. discharge p. A *branch* to carry *soil water* or *waste water* to the *stack*.

branch vent, b. ventilating pipe, anti-siphon p., trap v. A pipe which admits air to the downstream side of the water *seal* in a *trap* at the start of a drain. Usually it is run upwards to the *stack vent*, which lets in air through its top end. The branch vent prevents *unsealing* of the trap as a flush of water passes down the *soil stack*, e.g. as in a *one-pipe system* of sanitary pipework.

brash timber, brashy t. Timber

which breaks with small resistance to shock and little or no splintering. This 'short grain' may be due to fungus.

brass A metal alloy made from a mixture of copper and zinc, often with small amounts of nickel, lead, iron, etc., used in traditional builder's *hardware* and some pipe *fittings*. Brasses are mostly easily formed, strong, and corrosion-resistant, but are best not used with other metals, which could cause *dezincification*. Alpha brasses have up to 37% zinc and are softer than beta brasses with 37 to 46% zinc.

Brazilian mahogany, American m. (Swietenia mahogoni) True *mahogany*, a close-grained red timber with fine silky texture, hard, and with very little *moisture movement*. It is stronger than *oak* but slightly easier to work and used for the very best *joinery*.

brazing A generally simple, inexpensive way of joining metals using a copper–zinc filler metal (hard solder) which melts above 500°C, normally with an *oxy-acetylene flame*. The two surfaces need suitable *preparation*, usually limited to thorough cleaning. A *flux* is also needed. Brazing, particularly *silver brazing*, is widely used for *capillary joints* in copper tube or brass pipe. Brazed steel joints are not as strong as welded ones.

BRE *Building Research Establishment*.

break (1) To cut or interrupt, the opposite of *bridging*. A *capillary break* stops rising damp, a *switch* breaks an electrical circuit, *ground breaking* is the start of excavations, and *demolition* is the breaking-down or breaking-out of work.

(2) To leave an open space, as a *fire break*, or any change, discontinuity, or offset.

break draught damper A flue *damper*.

breaker A power tool for breaking rocks, concrete, or roads, by forcefully hammering a pick or similar tool. It can be a hydraulic, pneumatic, or electric hand-held *jackhammer* or mounted on a *backhoe*.

breaker slab, cable cover, warning tile A clay or concrete slab in the *backfill* over buried cables, to indicate their presence, in the same way as *tracer*.

break-glass unit A manual *fire alarm* call point with a breakable glass cover. Replacement glass should be similar to the original.

breaking-down The first of two stages in *conversion* of timber, in which the log is sawn into large squared *flitches*.

breaking of seal The *unsealing of traps*.

breaking-out See **break out**.

breaking the bond Inserting a *bat* in brickwork to give *broken bond*.

break joints *Staggered joints*.

break out To cut away part of a building *fabric* to make a secure connection with new work, e.g. the *toothing* of brickwork or the exposing of reinforcement bars.

breast (1) **chimney b.** A projection of a wall into a room, containing the flue and hearth of a fireplace.

(2) The wall under a *window sill*, down to the floor.

(3) (Scotland) The *riser* of a stair.

breast drill A large hand *drill* operated with both hands, with an extension and plate for thrust from the user's chest.

breather hole A tiny hole from the air space between *secondary glazing* out to the 'cold' side, where the air contains less moisture in winter, even at a higher *relative humidity*. As atmospheric pressure rises and falls, the air space gradually fills with dryer air, reducing *condensation*, but dirt and insects may enter unless the holes have a filter, e.g. porous tape.

breather membrane, b. paper, insulation sheathing A *building paper* that is microporous to allow *ventilation* and the escape of water vapour. It is put on over the outside of heat *insulation* and under the external *cladding* of a timber-framed wall (or roof). It also helps prevent *driving rain* from passing inwards. Breather membranes must be on the 'cold' side of the wall, that is on the opposite side of the framing to its *vapour barrier*. Some are made for rough handling and so can be used as 'housewrapping' to keep out the wind and wet during construction. See BS 4601.

BRE Digests Brief and to-the-point reviews of nearly 400 building subjects published as leaflets by the *Building Research Establishment*, covering materials, building performance, structures and services, design and site procedures, defects and repairs. Before use, the latest Index, issued each year as a Digest and printed on yellow paper, should be consulted.

breech fitting A *junction* pipe fitting.

breeching A double inlet to a *dry riser* accessible from the street.

breeze block A building *block* formerly made with coke breeze from gas works. It is superseded by the *clinker block*.

breezeway A *covered walk* between buildings.

bressummer, breastsummer A long heavy *lintel* across an opening, to carry the wall above. It was usually in timber and the wall was of brickwork or masonry. Reinforced concrete or steel beams do the same job today.

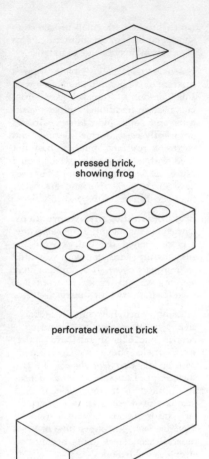

pressed brick,
showing frog

perforated wirecut brick

solid wirecut brick

Three types of brick.

brick, building b. A rectangular block no larger than 338 × 225 × 113 mm. Building bricks can be easily held in one hand and are usually proportioned for the lengths of each side (face, bed, end) to form multiples that work *bond* sizes

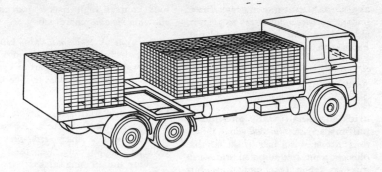

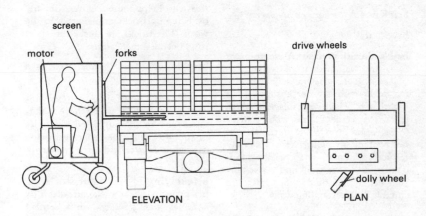

Brick deliveries.

when laid with mortar joints in walling. In hot climates bricks are often sun-baked (unburnt) from *adobe*. In temperate climates there is a long tradition of *clay bricks*, fired in a kiln, with fewer *concrete bricks*, *calcium silicate bricks*, and *firebricks*.

Building bricks vary widely in strength, durability, and appearance, from the most costly *engineering bricks* or *facing bricks* to *commons*. They are mostly used to build walls, and vary in type from *solid* to frogged, *cellular*, *perforated*, and *hollow*. Bricks shaped differently from the normal full brick are called *specials*. Bricks are also used as *pavers*. The delivery of bricks to building sites is usually mechanized, e.g. with self-offloading vehicles or transportable all-terrain forklift trucks, which handle shrink-wrapped or strapped packs. The packs can be split down into 'blades' for *brick trucks*.

In Britain today the standard brick is

215 × 102.5 × 65 mm in *nominal* size, making a *brick format* of 225 × 75 mm, although *modular bricks* are available. From 1936 until metrication the standard British brick was $8\frac{3}{4}$ × $4\frac{3}{16}$ × $2\frac{5}{8}$ in., making a format of 9 × 3 in. American sizes are less uniform than the British. See diagrams, pp. 48, 49.

brick-and-a-half wall, one-and-a-half-brick w. A *solid wall* with a thickness of one and a half brick lengths, plus one joint. In traditional brickwork this was $13\frac{1}{4}$ in. (340 mm) thick, but today is a 327 mm solid wall.

brick axe A *bricklayer's hammer*.

brickbat A *bat*.

brick damp course A *damp-proof course* in a wall made from two courses of *damp-proof course bricks*, or *engineering bricks* with a water absorption of less than 7%, laid with *staggered joints* in 1 : $\frac{1}{4}$: 3 cement : lime : sand mortar (designation 1 mix). The mortar should not contain any *plasticizer* as this may reduce adhesion. As brick damp courses are more rigid and resist tension better than other types, they are suitable for freestanding walls.

brick elevator Mobile *contractor's plant* (*C*) used for raising building materials to a *scaffold*. It has a rubber belt and steel *cleats* to stop bricks slipping at steep angles.

brick format The working size of brickwork, the *nominal* dimensions of a brick plus its associated mortar joints. With UK metric bricks and 10 mm wide joints the format is 225 × 75 mm (horizontal × vertical). The layout on plan of brickwork walls and openings is set out to multiples of half the format, or 112.5 mm, to avoid *broken bond*. Vertical format is based on the height of full *courses*. Four bricks long thus makes 900 mm and

four courses high makes 300 mm, allowing *modular* coordination.

bricklayer A *tradesman* who builds and repairs brickwork (bricklaying), on large projects working as part of a *gang* under a foreman bricklayer. See diagrams, pp. 51, 281.

bricklayer's hammer, brick axe (USA **ax, axhammer**) A small *hammer* with a sharp *cross-peen* as well as the striking face, used for breaking and dressing bricks.

bricklayer's labourer A *skilled operative* who helps the *bricklayers*, mainly by bringing bricks and mortar to the work. The bricks are stacked in short rows between *spot boards*.

bricklayer's line A *line* used as a guide during *bricklaying*.

bricklayer's scaffold A *scaffold* supported by *putlogs*, which have a flattened end to fit into raked-out holes in the *bed joints*, the other end being carried on *ledgers* held up by the *standards*.

bricklaying The skilled *trade* of laying bricks in *courses* to make *brickwork* by *bedding and jointing* with cement mortar. Joints are usually *shoved*, and the bricks kept true and plumb by working to a bricklayer's *line*, which is stretched between *profiles*, or between wall ends raised first by *racking back*. *Bed joints* and *perpends* should average 10 mm thick. Bricklaying can be *common brickwork* or *facework*. It is a *wet trade*. See diagram.

brick nogging Non-loadbearing brickwork infilling, usually between the *columns* of a concrete or steel-framed building, often laid *brick-on-edge*. *Wall ties* are needed as well as a *movement joint* above the top course.

brick-on-edge Brickwork, usually of

Bricklaying.

headers, laid on edge for a *capping* or *sill*.

brick slip, s. brick A *special* brick of less than normal width (20 to 50 mm wide instead of 102.5 mm), either made to size or, even better, cut with an abrasive disc from the same batch of *facings* as for the wall. It is used to mask the edge of a concrete floor slab in a brickwork wall or for a *boot lintel*, to which it is bonded with metal ties and resin mortar.

brick tie A *wall tie*.

brick trowel, bricklayer's t., laying t. A long tapering triangular *trowel* held in one hand for bedding and jointing bricks, which are lifted with the other hand.

brick truck A trolley with a tray and one or two wheels for picking up a stack or blade of *bricks* and wheeling them directly to the work, thus saving loading them into and out of a *wheelbarrow*.

brick veneer *Facings* laid *half-brick* thick to form the outer half of an external *cavity wall*, the inner half being *timber-frame construction*, with *studs* and a *dry lining*, used in North America and Australia.

brickwork (1) A wall made up of *courses* of bricks laid in *mortar* according to a *bond*, used for *cladding* and *infilling*. Brickwork, particularly of *clay bricks*, is strong, durable, and maintenance-free provided it is properly built. This entails correct choice of *mortar designation*, suitable *damp-proof courses* and *movement joints*, and of course skilled *bricklaying*. Since brickwork is permeable, external walls usually have a *cavity*. See BS 5628.

(2) The *trade* of bricklaying, forming a *work section* of a specification, often along with *blockwork*. In *quantity surveying* often shortened to 'bwk'.

brickwork beam A concealed *lintel* made by inserting *reinforcement* in the bottom bed course of brickwork over an opening, for clear spans up to 2.10 m.

brickwork chaser, keyway miller A power tool that cuts a neat *chase* in brickwork to receive electrical *conduit* or a pipe.

bridge, bridging A small area that allows a large flow of damp or heat through a material, often the result of poor *workmanship*. In a *cavity wall*, mortar droppings from bricklaying can lodge on *wall ties* and form a bridge for water penetration by *capillary action* as well as a *cold bridge*. In aluminium *patent glazing*, bridges are prevented by *thermal break* construction. See also *C*.

bridge board A *cut string*.

bridging (1) The spanning of a gap with *common joists*.

(2) The stiffening of adjacent wooden floor *joists* by a row of *solid bridging* or *herring-bone strutting* at right angles to the joists. When the floor is not stiff enough, bridging is used to make the joists deflect together when one is heavily loaded. Long spans have *double bridging*.

(3) The covering over of a gap in a *substrate* by a paint film.

(4) The by-passing of a *damp-proof course* (dpc), or of a *cavity*, which allows water to pass into a wall. This can

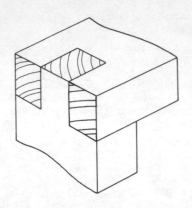

A bridle joint.

occur because of earth, rendering, or paintwork covering a ground-level dpc, or because of a *screed* at dpc level inside a house. The bridging of cavity walls may be due to mortar droppings on the *wall ties*. See also **bridge**.

bridging piece A short *bearer* either between or across *common joists* to carry a partition.

bridle (Scotland) A *trimmer joist*.

bridle joint A corner joint resembling a *mortice-and-tenon joint* and used for heavy framing. One member is cut at the sides to leave a central tongue which enters a notch or mortice. See diagram.

bright (1) A description of freshly sawn timber without discoloration.

(2) **b. steel** Steel with low surface oxidation.

brilliance (1) The cleanness and brightness of a *colour*.

(2) The clearness of a *varnish* or lacquer; the absence of opalescence and similar defects.

brindled bricks Clay bricks with stripes derived from the way they were stacked in the kiln, often used for *facings*.

bringing forward Making minor repairs to a *substrate* as part of surface preparation before repainting, such as patching the *plasterwork*, filling, and applying *priming* and *undercoats*. It is followed by the general application of the final coats.

brise soleil A French term for a permanent sun shade (horizontal or vertical blades, decorative screen walls) outside an external wall, used in tropical or Mediterranean countries to reduce glare and heat including *solar gain*.

British Board of Agrément The body which issues *Agrément Certificates*.

British Standard (BS) A document agreed between producers and users that describes how things should be done. It may include standard sizes of parts and products, details of widely accepted materials and workmanship, methods of testing, *codes of practice* (issued as a BS and called a BS Code), and glossaries of preferred terms. Standards allow coordination between different products and the general adoption of *reference specifications* by simply stating a BS number. Care should be taken not to infringe EC law by using BS numbers which would exclude an equivalent *European Standard*. Of the nearly 10 000 numbered and dated British Standards, many with several parts or sub-sections, some 1200 are revised annually by specialist committees. They can be bought from BSI sales in Milton Keynes or in Green Street, London W1. Many are held in local libraries, either printed, on microfiche, or on compact disc (see Preface for details). The main British Standards connected with building are mentioned in other entries in this dictionary.

British Standards Institution (BSI) The organization that prepares and publishes *British Standards* and *codes of practice*, as well as the English language versions of *European Standards*. The BSI also participates in the *ISO* for International Standards and cooperates with other national standards bodies: *AFNOR, ANSI, DIN*, etc.

British Thermal Unit (BTU, BThU) The amount of heat needed to warm one pound of water from 39° to 40°F, equal to 1.055 kJ or 0.252 kCal (1 BTU/h = 0.293 W).

broach (1) The pin in the keyhole of a lock with a hollow key.

(2) A pointed mason's chisel used for *wasting* stone.

broadloom carpet Fabric *floor covering* in rolls 2.50 to 4.00 m wide. Wide rolls reduce the labour of laying by eliminating some joints between rolls.

broken bond, irregular b. *Brickwork* or *blockwork* which is not entirely built to *format* and so has *bats* in the courses rather than all full bricks. It can occur from starting a corner with the *stretcher* along the wrong wall. In *facework* broken bond is unsightly, even with expensive *fair cutting*, and it should be eliminated by re-breaking or 'reversing' the bond below ground level with a bat.

broken white A white paint which has been toned down, usually to a creamy colour.

bronze A hard, corrosion-resistant alloy of copper and usually tin, sometimes with other elements. Aluminium bronze is mainly copper and aluminium, with iron, nickel, and manganese, and as strong as mild steel. Other types are *silicon bronze* and *gunmetal*. Bronzes are joined by *brazing*.

bronze disease Bright green spots of cuprous chloride on bronze from chloride attack, damaging the *patina*. It can occur if relative humidity is over 40% or from contact with wood, sea spray etc. and is removed by ultrasonic cleaning, chemicals or electrolysis.

broom-finish concrete A floor slab that is floated and trowelled smooth, then finished by drawing a stiff broom across it to make a non-slip surface.

brooming Scratching a floating coat with a stiff broom to make a key for plaster, or making a *broom finish* to concrete.

browning Undercoat plaster made from *gypsum* and sand, introduced in 1930, replacing lime : sand *coarse stuff*. These are both now rare in new work compared with pre-mixed *lightweight plasters*. Browning is applied in two or three coats, with mixes of $1 : 1\frac{1}{2}$ to $1 : 3$ plaster : sand.

brownstone A house faced with dark brown sandstone, formerly plentiful in America.

brush Paint brushes are made from synthetic fibres or animal bristles (stiff hairs) held on to a handle, usually with a metal ring or *ferrule*. The bristles of larger brushes may be set in synthetic resin. Brushes are used for the *application* of coats, the *cutting-in* of edges, or the *stippling* of finish coats. They should be thoroughly washed with *solvent* after use, but can be kept in water overnight.

brushability The ease with which paint can be applied by brushing. Brushable paints are not gummy, do not cause *ropiness* and enable a *live edge* to be easily picked up.

brushing (1) Wood *primer* should be brushed, not sprayed. Paint for top coats is spread evenly by *crossing* followed by *laying-off*.
(2) The removal of loose material from a surface by vigorous brushing with a wire brush, usually as *preparation* of a *substrate* for painting. Brushing is not suitable for *dusting*.

brush seal A long strip of bristles fixed to project from the edge of a door leaf, as a *wiping seal* for *weatherstripping* or for *smoke control*.

BS *British Standard*.

BSI *British Standards Institution*.

BTU, BThU A *British Thermal Unit*.

bubbling Bubbles of vapour in a paint film containing very volatile solvents. If they disappear before the film dries they are not a *defect*.

buck (USA) A *subframe* for a door or window.

bucket-handle joint, concave j. A durable *face joint* in brickwork, hollowed out by pushing along a *jointing tool* to compact the mortar.

buckle clip A sheet metal *cable clip* with a buckle end and a fixing nail.

buckshot A clayey sand with spots of darker clay. It collapses suddenly after several hours if left unsupported.

buff To polish or grind down a floor finish of *terrazzo* or screeded material. The process is derived from the high-speed buffing wheels of mechanical engineering, formerly of buffalo leather, which polish with slight abrasion.

buggy (USA) A *motorized barrow*.

build The thickness of a *film* of paint or a coat of filler compared with the usual or the total thickness. See **high-build coating**.

buildability Simplicity of design, enabling site work to be quick and easy. In general *brickwork* has good buildability. See **Good Building Guides**.

build and burrow A time-saving building technique in which *piles* (*C*) are first sunk for foundations and *substructure* columns, then the *ground floor* is built. Normal *bottom-up construction* of the superstructure then follows, simultaneously with *top-down* excavation and construction of the substructure, one *sub-level* at a time.

builder (1) A house builder, usually a skilled tradesman who works on site and does his own estimating and contract negotiations.

(2) A building company, usually a *main contractor* for large building projects and associated *external works*. The biggest builders employ hundreds of professional staff and site personnel to do each specialist job. In France, Germany, and Japan, builders usually do the detailed design.

(3) See **chartered builder**.

builder's equipment See **contractor's plant** (*C*).

builder's handyman A *jobber*.

builder's labourer A *semi-skilled* worker who does general work such as demolition, digging, unloading deliveries, cleaning up, and concreting.

builder's ladder A single-length ladder with round *rungs*, used by leaning against a wall and usually made of aluminium or fibreglass.

builder's level A spirit *level* or a *dumpy level* (*C*).

builder's lift A temporary *lift* for materials and passengers. It usually stands on its own base beside the new building, has a platform and cage rather than an enclosed car, and is larger than a *hoist*.

builder's line A *line*.

builder's rubbish Rubbish from the *building trades*. It does not include *spoil* from excavations.

builder's staging A mason's *scaffold*.

builder's tape A tape for *setting-out*, usually 30 m long and made of steel or PVC-coated fibreglass, which rolls up into a handy case.

builder's work drawing A drawing showing any *builder's work in connection* required by a specialist sub-contractor. To allow *coordination*, the drawings are given to the *main contractor* before structural work is started.

builder's work in connection (BWIC) Work in the *building trades* that needs to be done to help work in other *trades*, particularly the *mechanical and electrical* services. Builder's work in connection is usually done by the main *contractor* as *attendance*. It can include forming *penetrations* to run pipes, placing *concrete inserts* to carry fixings, laying blockwork to form a services *duct*, and removing *protection of finishings* at the end of the job.

building A building is any structure with a roof to provide shelter from the weather for occupants or contents, although it may also have other *elements*. Houses (residential work) are often designed by local builders, while larger projects with complicated *services*, such as office blocks, car parks, and airport terminals (industrial and commercial work), are designed by *architects* or other consultants.

Building Act 1984 The legislation that governs building in the UK, involving the *Building Regulations*.

building block A *block* of precast concrete, burnt clay, etc.

building board Rigid sheeting made from materials such as wood, gypsum, fibre-cement, flax, or cane fibre, or a *composite*. Examples are *chipboard*, *gypsum plasterboard*, and *hardboard*. It is usually cut on site to fit and is fixed with *adhesive* or *fasteners* such as nails, for use as *cladding*, wall *linings*, floors and decks, or *insulation*. It may have only one *face* and is usually thinner than other *panel products*.

building brick (USA) A *common* brick.

Building Centre An organization set up to provide information on building products, with branches in London (displays of finishings and fittings, and the Building Bookshop) and provincial cities.

building cleaning The removal of grime from external walls, often by *blast cleaning*, and the repair of the facings.

building code Local building laws in America. See **Building Regulations**.

building contractor See **builder**.

building control Laws and inspection procedures to ensure that buildings are built correctly. It is administered by the *local authority*.

Building Control Officer In London the fourteen Principal Building Control Officers must inspect jobs during design and construction to see that buildings comply with legal requirements for safety (stability, fire, etc.).

building element An *element*.

building equipment Building *services* and the internal fittings in a completed building.

building-in The fixing of a building *component* into a wall or other part of a building. Wall ties, air bricks, and brackets can be built in as work proceeds by bedding in mortar and laying bricks or blocks over and round them. Building-in can also be done by *casting-in*, by leaving a hole or *pocket*, or by breaking a hole or *chase* in completed work. The component can then be grouted into the hole or fixed into it with *dry-pack mortar*.

building inspector An employee in Britain of a *local authority*, building society, or insurance company who tells his employers whether a building complies with the law and advises them on its ratable value, fire risk, and mortgage value. He needs a wide knowledge of building construction, which may be shown by the Building Inspector's Certificate of the Institution of Municipal Engineers.

building line The line fixed, usually by the local authority, as a limit to building near a road. It may be behind the *frontage line* or sloped at a high level.

building management system (bms), b. automation s., intelligent b. A building system that uses an internal *communications installation* to operate or monitor building *services* such as air conditioning, lighting, lifts, and entry systems, usually working through *automatic controls*, using predictions made by computers that reprogram themselves to react to changes in climate and user requirements. See BRE Digest 289.

building paper Fibre-reinforced bitumen between layers of *kraft* paper, laid under concrete to prevent loss of cement into the earth and damage from chemicals in the soil. It is also used for many other purposes – to cover the *boarding* of a wall or roof, for example. In Britain it is made in rolls 1.8 m wide and 231 m long. It may also be called roofing paper, *breather membrane*, concreting paper, *sheathing paper*, waterproof paper, etc., some of which may be less strong or less waterproof than that described above. Many types are now based on polyethylene or polyvinyl chloride instead of bitumen. See BS 1521.

building permit Authorization by the *local authority* to erect a building, essential before site work can start. It is given in reply to an application.

Building Regulations Laws in Britain made under the *Building Act 1984*. They control the layout and materials used in buildings to ensure the strength and durability of the fabric as well as the health and safety of the occupants, particularly in a fire. Details may be covered in the associated manual, or *Approved Documents*. The Building Regulations in Britain have superseded local byelaws. There are now only three main sets (England and Wales; Scotland; inner London). In America and Australia, Uniform Building Codes have similar aims. See **Approved Document**.

Building Research Establishment (BRE) The BRE of the British Department of the Environment does government research in three main areas: construction, fire, and timber. The headquarters at Garston (Watford, 30 km from London on the Ml) now includes the Timber Division, moved from Princes Risborough. The Fire Research Station (FRS) is at Borehamwood (10 km nearer London) and Cardington (40 km further north), where a former airship hangar is used for full-scale fire tests. The Scottish Laboratory is at East Kilbride, Glasgow. BRE Technical Consultancy advises building owners, contractors, designers, and materials suppliers on matters such as fire protection, new product evaluation, refurbishment, site investiga-

tions, and environmental performance. It can be contacted through the Central Marketing Unit at Garston, tel. (0923) 664 800. BRE's publications (usually inexpensive and held in public libraries) include many books and papers, the *BRE Digests*, *Good Building Guides*, and *Information Papers* as well as leaflets on news of research. BRE's databases, BRIX (for construction) and FLAIR (for fire), can be searched by any organization with the password, which can be obtained free from IRS/Dialtech, London. Videos in the UK series can be hired only from CFL Vision, Yorkshire, tel. (0937) 541010.

building rubbish *Builder's rubbish.*

building society An organization for financing building, in Britain usually backed by an insurance company. Most private building is financed by these societies. The home owner repays the society over a period, usually not less than ten years, a large part of the repayments being interest. A mortgage is required for security. The buyer must make a down payment of at least 10% (often 25%) of the purchase price of the house.

building surveyor A person trained in the techniques, costs, and law of building construction. He or she advises on alterations, building defects, *easements*, extensions, renovations, energy conservation, planning applications, improvement grants, maintenance, fire insurance, and structural surveys of buildings in use. Chartered building surveyors are Fellows or Associates of the Royal Institution of Chartered Surveyors. Others belong to the Incorporated Association of Architects and Surveyors or the Faculty of Architects and Surveyors.

building system Either traditional or *industrialized building* methods.

building trades The *trades* directly concerned with the *structure* and *finishings* of a building, such as *bricklaying*, *plasterwork*, and *painting*, but not the plumbing, electrical, and mechanical *services*. The building trades usually do the *builder's work in connection*, but the term may exclude *demolition* or *groundworks*.

building-up of prices The calculation of the elements of a price in an *estimate*.

built-in (1) Fittings recessed into the fabric of a building, or a structural *element*, e.g. *luminaires* in a ceiling or cupboards in a wall, are described as built-in.
(2) Securely attached by *building in*, often becoming a *fixture*.

built-up A built-up *assembly* is one made up of several *components*, usually glue-fixed, but sometimes screwed, nailed, bolted, or *welded* (*C*).

built-up roofing *Membrane roofing* of two or more layers of *bitumen felt*, usually to weatherproof a *flat roof*. It is laid with *staggered joints* and jointed with *bonding compound* or *sealing compound*, by *torching-on* or *pour and roll*. Different types of bitumen felt are used for the *base sheets*, *intermediate sheets*, and top or *cap sheets*. Careful workmanship is vital for long life. For 3-ply roofing, guarantees are often given for twenty years and for 2-ply, ten years.

bulk excavations *Groundworks* which remove large amounts of material and reduce the general level down to near the *formation*. They are made with large and efficient *excavators* that are not suited to digging out the *isolated excavations*.

bulkhead, turret A roofed box shape built above a roof to cover a water tank, lift shaft, stair well, etc.

bulkhead luminaire, b. fitting A robust and watertight wall-mounted lighting fitting with a cast-glass bowl behind a guard wire, made for *plantrooms*, etc.

bulldog plate connector A *toothplate connector*.

bullet catch A *ball catch*.

bulleted key A flat *key* with grooves.

bullet-head nail (USA **finish n.**) A *nail* with a small rounded head.

bullet-resistant door A steel or aluminium door with wood facings resembling an ordinary door but providing more *security*.

bullet-resistant glass High-grade *security glazing*, usually *laminated glass* 20 to 200 mm thick with many interlayers, or a glass/polycarbonate composite.

bullhead connector A *straight tee* fitting for a duct or a pipe.

bull header A *special* brick with a *bullnose* end.

bullnose The rounding of an *arris*; in general any rounded end or edge of a brick, a step, a joiner's plane, etc.

bunched wires, bundled w. Prefabricated electrical wiring made up for *trunking*.

bund An uninterrupted wall of earth (earth bund), blockwork (bundwall), etc., built round a storage tank containing oil or other hazardous liquid, to such a height that the volume of liquid it could contain is slightly larger than the maximum volume of liquid in the tank. If the tank leaks, therefore, the bund prevents dangerous spreading of the liquid. Surface-water drains are not provided within the bund; the water is drained by pumping.

bur A drill *bit* used for widening holes rather than drilling them.

burglar alarm An *intruder alarm system*.

burglar-resistant lock A *security lock*.

buried services, underground s. Pipes or cables buried in the ground, for drainage, electricity, gas, telecom, water supply, etc. Before work is started on site, each service company (gas, water, electricity, etc.) must be contacted and exact positions marked, e.g. with paint or wooden pegs, using location drawings and a metal detector. Excavation should be done very carefully, usually with hand tools – it is common to find pipes or cables out of position or not as described, resulting in injury or costly damage. Many services are marked with protective *tracer* or *breaker slabs* in the backfill. Temporary cables on building sites should be pegged.

burl figure *Burr* veneer.

burlap (USA) *Hessian*.

burner The part of a *boiler* or gas *torch* where oil or gas fuel is released, mixed with air, and burnt.

burning off Removing old paint by heating it with a *blowlamp* until it softens, then scraping it off. Using a *heat gun* is safer and releases fewer toxic fumes. Neither should be used to remove old *lead paint*.

burnishing Polishing metals for *finishing* or to *hone* a cutting edge.

burnt clay *Ceramics* after *firing* in a kiln, either *soft-burnt* or *hard-burnt*.

burnt lime *Quicklime*, a stage in the making of building *lime*.

burr (1) A jagged edge left after metal is cut, or a *fin* on concrete from a *formwork* joint.
(2) **burl** The curly, much valued *figure* got by cutting through the en-larged trunk of certain trees, particularly walnut. It is formed by the dark pith centres of many undeveloped buds.

burring reamer A tool that removes *burrs* left by a *pipe cutter*.

bus bars Bare copper or aluminium *conductors* fixed inside *trunking* which distribute heavy-duty electrical power to *tap-off units*.

bus coupler A simple *coupler* between *bus bars*. It can be opened or closed only with the power switched off.

bus duct, busway Electrical *trunking* for *bus bars*.

bush, bushing (1) An insulated tube to protect cables where they pass through a hole in a metal cover.
(2) A short screwed pipe fitting for connecting pipes of different diameters. Unlike a *reducer*, it is threaded inside on one end and outside on the other.
(3) A fitting on the end of a *conduit*, where the cables emerge.

bushfire A blaze in flammable eucalyptus forest, e.g. in Australia, a *fire hazard* to buildings. It occurs in dry and windy weather, spreading through the tree canopy as a main fireball, sucking smouldering leaves into the air, dropping them ahead of the fire front, and starting more fires in the understorey. Buildings are best protected by clearing away combustible material round the outside (although several rows of deciduous trees may stop a grass fire) and by having grilles over openings and gutters to keep out flying cinders.

bush hammer A hand-held machine (usually air-driven), or a mason's hammer, with rows of raised tooth-like pyramids on its face, used for *hacking* or *scabbling* the surface of fairly hard

butane

materials such as stone or concrete (French 'boucharde').

butane A *bottled gas* used as a fuel for a *blowlamp*. It boils at −0.5°C and cools on expansion, freezing its jets and regulators in frosty weather.

butt (1) To meet without overlapping, as in a *butt joint*.
(2) A *butt hinge*.
(3) The thick end of anything, such as a log or a *shingle*.

butter coat A soft, wet coat of *render* for a *dry-dash finish*.

butterfly valve A valve with a flat disc on a turning spindle.

butterfly wall tie A *wall tie* made of galvanized steel wire about 3 mm diameter, bent to form a flattened figure 8. This tie transmits less sound than the stronger, stiffer *vertical twist* tie (strip tie). See diagram, p. 500.

buttering The spreading of *mortar* on the ends of bricks to form *perpends* or on the backs of floor tiles before *bedding*.

butt gauge A tool used for marking the width of the *rebates* for a *butt hinge* in a door *stile* and *jamb*. See diagram, p. 67.

butt hinge The commonest *hinge* for doors. When the door is shut the two halves are folded tightly together. Each flap is usually *morticed*, one into the door frame and the other into the *hanging stile*. The ordinary steel butt is very cheap and durable, but ball-bearing butts are smoother running, noiseless, and longer lasting, though much dearer. See diagram, p. 229.

butt joint A *joint* between two square-ended pieces of the same thickness without overlap.

button (1) A *push button* for a doorbell, to control a *lift*, etc.
(2) A small piece of wood or metal loosely held by a screw so that it can be turned and thus hold a cupboard door shut; usually smaller than a *knob*.

button-headed screw, half-round s. A *screw* with a hemispherical head.

butt veneer *Veneer* with the strong curly figure from a tree trunk. See **burr**.

butyl sealants Rubbery *one-part* sealants which dry by *solvent release*. They stick to most surfaces without a primer, but are not suited to severe conditions.

buzzer A *sounder*, often for an alarm.

buzz saw A *circular saw*.

byelaw, by-law, building b. Regulations that formerly controlled British building and were made by each *local authority*, but are superseded by the *Building Regulations*. There are other types of byelaws.

by others Work by others is work in another *trade* or another contract, beyond the *limit of works* or done before or after, e.g. *preliminary work*. To the plastering trade a blockwork wall is a *background* 'by others'. A *cable tail* left coiled 'by others' is for the wiring-in of electrical equipment or fittings.

bypass An arrangement of pipes (or conduits) for directing flow round instead of through the normal pipe, such as that leading to the *water meter*.

C

cabin A small building used for *site accommodation*, often on *jack legs*.

cabinet conditioner A packaged *air conditioner*.

cabinet work Very accurate *joinery* work in wood.

cable One or more electrical *conductors* covered with *insulation* (the core), inside a common insulating sheath, sometimes *armoured*. Cables for *wiring* are run overhead, underground, in *trunking*, on *cable tray* and cable *ladders*, or drawn in to *conduits* and *cable ducts*. Cables can be *flame-retardant*, *twin*, *festooned*, or *mineral-insulated*.

cable chute A cover over a cable on a wall, usually a steel channel with outward flanges, to give *mechanical protection*.

cable clip, c. cleat A *fixing* usually with a plastics hook and a nail (or masonry plug) that is driven into a wall to support a cable run.

cable cover A *breaker slab*.

cable duct, pipe d. A pipe into which electrical cables are *drawn in*, usually plastic drain pipe larger than 50 mm dia. It can be buried under a road, built into concrete, etc., like *conduit*.

cable gland A seal used where a cable enters a casing, or put on a cable end, particularly with *mineral-insulated cables*, to keep water out.

cable sleeve A *hook intake*.

cable tail A length of cable left coiled up for *wiring in* to a fitting.

cable tray A length of perforated galvanized steel sheet with a turned-up edge that supports cable runs, which are 'laid in' from the side. Cable tray requires earth *bonding* and may have a cover. Spare space should be allowed for extra cables.

cableway Electrical *trunking*. See also *C*.

cadmium plating A corrosion protective *finish* for steel articles such as wood screws. After *chromating* it takes paintwork better.

cage of reinforcement *Main bars* surrounded by *binder bars*, forming a grid that resists shear forces in the concrete and holds bars in place during concreting.

calcium carbonate ($CaCO_3$) The chemical name for chalk, limestone, marble, spar, and so on. It can act as a *flame retardant*.

calcium chloride ($CaCl$) An *accelerator* now rarely used in concrete, mainly because it corrodes steel.

calcium hydroxide ($Ca(OH)_2$) Slaked lime, sold as *dry hydrate*.

calcium oxide (CaO) *Quicklime*. When water is added it becomes *calcium hydroxide*.

calcium plumbate primer A *lead paint* formerly used on galvanized steel.

calcium silicate A strong *binder* found in Portland *cement* but also made by heating lime with silica sand. See below and *cement compounds* (*C*).

calcium silicate board Tough, general-purpose *building board* made in a similar way to *calcium silicate bricks*. It is *asbestos-free*, non-combustible and resists fire, water, and most chemicals. Lightweight types with *vermiculite* fillers are used for fire protection *encasement* of structural steel.

calcium silicate bricks Low-cost bricks made from sand (sand-lime brick), crushed flints (flint-lime brick), or other silica-bearing rocks, mixed with slaked lime, then pressed to shape and hardened by heating in a steam *autoclave* for several hours, forming *calcium silicate* between the grains. They are true to size and either smooth or textured. Different strengths are available. Manufacture can be accurately controlled. The bricks are as dense as most clay bricks, are frost-resistant, do not suffer from *efflorescence* and have good resistance to *sulphate attack*. A combination of contamination by road or sea salt plus frost can cause *spalling*. Their *moisture movement* is high: on drying they shrink about 0.35% (1 in 300), so that *movement joints* in brickwork should be provided every 7.5 m horizontally and at each storey.

calcium sulphate (USA **c. sulfate**) (CaSO₄) The mineral anhydrite, which has the same chemical composition as calcined *gypsum*.

calcium sulphate hemihydrate See **hemihydrate plaster**.

Calder coupling (USA) A stainless-steel *sleeve coupling* with a rubber gasket for connecting socketless cast-iron pipes.

calendering Squeezing sheet materials between rollers, down to 1 mm thick for some materials, with close control. Calendering can give a smooth or glossy surface, for PVC sheeting or *linoleum*, or a rough surface, for bitumen-polymer *damp-proof courses*.

calk (USA) *Caulk*.

call button A control push button for a *lift*, either in the car or on a landing.

call for bids (USA) See **tendering**.

call point A manual *fire alarm* button, usually a *break-glass unit*.

calorifier An industrial hot-water service which has a water-to-water *heat exchanger*, used for heating low-pressure hot water. Calorifiers have an insulated steel shell, inside which there is a submerged coil of pipe with steam or hot water passing through it, supplied from a central boiler plant. Calorifiers can be large storage or smaller non-storage types, depending on frequency of demand and boiler output rates, or can be built into the top of a *combination boiler*. A domestic *indirect cylinder* is a small storage calorifier.

camber (1) A slight upward curve in a structural member in the opposite direction to the sag, so that the member is flat in service. The camber of a wide-span reinforced-concrete *waffle slab* can be as much as 40 mm. It is put in during final levelling of the *formwork* before concreting. Camber resulting from *prestressing* (*C*) is usually called 'hog'.

(2) *Bowing*.

came An H-shaped strip of metal that fixes each piece of glass to the next one in *leaded lights, copper lights*, or stained-glass windows.

candela (cd) The *SI unit* of *luminous intensity*.

cane-fibre board, bagasse b. An *insulating board* made from the cane trash residue from sugar production

canopy An *open roof*, or a kitchen *hood*.

cant (1) A flat surface not at right angles, a *splay*.
(2) To tilt, *splay*, or *bevel*.

cant brick A *splay brick*.

cant stop brick A *special* brick to make a neat end to a *splay*.

cant strip (USA) A *tilting fillet*.

cap (1) The top part or head of a component. See **newel cap**.
(2) The *coping* on the top of a wall.
(3) The connector of a lamp, either *Edison screw* or *bayonet*.
(4) A cover, with internal threads, screwed over the end of a pipe. It thus closes, or caps, the pipe.

capillary action, capillarity The tendency of water (or any fluid) to be sucked into a narrow space such as the tiny interconnected holes of a porous material, or any close joint. Capillarity causes *rising damp* in walls and the *penetration* of rain across *bridging* in a cavity wall or under a window.

capillary break A gap, space, or cavity between two surfaces large enough to prevent *capillary action* of water. A *feint* in the edge of metal *flashing* is a capillary break.

capillary joint (USA same as **sweat j.**) A neat, compact joint in *copper pipe*, *stainless-steel plumbing*, etc., made by putting the end of a pipe into a fitting very slightly larger in diameter, after cleaning the contact surfaces and applying a *flux*. The pipe joint is then heated with a *blowlamp*, etc., until the *solder* or *brazing* material melts and flows in evenly round the space by *capillary action*. In drinking-water pipes, *lead-free solder* is required. Heating must be done carefully. The strongest joints are made with *silver brazing*; joints made with solder can come apart if the pipe freezes and are forbidden within the thickness of a wall.

capping (1) A decorative or protective cover on an angle, end, top, etc. It may be of a different material from the general area, or of the same material used differently. See **coping**.
(2) In *supported sheetmetal roofing* a metal strip as a *joint cover*, e.g. over a *wood roll*, either separate from, or welted to, the edge of the roofing sheets, which are dressed up the sides of the roll. Capping bottom edges are dressed down flat to form a *splash lap*, or if sloping or vertical they end in a *feint*.
(3) A metal section fixed to the outside of some *patent glazing* bars to hold the glass and protect the stem of the bar from the weather.
(4) Closing off the end of a pipe with a *cap*.

capping strip A *joint cover*.

cap sheet A *bitumen felt* used as the last sheet in a *built-up roofing* membrane. The top side is *self-finished* for *solar protection*, e.g. with *aluminium foil* or *mineral surfacing*, and the under side is surfaced for bonding to an *intermediate sheet*. The sheet is usually reinforced with fibres for resistance to indenting, puncturing, and tearing.

capstan head A *crosstop*.

capstone The *coping* on top of a wall.

caption A text forming the title to an illustration. Also a brief note on a *drawing* describing the materials used or how the work is done.

car The enclosed travelling platform of a *lift* or elevator.

car apron A projection of the floor of a *lift car* under the door. It meets a similar projection from the floor slab of the landing.

carbide tips, tungsten-carbide t. Tips on the cutting edge of many woodworking *edge tools*, hard enough to cut soft metals like aluminium, soft bricks, and some wallboards. Used on timber, carbide-tipped *circular saw* blades stay sharp for about three times as long as a normal steel blade but need to be sharpened in a special machine.

carbonation, carbonating Absorption by lime of carbon dioxide from the air, converting it to hard, stable *calcium carbonate*. Non-hydraulic lime sets in this way and the process is quite slow; old walls built with lime mortar took at least two years to harden, but after a century or more became very strong. Air pollution and exhaust fumes also encourage carbonation. Concrete is carbonated by conversion of its *free lime*, slowly from the outside inwards. It increases hardness, causes some irreversible shrinkage, and reduces concrete's alkalinity, allowing steel *reinforcement* close to a bare outside surface to corrode.

carbon dioxide (CO_2) An inert gas used as a *medical gas*, as an *extinguishing medium*, and for shielded arc welding.

carbon dioxide fire extinguisher The reduction of the oxygen content of the air from 21% to 15% smothers most flaming fires (although not glowing wood), but to achieve this involves about 30% CO_2 in the air, which is fatal. Even 5% can lead to shortness of breath and a slight headache (removal to fresh air usually results in full recovery). But the danger is less than the dangers of fire, and thousands of CO_2 inert-gas auto-suppression *fire-extinguishing systems* have been installed by industry since 1929. Several systems exist. For enclosed spaces the preferred method is total flooding. *Mechanical ventilation* should stop before, or simultaneously with, the discharge of CO_2. Provided that losses of CO_2 are not high, this method can extinguish deep-seated fires, and is also suitable for electrical fires, fires in fuel tanks, and other fires where water cannot be used. Where total flooding is impracticable, local application or manual hose reels are used. See BS 5306 part 4.

carcase, carcass The loadbearing *fabric* or structure of a building, particularly for *timber-frame construction*. See diagram.

The plan and section of a house carcase, omitting doors, windows, and plaster; and floorboards in plan.

A	= opening trimmed for stair	K	= herring-bone strutting
B	= trimming joist	P	= wall plate
C	= trimmer joist	PT	= plain tiles (resting on battens)
D	= trimmed joists	R	= rafter
dpc	= damp-proof course	RB	= roof boarding (covered with
E	= common joists		sarking felt)
F	= footing	RP	= ridge piece (or ridge)
G	= opening trimmed for fireplace	RT	= ridge tile (set in mortar)
H	= cavity wall	T	= soffit board
J	= ceiling joist	TT	= tusk tenon

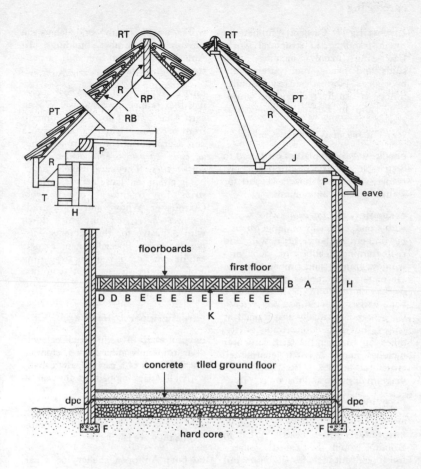

RT

RT

R · RP

RB

PT · PT

P

R

T

H

R

PT

R

J

P

eave

floorboards

first floor

B A

D D B E E E E E E E E

K

concrete

tiled ground floor

dpc

dpc

F

hard core

F

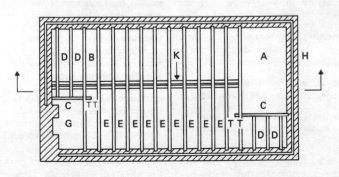

D D B

K

A

H

C

T T

C

G

E E E E E E E E E E T T

D D

carcassing (1) Carpentry timbers in rough framing and structural work. Carcassing mainly includes wall studs and plates, roof rafters, and floor joists. Normal carcassing-quality timber is usually redwood or whitewood and in Europe is not *stressgraded*.
(2) The *roughing-in* of pipework.

cardkey A coded plastics card used to operate an *entry system* through a card reader, which may have a keypad for security identification numbers.

carpenter A *craftsman* who works with wood, especially on large *carcassing* timbers for house floor, wall, and roof framing, loadbearing door and window frames, and timber flooring. He or she also applies *dry linings* or works as a *formwork carpenter*, using a wide variety of hand and *power tools*. Carpenters are usually also trained as *joiners*, but follow one trade or the other. In America the term carpenter includes joiner. (French 'charpente', structural timber or steelwork.) See diagram opposite and that on p. 346.

carpenter and joiner The title of a fully qualified *tradesman*.

carpenter's finish (USA) What in Britain would be called 'joiner's finish', since it includes all *joinery* but not roughly finished work such as rafters.

carpenter's hammer A *claw hammer*, but see also **joiner's hammer**.

carpenter's hardware Rough builder's hardware used in *carcassing*.

carpenter's helper, c. mate A *builder's labourer* who helps a carpenter.

carpentry The craft of cutting and joining timber to make structural frameworks, usually rough work such as the *carcassing* of timber buildings. In America (but not in England) carpentry includes *joinery*.

carpet Warm, soft textile *floor coverings* that reduce impact noise. Woven carpet has a *pile* of fibres projecting up from a *backing*; felted carpets have a *nap*. Carpets can come in rolls, either as *body carpets* or *broadloom*, or as *carpet tiles*. They are usually *fitted*, being either laid loose between *gripper strips* for traditional woven wool (Axminster, Wilton) and heavy-grade *contract carpets*, or fully stuck down with adhesive for the lighter *needle-punch* carpets. Specially made carpets can be *antistatic* or resistant to staining and cigarette burns, but common synthetic-fibre carpets are easily damaged by heat.

carpet gripper A *gripper strip*.

carpet strip A strip fixed below a door, originally in hardwood, of about the thickness of a carpet, now usually a metal strip clipped to the carpet edge.

carpet tile A square of *carpet*, usually 500 × 500 mm. In lifts and computer rooms they are laid loose.

carport An open shelter for a car, usually near a house. It has a roof, but is not fully walled.

Carrara white High-quality white marble from Carrara in Italy, used as a finishing. It is among the harder, more durable types of marble.

carriage (1) **rough string** An inclined timber or steel section parallel to the two *strings* against the underside of wide *stairs* to support the treads in the middle.
(2) **bogie, end c.** The unit containing the wheels of a travelling crane.

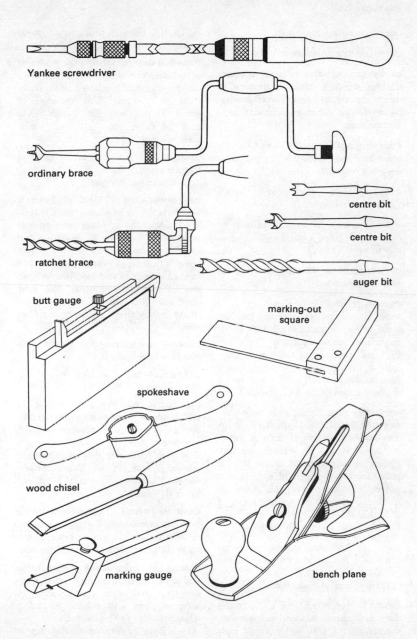

Yankee screwdriver

ordinary brace

ratchet brace

centre bit

centre bit

auger bit

butt gauge

marking-out square

spokeshave

wood chisel

marking gauge

bench plane

Traditional carpenter's tools. See also power tools, saws, and joiner's tools.

carriage bolt

carriage bolt A *coach bolt*.

carrying arm An *outreach arm*.

carrying capacity The maximum current which a cable is allowed to carry, limited, to prevent overheating, by a *fuse* or other overcurrent protection gear.

carrying out Doing the work of erecting a building, usually in the *performance* of a contract.

carry up To lay *brickwork* or *blockwork* up to a given height.

carting away Removing rubbish or excess spoil from a building *site*, usually to a public tip.

cartridge filter A filter with a disposable filter *element*.

cartridge fuse A *fuse* enclosed in a tube of insulating non-combustible material (glass, ceramic) filled with sand. Since there is no danger of fire when the fuse melts, it is safer and more reliable than rewirable fuses. It has a *high breaking capacity* – the fuse melts at about twice its rated current.

cartridge tool, fixing gun, powder-activated tool, stud gun A pistol for driving *shotfired fixings*, usually indirect acting with a low-velocity captive piston. A safety guard is needed. It may have ten-shot cartridge-discs or a power adjustment system. Care is needed to prevent fixings flying out of the other side of soft or thin materials. *Ear protectors* are usually compulsory.

case (1) An outer envelope, for instance the *facings* of a building or the metal cover of a *lock*.
(2) (Scotland) A *cased frame*.

cased frame, box f. The hollow, fixed parts of a traditional wood *sash window*, containing the sash weights and pulleys, and bounded by visible boards

called the outside lining and inside lining.

case-hardening During the *seasoning* of timber, the outer skin often dries out, shrinks, and hardens more quickly than the inner part. The stretching of the outer part may also cause *checks* and the differences lead to *warping* after sawing. See also *C*.

casement (1) A hinged or pivoted framed opening *light*.
(2) A *casement window*.

casement door, French d., French window A pair of glazed (usually outside) *double doors*, hung on opposite sides of a *frame* in an opening and meeting in the middle. In Britain they open outwards, in France inwards, but in cold countries such as Austria with *coupled windows* the outer pair may open outwards and the inner inwards. Being double they usually have a *Cremona bolt*.

casement fastener For single casements, a *sash fastener*.

casement stay A bar used for holding open a *casement*.

casement window A window in which one or more *sashes* are hinged to open. Generally, as in door hinges, the pins are vertical, that is the sashes are *side-hung*. In Britain they usually open outwards and on the Continent inwards. They are simple and cheap, but easily damaged by the wind.

cash flow chart The *main contractor's* estimate of spending for a large project, usually shown as a graph, given to the *client* for planning *monthly instalments*.

casing (1) **casework** A *joinery* frame and *dry lining* hiding rough work, such as a structural beam, stair framing, or services pipes and cables. A casing often provides *fire protection*.
(2) A *lining* or door or window frame of light timber.

(3) Formwork *boxing* or concrete *encasement*.

casing bead An *angle bead*.

cast Thrown or projected on to a surface, e.g. *dashed* or *roughcast*.

casting Making a component by placing fluid material in a mould and allowing it to harden, usually by cooling or by the setting of a *binder*.

casting-in Concreting by the *structural trades* round such components as electrical *conduits* or *concrete inserts*, as a way of *building in* to a concrete slab or wall.

cast-iron pipes *Ductile iron* drain pipes or pressure pipes, usually *centrifugally cast*, with *spigot-and-socket joints* or *sleeve couplings*. Cast-iron pipes should always have an outer coating of bitumen and they are often lined inside with cement mortar. See BS 437 and BS 4622.

cast stone, reconstructed s. Natural-stone *aggregate* bonded with cement, used like solid stone, for masonry, roof tiles, etc. It can be surfaced either with a *face mix* of granite chips, or with York stone, sandstone, or slate over the backing. Cast-stone lintels of sandstone and lime/pozzolan cement, used in AD 1138 to repair fortifications at Carcassonne, have lasted as well as natural stone. See BS 1217.

catch (1) A simple door *fastener* for holding a cupboard closed. Some, e.g. *magnetic* and *ball catches*, allow the door to be pulled open.
(2) A pawl or safety catch on a crane hook.

catch bolt A *return latch*.

catchpit, grit trap, sump A pit before the inlet to a drain pipe where sand, grit, and other detritus settle. It should be regularly cleaned out.

cat ladder A ladder or a board with cleats fixed to it, laid over a roof slope to give access for workmen to repair the roof without breaking the tiles.

catwalk A gangway for access round the upper walls of high buildings, such as steelmills, giving access to the roof and eaves for maintenance work.

cauldron A *bitumen boiler*.

caulk, calk (1) A *fishtail fixing lug*.
(2) To fill a joint with *caulking*.

caulked joint A *lead-caulked joint*.

caulking (1) Making a tight joint by forcing in *jointing* material or a *sealant*.
(2) Splitting and twisting the ends of a metal bar to make a *fishtail fixing lug*, which gives good *bond* to mortar or concrete.
(3) See **cocking**.

caulking chisel, c. iron A plumber's tool for compacting the lead in a *lead-caulked joint*.

caulking gun, pressure g. A hand-operated pistol which is used to force gun-grade *sealant* out of the nozzle of a cartridge.

caulking hammer A medium-weight hammer used to strike a *caulking chisel*.

cave-in The collapse of the sides of excavations, which should be prevented by *earthwork support*.

cavity A space between solid objects, or a *void*. In a *cavity wall* it is usually 50 mm wide (or 25 mm with *cavity insulation* in a sheltered location). In *double glazing* it is the space between the two *panes*, filled with air or gas. See **drained cavity**.

cavity barrier A *fire barrier*.

cavity batten A temporary *batten* (or plastics tray) inside a *cavity wall* to

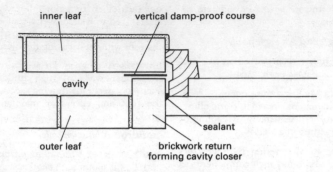

inner leaf

vertical damp-proof course

cavity

sealant

outer leaf

brickwork return
forming cavity closer

Cavity closers.

catch mortar droppings during brick-laying. It rests on the highest row of *wall ties* and may have a recovery cord on each end.

cavity closer Material to close the cavity of a *cavity wall* at a door or window opening. Insulated cavity closers and *combined lintels* both reduce *cold bridges* compared with a solid end made by forming a *return* in the brick-work outer leaf, which needs a vertical *damp-proof course* where it meets the inner leaf. See diagram.

cavity fill, filling, infilling *Cavity insulation* put into an existing wall (fill-ing the cavity) to reduce heat losses. The most-used material is urea-formal-dehyde foamed in-situ, which must be carried out to Schedule 1 of the *Build-ing Regulations* by a person with a BSI Certificate of Registration of Assessed Capability. The wall must comply with BS 8208, and the foam with BS 5617. The foam fills the cavity, restricting ventilation and drainage, and tends to crack after installation, providing a path for *driv-ing rain* and creating *dampness*. It also releases formaldehyde, a poisonous gas, which must be kept out of the building by a brick or block inner *leaf* with all holes sealed. Poly-urethane foam fill is also used, as is *loose fill* of dry or adhering beads of polystyrene.

cavity filler A *fire stop*.

cavity flashing A *damp-proof course* that crosses a cavity.

cavity floor An *access floor*.

cavity inspection (1) While *cavity walls* are being built they need to be inspected to make sure that there are no *bridges* from mortar *droppings* or other causes of dampness. Inspections can be made from the bricklayer's scaf-fold, and any remedial work is done immediately. Good site practice in-cludes the use of *cavity battens*, or the *cleanout of wall cavities*.

(2) In a completed building, if a wall shows signs of *dampness* (and other causes have been eliminated), the cavity can be inspected by removing a brick from a corner of the wall. Any *droppings* or other bridges can be knocked away with steel rods.

cavity insulation Usually boards or sheets of expanded polystyrene *insula-tion* fixed to the outer face of the inner

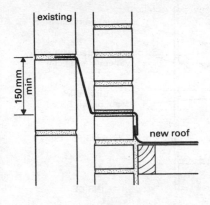

150 mm min

existing

new roof

A cavity tray.

leaf of a *cavity wall* during construction, leaving the normal cavity width of 50 mm. Cavity insulation is an economical way to achieve the *U-values* required by the *Building Regulations*, but requires careful construction to avoid rain penetration at *perpends* or *cold bridges* (BRE Digest 277). Insulation of existing walls is by *cavity fill* (BRE Digest 236).

cavity party wall A *double-leaf separating wall.*

cavity tie A *wall tie.*

cavity tray A *damp-proof course* that crosses the 50 mm wide cavity of a *cavity wall*, stepping up at least 150 mm between the outer and inner leaves, to form a gutter which leads to a *weephole* in the outer leaf. Used at an *abutment* where brickwork rises above a roof slope, it keeps the wall below dry by draining rainwater back through the outer leaf. The tray is usually built into the bed courses of the outer leaf, but can be simply turned up to rest against the inner leaf. The outer edge of a cavity tray may be joined to a lead *counter-flashing*, which is later turned down, outside the wall, over the roof

upstand flashing when the roof covering is completed. Trays are often preformed of heavy plastics to many shapes: stepped, staggered, arched, stop ends, etc. See diagram.

cavity wall, hollow w. (1) A warm, dry external wall with a 50 mm wide cavity between two leaves. The cavity prevents rainwater from reaching the inner *leaf* and provides some heat *insulation*. The outer leaf is usually a *half-brick wall* of clay *facing bricks*. Inner leaves depend on local climate; in Britain they are usually of insulating *blockwork*. The two brick or block leaves are held together by *wall ties* at 600 mm horizontal spacing and 450 mm vertical spacing, unless they are omitted for *sound insulation*. Rainwater may saturate the outer leaf, even penetrate the porous brickwork, but cannot cross the wall ties, and is allowed to escape through *weepholes*. Cavity walls need a high standard of *workmanship* and *cavity inspection*. See diagram, p. 72.

(2) A *double-leaf separating wall.*

ceiling The visible upper surface of a room, which can be an open slab *soffit*, or any type of *dry lining* fixed to floor joists or *ceiling joists*, or a *suspended ceiling*, or a *stretched covering*. A *rooflight* can form part of a ceiling. Ceilings are usually painted white or a light colour.

ceiling binder A running tie between ceiling joists or *trussed rafters*.

ceiling diffuser An *air terminal unit* for supply air.

ceiling fan A slow sweep fan with blades 900 to 1500 mm wide, hung from a vertical downrod, with a wall-mounted speed controller, usually reversible so that it can blow downwards in hot weather or upwards in cold weather to recirculate warm air from near the ceiling.

ceiling hangers

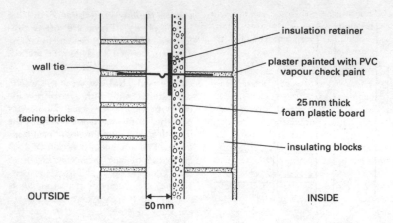

insulation retainer

plaster painted with PVC vapour check paint

25 mm thick foam plastic board

insulating blocks

wall tie

facing bricks

OUTSIDE

50 mm

INSIDE

A cavity wall.

ceiling hangers Steel rods attached to *primary fixings* in the underside of a slab to carry a *suspended ceiling*. They are usually adjustable.

ceiling height, headroom The height of the ceiling *soffit* above *finished floor level*.

ceiling-height unit A door frame, wall panel, etc., from floor to ceiling.

ceiling insulation, loft i. Thermal *insulation* for the *pitched roof* of a house, which is the area of greatest heat loss. The insulation can be a quilt, bats or loose fill and is usually placed on top of the *ceiling*, over or between the *joists*. This creates a *cold roof*, and to prevent *condensation* a *vapour barrier* is needed underneath, usually above the ceiling plaster. The insulating material should be stopped short of the *eaves* to allow ventilation. Before ceiling insulation is placed, a check should be made that no electric cable will be covered and cause a dangerous *temperature rise*. See diagram, p. 147.

ceiling joist A joist which carries the *ceiling* beneath it. It can be part of the roof or run separately below the *floor joists* of an upper storey. A traditional ceiling is carried on the underside of floor joists, rather than on separate ceiling joists, but *sound insulation* can be improved by this *discontinuous construction*.

ceiling outlet A *lighting point* on a ceiling for wiring in a cable to a *pendant fitting* or a surface-mounted *luminaire*. It may have a *rose*.

ceiling strap A *ceiling hanger*.

ceiling switch An electric light switch with a pull cord from the ceiling.

cell One of the small, often microscopic, units that make up the structure of wood and other plant tissues.

cellar A room or rooms of which more than half is below ground level, usually reserved for storage or for the *central heating* boiler, whereas a *basement* is a *habitable room*.

cell ceiling, egg-crate c. A *suspended ceiling* of closely spaced vertical blades forming a square grid, which hides air conditioning and lighting systems while acting as a diffuser for both. If *sprinklers* are mounted on a cell ceiling they need a backing plate to collect the hot buoyant smoke.

cellular (1) **Cellular materials** have many tiny internal voids, which make them *lightweight* and improve their thermal *insulation* but reduce strength and resistance to impact damage. They can be aerated, expanded, or foamed, and have open or closed cells. Cellular materials rarely have large voids, i.e. they are used to make 'solid' components.

(2) **Cellular construction** (hollow construction) has a *core* that contains large voids, usually made of solid materials, such as clay, wood, dense concrete, or cardboard. The core is enclosed inside *facings* and *lippings*. The resulting component is lightweight, but thin facings may show *core patterning*.

cellular brick A *brick* with at least 20% of its cross-section in the form of holes, which are closed at one end. They are not made in Britain. See **hollow clay block**.

cellular concrete *Aerated concrete*.

cellular-core door A light-duty *internal door* which does not provide much *sound insulation*. It has *flush* facings of plywood or hardboard over a *core* made of strips of cardboard in a honeycomb shape, or paper coils, or timber strips. Solid timber is used for the *stiles*, *rails*, and *lock block*.

cellular glass *Foam glass*.

cellular lighting diffuser A *diffuser* with a fine grid of vertical blades.

cellular plastics Materials usually described as *expanded plastics*.

cellulose enamel *Lacquer*.

cellulose paint Very flexible, tough paint, based on *nitrocellulose* (*C*), with

alkyd resins and solvent *thinners*. Cellulose paints dry by *solvent release* and are best applied by spraying, which can be difficult on site but is common for factory-finished units.

CE mark The European Community *conformity mark*, based on the letters CE, placed on a product that conforms to the *Construction Products Directive*, backed by a *quality management* system and attestation by the maker or a certifying body.

cement (1) A *binder*, usually meaning *Portland cement*, which sets to hold together the materials of *concrete, mortar, render, building boards*, etc., into a mass which then hardens and gains strength. Cement for use on site is delivered in bags or in bulk. It should be stored in dry conditions to prevent *warehouse set* and used within four weeks. It is usually added to the *mix* last and measured by weight or by the full bag, for accurate *batching*. Fresh wet cement is highly alkaline and can damage the skin or eyes; and most materials made from cement become alkaline when damp. Special cements include: high-early strength, masonry, sulphate-resisting, supersulphated, and white cement. See also *C*.

(2) Any strong *adhesive*, often a synthetic *resin*, or the natural bond or matrix that binds a material together, such as that which binds sand grains into sandstone.

(3) A common term for a cement *screed* or any mortar or concrete.

cement asbestos *Asbestos cement*.

cement content, c. factor The weight of cement per cubic metre of concrete. The addition of more cement to the *mix* improves strength and durability but also increases *drying shrinkage*. Structural concrete has a high cement content and *dry lean* a low one.

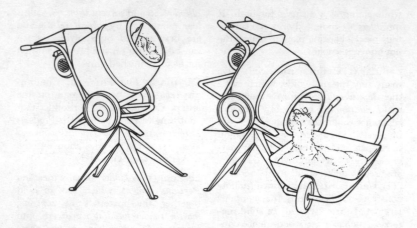

A cement mixer.

cement fillet, mortar f., weather f. A continuous triangle of *mortar* in the corner of an internal angle at an *abutment*. It is not good practice – the recommended detail is a *soaker* flashing.

cement:lime:sand mortar *Composition mortar*.

cement mixer A fat pear-shaped drum with an open top, turned slowly on a shaft, to mix the cement, aggregate, and water of *mortar*. The mix is emptied by tilting the drum. A *concrete mixer* (*C*) is bigger. See diagram.

cement paint (1) A mixture of cement and water which when applied to the outside of concrete, masonry, or brickwork improves its resistance to rain penetration. Ordinary cement paint is not very pleasantly coloured, but it is cheap. Coloured cement paints are very much dearer. Cement paint must not come into contact with *gypsum plasters*.
 (2) An *alkali-resistant paint* which can be applied over *masonry*, concrete, render, etc.

cement rendering, c. plaster
Plasterwork based on cement mortar, which is water-resistant but often not waterproof, making it suitable for *wet areas* or as *external rendering*.

cement-rubber latex A tough, flexible material used for *jointless floors*, *anti-graffiti treatment*, etc. The aggregate may be marble or other stone chippings or cork or wood chips. Such a floor can be ground and buffed like *terrazzo*. It can be as thin as 6 mm, or even thinner if used as *levelling compound*.

cement screed A *screed* of cement mortar, the traditional type.

cement slurry A liquid mix of cement and water, of creamy consistency, used for injection or to wash over a wall. See also **grout**.

cement-wood flooring *Jointless flooring* made from Portland cement, sawdust, sand, and pigment, which can create a non-*dusting* and quiet floor suitable for offices or living rooms, although it is unsuitable for wet areas such as bathrooms, kitchens, and workshops.

CEN *European Committee for Standardization.*

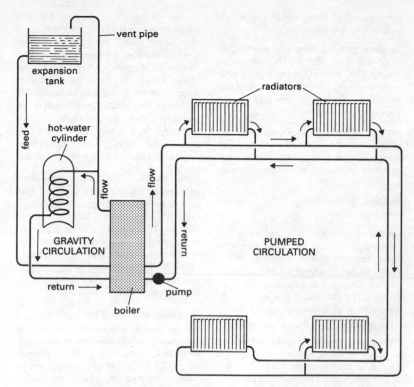

Central heating, two-pipe system. The hot-water supply, which comes from the cylinder, is not shown. (With acknowledgements to *Which?*, Consumers' Association, 1989.)

centers, centering, centres (1) Curved temporary supports for an arch or dome during construction.

(2) *Floor centres*.

central heating The heating of buildings using one energy-efficient central *boiler* (or a *district heating* system) and pumps to transfer heat to radiators by hot water in a pipe circuit. Alternatively *warm air* may be directly supplied through ducts. Radiators give a more comfortable heat than ducted warm air and at a lower temperature, costing less for fuel, although they take longer to warm up. They can be combined in a *split system*. Hot water is a simple *heat-transfer fluid* which is easily moved by *circulating pumps* and works best in a *sealed system*. Pump-assisted circulation also has the advantage that boilers no longer have to be in the cellar, as was the case for *gravity circulation* systems. In domestic installations the water is run in *small-bore* or *microbore* copper pipes, while commercial *low-pressure hot-water* systems are run in black mild-steel pipes. Apart from space heating, the boiler is often also used for production of *domestic hot water*. See diagram.

centreline A line for *setting-out* that runs lengthways through the centre of a symmetrical part, either the whole building or a component.

centre-opening door, bi-parting d. A double door that slides apart from the middle. For *lifts* it is made in four parts, with the inner halves of each leaf carried on the lift car, meeting matching outer halves at each landing.

centre planks The planks next out from the *heart planks* of a log that has been *flat sawn*, thus partly *quarter sawn* but free of heart pith. See diagram, p. 182.

centres Either *centering* or spacings *centre-to-centre*.

centre-to-centre The spacing or layout dimensions of components measured between their *centrelines*, or the span of a beam between the centrelines of the supporting columns. Components at regular spacings are said to be 'on centres' or measured 'in and over'.

centrifugal fan A quiet, efficient *fan* used in *air conditioning*. It has a *runner* or impeller with *blades* spinning in a casing, usually with *double inlets* and a V-belt drive from an electric motor.

centrifugally cast pipe, spun p. Precast concrete or cast-iron pipes made in a spinning outer mould. During setting they shrink inwards, but cracking is avoided as there is no inner mould.

ceramic-disc valve A tap with one fixed and one movable hard, smooth ceramic disc, with matching holes. The tap is opened by turning one disc until the holes are opposite those of the other. They are easy to operate, dripfree, and durable.

ceramics Burnt clay ware (Cerami was the potter's district in ancient Athens), usually meaning high-quality materials made from refined clay powder, dry-pressed to shape and fired

at 1000 to 1200°C (or higher for *fully vitrified ceramics*), but strictly including *clay bricks* (fired above 900°C), clay blocks, *vitrified clayware* pipes, and *terracotta*. Typically they are stiff but brittle and have excellent durability, water resistance, and *dimensional stability* (apart from their early *moisture expansion*). They can include cermets (ceramics in a metal matrix), used to make points for cutting tools. But sundried clays like *adobe* are not ceramics.

ceramic tile Attractive, water-resistant, hard-wearing floor tiles or wall tiles, of accurate shape and usually much thinner than *quarries*. They are made (to BS 6431) from refined powder pressed to shape dry, then fired until *vitrified* or *fully vitrified*, and often *glazed*. Different *adhesives* and fixing methods (see BS 5385) are used for kitchens, showers, and swimming pools.

ceramic veneer *Terracotta* external wall facings 30 to 63 mm thick, of large face sizes, hand-moulded or machine-extruded, and of very wide colour range. The thinner slabs are fixed by *bedding* in mortar, while the thicker slabs have 6 mm dia. anchor bars.

certificate (1) A statement signed by the *architect* that a certain percentage of work has been completed in a building *contract*. The builder is entitled to payment of this percentage, less *retention*. Certificates are issued *monthly*, then at *practical completion* and *final completion*. The *client* should not try to influence the architect's decision on the amount of a certificate.

(2) Any *certification* document.

certificate of practical completion A *certificate* issued when a building has been substantially completed, but with minor *defects*. With this certificate the *defects liability period* starts, allowing release of half of the *retention* money.

certification (1) A statement that an inspection carried out by an official body has satisfied certain criteria, for purposes such as *quality assurance*, *appraisal*, or *fire protection*. In Britain certification bodies are given approval by the National Accreditation Council for Certification Bodies.

(2) The issue of a *certificate*.

cesspit, cesspool (1) A *rainwater header*.

(2) A tank containing *sewage* (*C*); for details see *C*.

chain-link fence A woven-wire anti-intruder fence.

chain tongs, c. vice A plumber's heavy pipe grip with a chain that jams steel pipe against serrated jaws, to hold the pipe while it is being cut or threaded.

chair rail A protective board on a wall, level with the top of a chair back.

chalking The degradation of a *coating* such as paintwork due to decomposition of the *binder*, leaving a surface layer of pigment which can be easily removed. Chalking mainly occurs outdoors, with unsuitable paints. Washing may restore the original appearance or provide a surface for repainting.

chalk lime High-calcium *lime*, or fat lime.

chalk line, snapping l. A string *line* coated with chalk that is stretched tight between two points on a flat surface, then plucked to mark a straight line, used for *setting-out*. The line is usually rolled up into a case containing blue *whiting* (chalk powder).

chamfer A cut corner between two surfaces at right angles, made by removing the *arris* at a 45° angle. A chamfer is a symmetrical *bevel*. See **splay**.

chamotte *Grog*.

change of air An *air change*.

change order (USA) A *variation order*.

channel A U-shaped section, such as a half-round or three-quarter-round clayware pipe between drain pipes in an *inspection chamber* (*C*).

channel bend, slipper b. An open channel which is curved on plan, to guide the flow from a *branch* pipe in a *foul water* manhole, usually made of burnt clay; to standard angles and half-round or three-quarter-round.

channel insert A *fixing channel*.

chapping The flaking or roughening of the surface of a *casting*, from setting shrinkage dragging it against the mould. The chapping of concrete wall or column tops is common; for *direct-finish concrete* it can be reduced by *revibration* (*C*). Chapping is usually not permitted in *cast iron*.

charge hand, leading h. A *tradesman* in charge of a small gang of tradesmen and their *labourers*, the next grade below *trade foreman*.

chartered builder (MCIOB, FCIOB) (UK) A professional builder trained to university degree level in building technology, construction methods, project management, quantity surveying, building law, economics and finance, contracts administration, industrial relations, professional practice, site surveying, etc., who has been admitted as a member or a fellow of the *Chartered Institute of Building*.

Chartered Institute of Building (CIOB) (UK) A professional institution for all in building, founded as the Builder's Society in 1834 and granted a royal charter in 1980. Membership is open to professionals, technicians, and students, through several different routes, according to academic level and practical experience.

Chartered Institute of Building Services Engineers (CIBSE) (UK) A professional institution for designers of building *services*, formed by the amalgamation of the Institution of Heating and Ventilation Engineers (founded 1897) and the Illuminating Society (founded 1909) under a royal charter granted in 1976. The CIBSE Guide gives data on the indoor environment, equipment, and controls.

chase A groove in a wall (or a floor) usually for the *building-in* of services, such as pipes, conduits, or cables, or as a *raglet* for a flashing. Chases can be cut into *brickwork* or formed in the surface of concrete. Pipes in a chase should be encased in flexible conduit or laid in sand to allow for *thermal movement*. After services are installed, chases are filled with mortar for finishings to be applied over them.

chased Built into a *chase*.

check (1) A crack in converted timber along the grain and across the rings, not passing right through the wood, caused by *shrinkage* during seasoning. It is a *gross feature* in structural timber, but fine checks may add character to a *veneer*.
(2) A device that slows or damps down a movement, e.g. of a *door closer*, usually with hydraulic or air action.
(3) The inspection or verification of work to ensure that it complies with the *specification* or that the *setting out* is right.
(4) A *water check*.
(5) (Scotland) A *rebate*.

checked back Recessed, *rebated*.

checkerplate *Chequerplate*.

check fillet A small kerb to direct the flow of rainwater on a roof.

checking Usually *crazing*, but see **check**.

check lock A *snib* or privacy latch.

check out To form a *rebate*.

check rail A *meeting rail*.

check throat A groove near the outer bottom edge of a *sill* to stop rainwater running back inwards along the *soffit* and wetting the wall.

check valve See *C*.

cheek The side of a *dormer*, *mortice*, *tenon*, etc.

cheek cut (USA) In *roof cutting* a bevelled end to a *rafter* to fit against the side of a *hip* or *valley* member.

chemical conversion coating A *pretreatment primer*.

chemical injection An *injection damp course*.

chemically resistant floor tiles Special *ceramic tiles* with high resistance to chemicals, used for laboratory floors. They need to be bedded and jointed in a chemically resistant cement, based on bitumen, resin, or sulphur, but not Portland cement, which has poor resistance to chemicals.

chequerboard slabbing One method of *alternate bay construction*.

chequerplate, checkerplate Pressed steel *tread plate* for industrial flooring.

chill room, chiller A *coolroom*.

chilled counter A *servery* unit where cold food is kept cold.

chilled water Cold water (4° to 7°C) from a *chiller*, supplied to the *cooling coils* of an air-conditioning system.

chiller, water c. set (UK **refrigeration unit**) A machine which uses the *vapour compression cycle* to produce large quantities of *chilled water* for air-conditioning systems. The waste heat from chiller sets is usually disposed of by piping hot water to a *cooling tower*. Their cooling capacity is measured in kilowatts (in France in negative calories or 'frigories') or in the USA in *tons of refrigeration*.

chilling The deterioration of *paints* or *varnishes* that have been kept at low temperatures.

chimney Any structure containing a vertical *flue*. Open *fireplace* chimneys have a *throat* and *smoke chamber* leading to the flue, which projects above the roof to a height given by the *Building Regulations*. Industrial chimneys are usually freestanding on a concrete base beside a boilerhouse. They can contain several flues and are prefabricated and erected by specialists.

chimney block A *flue block*.

chimney breast The chimney wall that projects into the room and contains the fireplace and flues.

chimney cap, c. pot A *flue terminal*.

chimney gutter A *back gutter*.

chimney shaft The part of an industrial chimney that stands free of other structures.

chimney stack The brickwork containing one or more *flues* projecting from a roof, usually carried above the *ridge*.

chipboard, wood c. A *building board* made by bonding wood chip with synthetic resin *adhesives* (or cement) and either *extruded* or *platen-pressed*. Chipboard has a flat accurate *face* each side or can be *melamine-surfaced*. The standard grade, weaker than solid timber and with similar *moisture movement*, is used for joinery and *dry linings*. Structural-grade chipboard must be kept dry, at below 18% moisture content. Special grades are made for flooring or roof decking, some of them moisture-resistant. They can have tongued and grooved edges, which also improves the board's *fire resistance*.

Chipboard is the commonest *particle board*. See BS 5669.

chipboard screw, twin-thread s. A thick, twin-thread screw for use in *chipboard*.

chipped grain, torn g. Shallow holes in the surface of *surfaced* timber resulting from wood fibres being partly torn out by planing knives going against the grain.

chippings Crushed stone from 3 to 25 mm in size, for *roofing* or *dashed finish*.

chisel (1) A tool with a flat steel blade, sharpened across its tip to a straight cutting edge, held by a *tang* to a handle of tough plastics or wood. Chisels are used in woodworking to make *mortices* and *rebates*.

(2) All-steel chisels, made to be struck with a *club hammer*, include the *caulking chisel* and the *cold chisel* (*C*).

chisel knife A narrow *stripping* knife with a square edge not wider than 4 cm, used for removing paintwork.

chisel-set teeth *Carbide tips* on a circular saw shaped like a wood *chisel* and slightly wider than the saw blade.

chlorinated rubber paints Blended paints that contain natural *rubber* treated with chlorine, often containing paraffin waxes and *alkyd resin* binders plus a *solvent*, which evaporates and allows *curing*. They are hard (not rubbery) and water-resistant, and some are highly resistant to chemicals (acids, alkalis), making them suitable for *tanking* or lining swimming pools. Special types also contain *flame retardants*, but others may give off dangerous fumes if burnt. Dry conditions and adequate ventilation are needed during curing, as the solvent is slow to evaporate, flammable, and toxic. Frost does not stop curing, but slows it down.

chlorofluorocarbons(CFCs) Chemicals which are *vaporizing liquids* at normal temperatures, used in *refrigerants* and as blowing agents for foam plastics. Alternatives such as hydrofluorocarbons (HFC) are finding increasing use, as released CFCs severely deplete the earth's ozone layer. Damage is worst in January, when the earth is closest to the sun by about 3%, receiving 11% more radiation. The biggest hole in the ozone layer occurs over the south pole each southern summer. CFCs should be factory recycled, never burnt or dumped. See BRE Digest 358.

chroma Colour *saturation*.

chromating (1) Priming with paints based on chromates of zinc or lead, which are both rust *inhibiting pigments*.
(2) A clear golden protective coating for galvanized steel formed by dipping the articles in a hot solution of dichromate, or chromic and nitric acids.

chromium plating, chroming An abrasion-resistant *metal coating* consisting of an electroplated surface of chromium. When put on to iron or steel, the chromium adheres better if deposited on nickel previously electrodeposited on copper. Electro-deposited chromium is almost as hard as diamond and is also used for *hard plating*.

chuck The device on a *drill* that grips the *bit*. It usually has three self-centring jaws tightened by a key.

chute (1) A vertical pipe to guide falling loose materials, such as *refuse* or *linen*; or a rubbish *dropchute* (*C*).
(2) A steep open channel in a *drain*, or for handling *concrete*.
(3) A *cable chute*.

CIB (International Council for Building Research Studies and Documentation) Based in Rotterdam, CIB draws its members from the world's major research organizations, university faculties, public authorities, and private researchers. The CIBDOC database is accessed through the EURONET/DIANE host system.

cill An alternative spelling of *sill*.

cinder block A clinker *block*.

circuit (1) A continuous conducting path out to a load and back to an electrical power source; also the *wiring* with its two conductors. Examples: *outgoing* circuits, *branch circuits*, and *ring mains*.
(2) **circulation** Pipework through which a liquid circulates, going out through the *flow pipe* and back in the *return pipe*, e.g. the water in a *heating* system. A circuit may need *balancing*.

circuit breaker A *switch* which trips open automatically if there is *overcurrent* or other dangerous condition such as *earth leakage*. As *protection gear* it breaks the circuit at 1.5 times rated · current or less. It offers closer circuit protection than *fuses* and is simply reset by hand after the overload is removed. A single large *moulded case* type usually supplies several smaller *miniature circuit breakers*. European standard tests are given in BS EN 2495.

circuit protective conductor Formal term for the *earth* wire.

circuit protector Small electrical *protection gear* for a single circuit, such as a *miniature circuit breaker* or a *fuse*.

circuit vent (USA) A *branch vent* that runs upwards from before the last *trap* in a *branch* and connects to the *vent pipe*.

circular on plan Curved work, such as cylindrical walling, or a *dome*, involving extra *labours* and *setting-out*, or *special* bricks.

circular saw A *power tool* with a circular steel blade turned by an electric motor, hand-held or mounted as a *mitre saw*. The teeth round the rim

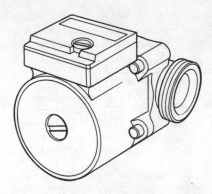

A circulating pump.

move at about 50 m/s and usually have *carbide tips*. Different blades are made to *rip* or to *cross-cut* wood, or the blade can be replaced with a *grinding disc*.

circular stair A *stair* with curved flights, or a *spiral stair*.

circulating pump, circulator A centrifugal pump, used in domestic central heating, which enables *small-bore* pipes to be used instead of the larger pipes of *gravity circulation*. The pump with its electric motor is a quiet, cheap, compact unit, easily replaced. Submerged rotor pumps usual in *sealed systems* are glandless and should not leak. See diagram.

circulating water The water contained in the closed circuit of a *central-heating* system.

circulation (1) In *planning*, the proper arrangement and proportioning of areas and spaces to facilitate movement of people from room to room, their access, communication, egress, or escape.
(2) The flow of water in a *circuit*, often called 'recirculation'.

circulation area, circulating a. A

passage, corridor, hall, or stairway used for *circulation* from room to room.

circulation pipe A *flow pipe* or *return pipe* in a circuit.

CI/SfB classification An arrangement of the information subjects in the building industry, promoted, administered, and developed by an agency of the RIBA for the building industry as a whole. It is based on the *International SfB* system and is set out in two manuals called the CI/SfB Construction Indexing Manual, one for general information (office libraries, contents of documents, etc.) and the other for project information (drawings and other *contract documents*).

cissing Mild *crawling*.

cistern A water tank with an open top, thus at atmospheric pressure, although it may have a loose cover. Cisterns are usually supplied through a *ballvalve* and may need a *tray* or *warning pipe*. They are used as feed and expansion tanks of *open-vented* low-pressure *hot-water systems*, for the storage of cold water, or where *direct supply* is not allowed by water authorities, e.g. to flush *water closets*. Water supply and feed cisterns, fitted in the roofs of British houses, are likely to become obsolete as *unvented-system* hot-water services and *sealed-system* central heating are adopted. See **dual-flush cistern.** See also diagram, p. 82.

City and Guilds of London Institute (C&G) The main awarding body for technical qualifications in the UK. Candidates are from approved training centres; C&G does not run courses itself. Its certificates are widely recognized in the traditional *building trades* and are part of the *National Vocational Qualifications* system.

clack valve A *check valve* (*C*).

cladding (1) (USA) **siding** A weather-

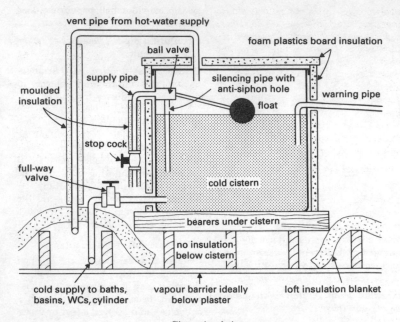

vent pipe from hot-water supply

ball valve

foam plastics board insulation

supply pipe

moulded insulation

silencing pipe with anti-siphon hole

warning pipe

float

stop cock

full-way valve

cold cistern

bearers under cistern

no insulation below cistern

cold supply to baths, basins, WCs, cylinder

vapour barrier ideally below plaster

loft insulation blanket

Cistern insulation.

tight skin covering an external wall, as part of the building *enclosure*. It is usually functional, but can include decorative *façade* cladding, and may have *insulation* or give some degree of security. Cladding is *non-loadbearing* and has to bear wind loads, impact damage, and temperature extremes. Long panels must allow free *thermal movement*. It can be of any durable material, such as *profiled sheeting*, *brickwork*, or *weatherboard*, or a system such as *curtain walling*, *patent glazing*, *cladding panels*, insulating *over-cladding*, or thin *stone facings*. Doors and windows may be included, but solid masonry is not cladding. Compare **lining**.

(2) A *finishing* or protective shell.

cladding panel Prefabricated factory-made cladding units, usually involving no *wet trades* on site. Most are *composite*, with an outer layer to resist knocks and weather. They include heavy concrete *large-panel systems*, pre-laid brickwork, and lightweight sheeting, which should have tamper-proof fixings.

cladding rail A *girt*.

claim (1) **contractual c.** An application for an *extension of time* and/or payment by a *contractor* (or *sub-contractor*) to recover costs not allowed for in the *tender* but for which he has an entitlement under the *contract*. Claims are often complex and can be based on direct or indirect factors. They should be prepared carefully to avoid *disputes*. See **variation**.

(2) A *monthly statement*.

82

clamp (1) A device for holding things together, often relying on a screw to tighten them. *Formwork* clamps are used while concrete is hardening, one of the most versatile being the *column yoke*. Clamps for services *fixings* usually grip from both sides and are stronger than *clips* or *cleats*. See also **cramp**.

(2) *Clay bricks* and fuel specially stacked in the open, or under an open-sided shelter, then set alight and burnt.

clamping plate A metal timber *connector*.

clamping time The length of time for which a joint needs to be held under pressure while an *adhesive* is setting, according to the amount of stress and the speed of *curing*.

clarification Additional information requested when the *contract documents* do not show exactly how something is to be built or when documents of equal *precedence* disagree.

clashing, interference The situation occurring when building *services* by different trades are both designed to fit into the same space, owing to lack of *coordination*. It is not unusual to change the detail layout of services on site, particularly in confined areas such as the space above a *suspended ceiling*, which apart from air-conditioning ductwork may also contain *luminaires*, pipework, and electrical *trunking*.

Class 0 (UK) A classification for *fire hazard* under the *Building Regulations*, given to materials tested to BS 476, which are: 'non-combustible' according to the *non-combustibility test*; have Class 1 *flame spread*; and have a low *fire propagation index*.

claw A curved split *peen* on a *claw hammer* or a *pinch bar*, used for pulling out nails.

claw-bolt lock A lock for a sliding

door, with a claw that grips the striking plate.

claw hammer, carpenter's h. A hammer with the usual flat face for driving nails and a *claw*, used for framing and *formwork* carpentry.

clay (1) As a *foundation* material clay has several disadvantages, notably its high *moisture movement* necessitating the use of *deep foundations* (*C*), etc. Clays vary widely in properties; some have more moisture movement than others, being affected by *trees* (such as poplars) and seasonal wet and dry cycles, which may lead to *cracking in brick walls*. See BRE Digest 343.

(2) The raw material for *ceramics*. There are many different types of clay.

clay brick The traditional building *brick*, made by shaping clay, then firing it in a kiln (or a *clamp*) until *hard-burnt*. The oldest types are handmade wet *moulded bricks*, and later *stock bricks*. The introduction of *pressed bricks* led to much stronger *engineering bricks*, now partly replaced by cheaper *wirecuts*. Clay bricks in general have excellent durability and low *moisture movement*, but all have some *moisture expansion*. See **Fletton brick**, **Keuper marl brick**, **London stock brick**, **Staffordshire blue brick**.

clay tile Any *floor tile*, *wall tile*, or *roof tile* made from burnt clay. Manufacturing processes vary in complexity, from extensive treatment and refining of the material for *ceramic tiles* to simpler shaping and firing for *terracotta* and *quarries*.

clayware *Sanitary fittings* and pipes made from burnt clay, which look good but are easily chipped or broken.

clean aggregate Granular material which is free of clay, silt, and organic matter, thus suitable for making concrete, etc.

clean earth An independent electrical *earth* system to protect electronic equipment (computers, telephones) against interference from inside a building.

cleaner's sink A *sanitary fitting* with a large deep *bowl* for rinsing cleaner's buckets.

cleaner's tap A draw-off *tap* for use by a cleaner.

cleaning Removal of rubbish; clearing of obstructions in pipes; preparation of a *substrate* for painting or plaster; bottoming or trimming sides of excavations; tidying after completion of work; blast or flame cleaning of steel, etc.

cleaning-down (1) The rubbing-down and removing of staining from *direct-finish concrete* or facework to make it ready for *handover*.
(2) **facelift, regrating** The renovation of external walls, patching, and rubbing of stonework, repointing, replastering, and repainting. See BS 6270.

cleaning eye An *access cover*.

cleanout (USA) An *access cover*, soot door, etc.

cleanout of wall cavities Some specifications forbid the cleaning-out of wall *cavities* and insist they they be kept clean during construction with *cavity battens*. Alternatively, each fifth brick in the lowest *course* can be laid in sand instead of mortar and removed after the wall has been completed, and all *droppings* and rubbish hosed out. The bricks are then replaced in mortar, which has to be done with care to make the joints look like the others in the wall.

cleanout trap A panel at the bottom of deep *formwork* through which rub-

bish can be removed before concrete is placed.

clean timber Timber free from knots (BS 6100). See **clear timber**.

clean up *Clearing site on completion.*

clear (1) Flawless, as of timber without *defects*.
(2) To unblock or clean a pipe or *drain*.
(3) Transparent, as of finishing, glazing, anodizing.
(4) Uninterrupted distance, span, height, the length of *clearance*.

clearance The distance between objects, e.g. the height or width of a stairway or passageway, or in joinery the 'play'.

clearing and grubbing The cutting-down of trees and removal of stumps before commencing *groundworks*.

clearing away The removal of excess spoil *off-site*.

clearing site before commencement The removal of obstructions from the site before commencing work, mainly the demolition of existing structures.

clearing site on completion, c. up The removal of *temporary works*, site sheds, rubbish, etc., after completion of work.

clear timber, free stuff Timber which is free from visible *defects* and imperfections (BS 6100). Since knot-free timber is generally only found in the lower part of the trunks of closely grown trees in virgin forests, clear and clean timber is now scarce in long lengths. However, short lengths (about 300 to 400 mm long) can be *finger-jointed* to give long boards suitable for *fascias*.

cleat (1) A batten or other small piece

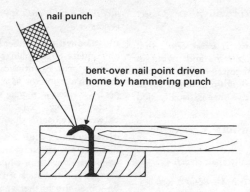

nail punch

bent-over nail point driven
home by hammering punch

A nail punch in use for clench nailing. A sledge hammer is held behind the nail head. The
nail tip should be bent over parallel to the wood grain.

of timber or metal fixed so as to reinforce a component or to stop objects
sliding along a surface.

(2) A timber *plugging* in a wall for
driving in a screw fixing.

(3) A *fixing* that supports pipework,
cables, or other *services* runs on a wall.
It is often held by screws or made in
two pieces. See also **clip**.

(4) A roofing *angle cleat*.

(5) The horizontal *rail* or *ledge* of a
timber door.

(6) A *tingle* or roofing cleat.

cleavage The natural layering in materials which make them easy to split in
one direction. Wood has cleavage along
the grain, slate is *riven* along its cleavage planes.

cleaving The splitting of a material
along its *cleavage* planes.

clenching, clinching Driving nails
through timber until the head is embedded on one side and the point projects
out of the other side, then bending the
point over, and possibly also driving it
into the surface with a nail punch.

Special *duckbill nails* are made for
clench nailing. See diagram.

clerestorey, clearstorey Windows
high in a wall to let in *daylight*.

clerk of works, c. of the works The
full-time site representative of the
client, whose main duty is *inspection* of
the work as it proceeds. He may also
agree existing *ground levels* recorded
by the contractor before groundworks
are begun, and the hours of *daywork*.
His approval in writing is often required before covering up work, e.g.
before the concreting of foundations.
A clerk of works is usually a *tradesman* of considerable experience near-
ing the end of his active career. He is
paid by the client and usually works
under the instructions of the *architect*
but takes no direct part in the running
of the job.

client The person or organization for
whom a building is being designed and
built, usually the *owner* of the land,
and of the building once it is completed. In a contract the client is called
the builder's *employer*.

client's representative A *project manager*.

climbing crane A *tower crane* that can lift itself on rams and swing in extra lengths of tower one by one. It can be freestanding up to about seven storeys, but above that is usually tied to the building.

climbing formwork A *jump form* or a *slipform* (*C*).

clinching Usually spelt *clenching*.

clink (Scotland) A *double-lock welt*.

clinker Fused ash from furnaces. If properly burnt and containing very little unburnt coal it is excellent *hardcore* (*C*) or *lightweight aggregate* for concrete or precast concrete building *blocks*.

clinker block, cinder b. A cheap, strong, lightweight building *block* made of *clinker* concrete. If kept dry, clinker blocks have very little *moisture movement*, but laid wet in a wall they can cause the plaster over them to crack badly as they dry out and shrink.

clip A metal *fixing*, attached by tightening a screw or driving in a nail, by hooking or a snap-fit arrangement, used as an accessory for fastening cables, pipes, scaffold tubes, roof tiles, etc. Clips are similar to *cleats* but usually simpler.

clip anchor An *angle cleat*.

cloak A waterproof covering of *supported sheetmetal roofing* over a projection above a general roof surface, to protect against rain, snow, etc. But see **overcloak**.

clock A *receiver clock* for *time distribution*.

clock timer A *timer* that switches *circuits* on or off at regular times, usu-

ally with a *solar dial*. It is a simple *programmer*.

close boarding Timber *boarding*, usually square-edged planks laid side-by-side to form a continuous surface for fixing insulation, or as a deck for *supported sheetmetal roofing* and roof tiles, or for *fencing*. Gaps usually appear between the boards due to shrinkage.

close-contact glue An *adhesive* used to make strong joints in wood, which need heavy pressure while the adhesive sets. It will not stick if the surfaces to be joined are further apart than about 0.15 mm, unlike *gap-filling glue*.

close-coupled cistern A low-level *water closet* cistern attached to the rear of the *pan*. The two may form a matching suite.

close-couple roof A traditional framed roof with common *rafters* joined at the wall-plate level with a *tie beam*, used for spans up to about 4 m.

closed shop A site where only members of a union can work.

close-grained timber *Narrow-ringed timber*.

close nipple A short *nipple*. The two pipes being joined are therefore close to each other.

closer (1) **closure piece** Material used to fill a gap, e.g. at the end of a *cavity*.

(2) A *special* brick or block made or cut narrower than usual to complete the *bond* at the end of a wall, e.g. a *queen closer* or a *king closer*.

(3) A *door closer*.

close string, housed s. A stair *string* with rebates cut into the inside face, into which the treads (also risers if any) are housed. Compare **cut string**.

closing in Working towards the *enclosed stage*.

closing device Hardware used as a

fastener to secure joinery in a closed position but not for *locking* it.

closing face The side of a *door leaf* which closes against the frame.

closing jamb A *shutting jamb*.

closing of excavations Final backfilling over services or round foundations, to reinstate the *original ground level*.

closing stile A *shutting stile*.

closure piece A *closer*.

clout nail, felt n. A stout galvanized-steel *nail*, 10 to 65 mm long, with a large flat head, for fixing bitumen felt, metal roofing, plasterboard, etc.

club hammer, lump h., mash h. A robust hammer with a fairly rectangular head of about 1 kg and a handle about 250 mm long, used in one hand. See diagram, p. 281.

coach bolt, carriage b. A *fastener* for heavy timber sections with a head that is rounded on top and enlarged underneath to a square section, enabling it to grip the wood without turning as the nut is tightened. See diagram, p. 165.

coach screw (USA **lag bolt**) A large *gimlet-pointed* screw for making *fixings* in wood, by turning the square head with a spanner. A *pilot hole* should be drilled for it. See diagram, p. 165.

Coanda effect The tendency of a stream of air to cling to a surface when blown parallel to it. Coanda-effect ceiling diffusers are used for *supply air* outlets. They often produce a dirty stain on the surrounding ceiling.

coarse-grained timber *Wide-ringed timber*. Not the same as *coarse-textured timber*.

coarse stuff (1) Lime:sand mortar, a traditional material for base coats of plasterwork, no longer used in new work.

(2) Lime:sand *ready-mixed mortar* for bricklaying.

coarse-textured timber *Wide-ringed timber* with a *texture* of relatively large wood elements (coarse grain or open grain) or unusually wide *annual rings* for the type of wood. Such timber may show *whitening in the grain* if varnished.

coat A single layer of any material (paint, plaster, render, asphalt) applied at one time, to a stated thickness, to make a *coating*.

coated felt A *mineral-surfaced bitumen felt*.

coating A layer of material applied in one or more *coats*, to form a *finishing* that can be decorative or protective. Common coatings are *paintwork, galvanizing, organic coatings*, and *anodizing*.

cob Clayey soil and straw or fine roots, shaped in place, to make the walling of an *earth building*.

cock A simple *stopvalve* for cutting off a supply of water, gas, or other fluid in a pipe. It is opened or closed by a quarter turn, exposing a *full-way* hole through a tapered plug.

cocked hinges Door hinges fixed askew to raise the door when it is opened, to clear carpets and to make it close by itself. *Rising-butt hinges* do the same job more expensively but more elegantly.

cocking A traditional joint made by cutting two mortices in a *wall plate* and one in the bottom of a beam, forming a *cog*, before *truss clips* became usual.

cocking piece, sprocket A narrow board nailed to the ends of rafters at the *eaves* to give an eaves overhang a slightly flatter slope than the rest of the roof. For *single-lap tiles* or slates, BS

5534 does not recommend this practice.

code of practice (CP) A publication describing accepted *good practice* in a trade, specialist work area, etc. Many former BSCPs are now referred to as BS Codes and published as *British Standards*. Codes of practice are also issued by research bodies, professional institutions, and trade associations. *See* **Eurocode**.

coffer A deep recess in a *soffit*. See next entry.

coffered slab A deep reinforced-concrete *suspended slab* cast over *pan forms* between which the main bottom bars are laid, forming a grid of ribs. Large spans are possible because of the reduction in *dead weight* (*C*), but the formwork needs *camber* in both directions. See **waffle slab**.

cog A small projection or *nib*.

coil (1) **battery** A curved pipe containing hot or cold flowing water, a *heat exchanger* from liquid to air, used in air conditioning as a *hot coil* for heating or a *cold coil* for cooling.
(2) Electrical windings of a *relay* (*C*), *contactor*, or transformer.

coil heating *Concealed heating* with piped *low-pressure hot water* or electrical *mineral-insulated cables*, common in *underfloor heating*.

coincidence effect A reduction in the *sound insulation* of a sheet material from flexing waves which pass through it more easily than does vibration at right angles to the surface. For a lightweight *double-leaf separating wall* or for *double glazing*, a sound at the 'coincidence-dip' frequency and one octave above may be 5 to 10 *decibels* louder on the other side than other sounds. Thick materials and *damping* reduce the coincidence effect. See BR Digest 143.

coir Natural fibres from the outside of coconuts, used to make doormats.

cold battery A *cold coil*.

cold bridge, heat b., thermal b. A piece of metal, such as a pipe or wall tie (or any other conducting substance), that passes through a wall and carries heat through it. This means that the inner surface of the wall round the pipe or wall tie will in summer be warmer and in winter be so much colder than the remainder of the wall that *condensation* may occur there. *Bridging* of this sort is particularly undesirable in a *cavity wall* or any other type of *discontinuous construction*.

cold cathode lamp A *fluorescent tube* without a heated cathode. It operates at high voltage with low light output and is used for long, complex shapes.

cold coil, c. battery A *heat exchanger* in an air-conditioning system, fed with *chilled water*, to produce cold air. It may need a *condensates pan*.

cold cupboard A *servery* cabinet kept at about 5°C.

cold deck A flat roof construction that works as a *cold roof*, no longer recommended by the *Building Research Establishment*. See **warm deck**.

cold galvanizing A protective coating of *zinc-rich paint* used on steel.

cold roof A roof with its insulation near the ceiling and a ventilated space above it. It can be a *pitched roof*, or a *flat roof*, which is also called a *cold deck*. The main risk of a cold roof is that warm moisture-laden air from the house will enter and as it cools form *condensation*. This is prevented in two ways: by providing a *vapour barrier* at ceiling level (which may interfere with ceiling finishings) and by *ventilation* of the ceiling cavity. See **warm roof**. See also diagram, p. 147.

cold store Any refrigerated and insulated walk-in for storage at low temperature. It may be either a *coolroom* or a *freezer*.

cold-weather working See **frost protection**, **winter working**.

collapse The excessive and irregular shrinkage of *kiln-dried* hardwoods, dried too quickly or too much, usually with some weakening of the wood cells, although the overall shape is recovered during *reconditioning*.

collapse grading The *stability* rating of a *fire door* given by testing.

collapsible form *Formwork* that can fold or telescope inwards to allow *stripping*.

collar (1) A ring or *fillet* of asphalt built up round a vertical pipe passing through an asphalt roof to ensure a watertight joint at the pipe.
(2) An enlargement outside a pipe or a reduction within its bore. It is often made to bear on another collar to ensure a tight joint between pipes, as in a *union*.

collar beam, span piece A horizontal tie beam, as in a *collar-beam roof*.

collar-beam roof A traditional framed roof with *common rafters* joined half way up their length by a horizontal *tie beam*, to give more headroom than a *close couple*.

collar boss A pipe fitting for a plumbing *stack*, with *bosses* that can be drilled out for future connections, such as an extra bath or basin *waste*.

collection In quantity surveying a list of figures and their sum, such as preliminary calculations from *taking-off*, to record minor dimensions and their sources or the sub-total of a page or *trade*.

collector (1) (USA) A drain pipe connected to a *sewer* (C) or storm drain.
(2) A *solar collector*.

collusive tendering (USA **bid collusion, b. rigging**) In competitive tendering, an illegal private agreement between tenderers on the prices they each submit; an interference with free competition.

colonnade A row of similar columns joined at the top by beams. Compare **arcade**.

colour (USA **color**) Any colour can be fully described by its *hue*, *lightness*, and *saturation*, e.g. using *Munsell references*. Colour in *paints* comes mainly from the *pigment*. Black is considered a colour for building purposes, but white is not. See BS 4800 and BS 5252.

colour chart, c. schedule A list of rooms and the colours to be used for decoration.

colour coating *Organic coating*.

colour coding Identification colours for services, *safety signs*, etc., e.g. red for a fire main, green/yellow for an earth wire. See BS 1710.

column An upright structural member, square, round, or rectangular, of reinforced concrete, timber, brickwork, or blockwork, including steel *stanchions* (C). Columns carry vertical loads (weight) in compression. They can be *isolated*, *engaged* in, or *attached* to a wall. See diagram, p. 90.

column clamp Bars, bolts, or *column yokes* put round *column forms* and tightened to prevent leakage during concreting.

column form, c. box *Formwork* for a reinforced concrete *column*.

column splice A *field splice* in a steel column for a tall building, joined

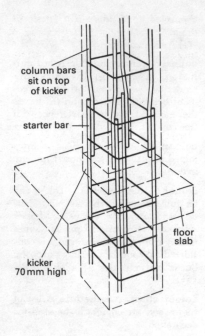

column bars
sit on top
of kicker

starter bar

floor
slab

kicker
70 mm high

A column.

by bolting at about every third storey.

column starter A *starter* bar for a reinforced concrete column, long enough to pass through the floor slab and *kicker* and to lap with the column bars. Column starters in foundation *bases* can easily shift during concreting and their *setting out* should be re-checked at kicker level, before the *ground slab* is cast, to ensure correct position when the column is built.

column yoke A set of four interlocking steel bars with hook ends, tightened by hammering wedges through slots, used as a *column clamp*.

comb (1) **c. steel** A replaceable toothed blade for a *comb hammer*.

(2) A paint grainer or glue applicator.
(3) A *scratcher*.

combed joint, cornerlocked j., laminated j. An angle *joint* in which the ends of the pieces are slotted to mate with each other.

comb hammer An air-driven power tool with a slotted *peen* that grips a *comb*, or a *patent axe* used for hacking or *scabbling* concrete or stone.

combination boiler A *boiler* with a *calorifier* for domestic hot water production mounted above it, inside the same casing. The waterways are directly connected, giving rapid heating and a small calorifier.

combination pliers, combinations (USA) *Footprints*.

combination tank An *open-vented* hot-water *cylinder* with a *cistern* built over it. It is less prone to freezing than a separate cistern.

combination tap assembly A *mixer*.

combined decking and reinforcement *Composite decking*.

combined extract and input *Mechanical ventilation* that combines *extract systems* and *input systems*. It is more complicated than either, as the volumes of air handled have to be matched.

combined lintel, l. damp-course A *lintel* of *powder-coated* galvanized steel or stainless steel shaped to act as a *damp-proof course* for a *cavity wall*. Some local authorities also require a conventional damp-proof course.

combined system A system of *internal pipework* in which *soil water* and *waste water* are carried in one drain pipe. See also *C*.

combplate The toothed plate at the step-on and step-off of an *escalator*.

combustible Liable to burn, e.g. timber, plastics, and synthetics, even after fire- or flame-retardant treatment. Testing of building products for *fire hazard* starts with the *non-combustibility test*. If they are combustible, they are subjected to further tests for *ignitability* and *flame spread*.

combustion gases Usually colourless toxic gases such as carbon monoxide, carbon dioxide, hydrogen chloride, and hydrogen cyanide, given off by a fire, often before the visible *smoke*. They can kill people by asphyxiation or poisoning.

combustion gas detector A *fire detector* that reacts to *combustion gases*, thus giving warning in the very early stages of *fire development*.

comfort index A single figure that takes into account air temperature, *mean radiant temperature*, air velocity, and *relative humidity*, used in the design of air-conditioning systems. Several different comfort indexes are in use, which all give fairly similar results: *equivalent temperature*, *environmental temperature*, globe temperature, and dry resultant temperature. Figures from the USA *effective temperature* are slightly different. The most suitable comfort index for a particular room or space differs with age, sex, type of clothing, and activity, for instance: swimming pool 26°C, hotel bedroom 22°C, living room 21°C, laboratory 20°C, office 19°C, shop 18°C, gymnasium 16°C, warehouse 13°C, etc.

comfort zone A range either side of the ideal *comfort index* within which most people still feel comfortable. For 70% of people the temperature should be within about 2.2°C of the comfort index.

commencement The start of building work on site, usually the date when *possession* of the site is given to the *contractor*. It marks the beginning of the *time for completion*. Each *trade* also has a commencement date. Before commencement an *advance* may be paid for *preliminaries*.

commissioning Starting a completed building *service* and checking that it works correctly. The process may extend over quite some time, both before and after *handover*, and can consume much power and water, particularly for complex mechanical and electrical installations. It includes *drain tests*, *air balancing*, and checks for *equipment noise*, as well as instrument measurements of performance levels against figures given by the *specification*. When commissioning is completed it is usual to give the operating and maintenance manuals and *as-built drawings* to the building owner, unless there is a maintenance contract with the equipment supplier or installer.

common, c. brick (USA **building b.**) A *brick* suitable for general purposes but without the attractive appearance of a *facing*.

Common Arrangement (UK) A *sequence of trades* agreed between representatives of leading organizations in the building team (ACE, BEC, RIBA, RICS). It has twenty-four sections (A–Z except I and O) which are subdivided into 300 Common Arrangement work sections (CAWS), such as G 20 or H 21. The Common Arrangement is used in the *Standard Method of Measurement* SMM7, and some *British Standards Institution* publications, and could be used for *specifications* on large projects with many *sub-contractors*. See table, p. 400.

common bond Either *stretcher bond* or, in America, *garden-wall bond*.

common brickwork

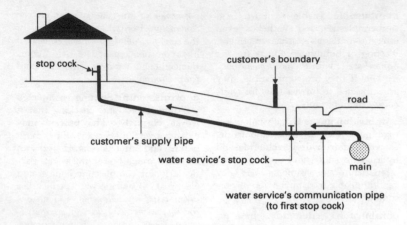

stop cock

customer's boundary

road

customer's supply pipe

water service's stop cock

main

water service's communication pipe
(to first stop cock)

A communication pipe is a service pipe for gas or water supply from a main.

common brickwork Brick walling which is to be plastered. The bricks may be *rough cut* and mortar joints are struck off or *raked out* for key.

common grounds, rough g. Timber battens fixed directly on to a wall, usually nailed into *pluggings*, to carry a *dry lining*.

common joist, rafter, stud, etc. One of several framing timbers that share loads with others running parallel to them at fairly close spacings. Common floor joists are usually 450 or 610 mm apart and tied together by both the *flooring*, nailed directly to them, and any *bridging* or *bearers*. Common ceiling joists have a *hanger* for load sharing, while common rafters and *trussed rafters* have tiling or sheeting *battens*, and common studs have *noggings*. Both *intermediate* and *trimmed* timbers can act in common.

common trench A services trench shared by a gas pipe and an electrical cable, or by a soil drain and a waste drain, which are usually kept at least 150 mm apart.

common truss A *trussed rafter*, but see **common joist**.

common wall (USA) A *separating wall*.

communicating door An *internal door* between two rooms.

communication pipe The pipe between the supply main and the consumer's *stopvalve* or the boundary, whichever is nearest the main. It is the part of the *service pipe* which belongs to the water or gas service company. See diagram.

communications installation Part of the *electrical services* installation for alarms (fire, intruder), data, fire detectors, telephone, time distribution, sound distribution, and remote controls. In a *building management system* there may be a data highway or a local area network. Wiring for communications is usually separated from other electrical services.

compaction of concrete Driving out, usually by *vibration*, air that has entered concrete during *placing*. The *formwork* is then completely filled and the *reinforcement* surrounded, leaving no *blowholes* or *honeycombing*.

compartment, fire cell One of the spaces, separated by *fire walls* and floors, into which buildings (except for housing) are divided in order to limit the spread of fire. The maximum volume of a compartment depends on building type and is stated in the *Building Regulations*. In some buildings they allow the compartment size to be doubled if *sprinklers* are installed.

compass saw, keyhole s., locksaw A handsaw with a blade which tapers to a point, usually made as a *padsaw*. It is used for cutting round sharp corners. See diagram, p. 388.

compatibility Chemical suitability of coatings (paintwork, plasterwork) or of *adhesives* to the backing or substrate. If parts of a *paint system* are incompatible, the mixture may result in low adhesion, poor *gloss*, pinholing, slow drying, greasiness, or *crawling*. Impregnation of timber with *preservatives* may lead to incompatiblity with paints when *overpainting*, or with adhesives or glazing compound. Guidance on paint makers, systems, and types is given in BS 8000.

completion The end of work on a *contract* or sub-contract, or for a trade or activity, the *stage* immediately before. At the end of a project a pre-completion inspection is made, followed by *practical completion* and later *final completion*. Progress towards completion is usually given as a percentage, e.g. 'brickwork 50% complete'.

completion bond A *performance bond*.

completion date The last day for *completion*, calculated from the date of *commencement* plus the *time for completion*, to which any *extensions of time* are also added. If work goes beyond this date, *damages for delay* may be payable.

component (1) A *product* made as a separate unit to serve a particular function, e.g. a *door set*. Components can be *modular* sizes, with *dimensional coordination*, as when they are to be used in *assemblies* to make up larger *elements*. They include parts of the electrical, plumbing, or gas *services* installation.
   (2) A part of the *daylight factor*.

composite, c. construction, mixed c., sandwich c. A component made of two or more materials, combined to take advantage of their properties. The materials in large composites should have similar *thermal movement*.

composite board, sandwich panel A panel for *cladding* or other purposes, often with an insulating core inside a loadbearing protective skin.

composite decking, participating d. Steel sheeting with a deep dovetail profile, used as both *permanent formwork* and *reinforcement* in a *composite floor*. See BS 5950.

composite floor A steel and concrete *suspended floor*, used for rapid construction of multi-storey buildings. The *composite decking* is laid on top of steelwork beams and covered with a topping of normal dense or lightweight concrete. Additional *fabric* reinforcement is embedded to reduce shrinkage cracking and ensure some strength in a fire, making *fire protection* insulation of the *soffit* unnecessary. This also leaves exposed the dovetail slots in the decking, allowing them to be used for fixing hangers for *suspended ceilings* and services. See also *C*.

composite wood board *Blockboard* or *laminboard*.

composition flooring A floor finishing made up of cement, wood particles, gypsum, calcium carbonate, pigment, and linseed oil. It does not suffer *dusting*, wears well, comes in many different colours, and can be laid either *jointless* or resembling *wood-block flooring*. It is suitable for laboratories, workshops, hospitals, or churches.

composition mortar, compo. Cement:lime:sand mortar, in proportions from $1:\frac{1}{4}:3$ (designation 1) to $1:2:9$ (designation 4). It has good *plasticity* and brickwork laid in compo has more resistance to cracking than cement *mortar*, but it sets more slowly, especially in cold weather. See **mortar designation**.

compound beam, built-up beam A beam built up of several timbers, held together either with *connectors* or by nailing, bolting, adhesive, or any other method. See **glued-laminated timber**.

compound walling Walls laid in two or more skins of different materials.

comprehensive contractor's liability insurance Cover for the risks during construction, including public liability, fire, and theft, which is usually required by the *conditions of contract*.

compressed air Air at about 7 *bars* pressure, delivered from a *compressor* through pipes and hoses, mainly used on building sites for *pneumatic tools*, which are rugged and powerful for their weight, and for *air-blast cleaning*.

compressed granulated cork *Cork board*.

compression glazing Window *gasket glazing* in which a preformed gasket is compressed by a bolted or clip-on metal beading, a wedge insert, etc.

compression joint, c. coupling, c. fitting A joint in *light-gauge copper tube* or *plastics pipes*, made by screwing together the ends of the two pipes to be joined by means of large nuts outside them. The joint can be *manipulative* or *non-manipulative*; both are bulky but generally reliable.

compressor A machine for producing *compressed air*, usually as part of a set with a *receiver*, or for lower pressures a *blower* or a *fan*.

Compriband Bitumen-impregnated polyurethane foam *sealing strip*.

computer Computers are used in *building management systems* and for many tasks in the building process, such as *quantity surveying*, *estimating* and the *programme*.

computer room A room for mainframe computers and other data-processing equipment, usually with packaged air conditioning, fire-resisting walls, an *access floor*, and *antistatic* finishings.

concave joint A *bucket-handle joint*.

concealed gutter A *box gutter* behind a roof *fascia* or *cornice*. At a normal viewing angle, from the ground, the gutter is hidden.

concealed heating Heating in a floor or ceiling with *coil heating* embedded in the concrete floor slab, out of reach of small children.

concrete A mixture of coarse and fine *aggregate* (sand and stone), *cement*, and *mixing water*, plus any *admixtures*, that sets within a few hours of mixing. Dense concrete is a versatile, cheap material, strong in compression and with low *moisture movement*. Walls and floors have good *fire resistance* and no *fire hazard*. Concrete is highly alkaline, which prevents the rusting of steel *reinforcement*. It is attacked by *sulphates*, many acids (acid rain, lactic acid), as well as by *salt air* or seawater. *Carbonation* can lower the alkalinity that prevents rusting. Ordinary concrete is unsuitable for severe exposure. Normal structural concrete for *in-situ* work is usually *readymix*, delivered fresh to enable *placing* in *formwork* before it sets, although silos containing dry mix, with a screw-feed water mixer, are also used in Germany. The use of *plasticizers*, a low *water/cement ratio* (*C*), and good compaction improve *durability* while reducing concrete's irreversible *drying shrinkage* and *creep* (*C*). The gain in strength with time is fairly slow; it needs *curing* and protection while it is *green*. A typical structural concrete (grade 25, or C 25) is designed for a compressive strength of 25 N/mm² (newtons/square millimetre) at twenty-eight days, although much higher strengths can be achieved. For in-situ structural concrete, *cube tests* (*C*) are needed, but they cannot be made for some days. Concrete is mainly used in buildings for structural walls, columns, and slabs, or for fire *encasement*. On major projects concrete work is often the largest *wet trade*. If used as a background for *plaster*, smooth dense concrete walls and *soffits* usually need *spatterdash* or much surface *preparation*. See **lightweight concrete**.

concrete bed A *ground slab* for a bedded *finishing*.

concrete block A building *block*.

concrete bonding plaster *Bonding plaster*.

concrete brick A *brick* made from sand and cement, usually by an *immediate demould* process. Concrete bricks can be obtained in many different colours. Though inexpensive they are not as popular as *clay bricks*.

concrete cancer A vague term which may mean failure from rusting *reinforcement* or from *alkali – aggregate reaction*

(*C*). See BRE Digests 263, 264, and 265.

concrete chute A semi-circular steel channel used to guide fresh concrete during *placing*, e.g. from a delivery *readymix* truck or a crane *skip*.

concrete cleaning *Cleaning down*.

concrete gang A *gang* of *concrete workers*. Its size can change from hour to hour.

concrete grade, c. group One of the different *mixes* of concrete, depending on use, such as normal concrete for strength, *dry lean* for blinding, etc.

concrete grinder A *floor grinder*.

concrete insert A *primary fixing*, such as a *fixing channel* or *dowel*, nailed or bolted against *formwork* before concreting and left in the face of a wall or slab during *stripping*. Plastics foam is often inserted to keep concrete out of the insert.

concrete masonry unit A building *block* or a *concrete brick*.

concrete mixer A large *cement mixer*. See also *C*.

concrete nail (USA) A *masonry nail*.

concrete pump *Contractor's plant* (*C*) for placing concrete, which is discharged from a *readymix* truck into a hopper. Mobile concrete pumps have an articulated boom to a hose which is moved around as needed. Only two men are necessary to operate a small pump, one at the hopper, the other on the hose. Static concrete pumps are used on large projects but need *duplication*. They can be located in a future delivery bay area or under a chute below a *façade retention* framework. Concrete pumps can place up to 100 m³ per hour but need *pumping-grade concrete*. See diagram, p. 96.

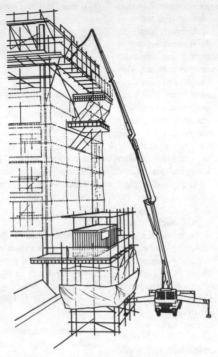

A concrete pump.

concrete roof tile A coloured tile with a smooth or granule finish, made in a wide variety of profiles, usually single-lap *interlocking*. Concrete roof tiles have good frost resistance and should last fifty years or more.

concrete slab A *ground slab*, a *suspended slab*, or a *paving slab*.

concrete worker A semi-skilled *operative* who does *concreting* as well as *screeding* or rough surface finishes.

concreting The placing of fresh concrete in-situ either on the ground or into *formwork* or into an excavation for *trench-fill foundations*, etc. Concreting involves handling equipment such as *skips* or a *concrete pump*. The mix should have good *workability* and not suffer *segregation*. Placing is followed by *compaction*.

concreting paper *Building paper*.

condensates pan A tray under a *cold coil* for the drainage of water from *condensation*.

condensation Water or other vapour in warm air ·turns to liquid when the air temperature drops below the *dewpoint* (*C*) of the liquid. This can happen merely from contact with something cold, with release of *latent heat*. Indoor air carries water vapour, released from people breathing, cooking, washing, etc. This water vapour is normally removed by *ventilation*, but if

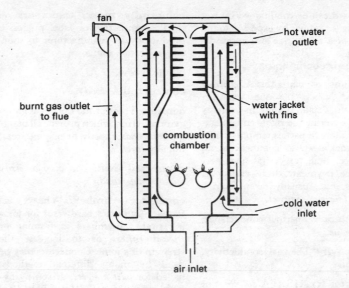

fan

hot water outlet

burnt gas outlet to flue

water jacket with fins

combustion chamber

cold water inlet

air inlet

A domestic condensing boiler. The condensate is not highly acidic and can go straight into household drains. No changes are needed to radiators or controls. Condensing boilers help the environment, lessening global warming and acid rain.

not surface condensation occurs, seen as dewdrops, *dampness*, or on windows, walls, etc. Condensation need not be visible. *Interstitial condensation* occurs inside walls or *insulation*, or in a *cold roof*, or in *double glazing*. In all cases prevention involves excluding moist air from the 'warm' side by a *vapour barrier*, with ventilation or *breather holes* to the 'cold' side. Reducing ventilation of habitable rooms to save energy can lead to condensation, or worse effects – a *sick building*. Very damp places like laundries or cold stores may need *dehumidifiers*. Fat vapour from frying can condense inside brickwork or plaster, even making clothes in a next-door cupboard greasy. See BRE guidebook BR 174, videos AP 21 and AP 24. See also **flue condensation**.

condensation gutter A narrow chan-nel inside a window across the foot of the frame, with a small hole to the outside for drainage of condensed water.

condensing boiler A highly efficient gas-fired *boiler* that cools the flue gases to 50–60°C, causing *condensation*, thus extracting about 86% of the heat available in the fuel, instead of 70% for a conventional boiler. It needs a drain for the mildly acidic condensate and a fan to improve the *forced draught*. A visible plume of water vapour is given off from the flue, which needs a high exit velocity (BRE Digest 339). See diagram.

conditioner An *air conditioner*, such as a window unit for one room.

conditioning (1) Wetting or drying timber, hardboard, etc., to bring it to a *moisture content* suitable for use.

Hardboard can be conditioned by wet sponging, or timber by exposing to the dry air of a room. See also **reconditioning**.

(2) See **air conditioning**.

conditions of contract A *contract document* which describes contract matters in detail. It may have many clauses, covering the rights and obligations of the two parties (employer and contractor), as well as: commencement, insurance, time for completion, performance, payments, delay, variations, disputes, etc., usually taken from a *standard form of contract*.

conductance Thermal conductance, the *C-value*.

conductivity Thermal conductivity, the *k-value*.

conductor (1) (USA) A *downpipe* or rainwater pipe.

(2) A lightning protection *down conductor*.

(3) **lead** (pronounced 'leed') An electrical cable, or its copper or aluminium core.

conductor head A *rainwater header*.

conduit Small-diameter plastics or metal tube (up to about 50 mm dia.) inside which electrical cables are run, using the *draw-in system*. Conduit can be rigid, flexible, or pliable, and cast into concrete, built into brickwork, or surface-mounted. Wiring inside conduit is usual on large projects or where more protection is needed than that given by *sheathed wiring*.

conduit bushing A short insulating sleeve in the end of a *conduit* to protect the cable insulation.

cone tie A *form tie* with metal or plastic cones on each end that are unscrewed after formwork is stripped. After patching the tie ends have correct *cover*.

conformity mark, trademark A mark, stamp, tag, or label put on a product by its manufacturer to show its *certification* status.

congé A *cove*.

conical roll A *wood roll*.

conifers Trees of the botanical group gymnosperms, which provide all building *softwoods*, mostly fir or pine trees, but also *larch*.

connected load The *after-diversity maximum demand*.

connector (1) **timber c.** A heavy steel *fastener* (builder's hardware) for joining timber members in framing or *trussed rafters* or to increase the strength of a joint. Connectors may be fixed by many different methods: nailed for *joist hangers*, driven by hydraulic press for *nailplates*, bolted for *toothplates*, *split rings*, and *shear plates*, etc. Most are *galvanized*, but rusting can occur from unsuitable timber *preservatives*. See diagram.

(2) A *tubular* pipe fitting with a *long-screw* that has an ordinary *taper thread* at one end and a long *parallel thread* at the other, with a *backnut*. The length of the parallel thread allows the *coupling* to be screwed back on to it completely so that the connector can be removed and the pipe run unmade at will. A *double connector* is more compact. See diagram, p. 177.

conservatory A room with a translucent or transparent roof, intended for growing plants, but often used as occasional living space. It can benefit from *passive solar heating* if on the sunny side of a building, but is expensive to heat normally. If smaller than 30 m² it requires no *building permit* in the UK.

consistency The *body* or stiffness of fresh paint, filler, concrete, etc.

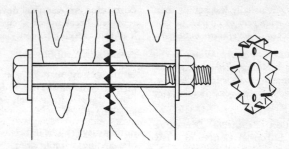

A toothplate timber connector held by 12 mm dia. bolt and nut and washers.

console (1) A control panel, often with a keyboard.

(2) A roof *monitor*.

consolidating rammer, jumping jack, stomper A motor-driven air ram with a flat end plate, held upright by two handles, used for the *compaction* of *fill*. See diagram.

consolidation *Compaction* or *dynamic consolidation* (C).

A consolidating rammer.

constant-volume box An *air terminal unit* which adjusts outgoing air temperature, rather than air volume, to maintain conditions in the space served.

construction The activity of erecting buildings and other structures, including commercial, industrial, and residential work.

Construction Industry Training Board (CITB) (UK) An organization set up to improve the quality of training in the construction industry. It is an awarding body for *National Vocational Qualifications*, trains *scaffolders*, etc.

construction joint A monolithic, strong joint in *reinforced concrete*, connecting work done on two different days. The first section has *starter bars* left projecting so that the new reinforcement laps with the old. Vertical concrete surfaces may have a *joggle joint*, or be cast against mesh, to give a good *key*. Horizontal concrete surfaces have to be cleaned of any *laitance* or weak concrete, usually by *scabbling*. *Formwork* should be tightly clamped at construction joints and often has a foam strip to prevent leakage.

construction management (USA) See **management contract**.

construction manager A contractor's manager, usually less senior than a *contracts manager*, in charge of one large project or of several *site managers*.

construction moisture Water in the fabric of a building, from *wet trades*, such as concrete work and brickwork, which needs *drying out* before dry trades can start.

Construction Products Directive (CPD) A European Community law aimed at removing barriers to trade between member countries. It requires construction products to meet six essential requirements in service: strength and stability; fire safety; health, hygiene, and the environment; safety in use; protection against noise; energy, economy, and heat retention.

construction time *Time for completion.*

consultant A person paid to give advice, for instance a registered *architect*, a chartered *consulting engineer* (*C*), *surveyor*, etc., who acts on behalf of a *client*. A consultant may design and prepare *contract documents* for a building project, call for *tenders*, and carry out the *supervision* of the work.

consumer unit A neat cabinet with equipment to control electricity supply to a house. Cables run to it from the *intake unit* and meter. It usually contains an *earth bar* and the *fuseboard* leading to each outgoing circuit.

contact adhesives *Adhesives* which stick immediately the two surfaces are brought together, usually *latex cement* in a fast-drying solvent. They are applied 'two-way dry', that is spread on both surfaces and allowed to dry briefly before the parts are brought together. Firm but brief *pressure* may be required, but not clamping. The parts must be accurately aligned before they touch.

contactor An electrical device for the repeated breaking of a circuit under load (i.e. carrying current).

container A rubbish *skip*.

contingency sum A *provisional sum* in *bills of quantities* for unforeseeable work, e.g. *risk* items such as pumping after storms.

continuity (1) The unbroken coverage of a coating, without *skips*.
(2) The continuous effective contact of all parts in an electrical circuit, to give high conductance (low resistance).
(3) See *C*.

continuous corbel A long *corbel* for a *movement joint* in a floor slab.

continuous mechanical ventilation (CMV) *Mechanical ventilation* with slow, quiet electric fans that run all the time; usually a ducted *extract system*.

contract A contract is formed when an offer by one person is accepted by someone else. Building contracts are made between the *contractor* who is to carry out the work and the client or *employer* who is to pay for it. Pre-contract work for large projects usually includes *estimating* from drawings and bills of quantities, to build up a *tender*, which is usually fixed-price. Not all contracts are won by tendering – they can also be *negotiated*, for *cost reimbursement* or a *target price*, or even simply for *management*. Work on site is done in *performance* of the contract. After *completion*, many contracts lead to post-contract work, such as settling the *final account*.

contract bond A *performance bond*.

contract carpet Heavy-duty *carpet* for non-domestic traffic conditions.

contract documents The working *drawings*, *specifications*, and *schedules* for doors, finishings, etc., which show the building work to be done, plus the *conditions of contract* and contract *agreement*, prepared by an *architect* or consulting engineer. The *bills of quantities* can be a contract document, as can the builder's *programme*, but usually they have low *precedence*.

contractor A person (or company) who agrees to do something in return for payment. A building contractor agrees to carry out building work before a stated *completion date*, usually acting as the main or prime contractor, *subletting* specialist work to *sub-contractors*.

contractor's plant See *C*.

contractor's surveyor A *quantity surveyor* employed by a contractor, either for *estimating* or for the financial control of work on site.

contracts manager A person in charge of one large project or several small projects, usually located in head office and senior to a *construction manager*, although titles vary between contractors. Contracts managers have long experience in controlling building work and deal with the architect or engineer (or other client's representative).

contract sum, c. price The amount of money to be paid by the *client* to the *contractor* in return for putting up a building. The *agreement* gives details of payment, usually *monthly instalments* plus a *final account*, which takes into account any *variations*. See **claim, lump-sum contract**.

contract time *Time for completion*.

contractual foundations Simplified foundations used for estimating and programming works when real ground conditions are uncertain. After *good ground* is exposed the foundations are designed in detail and a *variation order* issued.

control gap A *movement joint*.

control gear The *starter* and choke or *ballast* used to strike the arc and limit current in a *discharge lamp*. It may consume power and produce heat.

controls See **automatic controls**.

control valve A *discharge valve*.

convection The upward flow of a heated fluid as it expands and becomes less dense, taking heat with it. The space it originally occupied is then filled by denser cooler fluid. This is the cause of the *stack effect* in chimneys or lift shafts, etc., in a fire. Heated air may rise in the centre of a room, to drop back down the cool walls.

convector A low-temperature *heater* which draws in cold air through the bottom of its casing, passes it over a heating element or coil, and releases warm air through a grille in the top. It may be a *fan convector*. See diagram, p. 102.

conversion (1) The work of making timbers out of logs by ripsawing, splitting, or hewing parallel to the grain. Conversion has two stages: *breaking down* the sawlogs into *flitches*, then *resawing* into *square-sawn timbers*.

(2) Building alterations or renovation, often for a new type of *occupancy*.

(3) The highly undesirable weakening of concrete containing *high-alumina cement* (*C*).

conversion coating A *pretreatment primer*.

cooktop A *hob unit*.

cooler A *coolroom*.

cooling coil A *heat exchanger* supplied with *chilled water* from the air-conditioning plantroom, used to cool *supply air*. It usually has a *condensates pan*.

cooling tower A device for cooling large quantities of warm water by spray-

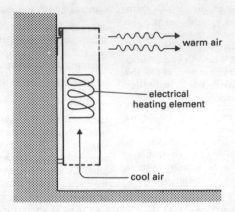

An electrical convector.

ing it over baffles past which a rising stream of air is drawn, guided by an outer casing. Some water evaporates, cooling the remaining water. The cooled water falls into a tray and is recirculated. Cooling towers are used to dispose of waste heat piped from air-conditioning *chillers*. Mechanical draught models with plastics internal baffles are usual in new work and are easily identified by their squat bottle-shaped casing, tapered at the top for the fan. They can be mounted at ground level or on a rooftop. Tall chimney-type natural-draught cooling towers are now mainly limited to large installations, such as power stations. Cooling towers should not allow spray drift, should be cleaned and disinfected every six months, and have water treatment against corrosion and bacteria to prevent *legionnaire's disease*.

coolroom, chill room, cooler A walk-in *cold store* kept at +2 to +4°C.

coordination The planning of complex projects so that components fit together properly and work is carried out in a suitable sequence for each of the *trades* involved. On large projects this can include using *background drawings* for the building services, to prevent *clashing* and to show *builder's work in connection*.

coordination meeting A monthly meeting on site, attended by the general foreman and trades foremen, to discuss the *programme* position and see what action each *trade* needs by other trades to maintain progress. It is chaired by the *site manager*, who takes *minutes*.

cope (1) **cap** To cover a wall with stones, bricks, or precast slabs which usually overhang to protect the wall from rain.
(2) A *coping*.
(3) To cut a *coped joint*.

coped joint A joint made by cutting one piece to fit against another, usually marked by *scribing*.

coping (1) **c. stone, cope** The top of a wall made with a row of stones, *specials*, a brick *capping* (on edge or on end and often over a *tile fillet*), interlocking blocks, metal, plastics, etc., usually overhanging, to protect the top courses against rain and frost. They can be saddleback, wedge, or parallel types. See BS 5642. See also diagram, p. 383.

   (2) Cutting a *coped joint*.

coping saw A saw for cutting round sharp curves. It has a thin, narrow, replaceable blade usually 150 mm long, held from each end in a U-shape frame. The blade can be released from the frame to insert it into a drilled hole from which sawing begins. Many blades are available for wood or ceramics (including *sawfile* blades); they are also used in *fretsaws* and *scroll saws*. See diagram, p. 388.

copolymer, copolymeric material High-performance *plastics* made by blending two or more *monomers* to make a better *polymer*, e.g. *styrene/butadiene copolymer* is a mixture of styrene and butadiene, not of polystyrene and polybutadiene.

copper A metal that is durable, malleable, and easy to work, although it hardens quickly when worked and needs *annealing*. It also has good electrical and thermal conductivities. Used by the Egyptians 5,000 years ago, copper pipe is now common. In monumental *roofing*, copper has a prized green *patina*, but rainwater from copper on to other metals, especially aluminium, can cause severe corrosion.

copper-chrome arsenate (CCA) A timber *preservative* that reacts with the cellulose in wood to form compounds lethal to insects and fungi but safe for humans, as they do not dissolve in water. CCA application is a factory process using *vacuum-pressure impregnation* done to BS 4072.

copper-faced felt, c.-covered roof shingles *Bitumen felt* with a thin surface layer of copper. It is *self-protective* and often laid as *shingles*, looking like the much more expensive *copper roofing*.

copper fittings *Pipe fittings* made for *copper pipe*, many of them not themselves made of copper but of brass or gunmetal. *Capillary joints* and *compression joints* are made of any of these metals. *Silver brazing* can eliminate the need for bought fittings, because the capillary joints can be made by expanding the tube after annealing, using special hand tools. See diagram, p. 104.

copper lights, electro-copper glazing Panels made from pieces of ordinary glass held between copper *cames*. The glass pieces are cut and ground to exact size, laid in a frame between strips of copper, and clamped together. Electrical connections are made using solder. The assembly is then placed in an electrolyte bath and a current is passed to deposit more copper on the strips, filling any gaps and forming H-shape flanges. The glass is held tightly enough for use as *fire-resisting glass*.

copper pipe Pipe from 3 to 100 mm bore, with walls 0.6 to 3.3 mm thick, used for *plumbing* installations such as *internal pipework*, water supply, and *small-bore systems*. Joints are made with *copper fittings*. Copper and lead pipes should not be joined – the *dissimilar metal contact* produces corrosion. Phosphorus-deoxidized copper pipe buried in a concrete floor can crack in the presence of ammonia from *latex cement* flooring adhesives. See **light-gauge copper tube**.

copper roofing *Supported sheetmetal roofing* of copper, generally from 0.4 to 0.5 mm thick, but in the best work 0.6 mm and in a low-pollution area as little as 0.3 mm. It can be laid in large areas, but no single sheet should exceed 1.3 m². Compare **copper-faced felt**.

corbel

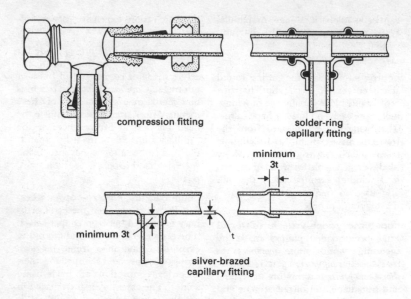

compression fitting

solder-ring
capillary fitting

minimum
3t

minimum 3t

t

silver-brazed
capillary fitting

Copper fittings.

corbel Brickwork, masonry, or concrete projecting from a wall face, usually as a support for a beam or roof truss, or a floor slab *continuous corbel*.

corbelling Brickwork projecting successively more each course to support a roof, chimney stack, *oriel window*, *gable springer*, etc.

corbel piece A *bolster*.

corbel WC pan A *wall-hung WC pan*.

cordless drill/driver An *electric drill* with its own batteries, either internal or as a plug-in 'power pack', which can be recharged many times. It takes either drill or screwdriver *bits*. As the battery voltage is low, any shock would be harmless.

cordless screwdriver A cylindrical *electric screwdriver*, similar to a *cordless drill/driver* but smaller and without a pistol grip. See diagram, p. 346.

cord switch A *pull-cord switch*.

core (1) The inner part of any *composite* material. Timber doors and relocatable partitions have a solid, semi-solid, or cellular core. Electric *cables* may have several conducting cores.

(2) One *conductor* in an electrical cable, with its own *insulation*.

(3) A *services core*.

core patterning, ghosting A pattern showing through on thin door *facings*, showing its *cellular* core and blocking.

core rail, handrail c. A steel *flat* beneath a *handrail*. For timber handrails the core rail is small, as the timber takes most loads. Larger-core rails are used to support extruded plastics handrails.

CORGI Confederation for the Registration of Gas Installers. See **gas**.

cork The bark of the cork oak grown

in Mediterranean countries and North America, mostly *granulated* for use loose or for making tiles.

cork board Granulated cork which has been compressed and baked to form slabs that are sliced to make *cork tiles* or *insulation slabs*.

cork carpet A resilient *floor covering* which is a type of *linoleum* containing granulated cork bonded with oxidized linseed oil and resins on a canvas jute backing.

cork tile A floor tile made from granulated cork compressed and baked at high temperature, when the natural resins bond it into a block. Tiles are cut from the block, usually 5 mm thick. Cork tiles make a very quiet, warm floor that is comfortable to walk on and wears well, although they need to be treated with a *sealer* after laying unless they are supplied pre-sealed. They should be protected from *indenting*. See also **rubber and cork tile**.

corner, angle A change of direction, usually of something standing up vertically, as does a wall. The line of a corner can be a sharp *arris*, or rounded or cut.

corner bead, c. guard profile, c. moulding An *angle bead*.

corner block (1) A slotted block which grips a wall corner when held in place by the taut bricklayer's *line*, used on corners raised by *racking back*.
(2) A *quoin block*.

corner brace A *knee brace*.

cornerlocked joint A *combed joint*.

corner tape *Scrim* for the internal corners between sheets of *gypsum plasterboard*.

corner trowel An *angle trowel*.

cornice A decorative moulding at the top of a wall, plain or with *enrichments*. External cornices throw rain drips away from the outside of the wall and help protect the roof against fire from below. Internal cornices hide the joint between the wall and the ceiling.

corridor A wide long access or *circulation area* inside a building.

corrosion Loss of surface metal, caused by chemicals, *dissimilar metal contact*, or stress. Contact with water that contains dissolved oxygen is needed for corrosion, so it is less likely inside a building, except within steel pipes. Corrosion outdoors is a major cause of economic loss in industrial countries and has led to the use of protective coatings, corrosion-resistant materials, etc.

corrosion inhibitors Chemicals which bond to a metal surface and reduce contact with corrosive media or prevent oxygen passing. Corrosion inhibitors include zinc compounds, sodium nitrate or chromate, some phosphates, and many complex organic compounds. They are used as *inhibiting pigments* in metal primers and in heating water. See next entry.

corrosion treatment of heating water The addition of *corrosion inhibitors* to the water *heat-transfer fluid* of a *low-pressure hot-water* heating system that has a *sealed system* of distribution pipes in *black steel*, to reduce corrosion by forming a film on the inside of pipes. See **barrier effect**.

corrugated sheeting A *cladding* material formed into a wavy shape, to stiffen it, either in metal, such as aluminium or steel, usually with a protective *organic coating*, or clear or translucent PVC and GRP. Corrugated roofing is often in a single run, full length from *ridge* to *eaves* to eliminate endlaps, but it still needs sidelaps of at least one and a half corrugations, unlike *profiled sheeting*.

Cor-Ten steel See **weathering steel** (*C*).

corundum, emery (aluminium oxide) A very hard black mineral, mostly used as a granular *abrasive* and in non-slip surfaces such as stair *treads*.

cost breakdown *Extended prices* for each item in *bills of quantities*.

cost estimate See **estimating**.

costing Calculation of the cost of work, from the *quantity* done with a certain amount of labour and materials. This actual cost is then compared with the tender *unit prices* as a basis for site *financial control*.

cost-plus-fixed-fee contract A similar type of *contract* to a *cost-plus-percentage contract* but with payment of the contractor's *overheads* and profits by a fixed fee.

cost-plus-percentage contract A *contract* used only for very urgent work which cannot await the completion of drawings. The *client* pays the *contractor* the full *flat cost* of the work, plus a percentage of these for *overheads* and profit. This form of contract gives the contractor a direct incentive to enlarge the scope of the work and to delay completion. For this reason it is usually done under close, strict site supervision. See also **daywork**.

cost price The *flat cost*.

cost-reimbursement contract, 'do-and-charge' Any *contract* that is paid on the basis of *cost-plus-percentage*, fixed fee, or *value cost*. Whenever time is available before the contractor starts work on site, a *fixed-priced contract* is used and this type of contract avoided.

cottage roof Roof framing without *principal rafters*, having only *rafters* running from the *wall plates* up to the *ridge* and joined with *ceiling joists*.

cotter, key A steel wedge driven into a joint to tighten it.

council (UK) A *local authority*.

counter-batten A wooden *batten* fixed over *sheathing felt* to the *roof decking* and under roof *tiling battens* to allow any rainwater below the slates or tiles to run away. Counter-battens run parallel to the *rafters*.

counter-flashing, cover-f. A sheet metal *flashing* built into a *bed joint* of brickwork or blockwork, to be turned down later over an *upstand flashing*.

counter-floor A timber *sub-floor* to carry *parquet*.

counterjib A back half of a tower crane *saddle jib*, which carries the counterweight.

countersink A conical enlargement round the outer end of a hole drilled in metal or wood to allow the head of a screw to sink below the surface. By extension, a drill *bit* with a conical cutting edge to make a conical hole for a countersunk screw head.

countersunk fixing A hole drilled wide enough at the surface for a screw head to sink below the surface and to be hidden by a wooden *pellet* glued over it or by filler.

countersunk head screw A wood screw with a conical underside and a flat (or raised) top that usually ends up *flush* with the surface after it is driven.

coupled window An outside window with two *sashes*, one in front of the other, often hung on special double hinges for the sashes to open or close together. The *cavity* between the sashes

creates *double glazing*. The inner sash
should have vapour-tight edge seals to
prevent *condensation* on the inside of
the outer panes. The outer sash may
be removable in summer. Its *sound
insulation* is good, about 35 decibels.

coupler, isolator An electrical connector used to make or break a circuit
when no current is flowing, unlike the
more costly *switch*; often a simple metal
link between *bus bars*.

couple roof A traditional pitched roof
with *rafters* and no *tie beam*, used for
short spans up to 3 m. See also **close-
couple roof**.

coupling, coupler (1) A *pipe fitting*
for making joints, whether in copper,
plastics, or steel. It may be a *capillary
joint*, a *compression joint*, a *sleeve coupling*, or screwed, or a *connector*.
   (2) In *tubular scaffolding*, a piece
which clamps two or more tubes together, by a bolt, drive-in wedge, or
finger-operated clip.

course A parallel layer of bricks,
blocks, tiles, etc., usually in a horizontal
row of uniform *gauge*, including any
mortar laid with them. The term is
also used for any layer of material,
such as damp proofing. See **format**.

coursed squared rubble (USA
random ashlar) *Squared rubble* built
into occasional courses.

court, courtyard An unroofed *circulation area* inside a building. See also
atrium.

cove, congé A concave moulding
joining a wall to a ceiling or floor. See
diagram.

coved tile An L-shaped tile *fitting*
with a *cove* between floor and wall.

cove lighting *Indirect lighting* from
above a fixed *cove* or *cornice*. The light
is thrown up to the ceiling and reflected

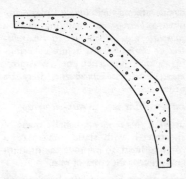

A precast plaster cove for a ceiling–wall
junction, in 10 m lengths.

in a pleasant, diffuse way with no
glare.

cover (1) A lid or cap that can be
removed for access or inspection.
   (2) A decorative moulding, fillet, or
plate to hide a joint or fixings.
   (3) The thickness of one material
placed over another, usually to protect
it, such as concrete over reinforcement
or backfilling over buried pipes or
cables.

covered walkway, breezeway An
open-roofed pathway between two
buildings.

cover flashing A *counter-flashing*

covering Sheet materials, *building
board*, felts, and textiles used as a *finishing*, *cladding*, or *lining* over a loadbearing structure or over a supporting
framework or *core*. Coverings for
floors, walls, partitions, and roofs are
made for different service conditions.

covering capacity In painting, an
ambiguous term – see **hiding power,
spreading rate**.

covering-up Work which hides previous work. It can include *backfilling*
excavations for foundations or buried

drains, fixing *dry linings*, plastering over a *background*, painting over a *substrate*, etc. Prior notice to the *architect* or *clerk of works* before covering-up allows *inspection* or *remeasurement*, otherwise *uncovering* may be ordered.

cover pillar tap An *easy-clean tap*.

cowl A *flue terminal*, usually made of sheetmetal, with static louvres or a rotating head, to improve the draught and prevent the entry of rain.

crab A portable winch for lifting heavy loads, or a tower crane *trolley*.

cracking in brick walls Cracks in brick walls result from *movement*, either in the foundations or the wall itself, but rarely indicate that the building is about to fall down. The causes include incorrect *movement joints, moisture expansion* of clay bricks, rusting of *wall ties*, or a heavier load on one wall than on another with the same *strip footing*, or the presence of wall openings or changes in wall thickness. *Differential settlement* (C) particularly of houses with *shallow foundations* (C) can be a cause if the foundations are on clay or silt, which swells as it gets wetter. Different *moisture movement* in the ground can also be caused by differences between inside and outside, or by different garden watering, or by the felling of a tree, which allows the soil to get wet and expand. Some foundations can be deepened by expensive specialist techniques like *underpinning* or a *diaphragm wall*. See BRE Digest 361.

cracking in paintwork The breakdown of a paint film, with cracks through at least one coat. Cracking of sheltered or internal paintwork is usually the result of defects in *application*, such as insufficient drying before re-coating. Some of the various types of cracking are *crawling, crazing*, and *hair*

cracking. The cracking of external paintwork is often due to long exposure to the weather.

cracking in plasterwork Fresh plaster made with *cement* can easily crack from initial *drying shrinkage* if *curing* is not continued until the plaster has gained adequate strength. Plaster can also be cracked by movement in a *background* such as a wall.

cradle, gondola The working platform of a *flying scaffold*.

cradling Strapping round *structural steelwork* beams or stanchions as a ground on which to fix *lathing* for plaster fire-protective *encasement*.

cradling piece A short *joist* from the wall each side of a *chimney breast* to the *trimmer joist*, carrying the floor boards.

craft A manual skill, requiring intelligence and care, that has been learnt, traditionally by *apprenticeship*. There are many among the building *trades*.

craftsman, craft operative A person skilled in a *craft*. In Britain he or she may have a *City and Guilds* certificate.

cramp (1) **clamp** A tool with an adjustable screw that is wound up to apply *pressure* to joinery parts while an *adhesive* is setting or during nailing, e.g. a *G-cramp, sash cramp*, or *floor cramp*.

(2) A metal strap built into a wall and holding a door frame or lining.

crane A machine with a hook hanging from the end of a steel wire rope attached to a *jib*, used for hoisting materials. The basic crane *movements* are lifting, slewing, and luffing (or trollying). Cranes have an alarm bell for the safe working load (SWL) and maximum load for each radius. Warning that a crane is working overhead can be given by flashing lights or, in the USA, by a *flagman*. To swing a crane jib over property without permission is an act of trespass. Building cranes are usually either mounted on a *tower* or *mobile*.

crank, cranked bar, offset A reinforcement bar with two slight bends in opposite directions to form a very flat S-shape, used at column splices.

crawling A *cracking* in gloss-finish paintwork topcoats, which shrink and reveal the *substrate*. It may be caused by inadequate surface *preparation*, such as sanding down a *gloss* surface or degreasing. Cissing is mild crawling.

crawl space Either an *underfloor space* or a services void under a ground-floor slab.

crawlway, creep trench A services *duct* large enough for a man to crawl along, with all its pipes or cables in place; often a *ground duct*.

crazing (1) Hair cracks in the surface of concrete, generally caused by excessive water in the mix or by steel trowelling too soon after laying. Map crazing is random crazing over the whole surface.
(2) A defect in a *final coat* of plasterwork or *rendering* in which many intersecting cracks form on the surface, usually caused by a weak *undercoat* from which the finish separates. The remedial action is to remove the bad patches down to the *background* and re-plaster.
(3) Bad *cracking* of paint, showing broad deep *checks*.

creasing A hood moulding that projects about 40 mm from an outside wall to throw rainwater away from the wall below, used under *copings* and above *flashings*. Tile creasings are made with sloping *plain tiles* bedded in

mortar to form a *drip*, finished with a *cement fillet* on top. See diagram, p. 383.

creeper (USA) A *jack rafter* for a *hipped end*.

creep trench A *crawlway*.

Cremona bolt A vertical bolt used to secure double *casement doors* or windows (i.e., double-door hardware). It has two half bolts that slide in opposite directions at the same time, one up, one down, to enter holes in the head of the door frame and in the floor. The half bolts are driven by a rack and pinon mechanism operated by a handle or a *panic* bar. They are fixed on one double-door *meeting stile*, which is shaped to fit the other meeting stile and hold both doors locked. They are a precision device that originated in Cremona, the city in Italy famous for its violin makers such as Stradivarius. Compare **espagnolette bolt**.

crew (USA) A *gang*.

cricket (USA) A *saddle* behind a chimney to throw off water to both sides.

cripple Any framed member shortened to make a *filling-in piece*, such as above an opening, or a *jack rafter* from valley to ridge.

critical activity Any work that directly affects the *completion date* of a project, such as foundations and structural trades. Prefabrication, by removing work to a factory, can be used to reduce the number of critical activities (see next entry).

critical path method (CPM), c.p. analysis (CPA), c.p. scheduling (CPS) The type of *network analysis* most common in the building industry, used to simplify the management of large, complex projects. CPM shows how to calculate the length of each series of *activities*; the longest series is the critical path, which determines the shortest time to complete the project. Computers are used to do the *forward pass* and *backward pass* and calculate each type of *float*, then print out a *programme*, usually as an *arrow diagram*. As work proceeds on site the programme can show quickly, and in an easily understood way, the tasks on which resources (materials, labour, equipment) should be concentrated. The programme can also be *updated*, which is only needed if *slippage* has changed the critical path. See diagram, pp. 352–3.

critical temperature The temperature at which a structural metal (steel, aluminium) softens when heated and can no longer carry a working load, usually well below its melting point.

crook (USA) The *warping* of timber, called *spring* in Britain.

cross A plumbing *fitting* in the shape of a cross.

cross-band, c.-banding, crossing In *plywood* or *composite board*, a layer with *grain* perpendicular to the core grain. It reduces shrinkage and cracking or expansion. In five-ply the cross-bands are the layers between face and core and between back and core (face crossing, back crossing). In *three-ply* they are the two outer veneers.

cross-bracing Tie rods, bars, etc., between the corners of a rectangular frame that stiffen it.

cross-bridging (USA) *Herring-bone strutting*.

cross-brushing *Crossing* with a brush.

cross-cut To cut with a saw at right

angles to the grain of wood, or a saw cut so made.

cross-cut saw A coarse saw with its teeth *set* and sharpened to cut across the grain of wood. The odd and even teeth are set in opposite directions and the tips sharpened to a fine point. See diagram, p. 402.

crossed-slot head, cross-slotted h. The most usual *recessed-head screw*. There are several different makes, each requiring a different *cross-top screwdriver*.

cross-garnet hinge A weighty *T-hinge*.

cross grain A *grain* in which the fibres do not run parallel with the length of the wood. It may be partly *end grain*, diagonal grain, *interlocked grain*, or alternating. All of these, the contrary of *straight grain*, make the wood difficult to work with *edge tools*.

cross-grained float A wooden *float* about 300 × 100 × 25 mm, similar to a *hand float* but thicker, with the grain parallel to the short side, used for *floating* coats of plasterwork. See diagram, p. 332.

crossing (1) Putting on a coat of paint with a brush (cross-brushing) or roller (cross-rolling) by a series of strokes each at right angles to the previous series, progressively *laying-off*, to give a smooth finish.
(2) *A cross-band*.

cross-joint (1) A brickwork *perpend*, or (USA) *head joint*.
(2) In *supported sheetmetal roofing* a lengthening joint between sheets, which are secured with a *seam*. This is more weathertight than a simple *head-lap*.

cross-lap joint A *halving* between two pieces of wood at right angles to each other.

cross-linked polyethylene (XLPE) A tough rubbery thermosetting *plastics* used for the electrical insulation of *wiring* and for sheathing *armoured cables*.

cross-peen hammer Any *hammer* which has, opposite the stricking face of the hammerhead, a wedge shape with its cutting edge at right angles to the handle.

cross-rolled paint Paint applied by roller with *crossing*.

cross-slotted head A *crossed-slot head*.

cross-talk (1) Talk in one room that can be heard in another room through air-conditioning *ductwork*. It can be reduced by an *attenuator* or a *shunt duct*.
(2) Interference between two different communications circuits, so that the signal in one circuit is overheard in the other.

cross-tongue (1) A strip of *plywood* or a slip of wood with diagonal grain, glued into a saw cut between two members to stiffen an angle joint.
(2) A *loose tongue*.

cross-top, capstan head A tap handle in the shape of a cross.

cross-top screwdriver A special screwdriver for a screw with a *crossed-slot head*. There are many different patterns and sizes.

cross-wall An internal structural wall at right angles to the sides of a building. It provides *bracing* (*C*) against wind loads.

cross-welt A *seam* between the head and tail of two sheets of *supported sheet-metal roofing*, usually a *double-lock welt* dressed down flat. Cross-welts are *staggered* and run parallel to the *ridge* or *eaves*.

crowd To pivot a *backhoe* bucket about the end of the digger arm, i.e. to dig.

crowd force The force that can be exerted by the teeth of a *backhoe* bucket.

crown The highest point of the inside of a *drain*, sewer, or *trap*. See also *C*.

crown course A top course of a wall, often *brick-on-edge*, or roof trim at the top of a slope, as an alternative to a *ridge capping*.

crown plate A *bolster*.

crown post Either a *king post* or a short vertical post near the middle of a hammer-beam roof. It may have a *bolster* under the beam.

cruciform Shaped like a cross.

cruck house The earliest English medieval timber house that has survived. Only a few remain. The roof is carried on pairs of timbers from ground level to ridge, somewhat like a modern *portal frame*. Timbers with the right curve (crucks) had to be chosen.

crushed aggregate Granular material made from natural rock that is broken and crushed, used for making *concrete*.

crutch handle A simple T-shaped tap handle, used for utilitarian purposes.

cube (1) The measure of volume used in *bills of quantities*, such as for concrete or groundworks; given in cubic metres (m³).
(2) A concrete test cube. See *cube test* (*C*).

cubicle A small room for privacy or isolation, such as a *shower room* or *water closet*.

cubicle switchboard A *switchboard*

with factory-made units that plug in to a wall rack, used for large installations. See BS 5486.

cubing (1) Determining volumes in cubic metres, cubic feet, etc.

(2) An approximate method of *costing* a proposed building, used in the earliest stage of its design, determined by multiplying the cost per m³ of recently completed similar work by the volume of the building in m³.

cunette channel (USA) A prefabricated manhole *channel*.

cup (1) The natural *warping* of a flatsawn plank of timber, caused by drying *shrinkage*, concave towards the bark.

(2) A *screw cup*.

cupboard Fittings, usually of *joinery*, either factory-made units or *built-in*.

cupboard latch A simple catch or similar *ironmongery* for securing a cupboard door.

cupro-alloy A metal alloy with *copper* as its main constituent, such as *brass* or *bronze*.

cup shake *Ring shake*.

curb, kerb A timber on a roof, sometimes to form a *roll*, such as at the break between the upper and lower slopes of a *mansard roof*.

curing, cure Improvement in the strength, hardness, durability, and other properties of a material after it has set, by chemical changes usually resembling those of setting.

For the *resin* binders in adhesives, sealants, and paints this involves *polymerization* and is often associated with *drying*. All *two-part products*, e.g. *polyurethane* or *epoxy*, undergo curing. Heating usually accelerates curing, notably for *thermosets*, leading to the *radio-frequency heating* of adhesives, although in some cases cooling may pro-

mote curing. Curing can also be caused by other processes or modified by chemicals. Moisture-curing sealants contain a dried curing agent which is re-activated by moisture from the air.

For *concrete*, curing (or maturing) involves keeping it wet and at a temperature near 20°C, or warmer for *accelerated curing* (*C*). Curing methods include spraying with water, covering to reduce evaporation, and insulating or heating. The main aim is to replace water used in early *hydration*, or lost by evaporation and into formwork, and to encourage chemical reactions in the cement by warmth. Curing improves concrete strength and durability while reducing *drying shrinkage*, cracking and dusting. Low temperatures slow down curing; the *degree days* since *placing* can predict the time to reach strength targets. Curing is also needed for cement-based *rendering*, floor *screeds*, etc.

curing agent An *accelerator*.

curing blanket, frost-protection quilt Thermal *insulation* placed over fresh concrete in cold conditions, particularly overnight, so that the *heat of hydration* is not lost but causes *self-curing*.

curing membrane A resin and solvent sprayed over a horizontal concrete surface soon after laying. It forms an elastic film that prevents evaporation from the concrete of water needed for *curing*. As the film is soft and gradually disintegrates, it does not interfere with the *bond* to a *screed*.

curling The lifting of the edges of an unbonded floor *screed* due to greater *drying shrinkage* in the top than the bottom. It can be prevented by long *curing*, but since that delays *drying of screeds*, it is usually overcome by reinforcing the screed with steel *fabric* or having a screed at least 100 mm thick.

current-balance circuit-breaker A *residual-current circuit-breaker*.

curtailment The bending of the end of a *reinforcement bar* to increase its *anchorage* into concrete, e.g. by a hook or a right angle.

curtaining, sagging A defect arising from excessive flow in a coat of *paintwork* during application, particularly on vertical work. It produces bow-shaped ridges, which look like hanging curtains. Curtaining may collapse into a *run*.

curtain wall Façade cladding with sheets of glass (or other materials) held in a metal frame, giving a decorative and durable external skin to a building while saving weight and space. In many ways it is similar to *patent glazing*, but is used in prestige situations like offices or public buildings. Solutions need to be adapted for local climate, cityscape, and distance to the sea. Complete systems are factory-made and quickly erected on site, usually by specialists, allowing the early *enclosure* of the building. Installation usually involves no *wet trades*, whether they are *unit systems*, placed by crane, or *stick systems* assembled on site from inside the building or from a *flying scaffold*. Some systems allow openings to be left in the façade, giving crane access for later trades. Most systems have aluminium bars, with grooves for *gasket glazing* and drainage, and in cold climates have *thermal breaks*. Weathertightness against *driving rain* is achieved by *pressure-equalized joints* and *drained and ventilated joints*. Loose objects on adjoining buildings, such as gravel *ballast*, even two storeys below, may need to be secured or replaced to prevent glass from being shattered in a storm.

curtilage The land occupied by a dwelling and its garden.

cushion (1) A seating, usually a firm strip of plastics or synthetic rubber, for glass along the full length of a *patent glazing* bar and capping to prevent rattles and vibration.

(2) A *pad*.

cushion action A buffer that slows a *sliding door* before the end of its *travel* to prevent damage or rebound.

cushioned flooring A resilient *floor covering* with a tough, flexible wear layer, a stabilizing interlayer, and a honeycombed backing, either in vinyl or *sponge-backed rubber*.

cushion edge, cushioned e. A slightly rounded edge to a ceramic *wall tile*.

cut (1) To divide into separate parts, or to shorten, e.g. by sawing.

(2) To add a *solvent* to a solution, diluting it.

(3) **short length** A piece of pipe shorter than a full length.

(4) *Cutting-in*, or a *ward cut*.

cut and fit A *scribed joint*.

cut-away front seat A *water closet* seat with an open front, used for public toilets.

cut brick A brick that has been *rough-cut* or *fair-cut*.

cut corner An angle with a wide *chamfer* making two 45° angles instead of one right angle.

cut nail A tapered *nail* of constant thickness made by cutting (shearing) it from steel plate.

cutout (1) An automatic device to switch off a circuit, usually an *automatic control*, e.g. a *pressure cutout*. Less often used to mean *protection gear* such as *fuses* and *circuit breakers*.

(2) The upper end of a *patent glazing* bar which is cut away for the *flashing*

to be dressed down flat over the glass and bar.

(3) A *penetration* cut through the *structure* of a building, usually more costly than one made during construction or prefabrication.

cutout box A *consumer unit* or a *distribution board*.

cut pile Carpet *pile* that projects as individual strands, made by weaving two carpets face-to-face, then cutting in between: e.g. *Axminster*.

cut stone *Dimension stone*.

cut string, open s. An *outer string* of a stair with its upper edge cut in steps so that the *treads* overhang it, used for the dignified stairs of the eighteenth century.

cutter block A steel block fixed to the spindle of a *spindle moulder, surface planer*, or similar woodworking machine. It carries two or more knives which shape or smooth the timber. All cutter blocks rotate at high speeds – at least 4000 rpm – and the knives usually have *carbide tips*. See **solid-moulding cutter**.

cutting and bending The prefabrication by a supplier of *reinforcement bars* to details shown on the *bending schedule*.

cutting gauge A tool used like a *marking gauge*, with a thin blade in place of the marking pin. It is used for cutting laths or rebating timber.

cutting-in The painting of a clean edge, usually a straight line, at the edge of a painted area, for example a paler area beside *highlighting*.

cutting-in brush A small brush for *cutting-in*.

cutting iron The sharp steel blade of a hand *plane*.

cutting list A list showing the sizes and sorts of timber needed for a job, or the reinforcement for a *bending schedule*.

cutting-off, trimming The removal of the top 600 mm or more of an *in-situ* pile or *diaphragm wall* (*C*), concreted to an extra height in muddy ground. This top piece of concrete, which is possibly dirty and weak, is discarded.

cutting torch, blowpipe A gas torch used for *oxy-acetylene cutting* of steel.

cutting waste An allowance made in *bills of quantities* for the difference between the size needed for a job and normal commercial sizes, wasted as *offcuts*.

C-value, thermal conductance The amount of heat that passes from one face to another of a building material. It is measured as the number of units of heat (watts) transmitted through unit area (square metres) for each degree of difference in temperature between the two faces. The C-value thus has the same units as *U-value* ($Wm^2/$ °C), although it is not the same thing, but lower by an amount given by the *surface coefficient*. Thus C-value is the conductivity or *k-value* divided by the thickness, and is the inverse of resistance or *R-value*.

cyclic movement, reversible m. *Expansion* of building elements followed by contraction as temperature and humidity conditions change back and forth, giving *thermal movement* and *moisture movement*.

cylinder (1) **storage c.** A closed circular tank for storing *domestic hot water* under pressure in a house *hot-water system*. Covered with insulation and a jacket, it should be as close as possible to the draw-off *taps*, so that the hot water comes through quickly and *dead* *legs* are kept short. Domestic cylinders are usually *storage water heaters*, with *direct* electric or gas heating, but if there is a boiler they can be *indirect*. The cylinder tank can be made of copper, or steel with a special corrosion-resistant inner coating. Pressure comes either from a feed and expansion *cistern* in open-vented systems or, increasingly, straight from the water mains in *unvented systems*. Stratification that separates hot and cold water is best in tall vertical cylinders, enabling the water to stay hotter between heatings. If horizontal, both cylinder ends are domed, if vertical the top is domed and the bottom is concave. Industrial hot-water systems usually have a *calorifier*.

(2) The round hole inside a *cylinder lock*.

cylinder lock, night l., Yale l. A *lock* with a cylinder containing a plug, which can be turned once the right key is inserted in a slot, raising *pin tumblers* or *disc tumblers* to the right height. They are used for an entrance door, are of either *rim* or *mortice* type, and are usually opened by a key from the outside and by a knob from the inside, or set with the *latch* permanently in the shut or open position by a catch inside the door.

D

dabs Blobs of liquid adhesive, mortar, or plaster used for fixing sheet or board materials, such as *dry linings, tiling*, etc. They are put on a wall or ceiling at intervals and squashed out flat, making individual points of firm *bedding*. Dabs are used in *spot gluing* and *dot and dab fixing*.

dado A border or panelling over the lower half of the walls of a room, from the top of the *skirting* up to the *dado capping*.

dado capping, d. rail, chair r. The *moulding* along the top of a *dado*.

dado trunking Surface-mounted *electrical trunking* run along a wall, similar to *skirting trunking* but usually near desk-top height.

damages for delay (USA **penalty clause**) Sums payable by the *contractor* for the late *completion* of a building, to compensate the *client* for loss. In *conditions of contract* it is usual to state *liquidated damages*.

damp See **dampness**.

damp course A *damp-proof course*.

damper (1) A flat metal plate or any device to control a flow of air (ventilation, air conditioning) or gas (smoke, flue draught, fire venting).
(2) A device to slow a moving part. It may act in one or both directions, e.g. a shock absorber on a spring or the *cushion action* of a sliding door.

dampness Dampness in buildings is usually first seen on walls and has many causes, including leaking roofs or pipe joints, overflowing rainwater *gutters*, entry of *driving rain* through cracks in walls or windows, *bridging* of cavity walls, *rising damp*, rain penetration of *solid walls, condensation, hygroscopic salts* in brickwork, and *dry rot* in timber. It is often associated with lack of *ventilation* and incorrectly installed *damp-proof courses, flashings*, or *vapour barriers*.

damp-proof course (dpc), damp course A strip of impervious material the same width as a brickwork or blockwork wall, or a *brick damp course*, to keep out moisture. A dpc at ground level, to exclude *rising damp*, is laid near the bottom of the wall, at least 150 mm above ground, and projecting 5 mm. At high level, e.g. above junctions of *parapet* walls with a roof and at the *lintel* above door or window openings, it diverts rain out of the *cavity*, to drain through *weepholes*. In *cavity walls* the outer edge of a dpc is two brick courses lower than the inner edge, to prevent entry of *driving rain* and both edges are built into the *bed joints*. Joints at laps and corners are covered with *bonding tape*. Dpc comes in preformed rolls, made of a variety of flexible materials such as *bitumen-polymer* or *pitch-polymer*, with or without fibres (in cold weather these types should be warmed before unrolling to avoid cracking), embossed polyethylene or thin, soft super-purity aluminium sheet. A ground-level dpc has been required by law since the Public Health Act 1875. For renovation this type can be made by *injection* or *sawn*. See BS 8215, **cavity tray, flashing**, and diagram.

damp-proof course brick A *clay*

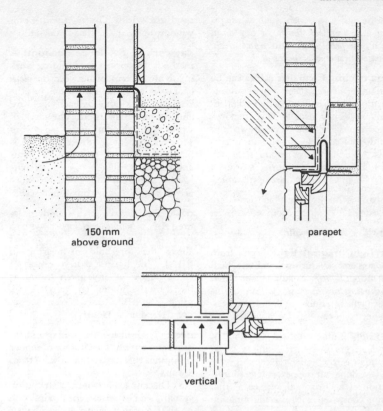

150 mm
above ground

parapet

vertical

Damp-proof courses.

brick with low *absorption*, used to build a *brick damp course*.

damp-proofing Putting a horizontal or vertical *damp-proof course* in a building.

damp-proof membrane (dpm) A wide layer of impervious material such as *mastic asphalt* or a plastics sheet *underlay* beneath a *ground slab* and up the outside of buried external walls, similar to *tanking* but not made to resist water under pressure. It can also be a *surface damp-proof membrane*.

dancing step A *balanced step*.

dap, dapping (mainly USA) A sinking such as those made for timber *connectors*, or a *housing* for a *ribbon board* in a *stud*.

Darby float A two-handed aluminium plasterer's *float*, about 1.30 m long, 120 mm wide, and 16 mm thick, used for levelling plasterwork, e.g. on ceilings.

dashed finish External plasterwork thrown on to a wall by hand or *roughcast applicator*. Dashed finishes can be either wet dash, such as *rough-cast* and *Tyrolean finish*, or *dry dash*, such as *pebbledash* and *spar dash*. Dashed first coat is *spatterdash*.

data cabling Computer cables can be run under an *access floor* or in a communications duct on *cable tray* or laid in *trunking*. They usually have factory-fitted plugs and cannot be drawn into conduits. Flat cables are run under carpets. A *building management system* may have its own *local area network* or even a data highway using fibre optic cable.

datum A mark from which *levels* are measured for a site, a room, etc.

day The *daylight width*.

daylight, natural light Light from the sun and sky during daytime, which for indoor *lighting* calculations is based on the *standard overcast sky*. Tinted windows which reduce daylight have been suspected of causing *sick buildings*.

daylight factor method A method of *daylight prediction* by taking a given point in a room (the reference point) and working out the percentage illumination of a horizontal surface at that point compared with the illumination which would come from the whole hemisphere of sky. The daylight factor is given by adding together the *sky component*, the *externally reflected component*, and the *internally reflected component*, less a *maintenance factor* to allow for dirt on windows. See BRE Digests 309 and 310.

daylight prediction, d. analysis A method of estimating *illuminance* levels inside a room from windows or other *fenestration*, e.g. the *daylight factor method*.

daylight protractor An instrument introduced by the *Building Research Establishment*, used for predicting *daylight factors*.

daylight width The *sight* width of a window opening that lets in daylight.

daywork (USA **force account**) A method of payment for building work involving agreement between the *clerk of works* and the *contractor* on the hours of work done by each operative and the materials used. The clerk of works signs the contractor's daywork sheets. Payment to the contractor is based on his or her expenses in labour and materials, plus an agreed percentage for *overheads* and profit. Daywork is *cost-plus-percentage* on a small scale.

dBA The *loudness* of a sound measured by a *sound-level meter* with a type A filter and given in *decibels* above the threshold of hearing (0 dBA). The figures closely represent the loudness of any given sound as heard by the human ear: 30 dBA in a library, 60 dBA normal conversation, 100 dBA a noisy factory.

dead (1) Unfit for use, said of mortar or plaster after its *initial set* or of render droppings that have been *screeded* from a wall.
   (2) Disconnected from a distribution system, said of an electrical wire with no source of power or a disused gas or water pipe; also applied to collection systems of drains or sewers. Before *demolition*, pipes can be checked for gas or water under pressure by removing a plug, drilling a small hole, etc. Drawings may show a pipe to be dead although it is really *live*.

dead bolt A bolt of a door *lock* that can be withdrawn or shot only by turning a key, as compared with a spring-loaded bevelled *return latch*.

dead end (USA **false exit**) A corridor or room with only one way out. In non-domestic buildings, it must have a *fire safety sign* to prevent it being mistaken for an *escape route*.

dead knot, encased k. A *knot* whose fibres are not intergrown with surrounding wood. It is easily knocked out and thus is a worse *visible defect* than a *live knot.*

dead leg A hot-water pipe leading to a tap and not part of a *circuit.* The hot water in a dead leg cools off between draw-offs, wasting water and heat, and increasing the danger of *legionnaire's disease.* Domestic hot-water *cylinders* are located near to the use points so as to limit dead legs to the lengths required by water companies. Large installations often have *secondary circulation,* but even they have short dead legs.

deadlight A *fixed light.*

dead lock A security *lock* which is worked from both sides by a key only. It usually has no door knobs and so may need a *door pull.*

dead shore A heavy upright timber (one of two) under each end of a *needle* used to uphold a wall above.

dead-soft temper The softness of *copper* sheet required for roofing work. Copper tends to get harder when bent or worked during placing. Working should be stopped before the copper breaks, and the metal *annealed* by heating.

deal (1) A piece of *square-sawn* softwood 47 to 100 mm thick and 225 to 275 mm wide (BS 6100).
(2) **red d.** *Redwood.*

death-watch beetle (Xestobium rufovillosum) A *borer* which burrows deeply into structural timbers, in particular the *sapwood* of English *oak,* and is therefore difficult to kill. It prefers decayed hardwoods, but will attack nearby softwoods. It leaves *shot holes* about 3 mm dia., and the adult makes a ticking noise.

debris-collection fan, protection f. A canopy projecting upwards and outwards from the face of a building under construction, renovation, or demolition, to catch falling objects which could injure people. It can be at first-floor level to protect pedestrians in the street, on top of a *flying scaffold,* etc.

deburring *Fettling.*

decay, fungal d., rot The weakening of timber which has been attacked by *fungus,* usually a serious *defect.* Most decay in buildings is *dry rot,* which despite its name requires a high *moisture content. Wet rot* is less common. Durable timbers have natural resistance to decay, without the need for *preservative* treatment.

decibel (dB) Sounds are measured by comparison of their intensities, starting from two sounds, one twice as loud as the other. This two-to-one ratio is called the Bel, but as that is too large a unit for normal use, the scale is further divided into ten diminishing logarithmic steps, or decibels. The *sound insulation* of a wall, floor, door, or window is always the same in decibels, irrespective of the *loudness,* which can be measured in *dBA.*

deck A platform, usually above ground and weather-resistant, or an unroofed foot-traffic area beside a building or swimming pool; also the structural base for a floor, an *access floor, flat-roof* covering, or *supported sheetmetal roofing.*

decking (1) The material used to cover a *deck,* such as timber *flooring,* panels of *building board,* or the *formwork* for a suspended slab.
(2) *Profiled sheeting* forming the underside of a *composite floor.*

decorative laminate See **laminate.**

deducts To save double *taking-off*, in *quantity surveying* items like brickwork are measured over all openings for doors or windows. Later, when taking off the doors and windows, a note is made to use the *dimensions* of the opening a second time, as a deduct from the *adds* for brickwork, thus arriving at the brickwork quantity.

deed An *agreement*.

deep freezer A *freezer* for storage between − 18°C and − 40°C.

deep-plan building Offices and commercial buildings with areas that are more than 4 or 5 m from a window. They used to have *light wells*, but today high levels of electric lighting (plus air conditioning) can make up for the loss of *daylight* (and ventilation), although this increases energy costs. Natural ventilation from one side only can give fresh air to a 10 m depth of office space.

deep-seal trap An *anti-siphon trap* with a *seal* 76 mm deep or more.

default The failure of one party to a *contract* to fulfil the agreement. This may involve *determination* of the contract. A *performance bond* or *retention* is usually demanded to guard against default by the *contractor*.

defect Defects may be due to human error, improper manufacture, the wrong choice of materials, lack of *compatibility*, poor *ventilation*, etc. A suitable *specification* plus supervision and *inspection* during construction are usual to prevent defects, but many still occur (see BRE Digest 176). Defects in a roof, building *fabric*, and foundations lead particularly to *dampness* and *cracking in brick walls*. Timber, being a natural material, often has irregularities or weaknesses which reduce its usefulness or strength, such as *decay*, insect damage, poor conversion or machining, *knots*, *shake*, *wane*, or *warp*. Some timber *gross features* make it unsuitable for structural purposes, others are only *blemishes*. Paintwork defects (see BS 2015) can be due to improper surface *preparation* or *application*. Less often they are caused by the *paint* and other materials not being properly made, stored, or handled.

Defect Action Sheets (DAS) Short papers warning of building problems that have occurred in design or on site, mostly in houses, published by the *Building Research Establishment*; e.g. the need for ventilation of the space in *warm roofs* or remedies for condensation and rain penetration. See **Good Building Guides**.

defects liability period, maintenance p., retention p. The time after a building is completed during which the builder has to fix anything that does not work according to the *specification*. The length of the defects liability period is given in the *conditions of contract* and is usually six months, starting from the date given on the *certificate of practical completion* and going until the date of *final completion*.

degreasing The removal of oil or grease before painting, by wiping the *substrate* with a *solvent*. Non-ferrous metals such as aluminium readily hold oil or grease. Even formed concrete surfaces may need degreasing to remove any mould oil or other *release agent* (*C*), so that plaster will stick.

degree-day value A figure describing the relative coldness of a site. In Britain, where it is assumed that buildings need no heating when the outside temperature is 15.5°C or warmer, the number of degrees that the average temperature is below 15.5°C on a given day counts as its degree-days. The values for a series of days can be added together to give a degree-day value for a particular period. The annual degree-day value is fairly constant for a particular site: London has a value of around 1950 and Aberdeen, also at sea level but 644 km further north, 3050. This figure is used by heating and ventilation engineers to calculate the annual energy requirement for *heating* a building as well as the sizes of boilers and radiators. Another use of degree-days is for *curing* times of concrete. French degree-day values are in degrees C below 18°C; US measurements are in degrees F below 65°F (18.3°C).

dehumidifier An air-conditioning unit which removes some moisture from the air by contact with a refrigerated cooling coil that chills the air below the *dewpoint* (*C*), less often by use of *desiccants*. Refrigerator types draw heat from the air and recover latent heat from the humidity, and that heat, which is given off by the refrigerator condenser coil, is used to reheat the outgoing air stream. See **vapour compression cycle**.

delamination The separation between veneers of *plywood* or between laminations of *glued-laminated* timberwork due to failure of the *adhesive*. See also *C*.

delay The late completion of work, which produces programme *slippage of dates*.

deletion A *variation*, usually to produce a saving.

deliquescence The liquefaction of *hygroscopic salts* as they absorb water.

deluge sprinkler A *drencher sprinkler*.

demand factor The *diversity factor*.

demolition Breaking up an existing building, usually to make way for a new building, with careful removal of debris and materials for re-use. Demolition is a skilled and dangerous *trade* carried out by specialist contractors; even a careful survey can fail to find problems. *Buried services* such as gas, electricity, and water have to be located if they are *live* and carefully exposed, then terminated or diverted. They may need temporary support, particularly pipes with *push-fit joints*. Demolition work may include *façade retention*, the *shoring* or *underpinning* of adjacent buildings, and *asbestos removal*.

demountable partition A partition that can be taken down and moved without involving *wet trades* or damage to finishes, and mainly used for office buildings. It is held in place by screw jacks or a *receptor* channel. There are many systems, the best being *relocatable*, but in general partitions which are easiest to move also give the lowest *sound insulation*.

dense concrete Normal *concrete*, not *lightweight concrete*.

dentil slip A piece of tile pushed into the mortar bedding under a *ridge tile* in the large trough of a bold pattern roof tile, to save mortar.

descaling (1) The removal of *mill scale* (*C*) as preparation for painting.

(2) The periodic cleaning of rust from inside *cast-iron* drain pipes.

description The part of an *item* in bills of quantities that briefly describes the materials, form, and location of a job of work in a way easily understood by an *estimator*. Descriptions are worded according to a strict set of rules given in a *Standard Method of Measurement*.

desiccant A chemical which absorbs so much water that it can be used to dry air, in a *dehumidifier* or elsewhere.

desiccation Thorough drying, using kiln heating or chemical *desiccants*.

designation See **mortar designation**.

design-build contract, package deal, turnkey project An agreement with a client by which a *contractor* takes full responsibility for design and construction. This can reduce the delays of coordination between consultants, but it increases the risk of unsuitable construction. Only conceivable with a sophisticated, conscientious contractor, it is less secure for the client than a *management contract*.

desludging The removal of *sludge* (*C*) from a *catchpit*, a *septic tank*, a boiler, etc.

desuperheater A unit for transferring heat from *superheated steam* to heating water. The water is sprayed into the steam, which gives up its great *latent heat* during condensation.

detail A large-scale *drawing* of a small part of the work.

detector A *sensor*, particularly one that warns of something unwanted, such as a fire or an intruder. It may be used to sound an *alarm*.

determination, termination The ending of a *contract* before work is completed, by either the *employer* or the *contractor*, for reasons stated in the *conditions of contract*, such as the work not having been done or not paid for. Determination is more final than *suspension of works*.

devil float, nail f. A plasterer's *hand float* with a nail near each corner, projecting about 3 mm, for *scratching* the surface of fresh plaster to make a *key* for the next coat.

dezincification The loss of zinc by chemical attack or electrolytic action.

It can occur in *brass* pipe fittings in contact with hot stagnant water containing chlorides, which selectively dissolve zinc. Dezincification-resistant brass is made. It also happens to *galvanizing* in pipework and cisterns if there is *dissimilar metal contact* with copper.

dft Paint *dry-film thickness*.

Diagrid floor A *waffle slab*.

diamond break stiffening Two slight folds made on the diagonals from corner to corner of a panel of sheetmetal, often used to reduce the drumming of air *ducts*.

diamond saw A *circular saw* with *abrasive* industrial diamonds set round its rim. It is usually guided by rails or wheels and can cut concrete or stone. Water is sprayed on to the blade during cutting. See diagram.

diamond washer A curved washer used with a *hook bolt* or other fixing through roof sheeting. It fits over the corrugations like the *limpet washer* and may be of aluminium, *powder-coated* galvanized steel, or plastics.

diaphragm tank A closed tank with its inner space divided in two by a rubber diaphragm. The upper space is filled with air or nitrogen, the lower space with water (or vice versa). Diaphragm tanks are used as *expansion vessels* or for pressure maintenance. See diagram.

diaphragm wall See *C*.

diatomite, diatomaceous earth A soil composed of hollow siliceous skeletons of tiny marine or freshwater organisms (diatoms) found in Denmark and elsewhere. It is used as an *extender* in paints, as a *lightweight aggregate*, and to make *flue blocks* for temperatures up to 1300°C.

die (1) At the upper and lower ends of

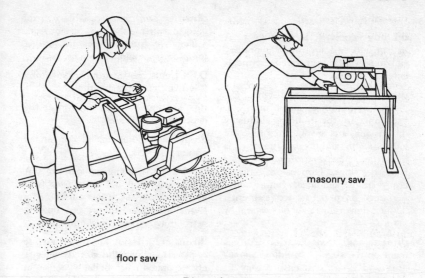

masonry saw

floor saw

Diamond saws.

a turned *baluster*, the enlarged square part that meets the handrail or plinth.

(2) An internally threaded metal block for cutting male threads on bars or

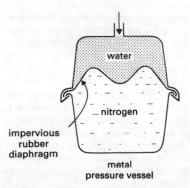

water from sealed system expands into tank on heating

water

nitrogen

impervious rubber diaphragm

metal pressure vessel

A diaphragm tank.

pipes, held in a *diestock*. Compare **tap**. See also *C*.

diestock A holder for *dies* with which screw *threads* can be cut by hand.

differs, key changes, variations The maximum number of different keys that can be made for a particular lock. Simple locks may have only 500 differs; *pin tumbler locks* have practically infinite differs.

diffuser (1) An outlet from a *register* or from any *air terminal unit*.

(2) A baffle to reduce glare from a *luminaire*.

diffuse reflection Light reflected equally in all directions from a surface with a matt finish, giving a gentle effect, unlike *specular reflection*.

digger A small *excavator*, such as a *backhoe*.

digging The loosening of ground by hand or machine. It may be separate from getting out the *spoil*.

diluent A *thinner*.

diluting receiver, laboratory r. A tank on a *waste* drain from a laboratory sink to prevent concentrated chemicals entering the sewer.

dimension A size or distance expressed by a number and a unit, e.g. 750 mm. In the building industry dimensions are given in order of length, width, and height, and work is built from *figured dimensions*, not *scaled* from drawings. Dimensions should always be checked on site before starting work on prefabricated elements. Timber dimensions may be *nominal* or *dressed*.

dimensional coordination Agreement on the sizes of building *components* and *elements* to ensure easy fit into a *modular system*, simplifying both design and building.

dimensional stability A material is dimensionally stable if it has no *moisture movement*, little temperature movement, no *creep* (*C*), and does not shrink or expand for any other reason. Stone, glass, and old bricks have good dimensional stability. Cement and wood products are generally unstable, but *plywood* is much more stable than wood.

dimension lumber (USA) *Lumber* that has been *square-sawn* to uniform sizes, but not *dressed*, from 4 to 12 in. wide and 2 to 5 in. thick.

dimension stone, cut s. Natural stone that is cut to shape, usually sawn into rectangular blocks, for use in *ashlar* stonework.

diminishing courses Courses of *slates* laid with the largest at the bottom and the smallest at the top.

diminishing stile, gunstock s. A door *stile* which is narrowed (diminished) from the *lock rail* upwards to give more space for *glazing*.

dimmer switch A variable control used to decrease the light output from a *luminaire*. Special types are needed for fluorescent and halogen lamps.

DIN Deutsches Institut für Normung (German standards institute).

dip The lower bend of a *trap* or the depth of its water *seal*.

dipper The outer half of the hinged arm of a *backhoe*.

dip pipe, trapping bend A downward-facing 90° *petrol bend* with its end below the water surface of a *petrol interceptor*, to prevent floating petrol moving between chambers.

direct cylinder A hot-water storage *cylinder* in which the stored water is not separated from boiler water as in an *indirect* cylinder. All the earliest low-pressure systems were of this type.

direct dialling-in (DDI) A *private branch exchange* with individual numbers for each extension line. An outside caller can dial them without going through the operator.

direct-finish concrete, in-situ f., off-form c. A textured surface to concrete walls produced by the *formwork* alone. Since *patching* is usually not allowed, shutters need to be strong, accurate, and tightly clamped to prevent *fins* or *offsets*. Concrete needs to be thoroughly compacted to avoid *honeycombing* or *chapping*, and *re-vibration* (*C*) may be recommended if rough board linings are used. After forms are removed the finished surface may need *protection* until handover. Rust stains from rainwater off bare steel are difficult to remove during *cleaning-down*, but can be prevented by coating the steel with cement *grout*.

direct labour Building *tradesmen* and

labourers employed by the *client* or his agent (engineer or architect) directly, without the mediation of a *contractor*. About one in every three *local authorities* in Britain makes use of direct labour.

direct supply Water taken directly from the mains, not from a *cistern*, and used for *drinking water*, *unvented* hot-water systems, and *flushing valves*.

direct transmission of sound *Transmission* of sound through a separating wall, floor, etc. It may be *airborne* for part of its path, but it excludes *flanking transmission*.

dirt-depreciation factor (USA) The *light-loss factor*.

dirty money Additional pay to a building operative for working in difficult or unusual conditions.

disabled facilities, handicapped f. Help for the disabled in public buildings, e.g. access ramps for wheelchairs, not obligatory in private houses. See BS 5810.

disappearing stair A *loft ladder*.

disc grinder An *angle grinder*.

discharge lamp An efficient light source from an electric discharge passing through a glass tube containing vapour (mercury, sodium) or gas (argon, krypton, neon). The colour of the light may be modified by the vapour or gases, or in *fluorescent tubes* by an inner coating. Except for *cold cathode lamps* they need a heated cathode and usually *control gear*. All can produce radio frequency interference or 'hash'.

discharge pipe A pipe that carries *foul* water from *sanitary fittings*. It may also be used to carry rainwater.

discharge valve, regulating v. A valve which regulates pipe flow, as opposed to the simpler *stopvalve*.

disconnecting trap, disconnector, intercepting t. A *trap* that is required if a rainwater drain is connected to a *sewer* (*C*), to prevent foul air passing into the rainwater piping, then out into the air and becoming a health hazard.

discontinuous construction, isolation Construction with breaks in the continuity of a structure to reduce the sound *transmission* through it, both airborne and *impact*. It is more complicated than *monolithic* construction and more expensive, but is cheaper and more effective than alternative means of *sound insulation*. It can improve by 20 dB the reduction given by the mass of the two leaves in a *double-leaf separating wall*, a *floating floor*, or an *isolated ceiling*. See diagram, p. 185.

disc sander A small *sander* with a stiff rubber disc on a rotating shaft, which fits into an *electric drill*. The sandpaper is fixed to the centre of the rubber disc. It is simple and inexpensive, but cannot reach awkward corners. Skill is needed to avoid leaving curved marks on the work and it is harder to operate than an *angle grinder*.

disc tumbler lock A *cylinder lock* with spring-loaded disc detainers which allow the lock to be turned only by the correct key.

dishing A shallow sinking with sloping sides, used in *built-up roofing* to form a *rainwater header*. The roofing felts are worked into the dishing.

disinfection See **sterilization**.

dispersion The division into fine droplets of the liquid in an *emulsion paint* or other *emulsion* (*C*).

dispute, difference Disagreement between the *client* and the building *contractor* over a contract matter (or between the contractor and a *sub-contractor*, etc.). The *conditions of contract* may allow disputes to be settled by *arbitration*, instead of *litigation*.

dissimilar metal contact, bimetallic corrosion, electrolytic c. The corrosion of metals with a conducting path between them while in contact with impure water containing oxygen. For instance copper in hot- or cold-water pipes causes the *dezincification* of galvanizing; lead caulking accelerates the rusting of wrought iron exposed to the weather. In most cases direct contact can be prevented by inserting an isolating tape or *underlay*. It can also occur from indirect contact between some metals, e.g. from rainwater dripping from a copper pipe or flashing, which corrodes uncoated aluminium roofing.

distance piece A block of resilient material such as plasticized *polyvinyl chloride* about 25 mm long, fitted into a glazing *rebate* to prevent displacement of the glass. Distance pieces are positioned at about 300 mm spacings and should never coincide with a *glazing block* or a *location block*. They are used with *glazing materials* that need time to *cure* or are *non-setting*. See also *C*.

distemper A heavily pigmented matt *water paint* bound with glue *size*, largely superseded by *emulsion paint*.

distribution Inside a building, the pipes, ducts, and cables which send the *services* to each final use point. There may be separate internal installations for hot and cold water, electrical power and lighting, supply and return air, communications, and medical gases. See also **drop system**.

distribution board, d. panel, d. fuseboard, sub-board A *switchboard* for power or lighting *branch circuits*, and

their *protection gear*, to serve a zone or area in a large installation.

distribution pipe A pipe from a storage *cistern* or *main* to a draw-off *tap*.

district heating A method of heating houses or flats in one part of a town from a central supply of heat. This generally involves using the waste heat from electric power stations. Distribution is underground, using medium- or low-pressure hot water or steam, carried either by pipe-in-pipe mains with factory-fitted insulation and a plastics outer casing, or lagged twin pipes laid in *ground ducts*.

district surveyor The former title of a London *Building Control Officer*.

diversity factor, demand f., use f. In pipework or power distribution not all the flow or load occurs at the same time. The *installed load* is multiplied by the diversity factor to give the probable maximum flow, or *after-diversity maximum demand*, which is much smaller than the *installed load*.

diverter A *three-way valve*, as in a *bath/shower mixer*.

divided responsibility A duty or obligation shared between two or more people. This is a common source of disputes. Ways of avoiding it include *supply and fix* contracts and clear agreements on the *limits of works*.

division wall A *fire wall*.

'do-and-charge' contract A *cost-reimbursement contract*.

docking saw An electric *circular saw* on a stand, used to cut joinery timbers to length, usually at right angles, but it may be adjustable for *mitres*.

documents See **contract documents**.

dog A steel U-shaped *fastener* with

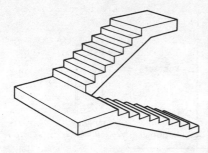

A dogleg stair.

spike ends used for fixing together heavy timbers, for instance a *dead shore* to a *needle*.

dog bolt A *hinge bolts*.

dog ear (Scotland **pig lug**) A box-like external corner formed by folding sheet-metal without cutting to avoid using a *gusset piece*.

dogleg stair, dog-legged stair A *stair* with two *flights* between storeys, a rectangular *landing*, and no stair *well*. For a timber stair the outer *string* of each flight is housed in the same *newel post*. See diagram.

dolomitic lime *Lime* (calcium oxide) with a high percentage of magnesium oxide – about 50% of each.

dome A hemispherical *vault*, circular on plan. The horizontal thrust from the dome must generally be carried by *reinforcement* in the *ring beam* at the *springing*. Domes are often crowned with a *lantern*.

dome-cover screw, d. top A *screw* with a small threaded hole in the middle of its head into which a stainless steel or chromed cover is screwed. It is often used for fixing a *water closet* to the floor.

dome end, domed e. The end of a hot-water *cylinder* or any *pressure vessel*

(C). Its shape can be dished-and-flanged or 'basket-handle'.

dome-head screw A *screw* with a head that has a rounded, raised top and a flat underside. The top stands slightly above the surface that the screw is driven into.

domelight A *rooflight* in one piece in the shape of a dome, usually about 1.50 m diameter and made of cast glass or moulded translucent *plastics*.

domestic hot water (DHW) Water for washing, usually at 60°C. Small installations have a storage *cylinder* or *instantaneous water heater*, while larger installations often have a *boiler*, with a separate *calorifier* and distribution with *secondary circulation*. All domestic hot water should be stored above 60°C (occasionally it should be raised above 70°C) and be above 50°C at the tap, to prevent *legionnaire's disease*.

domestic sub-contractor A *sub-contractor* chosen by the contractor.

domical grating A dome-shaped strainer over a *rainwater outlet*.

door (1) The *doorset* in a wall opening, for walking through; or on a cupboard. External doors are used for access and internal doors for communicating. Single and *double doors* are usually *side-hung*, but there are other types, e.g. *flexible*, *sliding*, *swing*, *revolving*, *folding*, and *up-and-over*. Special doors are used against fire or smoke.
(2) A *door leaf*.

door assembly A *doorset*.

door buffer, rubber b. A small rubber stud in the rebate of a door *jamb*, on the *closing face*, to reduce noise from the leaf slamming.

door buck, d. casing A *door lining*.

door closer, d. check A spring-oper-

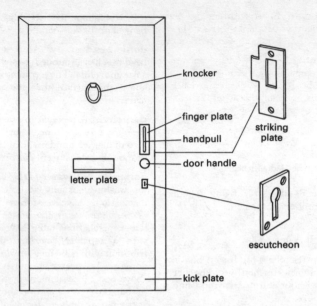

knocker

finger plate

handpull

striking plate

door handle

letter plate

escutcheon

kick plate

Door furniture.

ated device to keep a self-closing door shut until deliberately opened, usually with a one-way hydraulic or air *check* that acts during closing to prevent the door slamming, although 'back check' action can also cushion movement in the opening direction. Door closers can be mounted on the face of a door or concealed in the *transom*, door leaf, or floor. On *fire doors*, they should hold the door shut firmly enough to resist it being blown open by pressure from *flashover*. Old fire doors often have a strong spring and no check, with the result that they are heavy to open and slam shut.

door control A simple *entry system*, with an *electric striking plate*, usually linked to an *entry phone*.

door face The *lining* of a door leaf.

door frame A surround in which a *door leaf* is hung, heavier than a

door lining, and usually made of *solid timber* or pressed metal. It has two *jambs* and a *head*, usually about 100 × 80 mm and *double-rebated*, as well as a *spreader bar*. Frames are usually stood up in position, plumbed, and braced before *bricklaying* is started or the *screed* laid, and built in as work proceeds.

door furniture Decorative parts fixed to a door, including the *door handle*, key *escutcheon*, protective *finger plate* and *kick plate*, trim over *door closers*, etc. See diagram.

door handle A *lever handle* or a *door knob*.

door hardware All ironmongery for a door, including that needed for hanging, locking, closing, or weatherstripping it, as well as any decorative *door furniture*.

door head, h. member The top cross-member of a *door frame* or *door lining*.

door holder A *fusible link*, or an automatic magnetic or pneumatic device, that holds open a *fire door* until released by a *fire detector*, allowing the *door closer* to operate.

door jamb The upright side members of a *door frame* or *door lining*. For a single door, one is the *hanging jamb* and the other the *shutting jamb*.

doorkit All the pieces for a doorset, pre-cut ready for assembly on site, including the *door leaf, door frame, architraves*, etc.

door knob A *door handle* in the shape of a knob. It is more compact than a *lever handle*, but more difficult for children, the elderly, or the disabled to turn.

door latch A *latch*.

door leaf, door The openable part of a *door*. Timber doors have a *leaf frame* and panelling, or *flush* facings, or can have a *cellular core*. Door leaves can also be *all-glass, glazed, matchboarded*, or padded.

door lining (USA **d. buck, d. casing**) Cladding to the *reveal* of a door opening, installed after the adjacent walls are built. It is fixed using timber pluggings and packing pieces, and is less strong than a *door frame*.

door lock A *lock*.

doormat A mat with stiff bristles or ribs to collect dirt, used inside or in front of an entrance door; traditionally made of sisal and *coir*.

doormat frame, mat well f. A rectangular metalwork surround, usually in brass *angle*, forming the edge to a *sinking* in a floor *screed*. A doormat frame is best made too deep (so that the sinking is also too deep) rather than too shallow, and *packing* used under the *doormat* if necessary.

door opener An *entry system* with a *sensor* to detect people and operate an automatic door, often a double sliding door.

door post The upright *jamb* of a *door frame*.

door pull A handle for a swing door, or one with a *dead lock*. It indicates that the door is to be pulled and so is best not installed on a 'push' side.

door schedule A *contract document* with a numbered list of doors (e.g. D1, D2, D3), and details of location, type, hardware, door furniture, etc.

doorset A factory-made door unit, complete with its frame and leaf, architrave or cover moulding, stops and hardware, usually to a standard size.

door sill A door *threshold*.

door stile The upright side members of a *door leaf*. Compare **door jamb**.

door stop (1) A strip of wood nailed or glued to a door frame (or for expensive doors *rebated* out of the solid) on the head and both jambs, to prevent the door passing through the frame, and to make, as far as possible, an airtight seal.

(2) A small buffer fixed to the floor (floor stop), or wall, to prevent a door from opening too far or the door handle from striking the wall.

door strength Tests for door strength include: *heavy-body impact, hard-body impact*, torsion, closing against an obstruction, slamming, wrenching the handle, etc. See EN 108, EN 129, EN 162.

door switch A switch operated by the opening or closing of a door.

door threshold A *threshold*.

door tidy Because no *letter plate* achieves a finished appearance inside the door, a draught-excluding, spring-loaded metal plate, the door tidy, may be installed inside to cover the letter slot. It also improves *fire resistance*.

door unit A *doorset*.

doorway width The minimum clear width between the *jambs* of a door frame.

dormer, d. window A vertical window through a sloping roof, to give light into an *attic*, usually with a pitched roof, upright *cheeks*, and a front *apron*. Less often it is a *shed dormer* or an *internal dormer*.

dormer ventilator A small roof *ventilator* in the shape of a *dormer*.

dosy timber Wood which is beginning to *decay*. See **dote**.

dot (1) A small piece of solid material put on a surface.
(2) **lead d.** A *soldered dot*.

dot and dab fixing A method of fixing *dry linings* such as *gypsum plasterboard* to an uneven wall surface, using small pieces (75 × 50 mm) of insulating board (usually bitumen-impregnated fibreboard) of varying thickness as dots. They are bedded on the wall in smears of gypsum-based adhesive or *board-finish plaster*, placed at about 450 mm horizontal spacings and 1 m vertical spacings, and levelled with a straight edge. After the adhesive has set, *plaster dabs* are placed on them and the plasterboard is applied.

dote, doat The early *decay* of timber, indicated by dots or speckles, described as dosy, foxy, etc. Sometimes a *defect*.

dots and screeds A method of applying in-situ *plasterwork* or a floor *screed*, starting with *dots* at wide intervals, which are used as a guide to run strips.

The strips are then used as a *screed rail*. A flat surface is thus established, using the finishing material itself.

double-acting hinge A hinge for a *swing door*, such as a *floor spring*.

double bridging Two rows of *solid bridging* or *herring-bone strutting* between floor *joists*, dividing the floor area into three equal parts, usual for spans wider than 4 m.

double connector A short *connector* made of a piece of pipe with, at each end, a long *parallel thread*, each fitted with a back nut and a socket. It can be removed from a pipe run without dismantling the entire installation.

double curvature The shape of a surface curved in two directions, e.g. a sphere, *hyperbolic paraboloid*, etc. This can stiffen materials more than single curvature, e.g. a cylinder, corrugations.

double-decker lift A high-capacity passenger *lift* with two cars one on top of the other, serving different pairs of floors (e.g. odd numbers and even numbers) at each stop. About a third fewer lifts are needed, saving usable floor space.

double door A door with two *leaves*, the *active* and the *inactive*, hung from opposite sides of the same frame. Their *meeting stiles* come together.

double-door bolt A *Cremona bolt*.

double eaves course, doubling c. A row of *plain tiles, slates*, or *shingles* laid at the foot of a roof slope, over an *under-eaves course* of shorter tiles, etc., instead of a *tilting fillet*.

double glazing *Glazing* with two layers of glass or clear plastics separated by a *cavity*, usually a factory-finished *sealed unit* or windows with a *secondary sash*; less often a *coupled window* or a

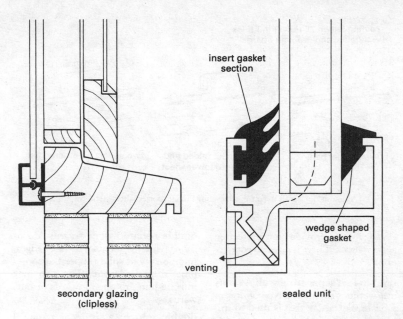

insert gasket
section

wedge shaped
gasket

venting

secondary glazing
(clipless)

sealed unit

Double-glazing.

double window. Double glazing is used for heat and/or sound *insulation* and differences in design depend on the main purpose. Both types are vapourtight round the inner pane, with the cavity, or the edge sealant, vented outwards.

Heat *insulation* of double glazing is at its best with a 12 to 20 mm cavity. A cavity wider than 300 mm allows convection currents to increase heat losses. Glazing is usually with flat glass or twin-walled *polycarbonate* sheet. Heat insulation can be improved by filling the cavity with argon and by applying a *low-emissivity coating* on the outer side of the inner pane.

Sound insulation can reach 30 to 50 *decibels* (dB) with double glazing, compared with only 20 to 35 dB for a single-glazed window. It improves as the space is made wider, with most of the benefit being gained at about 200 mm. Thick glass reduces *transmission* and the *coincidence effect*, and panes of different thicknesses reduce resonance. The frame should be heavy and effectively sealed round the edges. It may have an absorbent lining round the edges of the cavity. Filling the cavity with *sulphur hexafluoride* gas improves sound insulation for high-pitched speech but not for traffic noise. See diagram.

double-handed lock A *reversible lock*.

double-hung sash window A normal *sash window*.

double-inlet fan, d.-suction blower A *centrifugal fan* with an inlet each side.

double insulation Insulation within

double-leaf separating wall

roofing felt on 25 mm (1 in.) thick
tongued and grooved boards

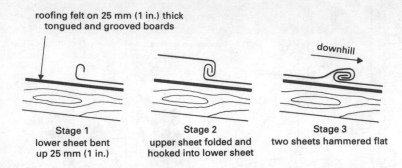

| Stage 1 | Stage 2 | Stage 3 |
| lower sheet bent up 25 mm (1 in.) | upper sheet folded and hooked into lower sheet | two sheets hammered flat |

A double-lock welt used for cross-joints in supported sheetmetal roofing.

electric tools to provide high protection against shock. There is no earth connection.

double-leaf separating wall A *cavity wall* without *wall ties*, a type of *discontinuous construction* that is used to improve *sound insulation* by reducing *transmission*. The wall leaves can be of masonry or lightweight timber *framed partitions* lined on one side with laminated plasterboard 40 mm thick. The cavity is 225 mm wide or more, and sound *absorption* inside it is slightly increased by *mineral wool* quilts 13 to 25 mm thick between the studs. The *fire stops* across the cavity must also be flexible, e.g. wire-reinforced mineral wool (BRE Digest 347).

double-lock welt, d.w. (Scotland **clink**) A joint in sheetmetal made by folding the edges together through 180° along one line, then folding again another 180°, to form a *seam* which is more secure than a *single-lock welt*, although it does not act as an *expansion joint*. They are mainly used to make *cross-welts* in *supported sheetmetal roofing*. See diagram.

double-pitched roof A *pitched roof* with two slopes away from a *ridge*.

double-rebated door frame A *door frame* which is rebated on both edges, traditionally of solid timber but sometimes of *pressed metal*. The door may thus be hung to open inwards or outwards.

double-return stair A ceremonial *stair* with one wide *flight* up to a *landing* and two narrower parallel side flights up to the next floor.

double Roman tile A *single-lap* standardized British clay *roof tile*, measuring ordinarily 420 × 360 mm × 16 mm thick. It looks like one *Roman tile* joined on each side to two half tiles, making two waterways. The edges are shaped to form 76 mm interlocking side laps. There are two nail holes but no *nibs*. Compare **Poole's tile**.

double-suction blower A *double-inlet fan*.

double-throw lock A *security lock* that requires two turns of the key to fully shoot the bolt.

double time *Overtime* paid at twice the usual wage rate, customary in the UK for work between 4 pm on Saturday and Monday morning, or later than three hours after normal finishing time from Monday to Friday.

double-triangle tie A wire *wall tie* of galvanized or stainless steel with each end bent to form a triangle, used for *cavities* up to 75 mm wide. See diagram, p. 500.

double vacuum treatment The *impregnation of timber*, usually with expensive clear organic *preservative*, by placing it in a tank from which air is sucked out, followed by flooding with the preservative and the return of air pressure. This forces the preservative solution deep into the pores of the timber. Air is sucked out a second time to recover excess amounts of preservative.

double welt A *double-lock welt*.

double window A window with two similar but separate *sashes*, hung in the same frame one inside the other.

doubling course A *double eaves course*.

doubling piece A *tilting fillet*.

Douglas fir (Pseudotsuga taxifolia), **Oregon pine** A durable *softwood* grown in Canada and the USA, available in long lengths. It is twice as stiff as other firs. It resembles European *redwood*, but is less dense.

dovetail A *joint* used in corners of drawers or fine joinery. The interlocking tenons (called pins) are fan-shaped, like a pigeon's tail, not straight as in a *combed joint*. Being thicker at their ends than at the root, they are not easily pulled out.

dovetail cramp A *cramp* for stonework which is a double-dovetail shape and may be of slate or metal. Leadclothed iron cramps were used by the ancient Greeks.

dowel (1) A short round wooden rod which is glued into drilled holes in two wooden parts to connect them by *dowel-*

ling. The dowel may be grooved to allow air and glue to escape as it is driven. A *treenail* is larger.

(2) A thick steel tube, threaded inside and cast into concrete *flush* with its surface. A bolt is screwed into the thread for use as an anchor or a lifting attachment. A bar fixed to the tube improves *bond* to the concrete and prevents it being pulled out. Other types of steel dowel exist.

dowel bar A piece of *plain bar* reinforcement rod across a *contraction joint* (*C*) in a large concrete *ground slab*. One half has a factory-fitted *bond breaker*.

doweller A *woodworking machine* for drilling *dowel* holes at a set distance from the edge of joinery members, often fitted to a sawbench and planer.

dowelling (1) Making a *joint* by drilling holes and putting in glued *dowels*. Strong joints can be made, even into *end grain*. See diagram, p. 134.

(2) Long pieces of round hardwood which are cut to length to make *dowels*.

dowel pin (1) A short round *wire nail* pointed at both ends.

(2) (USA) A headless nail with one point and a barbed shank. It is driven into a *mortice-and-tenon joint* to fasten it permanently.

dowel screw, handrail s. A wood screw threaded at each end.

down-collective control A basic type of *lift control* system, with operating buttons both in the car and on each landing. Calls from the landing buttons are not answered as the car goes up, but in order at each floor as it goes down.

downcomer A pipe in which water flows downwards, such as a *flush pipe* from a cistern or a rainwater *downpipe*.

down conductor A thick metal tape,

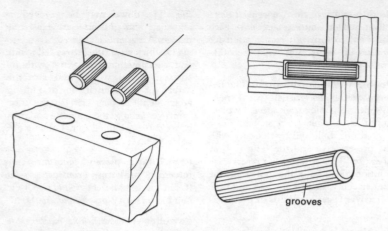

grooves

Dowelling.

cable, or rod of copper (or aluminium for an aluminium roof) going down the outside of a building as part of the *lightning-protection system*. It runs from the *air termination* to the *earth electrode*. The bottom end near ground level may be run inside a steel pipe for *mechanical protection*.

downlighter A narrow-beam *luminaire* near or recessed into the ceiling, aimed downwards to provide lighting without *glare*. See **high bay lighting**.

downpipe, downcomer, rainwater pipe (USA **downspout, conductor, leader**) A vertical or steep pipe which brings rainwater from roof *gutters* to the ground and into a *drain* or *soakaway* (C), as a successor to the gargoyle. The earliest pipes were of lead, which began to be replaced in Britain by cast iron in the nineteenth century. Today downpipes are mainly in plastics and snap fit with other *rainwater goods*. Accessories for downpipes include *rainwater outlets* and *rainwater headers*. Like all piping they must allow

for *thermal movement* without damage. See **offset**.

downstand A downward folded edge.

downstand beam A beam that projects below a slab *soffit*.

downtime A period during which an earth-moving machine is not available because of breakdown, refuelling, or maintenance. The opposite is *uptime*.

downwards construction *Top-down construction*.

dozy timber See **dote**.

draft (USA) See **draught**.

drafted margin A smooth uniform border 20 to 50 mm wide worked round the edges of the face of a stone.

draft stop (USA) A *fire stop*.

drag (1) A steel plate about 150 mm long by 100 mm wide, with plain or V-notched edges, used to level plaster surfaces, or to scratch the surface and give it *key* for the next coat of plaster, or to smooth the face of stone in *dragged work*. See also **French drag**.

(2) Resistance to the movement of a paint brush during *application* or *laying-off*, which can lead to pulling of the coat.

(3) See *C*.

dragged work Smooth-faced *ashlar* produced by a *drag*.

drain (1) A buried pipe that removes *foul water* from *sanitary pipework* and takes it to a sewer. Drains are usually laid as a line of *drain pipes* in the bottom of a *trench*, which is *backfilled* to give them a *surround* of *peagravel* or concrete. See BS 8301.

(2) To remove *foul water* or rainwater in gutters or drain pipes.

(3) A *gully* through which water enters a drain pipe.

(4) An open channel or gutter, an outlet, etc.

drainage A system of buried *drains*, or the *trade* that installs drains, work usually done by a *drainlayer*. Compare **plumbing**.

drainage easement An *easement* allowing drains.

drainboard A *drainer*.

drain chute A special *drain pipe* shaped to make *rodding* easy, tapered in its upper half, used on the outlet side of an *inspection chamber* (*C*).

drain clearing See **rodding, sink plunger**.

drain cock A *tap* placed at the lowest point of a water supply system through which the pipes can be drained.

drained and ventilated joint A stormproof joint between window glass and its frame, mostly used in *curtain walls* and *patent glazing*. The frame is made of metal profiles or *structural gaskets* with large voids, which connect to ventilation and drainage holes in the bottom. Air pressure inside the hollows is thus always the same as that of the outdoor air, so that driving rain is not forced through the outer joint. The little rain that does get through simply drains away. See **open-drained joint**.

drained cavity A way of keeping a basement or cellar floor dry. Hollow blocks are laid on the floor, sometimes covered with a *vapour barrier*. Any water below drains away either to the drains or to a sump. Compare **tanking**.

drainer, drainboard, draining board A sloping surface to allow water to drain into the bowl of a kitchen *sink*.

drainlayer, pipe layer A *tradesman* who lays drains.

drainline A row of pipes joined end-to-end to make a *drain*.

drain pipe A *length* of pipe made for use as a *drain* or *sewer* (*C*). Pipes, as well as their *joints*, may be either rigid or flexible. Flexibility is desirable in areas where the ground may move, e.g. near mines. Flexible pipes of *plastics*, ductile cast iron, and steel deform appreciably before collapse. Rigid pipes of *asbestos cement*, *vitrified clayware*, concrete, and grey cast iron break before their deformation becomes noticeable. Pipe joints made flexible by a *joint ring*, as in the *sleeve coupling*, allow rigid pipes to settle without breaking. But rigid pipes are not softened by standing in the sun, as PVC pipes may be. See diagram, p. 136.

drain plug, d. stopper An expanding plug which blocks the end of a drain run during a *drain test*. It is usually a *bag plug* or a *screw plug*.

drain rods Lengths of flexible *glass-reinforced plastics* or other material that

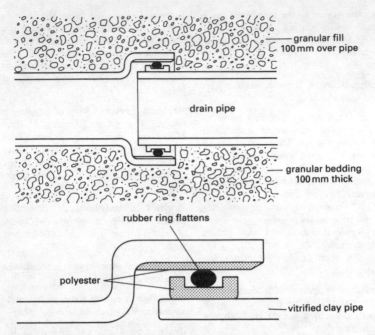

granular fill
100 mm over pipe

drain pipe

granular bedding
100 mm thick

rubber ring flattens

polyester

vitrified clay pipe

A vitrified clay drain pipe with push-fit joints on a granular bedding.

can be joined end-to-end and used for *rodding* drain pipes. See diagram, p. 374.

drain shoe A *special* drain fitting with both an *access cover* and an inlet for a *downpipe*.

drain test Drains are tested after laying, before the trench is *backfilled*. The main test, usually a *hydraulic test* (*C*), is for leakage and is made as each drain run is completed, with a check for straightness between manholes. At *commissioning* the main tests are for flow rate and freedom from debris by a *ball test*.

draught (USA **draft**) (1) The flow of air and burnt gases up a *flue* which occurs either because of their buoyancy or from the effect of a power-

driven fan, or both. The draught pressure is usually measured by a *U-gauge* in millimetres of *water gauge*.

(2) An air current that causes discomfort by chilling.

(3) The amount by which holes are out of line in *drawboring*.

(4) **draw** The inward taper or slope given to the sides of *formwork* so that it can be slid away from the hardened concrete.

draught stabilizer, d. diverter (USA **barometric damper**) A vertical metal plate in the sidewall of a *flue*, pivoted horizontally and counter-balanced so that the *draught* cannot be excessive for long. When the draught increases too much, the plate is sucked inwards, admitting cold air and thus reducing the draught. It

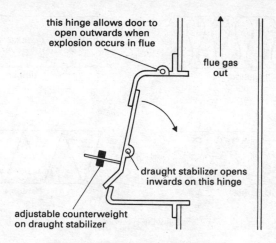

this hinge allows door to open outwards when explosion occurs in flue

flue gas out

draught stabilizer opens inwards on this hinge

adjustable counterweight on draught stabilizer

A draught stabilizer for an oil-fired boiler.

automatically closes when the draught diminishes, thus ensuring the correct draught. See diagram.

draught stop A *fire stop*.

draught strip, d. stripping *Weatherstripping*.

draw bolt A *barrel bolt*.

drawboring Drilling holes through a *tenon* and the morticed piece about 3 mm out of line (the draught) so that a tapered steel pin driven through the holes will pull the pieces tightly together. The steel pin is later replaced with an oaken *treenail*, making a joint without glue that can be dismantled and reassembled.

draw cable, d. wire (USA **fish tape**) A wire left in a *cable duct* or *conduit* during construction for a *draw-in system* of wiring or cables. Reels of draw line may also be blown through conduit using compressed air.

draw-in box, pull b. A box in a *draw-in system*.

drawings The main working documents used on site: plans, sections, and elevations – on a large project there are hundreds, grouped in sets. Small-scale drawings (such as floor plans) show the layout for *setting-out*, large-scale ones show *details*. Both may have *dimensions* and descriptive notes. Drawings may be either *contract documents*, such as the architectural, structural, and services drawings, or for *coordination* between trades, such as the builder's work, shop, and marking drawings. For complex services, drawings may go through many *revisions* during the contract, which are noted briefly in the *title block*. The last drawing revision should form the *as-built* set.

drawing symbols Marks on drawings (usually the plans) which indicate particular materials or components. See diagram, p. 138.

draw-in system A system of wiring *conduits* or larger cable ducts, usually cast into the concrete *structure*, through which wires can be pulled in from the

drawn glass

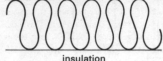

insulation

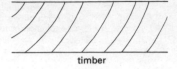

timber

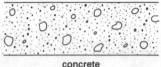

concrete

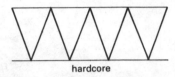

hardcore

Drawing symbols.

ends, or from *draw-in boxes*, using *draw cables*, a *snake*, *rodding*, etc. Wiring can often be replaced by using the old wires as a draw cable.

drawn glass *Sheet glass.*

drawn wire Mild-steel wire drawn through the hole in a die, which work-hardens it. It is used *as-drawn* to make *fabric* reinforcement.

draw-off pipe The pipe leading to a *tap*, usually a *dead leg*.

draw-off tap A water supply *tap*.

draw-off temperature The temperature at which *domestic hot water* leaves a storage cylinder.

drencher sprinkler, deluge s. A high-capacity outdoor *sprinkler*, used as a water curtain to prevent fire entering a building.

dress (1) To smooth timber, usually with a *plane* or *woodworking machine*.
(2) To work stones to their final dimensions by *wasting*, sawing, hammering, or chiselling.
(3) To fold and hammer down sheet material over a surface.

dressed dimension, neat size The finished size of joinery timber after planing. It may be 9 mm less than the *nominal dimension* in both directions, from removal of wood in sawing and planing, as well as *shrinkage* from a higher *moisture content*.

dressed stone *Dimension stone.*

dressed timber *Surfaced timber.*

dressing Shaping and finishing materials to their final dimensions or shape. See **dress**.

dressing compound (1) Bituminous liquid used hot or cold (cut back) to coat the exposed surface of *bitumen felt* and make protective *chippings* stick.
(2) *Levelling compound.*

dressings Smooth stone *masonry*, bricks, or *stucco* round doors or windows, or at the corners of a building, of contrasting type to the remainder of the facing brick, stone, or render. They may have *mouldings*. See **quoin**.

driers Compounds which encourage the oxidation and hardening of *drying oil* or hydrocarbon *solvents* of paints.

Those added by the user are mostly soluble resinates, linoleates, or naphthenates of lead, manganese, or cobalt, to BS 332.

drift bolt A steel pin usually not less than 22 mm dia. driven into holes bored 1.5 mm smaller as a fixing between heavy timbers. See also *C*.

drill (1) A hand or *power tool* that turns a *bit* held in the *chuck*, mostly used for boring holes. Drills are rated by the largest *drill bit* they can drive through a material, e.g. '10 mm steel, 15 mm wood'. See *electric drill*.
(2) A *drill bit*.
(3) To drill or bore a *penetration*.

drill bit, drill A *bit* for boring holes, driven by a *drill*, usually a *twist drill*. Drill bits can be *tungsten-tipped* or even *shanked*.

D-ring A pipe *joint ring* with one flat surface, which prevents it from rolling. The joint is made by sliding, and lubrication may be needed. See diagram, p. 252.

drinking fountain A small bowl with a water jet and a self-closing tap, made to prevent contamination from mouth contact or *back siphonage*. The water can be refrigerated.

drinking water, potable w. Water fit for human consumption. Contamination is prevented by *direct supply* from the main, avoidance of *back siphonage* and the use of non-tainting materials. Contact with coal tar or lead is now prohibited.

drip (1) A groove or rise in an under side to make water fall off. The outer edge of an exterior *soffit* has one to exclude rainwater, e.g. a *bead, downstand, drop apron*, or *throat*.
(2) On a flat or gently sloping roof which has a step perpendicular to the direction of flow, the lower sheet (*under-*

cloak) of *supported sheetmetal roofing* is turned up and over the step to bring it under the *overcloak*. This is laid later, dressed down over the riser and sometimes extending 40 mm on to the flat as a *splash lap*. Drips act as *expansion joints* in long panels.

drip edge A *drop apron*.

drip-free paint A *non-drip paint*.

dripping eave An *eave* with no gutter.

dripstone A stone *hood moulding*. A rectangular dripstone is a 'label'.

drip tray A *condensates pan*.

drive screw, s. nail A *fixing* like a nail, but with a square shank 2 mm or more thick, twisted several times in its length. It can be driven by hammer into wood or masonry, but is very difficult to withdraw.

driving rain, wind-driven r. In a storm, rainwater can enter a building, mostly through joints or cracks, forced through by the higher outdoor air pressure on the windward side of the building and emerging from the crack as a strong spray. The *perpends* in brickwork are a weak point, but rain can also find its way round or through *cladding, roofing*, and window frames, particularly at edges, joints, *ridges, hips*, etc. Driving rain should not be able to bridge a *cavity wall* or get through a *drained and ventilated joint* or similar *open-drained joints*, or under or round a *flashing*.

driving rain index A grading of site *exposure*, given by multiplying annual rainfall (mm) by average wind speed (m/s) and dividing by 1000. 'Severe' exposure is above 7 m²/s, 'moderate' above 3.7 m²/s, and 'sheltered' below 3 m²/s. See BRE Digest 127.

drop (1) A vertical ceiling panel at a change in levels.

drop apron

(2) **cable d., wall d.** A cable running down from the ceiling to a light switch, on the surface or inside a wall or partition; or its conduit.

(3) **d. panel** A *thickening* of the concrete at the head of a column in a *plate floor* (*C*) to resist the high shear loads at the column connection.

drop apron, drip edge In *supported sheetmetal roofing*, a strip of metal fixed vertically down to act as a *drip* at *eaves*, *verges*, or *gutters*, held by a *lining plate*.

drop ceiling A *false ceiling*.

drop channel A short but steep channel between the inlet and outlet of a *drop manhole*, usually for rainwater or *waste water*, but not *soil water* drains, which have a *backdrop*.

drop connection A *backdrop*.

drop end A *rainwater outlet* downwards from the bottom of an eaves gutter.

droplight A *pendant fitting*.

drop manhole A manhole with a large difference between the invert levels of inlet and outlet pipes. It may have a *backdrop* or a *drop channel*.

drop panel A *drop*.

droppings Unwanted fallen material which marks or damages other work or is no longer fit for use, such as bricklayer's mortar inside a *cavity wall*, paint on a floor, or *dead* plaster.

drop system A hot-water heating system which is fed from a sub-main at the top, with branches downwards (droppers) to the consumer points. It may have return pipes to form a *circuit*.

drop wire A *service cable* from the nearest pole of an overhead supply line connecting a house to the electrical mains.

drove (Scotland) A *boaster*.

drumming Noise from sheet materials that are struck or vibrated, such as stainless-steel *sinks* or sheet-metal *ductwork*. It can be reduced by a thick, soft coating on the unseen side, *diamond break stiffening*, etc.

drunken saw A *wobble saw*.

dry (1) The condition of paintwork after *drying* – either *dust dry* or *hard dry*, etc.

(2) See **dry construction**.

dry area An *area*. See also **wet area**.

dry concrete A *mix* with insufficient *mixing water* to give it normal *workability*, either one with a normal *cement content* or a *dry lean concrete*.

dry construction Building with no *wet trades*, as in methods using *dry linings, suspended ceilings*, or *precasting* (*C*), either with reference to complete *system building* or a particular building *element*. With full dry construction, a building is ready for occupation quickly but may be expensive. It may also be easy to take apart, thus offer less *security*. In spite of the name, these methods must not be used in a damp structure.

dry-dash finish The finish produced by throwing coarse aggregate on to a wet coat of *render* (butter coat) on a wall, where it sticks, giving an exposed aggregate finish. The finish can be *pebbledash, spar dash*, or *shingle dash*, and the dashing is either by hand casting or machine projection.

dry-film thickness (dft) The thickness of a *coat* of paint, usually measured in *microns*. Primer on steel has a dft of about 50–70 microns.

dry-gasket glazing *Gasket glazing*.

dry hip tile A roof *hip tile* that does not need traditional mortar *bedding*.

dry hydrate *Hydrated lime*, $Ca(OH)_2$, that has been slaked by the manufacturer with the calculated amount of water and sold in bags as powder. It can be soaked overnight to improve *plasticity*.

drying (1) The reduction of water content (moisture, vapour) in a material, causing reversible *moisture movement* or permanent *drying shrinkage*.

(2) The hardening of a coat of *paint* or varnish, making it safe from damage or enabling it to be *recoated*. Drying can be by *solvent release* from the *medium* or chemical *curing*, or a combination of both. Air drying is drying at air temperature; *forced drying* or *stoving* are at higher temperatures. Main stages in paint drying are: dust dry, tack free, dry to handle, and hard dry.

(3) *Seasoning* of timber.

drying of screeds *Screeds* to take a glued-down floor finish, such as thermoplastic or PVC tiles or sheet flooring, must be dry before the *adhesive* is applied or it will fail. Usually one month is allowed per 25 mm of screed thickness for *planning* purposes, but a moisture meter should be used for the final decision. See BS 8203. See also **curling**.

drying oil An oil such as *linseed oil* which forms a tough film when exposed to air in a thin layer, used as the base in *oil paints*.

drying-out After a building is *enclosed* and work in the *wet trades* is completed, the *construction moisture* has to be evaporated before *dry trades* can be started. An approximate rule of thumb is 'one month per 25 mm of thickness' for brickwork or concrete slabs, in good weather conditions and with adequate ventilation. In winter heating may be needed. Each kg of water to be evaporated needs 0.7 kW-h of energy (BRE Digest 163).

drying shrinkage Materials made with cement (concrete, mortar, render) shrink during their early life, starting even before *initial set*. In the first twenty-four hours shrinkage can be more than 1% in some circumstances, then a further 2 or 3 mm per metre up to twenty-eight days. This is much more than the effects of *moisture movement* or shrinkage from *carbonation*. Drying shrinkage is aggravated by a high *cement content*, which demands more water for hydration, by the loss of water to absorbent *formwork* and evaporation, or poor *curing*. The cracking it could cause in structural concrete is prevented by the use of *contraction joints* (*C*), *construction joints* between bays, etc. *Revibration* (*C*) may reduce visible effects, such as the *chapping* of column tops. For timbers, good *seasoning* reduces drying shrinkage.

dry joint A joint without mortar or adhesive. See *C*.

dry lean concrete A concrete *mix* with a low *cement content*, batched with less than normal water, giving a good (low) *water/cement ratio* (*C*) and low shrinkage. It is used for *blinding concrete*, loadbearing infill, and road bases, and can be compacted by roller.

dry lining (USA **dry-wall**) A lining to the inside of a building, made of any *panel product* or *tapered-edge plasterboard* or *wallboard* that needs no plastering except for narrow strips over joint *scrimming*. It is either nailed to *grounds* or to the *studs* of timber-framed walls, or bonded to *dabs* on brickwork. Dry linings reduce *construction moisture* and enable the building to be decorated and completed early. Fixings for heavy objects are made into the studs or with *toggle bolts*.

dry masonry Walling without *mortar*.

dry mix Usually *dry concrete*.

dry-pack mortar *Mortar* used for *pinning-up* or packing under baseplates, batched with very little water and rammed in to a narrow space with a stick. It has less shrinkage than normal mortar or *grouting*.

dry partition A *demountable partition*.

dry-powder extinguisher A portable *fire extinguisher* with a dry powder extinguishing medium, used for electrical and *flammable liquid* fires.

dry-press brick A *pressed brick* made by multiple pressing of semi-dry powdered clay or ground hard shale, enabling it to be placed straight in the kiln.

dry ridge tile A roof *ridge tile* with a clip fastening system. It does not need bedding in mortar.

dry riser (USA **dry standpipe**) A *fire riser* that is usually kept empty and thus cannot suffer frost damage from freezing water, as a *wet riser* can. Dry-riser inlets are usually located near the front entrance, giving the fire brigade easy access. It may have its own *booster* pump, or a *breeching* for connecting a mobile booster.

dry rot The *decay* of timber because of dampness feeding the fungus Serpula lacrymans; at least 22% *moisture content* in the timber is essential. At much below this level the fungus dies. Dry rot flourishes in dark humid conditions and its fruiting body (sporophore) can spread great distances. Its roots (hyphae) have threads like plant stems that transmit moisture, and they can force their way through the mortar of brickwork. If the timber cannot be kept dry, the rot is difficult to eradicate. Infected wood and the complete root system have to be exposed and burned or disinfected. See BRE Digests 299 and 364.

dry sprinkler A *sprinkler* supplied from a *dry riser*.

dry to handle A stage in *drying* of paintwork at which the component can be handled without damaging the coating.

dry trades Finishing trades that cannot be started until after the building is weathertight and *drying-out* completed. This is a major *stage* of construction and a builder aims to get dry trades in as soon as possible.

dry verge tile A *verge tile* fixed with clips, not mortar bedding.

dry wall (1) **dry stone wall** A wall built without *mortar*.
(2) A wall built by *dry construction*.
(3) **dry-wall** (USA) A *dry lining*.

dual-duct system Air conditioning with two separate *high-velocity* ducts, carrying hot air and cold air from the central plant to a *mixing box*, then low-velocity ductwork to the *supply air* outlets.

dual-flush cistern A water-saving *water-closet* cistern with a two-button *flushing mechanism*, one button giving a full flush of 9 litres and the other a half flush of 4.5 litres. In parts of Australia dual-flush cisterns are required by law.

dubbing-out The filling of hollows in a wall surface before *plastering*.

duckbill nail A chisel-pointed nail, easy to *clench*.

duckboard Timber boards nailed to joists, as a temporary floor.

duckfoot bend, rest b. A right-angle *bend* with a flat seating to carry the weight of vertical pipes or the thrust due to the change of direction of the water.

duct (1) **air d.** A large lightweight

pipe for *air handling* in air-conditioning or ventilation systems. Rectangular *folded sheetmetal* is used for low-velocity air and *spiral duct* for high-velocity air. Other materials include *glass-reinforced plastics*, plastics tubes, and *flexible ductwork*. Air ducts must have *fire dampers* at compartment floors or walls.

(2) **services d.** A passage of concrete, blocks, pipes, etc., in the building *fabric* or elsewhere, inside which *services* are run, such as pipes, cables, or air ducts, mainly for the plumbing, electrical power, and air conditioning. The ducts may be *shafts* for *risers* that run up a *services core* to the space above a *suspended ceiling*. In large buildings, services ducts need to be sealed to prevent *transmission* of sound and have *fire stops* at compartment floors or walls. See **ground duct, main services duct**. See also *C*.

duct cover A steel plate, a concrete slab, or open metal flooring panels over a *ground duct*, usually set in *rebates* on each side, to finish *flush* with the surrounding floor.

ducted flue, appliance ventilation duct A large *flue*, such as a *U-duct* or *SE duct*, that supplies fresh air to a *room-sealed appliance* then takes away its burnt gas. Details are given in the British Gas publication 'Gas in Housing'. See diagram, p. 481.

ductile iron A type of *cast iron* that is stronger and more impact-resistant than *grey iron*. It is used for water supply and drain pipes of 80 to 700 mm dia.

ductwork Air ducts for air conditioning or mechanical ventilation.

dummy activity An activity represented by a dotted line on an *arrow diagram* programme using the *critical path method*, as a clear and simple way

of showing how progress on one task prevents completion of another task. No resources are needed for a dummy activity, nor does it take any time. See diagram, p. 352.

dummy frame A temporary frame the exact size of a *doorset*, put up in its place while a brickwork wall is being built. The opening is accurately formed for the doorset to go in later, undamaged by any of the *wet trades*.

dunnage Waste timber used for packing between layers of stacked materials.

duodecimal system The feet and inches system, or any system in which twelve small units make one large unit.

duplex apartment (USA) A *maisonette*.

duplex control A *lift control* system that sends only one car to answer a landing call, where there is a bank of two lifts.

duplication of plant The installation in pairs of essential plant, such as pumps or fans, and their electric drive motors. One is on *duty* and one is on *standby*, often with automatic changing-over at each start-up.

durability The ability of a material (timber, concrete, ceramics, adhesive, paint) to stay serviceable for a long time. The *fabric* and *cladding* of a building should have lifetime resistance to moisture in all forms (driving rain, damp) combined with heat and cold, solar radiation, and frost. Periodic repair or replacement is usual for finishings, services, etc. See BS 7543.

duramen The *heartwood* of a tree.

duration The estimated time (weeks, months) needed to complete an *activity* in the *critical path method*, usually worked out from experience.

dust Dust from building materials such as cement, gypsum, fibres, and wood can be a health hazard on site. It should not be inhaled or allowed into the eyes, nose, or throat. In general, contact with the skin is best avoided. Cement, plaster, and lime should be handled so that they are not spread as dust. The abrasive cutting of concrete, bricks, stone, mineral fibre sheet, etc. or work with *power tools* should be done wet if possible, or by machines with dust extraction, and outdoors in preference to indoors. Fine dust can be detected by looking through a dust cloud towards a high-intensity lamp, while shielding the eyes from direct glare. If dust is produced, protective clothing, face-masks, respirators, and washing facilities are required, as well as the separation of workers to reduce the numbers exposed. Smoking, which deadens the lung's dust filters, should be avoided. Dust from preservative-treated timber may be poisonous, especially that from *hardwoods*. Special requirements apply to work with *lead* or the removal of *lead paint* and *asbestos*. In the UK details can be obtained from the Health and Safety Executive.

dust dry (USA **d. free**) A stage in *drying* of paintwork after which dust will not stick to the fresh surface.

dusting (1) The loss of material from the weak surface of a concrete floor, usually from excess water in the mix, careless laying, or poor *curing*. Dusting in strong concrete may be reduced by a *dust proofer* or a *hardener*.

(2) The removal of dust, the last step in *preparation* of a *substrate* before painting. Wiping with a *tack rag* or vacuum cleaning are better than *brushing*, which moves the dust to somewhere else.

dusting brush A large soft painter's brush, formerly used for *dusting*.

dust proofer, anti-dust product A liquid, painted on to concrete floor surfaces to reduce *dusting*. The *hardener* types contain fluorosilicates or similar chemicals that react with free lime in the concrete to form hard, stable compounds. *Floor-sealer* types are based on *epoxy resin* or *polyurethane*.

duty (1) The load capacity, or wear and exposure resistance, of a building component for given service conditions.

(2) In *duplication of plant*, the duty machine is the active one, not the one on *standby*.

duty load The normal maximum load in service of a *lift*: the greatest number of passengers, or the weight in kg allowed in the car.

dwang (Scotland) *Strutting* or *noggings* between *common joists, rafters, studs*, etc.

dwarf support wall A low wall that supports ground-floor *joists*.

dwarf wall A low wall, usually no higher than about one metre.

dye A colouring material which, unlike a *pigment*, is soluble and colours materials by penetration. Dyes are used to make *lakes*.

E

ear A fixing *lug*.

earliest date, e. event time In the *critical path method* the earliest likely date for start or finish of an *activity*. It is calculated by a *forward pass* and used in ordering materials and planning.

ear protectors Ear muffs or ear plugs to prevent damage to the ears of people exposed to loud *noise*.

earth (1) (USA **ground**) The mass of the earth, which conducts electricity, taken as zero voltage and used for *earthing*. Earth cables, or the 'circuit protective conductor', have green/yellow stripes. See **clean earth**.
(2) *Topsoil*.

earth bar A brass bar in the *main switchboard* or *consumer unit* to connect all building *earthing* systems to the main earth supply.

earth bond *Bonding*.

earth building, e. wall construction A house of unburnt *adobe*, *pisé*, or *cob*, usually owner-built and maintained at little or no cost. Its thick walls give fairly good insulation against heat, cold, and sound, but may require a protective coating to prevent rain scouring it. Earth buildings are best suited to dry inland climates and need careful selection of materials and proper construction. In earthquake regions earth buildings are dangerous – the walls lack toughness and can allow a heavy flat roof to collapse on to the occupants.

earth colours Natural paint *pigments*, yellow to brown in colour, dug up from the earth, such as *ochre*, *sienna*, or *umber*. They may contain some clay.

earthed concentric wiring A *mineral-insulated cable* in which the outer conductor, a metal tube, is earthed. It contains the other conductor.

earthed system The connection of the *neutral* wire to *earth* in a transformer sub-station, used for *low-voltage* distribution. The *fuses* and *switches* are on the *live* wire.

earth electrode, e. termination Steel-cored copper-clad stakes driven into the ground, or copper tape, or a metal plate buried at a suitable depth, sometimes even underground metal water pipes located where the earth is always damp. It provides a low electrical resistance to *earth*.

earthenware *Ceramics* of lower strength than *vitrified clayware*.

earthing Connections between conducting objects and the *earth* to protect against shock from the electrical installation or from lightning. See BS 7430. Compare **bonding**.

earthing lead The conductor which makes the final connection to an *earth electrode*, usually through a test clamp.

earth-leakage circuit-breaker (ELCB) A *circuit breaker* to give reliable protection against electric shock from mains-voltage equipment, but which should not be used as a substitute for *earthing*. It can be in the *consumer unit* or plugged in before an extension cord. For portable *power tools* it should operate at a leakage current of 30 mA and stop supply in less than 30 milliseconds (0.03 sec.)

of a dangerous fault, usually by detecting *residual current*.

earth termination An *earth electrode*.

earthworks *Groundworks*.

earthwork support Strong sheeting held with props against the sides of an excavation to prevent the earth collapsing and injuring or burying workers. Support should be provided for all but the most shallow trenches (less than 1.20 m deep). Examples are *planking and strutting* and *trench boxes* (*C*).

earthy metalwork A *metallic part*.

ease (1) To sandpaper the sharp edge of an *arris*, giving an *eased arris*.

(2) To free a tight fit between two joinery parts, giving them play to slide or close properly, usually by planing some wood off the edges.

(3) To slacken the pressure on *formwork* after concrete has hardened, by unscrewing *props*, driving out *wedges*, etc., as a first step in *stripping*.

eased arris An *arris* that has been very slightly rounded by rubbing with fine sandpaper. It is softened to the touch, although the rounding may be difficult to see. The preparation of timber joinery for painting includes easing the arrises.

easement In law, a right which a person may have over another's land, such as the right to run underground pipes across it for *drainage*, or a *wayleave*.

easy-clean tap, cover pillar tap A simple basin tap with a smooth cover screwed on over the tap body, hiding the *gland nut*.

eaves, eave The lowest part of a sloping roof, or the area beneath if it *overhangs*, although eaves may be *flush*. Eaves may have a horizontal *fascia* which carries the *gutter*. See also **verge**. See diagram.

eaves board A *tilting fillet*.

eaves course (USA **starting c.**) The lowest course of *tiles* on a roof, over the *eaves*. It is the first course to be laid and the bottom edge of each tile is not laid over the top edge of another tile but is raised by a variety of methods, such as a *tilting fillet*, an *eaves vent*, or a special *eaves tile* to form a *double eaves course*.

eaves drip, roof d. The lowest point of a roof covering, where water falls off, usually into an *eaves gutter*.

eaves flashing A *drop apron* from an asphalt roof dressed into an *eaves gutter*.

eaves gutter A *gutter* along the eaves *fascia*, in plastics, aluminium sheet, etc., usually made to match other *rainwater goods*. It should be centrally under and not more than 50 mm below the *eaves drip*.

eaves lining A *soffit board*.

eaves overhang The distance that *eaves* overhang beyond a wall. In Continental Europe it is often 1 m or more.

eaves plate A *wall plate*.

eaves soffit The underside of overhanging *eaves*, lined or open.

eaves tile A short tile about 215 mm long used in the *under-eaves course* below an *eaves course*, thus forming a *double eaves course*.

eaves vent, e. ventilator A narrow opening under an *eaves course* or in the *eaves soffit* to allow ventilation air into a *cold roof*.

EC A *Eurocode*.

echo The repetition of sound by reflection from a wall such as the rear wall of an auditorium. Echo is reduced by *absorption*.

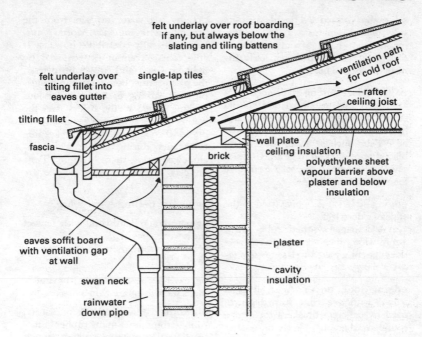

felt underlay over roof boarding
if any, but always below the
slating and tiling battens

felt underlay over
tilting fillet into
eaves gutter

single-lap tiles

ventilation path
for cold roof

rafter
ceiling joist

tilting fillet

fascia

brick

wall plate
ceiling insulation
polyethylene sheet
vapour barrier above
plaster and below
insulation

eaves soffit board
with ventilation gap
at wall

swan neck

rainwater
down pipe

plaster

cavity
insulation

A typical eave detail for single-lap tiles and a cavity wall. Ventilation openings must be provided at eaves and ridge, with a clear path between them.

economic thickness of insulation The thickness of thermal *insulation*, including costs of installation and protection, that can be justified on hot-water pipes by the saving in heating cost.

edge To work or shape the edge of a joinery component.

edge beam A beam that gives extra stiffness or *edge fixity* to the side of a concrete *suspended slab*.

edge bedding *Face bedding*.

edge distance The width that holes for bolts or similar *fixings* are kept back from an edge to avoid breaking it away.

edge fixity Structural *continuity* (*C*)

given to the edge of a *suspended slab* by an *edge beam* or wall.

edge form *Formwork* such as a wooden board, for the edge of a concrete slab. Its top is set at the level of the slab surface and used as a *screed rail*.

edge grain, comb g., vertical g. A grain seen in *quarter-sawn* wood. The best quarter-sawn oak shows *silver grain*.

edge isolation An *expansion strip*.

edge joint A joint in the direction of the *grain*, e.g. between two veneers. Compare **butt joint**.

edge nailing *Secret nailing* of floor boards, timber wall linings, etc.

edge-shot board A board with its side planed lengthways, to remove marks from hand sawing. It has a square edge and sharp *arrises*. See **laminate veneered board**.

edge tool A cutting tool, such as a knife or a chisel, usually made of special steel. The edge must be sharpened and honed once it becomes dull from working. Tools such as a normal all-steel bladed *saw* also need to be *set*. Circular saws and many other *power tools* have long-lasting *carbide tips*.

edging strip (1) A *lipping* fixed to the edge of a door leaf.
(2) A U-shaped synthetic rubber strip 4 mm wide which completely encloses the edges of a pane of glass for *secondary glazing*.

edging tool, edger A small steel trowel with one edge turned down, used by concrete finishers to form a rounded edge on a kerb or a stair *nosing*.

Edison screw cap A coarse-threaded cap for an *incandescent lamp*, to make electrical contact when screwed into the lamp holder.

effective storage The usable contents of a storage tank.

effective temperature (USA) A *comfort index* that takes account of radiant heat, air movement, temperature, and humidity.

efflorescence, bloom On the surface of *clay bricks*, powdery crystals that grow from the salts dissolved in the brick. It is unsightly, although usually harmless unless it lifts paint, plaster, or tiling. The salts come mainly from gypsum or pyrite in the clay and coal that bricks are fired with. Efflorescence is uncommon on mortar or *calcium silicate bricks* unless due to contamina-

tion by sea water or salts from the ground, such as *sulphates*. Mortar *plasticizers* generally contribute no salts. If efflorescence cannot be removed by light brushing and breaks up plaster, the plaster must be removed, asbestos-bitumen lathing or other impervious backing nailed to the bricks, and the wall re-plastered. It can be greatly reduced if not prevented by keeping bricks dry during laying, which includes covering them at night. Tests on clay bricks for efflorescence are given in B S 3921.

egg-crate ceiling A *cell ceiling*.

eggshell (1) Paintwork with one of the lower stages of *gloss*.
(2) A smooth matt finish to stonework.

egress The means of egress from building is the way out, e.g. in a fire.

elastomer Any synthetic or natural rubber, sometimes in the rubber industry defined as material resilient enough to be stretched to twice its length and, on release, to snap back to the original length. See next three entries.

elastomeric bitumen *Bitumen polymer*.

elastomeric polymers The materials of synthetic *rubbers*, such as *polychloroprene, polysulphide*, and plasticized *polyvinyl chloride*. They can be *copolymers*, e.g. *acrylonitrile butadiene styrene* and *ethylene propylene diene rubber*.

elastomeric sealant A liquid or paste that stiffens after application to become a rubbery *sealant* which undergoes *curing*.

elbow, knee A sharp right-angled pipe *bend*.

elbow board, e. lining The panelling below a window, on top of the wall.

electrical riser Vertical electrical distribution wiring that runs up a tall building, or the services *duct* which contains it.

electrical services, e. engineering, e. installation Building *services* for the distribution of *low-voltage* power and lighting, communications (telephone, data, alarms), lifts, and lightning protection. For safety and to minimize electrical interference, the cables for each type of service are run in separate *conduits* or *trunking*.

electric drills Versatile *power tools* with a *chuck* and a pistol grip. They are mainly used for drilling holes in materials like wood, steel, or masonry, and range in size from compact and lightweight units (including *cordless* types), through medium sizes with variable speed, reversing, or hammer action, to large types for concrete drilling. Drills are rated by the largest *drill bit* they can drive through a material, e.g. '10 mm steel, 15 mm wood'. A powerful drill with a blunt bit can set fire to wood or *breather paper*. See diagram, p. 346.

electric heating Space *heating* or the production of *domestic hot water* with resistances, using *coils* or panels, often off-peak *storage* type.

electrician A *tradesman* who instals or repairs *electrical services*, including *wiring* for circuits to machines, plant, or lighting.

electric motor Easily controlled, reliable electric motors are widely used in building *services* to drive fans, pumps, and blowers or to operate lifts, dampers, and doors. Since they may vibrate and be noisy, they often need *anti-vibration mountings*. Medium-sized and large electric motors may need to be supplied through a *starter* or from a special *motive power* circuit. Large units are run off *three-phase supply*.

electric panel heater *Panel heating* by electrical resistances. The panel may be surfaced with marble, plastics, etc., and can work at a high temperature of about 290°C, at a medium temperature below 120°C, or at low temperature below 81°C.

electric screwdriver, screwgun A *power tool* like a reversible *electric drill*, but with a screwdriver bit, used for its speed of operation and exact tightening force. It can also be a *cordless drill* or a *cordless screwdriver*. See diagram, p. 346.

electric storage floor heating *Underfloor heating* using off-peak power to warm a concrete floor overnight and radiate warmth during the day.

electric striking plate A *striking plate* with a remote control, usually for a *door control* operated through an *entry phone*.

electric tools Heavy-duty portable electric *power tools* can be run off a temporary site connection or a generator. Many 240 volt electric tools have *double insulation*, otherwise all exposed metal parts must be *earthed*. Often safety can be improved by supply through an *earth-leakage circuitbreaker*. Cords and plugs must be correctly connected, cable grips tightened over the outer sheath, and covers kept in place. Cords can be dangerous if damaged by being dragged over sharp edges or run over by vehicles. Tools, cords, and plugs need regular inspection and maintenance; damaged cords should be replaced – makeshift repairs are unsafe. In damp conditions 110 volt tools may be required, supplied through a transformer centre-tapped to earth, to limit any shock to a nonfatal 55 volts. Cables, plugs, transformers, etc., for 110 volts are coloured yellow. In wet conditions 25 volts are used. Small electric tools are often *cordless*.

electric water heater A hot-water *cylinder* or an *instantaneous water heater*.

electro-copper glazing *Copper lights*.

electrode boiler A *boiler* with submerged electrodes that heat the water as electric current flows through it. It is generally larger than domestic size and has the advantage that it cannot be overheated because of lack of water.

electrolier (USA) A *pendant fitting*.

electrostatic filter An efficient air-conditioning *filter* with a high-voltage direct-current field causing fine dust to coalesce into large particles, which are later caught by a *bag filter*. If there is a power failure the bag filter catches coarse dust.

element (1) **building e.** A major functional part of a building, made up from *components* or *materials*, e.g. the foundations, columns, floors, walls, and services. Both elements and components can be *modular sizes*.
(2) A *heating element*.
(3) A replaceable filter medium.

elemental bills of quantities *Bills of quantities* organized by dividing the work into building *elements*. The *items* therefore cover larger units with less detail, saving time in estimating.

elevated gravity tank A water storage on a roof, even a swimming pool, as a backup source of supply for a *sprinkler system* in a tall building.

elevation (1) In *levelling*, the height of a point above sea level *datum*.
(2) A *drawing* of something seen from the side, viewed as if projected on to a vertical plane, used to show the outside walls of a building.

elevator (1) (USA) A passenger *lift* or a service lift.
(2) A *brick elevator*.

elevonics (USA) Elevator electronic *automatic controls*. See **lift**.

elm A dull brown *hardwood* which *warps* badly if not carefully seasoned or *kiln-dried*. It should be kept either wet or dry but not allowed to alternate between the two. It has twisted grain and is even harder to split than *oak*, but in other respects is slightly weaker. It is cut as *burr* for veneers, or used in the solid for *wood-block flooring* and panelling.

embossed carpet Carpet with the *pile* cut to different lengths to make a decorative pattern.

emergency exit indicator lighting A *self-illuminating exit sign*.

emergency shutdown A *mushroom-headed pushbutton*.

emissivity The rate at which heat is given off from a surface compared with a black body, given as a decimal fraction. A dark *roofing* with high emissivity can become 5°C colder than the air on a cold night. Aluminium *reflective insulation* and special coatings have *low emissivity*. Compare **absorptivity**.

employer, client (USA **owner**) In a building *contract* the person who pays the *contractor*, usually the owner of the land and future building, of which the contractor has temporary *possession*.

emulsion paints Paints (also called 'acrylic', 'vinyl', 'latex', 'plastic', or 'water-based') which dry by the evaporation of water. They contain a synthetic *resin* in the form of a *dispersion* of tiny droplets suspended in water. During drying the resin droplets coalesce to form a film, which then undergoes *curing*. Many different resins are used, mostly *copolymers* such as acrylic/vinyl, acrylic/polyvinyl acetate, or acrylic/styrene. Application is by rolling or brushing, and they are suitable for *wet areas*. Though porous enough to allow some drying of new plaster (and allow *efflorescence* to pass without harming them) they should not be applied to a wall that is really wet. Alkalis from cement may discolour their *pigments*, but they resist alkali attack better than *oil paint*. Some have low vapour *permeability*. Many different colours are available, but oil-bound *stainers* or *oil paste* must not be used. They are usually matt, but some have medium sheen. If old emulsion is well cleaned, it can be re-coated without failure, using emulsion or *alkyd paint*.

EN A *European Standard*.

enamel, vitreous e. A very hard permanent finish attached firmly by firing an opaque *glaze* at high temperature, nearly as hot as for *porcelain enamel*. Glazes of *borosilicate glass* are used on cast-iron and pressed-steel articles like baths (usually white) as well as on flat glass. Enamel is very much more resistant to heat and wear and has better adhesion than *enamel paint*, but it chips when struck hard.

enamel paint Paint made with synthetic *resin* binders which dry to a smooth hard film resembling *enamel*. Enamel paints *flow* well but need good *undercoats* since they have little *pigment* and are not opaque. Compared with oil paints they are durable; many are suitable for *forced drying* or *stoving*. Stoved enamel paint is *heat-resistant*.

encapsulation See **asbestos encapsulation**.

encasement, fire protection A layer of *fire-resisting* material enclosing *structural steelwork*. It delays the heating of the steel and prolongs the time it takes to reach its *critical temperature*. Concrete, either cast in place after final plumbing of steelwork, or precast, with holes left for bolts, is the most usual encasement.

Dense concrete with *wrapping fabric* is needed for columns in traffic areas or if any *wall ties* or inserts are to be embedded. Steel floor beams may be encased with *lightweight concrete* or with other materials such as *sprayed mineral insulation*, plasterboard on *cradling*, or intumescent paint (formerly *sprayed asbestos* was used).

encasing (1) *Cladding* or lining that conceals services or structural framing. (2) *Encasement*.

enclosed stage, closing in The completion of a building *enclosure*, which allows *drying-out* and a start on *finishing trades*.

enclosed stair A *protected stair*.

enclosure (1) **envelope** The external weatherproofing of a building, its *cladding* and *roofing*.
(2) Any housing over electrical *accessories*.

encrustation *Furring*.

end bearing length The length of a beam that sits on its end supports. It must give sufficient bearing area.

end event The last event on a *programme* using the *critical path method*, or what is usually called *completion* of building works.

end grain The surface of timber exposed when a tree is felled or when timber is *cross-cut* in any other way. Fixings into end grain are easily pulled out unless made with large screws or *dowelling*. End grain should be sealed by double *priming*, preservative, or the adhesive in a joint.

end joint A *butt joint*.

endlap For *corrugated sheeting*, a *lap* in the direction of its length, usually 150 mm for 15° to 75° roof *pitch*, or 100 mm from 75° to vertical.

endlap joint A woodworking *joint* made by *halving* each member, used mostly for *angle joints*.

engaged column A structural *column* within a non-loadbearing brickwork or blockwork wall. The column and wall usually have the same *flush* surface on at least one side. The column is *attached* when seen on the other side.

engineered brick (USA) A *brick* which, with its mortar joint, measures when laid 200 × 100 × 80 mm (8 × 4 × 3.2 in.), that is five courses in 400 mm (16 in.) of height.

engineering brick (UK) A *clay brick* of high compressive strength and low *absorption*, e.g. *Staffordshire blue bricks* and some reds. Class A bricks are stronger than 70 N/mm² and have an absorption below 4.5%. Class B have a 50 N/mm² compressive strength and maximum 7% water absorption, and are often used for sewer *manholes* (*C*). See BS 3921.

engineering installations The mechanical and electrical *services* in a building, often grouped as a set of *trades*.

English bond A brickwork *bond* in which each *course* is either all *stretchers* or all *headers*. Despite its name, the Romans built walls in this way, as described by Palladio in 1570.

enrichment Decorative relief on the surface of a *moulding*, often representing leaves or a geometric pattern. It can be carved in stone or cast in *fibrous plaster*.

en-suite bathroom A bathroom which can be reached directly from a bedroom.

entry (1) A lobby or other *circulation area* used as a way into a building.
(2) An underground *duct* for an incoming service cable or pipe.

entry phone An *intercom* between the front door and a building occupant, including a circuit for remotely operating an *electric striking plate*.

entry system An electronic *door opener*, usually operated by a cardkey and card reader. Entry systems in *building management systems* can do security checks, call lifts, and switch on lights up to the person's destination, as well as record their presence for payroll purposes.

envelope The *enclosure* of a building.

environmental temperature A *comfort index* calculated from the average of one third of the air temperature (°C) plus two thirds the *mean radiant temperature*.

epdm *Ethylene propylene diene rubber*.

epoxy adhesive Very strong *two-part* glue, mixed in 50–50 proportions from kegs, tubes, or guns with two plungers. It must be put on a dry *substrate* cleaned of all oil and dust. Often a primer is needed, especially on steel.

epoxy mortar, e. resin/sand m. A *resin mortar* for patching shallow holes (3 mm) in concrete surfaces and repairing masonry. Their different rates of *thermal movement* prevent its use in bulk.

epoxy paint Colourful, tough paint made with *epoxy resin* binder, which resists chemicals (acids, alkalis), oils, and solvents, making it a useful *floor paint*. If used outdoors it suffers from *chalking*. The *substrate* needs to be dry and dust-free. Fumes from epoxy paint put on by spray gun are dangerous to breathe.

epoxy resin, epoxide r. An expensive *two-part* synthetic *resin* which cures to a strong, durable, water-resistant solid. Epoxies must be mixed accurately and used at temperatures above 5°C. They set in a time varying from twenty seconds to several minutes, quicker in warmer weather. They should be kept off the skin or washed off without delay, and the fumes avoided. Main uses are in adhesives, paints, and *resin mortar*.

equilibrium moisture content The *moisture content* at which timber neither gains nor loses moisture when subjected to constant humidity and temperature. In practice, joinery timber has *seasoning* to make it *air dry*.

equipment Building *services*, or *contractor's plant* (*C*).

equipment noise Disturbing sounds or *vibration* from building *services*, particularly motor-driven fans or pumps. Noise *transmission* into occupied areas can be reduced by several methods, including *anti-vibration mountings*, heavy walls for *plantrooms*, sealed holes, *flexible duct connectors*, and *attenuators*. A *fan* will emit 17 decibels more noise if its speed is doubled, but moves far more air, saving on space and cost. Similarly *high-velocity* air ducts are noisy but compact. Equipment with *modulated controls* tends to be less noisy than that with *on/off controls*.

equi-potential bonding Earth *bonding*.

equivalent temperature A *comfort index* that takes into account radiant heat, air movement, and dry-bulb (but not wet-bulb) temperature, and so, unlike the American *effective temperature*, equivalent temperature does not take humidity into account and is not therefore used in heavy industry where men sweat. See BS 5643.

erection The placing in its final location on site of any framework, main building *element*, tower, formwork, cladding panel, etc. Temporary bracing against the strongest likely wind should be kept in place until *plumbing* is completed and permanent fixings and bracing are installed.

escalation of contract prices (USA) *Fluctuations*.

escalator A moving stair between floors of shops, air terminals, etc., or in and out of underground stations. It saves people effort and moves crowds much more quickly than stairs or lifts but occupies more space. It needs *builder's work*, such as pits and electrical wiring, and a *fire shutter* if it passes through a *fire floor*. Packaged factory-finished units can be fitted in 48 hours, but need adequate site access and heavy handling equipment. See diagram, p. 154.

escape route The path along which building occupants can travel to reach safety in an emergency, e.g. a corridor or *escape stair*. The first part of the escape route may be unprotected for a distance of 9 m to 45 m, but the second part, the *exit*, must be a *protected route*. Wall finishings must not be a *fire hazard* and signs are required for exits and *dead ends*. Evacuation should be possible in under 2½ minutes.

escape stair, fire e., fire s. Part of an *escape route*, usually a stair completely surrounded by protective fire-resisting walls, reached through a lobby with self-closing fire-resisting doors. Alternatively an outdoor metal stair. For widths and details see BS 5588.

escutcheon A plate round a key hole, usually of metal. It may have a *key drop* and match other *door furniture*. See diagram, p. 128.

espagnolette bolt, shutter b. A verti-

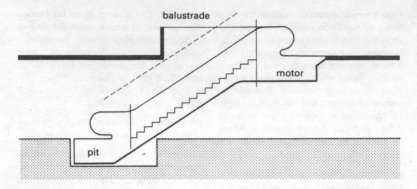

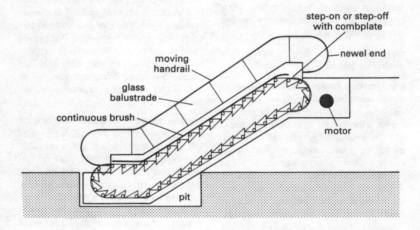

An escalator.

cal bolt for locking double shutters, or casement doors or windows, consisting of a long, turnable rod with off-centre hooks forged on both ends. The bolt is fixed on the *meeting stile* of one door leaf, which closes with the other leaf. A handle at mid-height is used to turn the whole rod, engaging the hooks in the door head and sill. The handle is then slipped behind a cleat to secure the shutter. The espagnolette bolt origi-

nated in Spain and is simple and robust compared with the *Cremona bolt*.

establishment charges *Overheads*.

estimate The probable cost of a building, usually arrived at using *bills of quantities*. A *margin* is added to give the *tender* price. The data in it may also be used for the *financial control* of building work on site.

estimating Determining the probable

cost of proposed building work, arrived at by *taking off* quantities from drawings or using those in *bills of quantities*, then multiplying by the *unit price*. Accurate estimating for use in *tendering* is usually based on the contractor's own *costing* system or a *price index*, while approximate methods rely on *cubing*. The cost of a contractor's estimating department is a major part of *overheads*.

estimator A person employed by a *contractor* to prepare building *estimates*, often a *quantity surveyor* working under the chief estimator.

etch primer A *pretreatment primer* used before painting non-ferrous metal surfaces, such as aluminium or *galvanizing*. It may be *two-part* and contain phosphoric acid to etch the metal and react with its other components, such as chromate *pigment* and butyral resin. One-part types are similar.

ethylene propylene diene rubber (epdm) A tough and weather-resistant synthetic *rubber*. One use is to make flexible preformed *flashings* where vent pipes pass through the roof. The 'm' in epdm means 'monomer', but the material is strictly a 'terpolymer'.

ethylene vinyl acetate (eva) A co-polymer *plastics* that resists severe exposure and low temperatures, used for glazing beads, pan connectors, etc.

eucalyptus, gum tree A tree with well-covered seeds, native to Australia but grown throughout the world in warm temperate climates. It is a *hardwood* with many species. Its timber is usually *kiln-dried* and *reconditioned* before use in buildings.

Eurocode (EC) A *code of practice* published by CEN (see next entry). It may interpret the *Construction Products Directive* and apply in the UK if mentioned in the *Building*

Regulations. ECs can start as a *prestandard* and later be converted to a *European Standard*.

European Committee for Standardization (CEN) (French, Comité européen de normalisation) The body that coordinates *European Standards* and *Eurocodes* working with the national *standards* bodies of eighteen countries in the European Community (EC) and European Free Trade Association. The aims of CEN are to liberate the buying of materials within the EC, and to help Europe as a whole to compete in world trade.

European Standard, Euronorm (EN) A *standard* agreed by all members of CEN (see above) and existing in three official versions, English, French, and German. It may start as a *harmonization document* or *prestandard*. Early ENs had separate numbers from their *AFNOR*, BS, or DIN equivalents, but many now have 'dual numbers' (BS EN or DIN EN or NF EN, plus figures), giving them status as a national standard ('norme' is the French and 'Norm' the German for 'standard').

evaporation (1) Loss of *vapour* from a liquid. The vapour removes *latent heat* during drying, with resultant cooling.
(2) *Solvent release.*

event The start or finish of an *activity* on a *programme* worked out by the *critical path method*. The earliest and latest event times are shown on an *arrow diagram*, but an event may have *lead time* in front of it. The beginning event is project *commencement* and the end event is *completion*.

even-textured, even-grained Of timber, having little variation in the size of the wood elements, e.g. timber in which there is little contrast between *springwood* and *summerwood*.

excavation (1) The work of digging holes in the ground to reach a *foundation* for a *service* or *structure*. It is part of the *groundworks*. Excavation is often done in two stages, *bulk excavations*, then *isolated excavations* for foundations and *trenches*. Excavators are used for most work, although hand excavation may be needed in congested spaces, round buried *services* or for *bottoming* and trimming the sides. Excavation produces *spoil*.
(2) A pit or trench dug in the ground. It may need *earthwork support*.

excavator A large machine with a digging *bucket* (*C*), usually worked by *hydraulic* rams.

excess current *Overcurrent*.

excess excavation *Excavation* deeper than required for the *substructure* and *working space* of a building. Under *foundations* it is usually made good with *dry lean concrete*.

excess spoil After completion of *groundworks*, any material from *excavations* that has not been used as *backfill*. Disposal can be on-site by *spread and level*, or off-site, usually by carting to a tip.

excess voltage *Overvoltage*.

execution (1) The signing of an *agreement* (execution by hand) or marking it with a company seal (execution under seal), thus formally recognizing the existence of a binding *contract*.
(2) The *performance* of a contract by carrying out the work.

exfoliated vermiculite An *ultralightweight aggregate* made from *vermiculite* which has been heated and thus permanently expanded to many times its original size. The bulk density is usually 65 to 200 kg/m³.

exfoliation The scaling of stone, caused by the weather.

exhaust Air rejected outdoors after it has been extracted from a room.

exhaust shaft A vertical *duct* to remove air.

exhaust system An *extract system*.

exit A corridor or other *circulation area* and the *exit door* for a fire *escape route*. Exit widths depend on the building *occupancy*. Details are given by *Building Regulations*.

exit door The last door on an *exit* route. It opens outwards and must not be locked while people are inside, only latched by *panic hardware*. In shops or cinemas, security may require an *alarmed door*.

exit sign A *safety sign* with 'EXIT' in white letters on a green background, either a stick-on plastics label or a *self-illuminating* unit. It may be accompanied by matching direction arrows.

expanded clay, bloated c., e. shale, e. slate (USA **haydite**) Hard porous pellets made by burning clay or similar materials in a cement kiln until it partly melts and is expanded by trapped air. Their bulk density is 320 to 1040 kg/m³. When used as *lightweight aggregate* they make quite strong concrete.

expanded perlite *Ultra-lightweight aggregate* made by heating *perlite* to form glassy spherical particles the size of small peas. It is used in insulating plaster and concrete, and its bulk density ranges from 80 to 240 kg/m³.

expanded plastics, foamed p. Cellular plastics foamed with heavy gases, formerly *chlorofluorocarbons*, mainly used for thermal *insulation* and supplied as a board, foamed in-situ, or sprayed, e.g. *expanded polystyrene*, *phenolic foams, polyurethane, polyvinyl chloride*, and *urea-formaldehyde*.

expanded polystyrene (EPS), PS foam A cheap cellular plastics made by foaming *polystyrene*, usually supplied as extruded board ('bead board') or beads for *loose fill* and weighing less than any other heat *insulation*, 16 kg/m³. Its maximum recommended temperature of use, 70°C, is low, but it is strong for so light a material, having a compressive strength of 207 kPa. Being moisture-resistant it is used for *inverted roofs* and *external insulation*. Wall and ceiling linings made of it melt at about 80°C, but are not a severe *fire hazard* if fully bedded with adhesive to a substance that will not burn. Such linings should be left unpainted or painted only with matt-finish flame-retardant paint or *emulsion paint*. Different grades are available for roof decks and under *floating floors*, while standard grade is used inside *composite* board and as pocket *formers* in concrete. It is softened by solvents such as *white spirit* and disfigured by boiling water.

expanded PVC, PVC foam An expensive *non-flammable* insulating material.

expanded vermiculite *Exfoliated vermiculite*.

expanding bit, expansion bit A wood-drilling *bit* with a single cutter which can be adjusted to varying radii.

expanding plug A *drain plug* for testing pipes.

expansion Increase in size. Water expands appreciably on heating and this can be seen in *domestic hot-water* systems (see entries below). Expansion of building materials that alternates with contraction (*cyclic movement*) is part of the wider subject of *movement*. The *moisture expansion* of ceramics and the *sulphate expansion* of mortar are irreversible.

expansion cistern An open *feed cistern*.

expansion joint (EJ) A *movement joint* mainly to allow *expansion*.

expansion pipe, vent p. In *open-vented* domestic hot-water systems, a pipe leading from the *cylinder* to a point over the feed and *expansion cistern* so that, if the boiler boils, the steam (or water) will discharge harmlessly into it. The boiling water expelled from the boiler may eventually fill the tank and flow out through the *warning pipe*.

expansion sleeve A *sleeve*.

expansion strip, e. tape, edge isolation, insulating s., isolating s. Resilient *insulation* used to fill the gap between a *partition* and a structural wall or column, or to separate a *glass block* from the structure and thus prevent damage to the glass. See **sealing strip**.

expansion vessel A closed *diaphragm tank* for the expansion of hot water, used in central-heating *sealed systems* and hot-water *unvented systems*.

explosive fixing A *shotfired fixing*.

export Carting off the site of *excess spoil* not needed for *groundworks*.

exposed aggregate finish A rough-textured surface on concrete made by removing a thin layer of cement and *fines* to reveal the coarse *aggregate*. Methods include the use of *retarders* on formwork, with early *stripping* then light brushing and spraying with water, washing with acid, *water blasting*, etc.

exposed metallic part, e. conductive p., earthy metalwork Any metal building component that could come into contact with the electrical system, such as pipes, ducts, door frames, baths, railings, cable tray, steelwork, etc. To protect against electric shock they are interconnected by earth *bonding*.

exposure The effects of wind and weather, rain, storms, frost, *freeze–thaw cycles*, sunlight, and its *ultraviolet* rays, etc., that can damage the roof and outside walls, or reduce their *durability* and increase heat loss. Mild, moderate, or severe exposure can be judged from the *driving rain index*. Very severe or extreme exposure is usually restricted to seaside or industrial situations.

extended price, extension The sum of money in a *priced bill* shown in the right-hand money column opposite each *item*, calculated by multiplying the *quantity* by the *unit price*.

extender (1) inert pigment A white powder added to a paint, often a crystalline mineral with low *hiding power*. It is added to adjust the paint's film-forming and working properties such as *thixotropy* (C). Most extenders are crushed as finely as *pigments*. The commonest are asbestine, *barytes*, *blanc fixe*, *diatomite*, kaolin, mica, silica, and *whiting*.
(2) A substance such as wood flour added to *glue* to dilute it and thus increase its spreading capacity.

extension (1) The enlargement of an existing building. Factory end walls are often built with future extensions in mind.
(2) An *extended price*.

extension bolt, monkey-tail b. A *barrel bolt* with a long tail so that it can be operated without stretching up or bending down.

extension ladder A telescopic *ladder* with a sliding extension which can be locked at a particular height. Domestic extension ladders are 5.20 to 6.40 m long, industrial types up to 10.70 m or more. They are usually made of aluminium or fibreglass.

extension of time Extra time added to the *time for completion* to give a later completion date because of *inclement weather*, *variations*, or for other reasons usually listed in the *conditions of contract*.

exterior paint, external p. A protective and decorative coating for use outside. Different types are made for wood, metal, masonry, etc. They have better resistance to weather and are usually more expensive than *interior paint*.

exterior plywood *Plywood* in which the *adhesive* (not the wood) is moisture resistant.

external angle A *special* brick, bent in the middle, for a *squint corner*.

external glazing *Glazing* on the outside wall of a building. It is usually *inside glazing*.

external insulation Foam *insulation* on the outside of a wall or roof to keep the building warm. It also ensures a more even temperature indoors because of the *thermal capacity* of the building *fabric*. It usually consists of boards of rigid and moisture-resistant *expanded polystyrene*, covered with *render* or any protective facing. A *vapour barrier* on the inside face of the wall or ceiling, *ventilation* of the insulation to the outside and a *fire barrier* are needed. Examples are wall 'overcladding' and *inverted roofs*.

external leaf The facing bricks forming the outer skin of a *cavity wall*.

externally reflected component In

the *daylight factor method* the light reflected towards a window from the ground or obstructions outside, commonly about one tenth the brightness of the *sky component*.

external plumbing *Rainwater plumbing*.

external rendering Cement *rendering*, smooth or *textured*, or *stucco*.

external wall A wall with one face exposed to the weather or the earth.

external works Construction works outside a building, a *trade* that groups roadbuilding, drainage, landscaping, street furniture, fencing, and the like.

extinguishing medium, extinctant Water, inert gas, vaporizing liquid, dry powder, or foam used in fixed or portable *fire extinguishers*. Water is the most common, being cheap, efficient, and readily available, but it is unsuitable for electrical and *flammable liquid* fires. Each medium has its uses and its own hazards in use; details are given in BS 5306.

extra, e. work Work which was not included in the original *contract* and has to be ordered by the *architect* or engineer in writing, generally by a *variation order*. It differs from *extra-over*.

extract air Air taken from a room for exhaust or recirculation.

extract fan, extractor f., exhaust f. A compact fan unit in a housing, fitted in a window, wall, or ceiling as a simple type of *mechanical ventilation*.

extractor hood A stainless-steel panel below a ceiling, down to near head height, to guide cooking fumes and vapour into the *extract system*.

extract system A system of *mechanical ventilation* that sucks air out of rooms through ductwork and outlets of different kinds, such as slots,

hoods, or grilles, and discharges it to the outside air, often with *heat recovery*.

extra depth The downward extension of an excavation to reach *good ground*, when old fill or poor ground that would not pass a *foundation inspection* is found. This deeper hole may be filled with mass concrete up to the original *formation* level, or the *contractual foundations* are lowered, or sometimes an extra basement is added.

extra-low voltage Electric current at less than 50 volts.

extra-over In *bills of quantities* many items have a *description* that is partly the same as others. To avoid repetitions, therefore, only the parts that are different are described. Thus 'breaking up old foundations' is commonly given as extra-over normal excavation. The estimator can re-use parts of the price already worked out.

extra space *Working space*.

extruded brick A *wirecut brick*.

extruded chipboard Wood *chipboard* made by squeezing through a slot, sometimes containing *voids*. It is not as strong as *platen-pressed* chipboard.

extruded gasket A preformed glazing *gasket*.

extruded section, extrusion Something made by *extrusion* (*C*), usually of plastics, rubber, or soft metal.

exudation Gummy resin droplets on the surface of *softwoods*, usually worst at knots or after warming.

eye (1) An *access cover*.
(2) An opening formed in a metal member, for example the eye of a hammer head into which the handle fits.

eyebrow dormer A *shed dormer* with

eyebrow dormer

a roof that is curved at each end to blend into the general roof with no sharp angles. An eyebrow dormer has no *cheeks*, and its roof is only slightly lifted above the main roof slope, with a central upward curve. It resembles an *internal dormer* apart from the curved roof.

F

fabric (1) The loadbearing *structure* of a building, its brick, concrete, or timber *carcase* or *frame*, as well as its *infill* panels. The term is often used to include cladding, windows, patent glazing, roof coverings or doors when they are attached to, or part of, the fabric, but not finishings or joinery.

(2) **mesh** Stiff steel wires welded in rows at right angles to make a grid. The wire can be as-drawn, with a smooth surface, or have bumps like a *deformed bar* (*C*). Different grades are made for structural concrete slabs, screeds, toppings for composite floors, house *footings*, and *wrapping* steelwork. See BS 4483.

fabric roll filter, roll-on f. A main *fresh-air filter* for an *air-conditioning* plant, made up of a long strip of fabric between two rollers. It is automatically advanced one complete panel at a time so that the partly blocked area does not overload the clean new area of fabric.

façade A front wall with formal decoration or any outside wall with high-class *cladding*.

façade panel A *cladding panel* made for a *façade*.

façade retention The temporary shoring of a valuable *façade* with a steelwork or scaffolding frame during partial demolition and rebuilding. The old floors and roof can be replaced with new ones. The frame is often wide enough to fit *site accommodation* and is protected from traffic damage. It may allow drive-through delivery of *readymix* to an underground *concrete pump*. See diagram, p. 162.

face (1) The visible surface of a building component (brick, door), the opposite of *back*.

(2) The side of *building board* with the best appearance. The face of plywood or blockboard is identified by the *face mark*. For *gypsum wallboard* an ivory colour indicates a face that can be painted without plastering. Grey indicates a face to be plastered. On *hardboard* one face is usually smooth and one textured.

(3) Either of the two broadest sides of a rectangular timber, or any side of a square timber.

(4) The front wall of a building, which must be behind the *building line*.

(5) A working surface of a tool, such as the flat face of a *hammer* or the front cutting edge of a *saw* tooth.

face bedding, edge b. Stone laid so that the *natural bed* is vertical. The stone is liable to flake away, as on the Houses of Parliament in London. Only arch stones are correctly laid like this.

face brick (USA) A *facing brick*.

faced wall A *solid wall* in which the *facing bricks* and *backing* bricks are bonded to act together under load, with no *capillary break* between them.

face edge, working e. (USA **work e.**) The first edge of a piece of wood to be straightened in *joinery* work, from which the other edges are measured.

face joint In brickwork the mortar of a bed or perpend *joint*, seen on the wall face, usually finished by *jointing*.

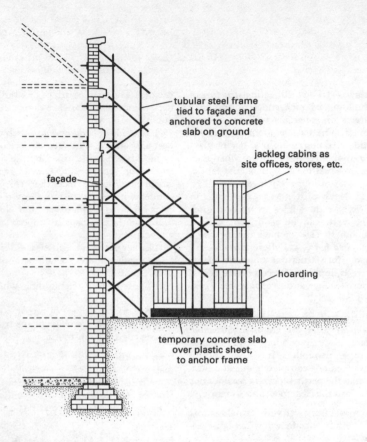

tubular steel frame
tied to façade and
anchored to concrete
slab on ground

jackleg cabins as
site offices, stores, etc.

façade

hoarding

temporary concrete slab
over plastic sheet,
to anchor frame

Façade retention during the demolition and rebuilding of a building interior.

facelift External *refurbishment*.

face mark A pencil X-mark on joinery or any similar way of identifying the *face* of *building board*, stone, etc. Other surfaces are trued from the face.

face mix A mixture of cement and crushed stone, usually 25 mm thick, placed on the surface of a mould for *cast stone* and backed with a cheaper, stronger mix, which is poured immedi-ately afterwards and therefore bonded to it.

face putty, front p. The triangular fillet of traditional glazier's *putty* on the outside of a window *pane*. See **back putty**.

facework, fair-face brickwork *Brickwork* that is laid to a high stan-dard of workmanship, to be of attrac-tive appearance without applied finish, usually in *facing bricks*. The work can include building a *sample panel*, blending different bricks, avoiding chips, achieving a uniform thickness of *bed joints*, *keeping the perpends*, avoiding mortar staining, and covering to keep rainwater out of the holes in *perforated bricks*. A wall of clay bricks *half-brick* thick can be built fair-face on only one side, because of inaccuracies or markings, but concrete *blocks* and *calcium silicate bricks* can be fair-faced both sides.

facing (1) Fixed indoor trim and linings, such as a *casing* or *architrave*, or a large area of wall *finishing*. The plywood skin on a door leaf is a facing.
   (2) External *cladding*, or a *facing brick*.

facing brick, facings (USA **face b.**) Durable bricks for external *cladding* of pleasing but not necessarily uniform colour and texture. They can be smooth or *textured*. Clay facings may be self-coloured, surface-treated with pigments or coloured by *flashing*. Concrete facings are often split to reveal the aggregate. See **facework**.

facing hammer (USA) A *bush hammer*.

factory-finished *Self-finished*.

factory-mixed *Premixed* or *ready-mixed* in factory conditions. The factory mixing of products allows tighter control of quality than site mixing.

Faculty of Building See *C*.

faience *Terracotta* that is fired twice, once to vitrify the *body* fully and then again with *glaze*. The body is highly porous but has low shrinkage during firing and so can be used for very large clay tiles. Named from Faenza in Italy.

failure The breakdown of a building product in service. See **defect**. See also *C*.

fair cutting The cutting of facing brickwork, always assumed to be *half-brick* thick (102.5 mm), therefore measured by length and not as an area. It can be done with an abrasive wheel, or less usually with a *bolster*, *trowel*, or *scutch* as in *rough cutting*.

fair-face brickwork *Facework*.

fair-face concrete Concrete with a *direct finish* of uniform colour, substantially free from surface defects.

fall A slope, e.g. of a *flat roof*, *gutter*, or *screed*, that allows water to flow away without *ponding*.

fall pipe A *downpipe*.

false body The high viscosity or *thixotropy* (*C*) of a *non-drip paint* when undisturbed. It is reduced by stirring.

false ceiling, drop c. A second ceiling below an existing ceiling, or a panel lower than the general area, usually carried on *ceiling joists* and lined with traditional materials such as plasterboard. It reduces the height of a room and may provide sound and heat insulation. It is usually not a *suspended ceiling*. See **isolated ceiling**.

false exit (USA) A *dead end*.

fan (1) **blower** A device driven by an electric motor with rotating *blades* that move air for *ventilation*. Centrifugal fans are mostly used for *air conditioning* and propeller fans for *mechanical ventilation*. Fans are usually stopped in a fire, except those for *pressurized escape routes* or for *smoke extract*.
   (2) A *debris-collection fan*.

fan coil unit An *air-handling unit* for air conditioning one room. It has a *coil* to heat or cool air, which is recirculated to the room by a built-in fan. Fresh air is supplied to the unit through ductwork. The unit has a filter and is supplied with heated or chilled water

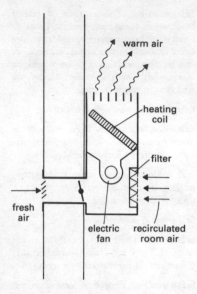

warm air

heating coil

filter

fresh air

electric fan

recirculated room air

A fan convector.

from a central plant. It is usually mounted in the ceiling and controlled by a local thermostat.

fan convector A domestic *warm-air heater*, usually fed with hot water pumped from a boiler. It has an electric fan that sucks cool air in at the bottom through a filter and heating *coil*, and blows warm air back into the room. Any fan convector, because of its fan, has a much higher heat output than a conventional radiator of the same size, although the electrical power demand of the fan is usually well below 100 watts. Though similar, *fan coil units* or *unit heaters* are larger. See diagram.

fang A *fishtail fixing lug*.

fanlight, transom window A window over a door within the main door frame, originally semicircular, now of any shape or replaced by an opaque overpanel.

fascia (1) **f. board, eaves f.** A board, or similar *trim*, set on edge along the *eaves*, to cover the rafter ends and carry the *gutter*. The fascia board's top edge is level with the top edge of the *tilting fillet* and may act as one. For a *cold roof* there may be an *over-fascia vent*. If ventilation is through the eaves *soffit*, the *underlay* is draped over the fascia and into the *eaves gutter*. See diagram, p. 147.

(2) The panel over a shop front, usually carrying the shop name.

fastener (1) Any device for joining units to each other. General-purpose fasteners such as *screws, bolts, nails,* and *staples* are usually of mild steel, sometimes protected against *corrosion*, or of stainless steel. See **fixing**. See also diagram.

(2) Door or window *furniture* that secures a *leaf* or *sash*.

fast to light Of a *colour*, unaffected by light or *ultraviolet*, not fading or discolouring.

fast track procedures Acceleration of the lengthy process of building by making only the decisions needed for immediate work on site. It requires close cooperation between the client, architect, and contractor, as well as careful programming and control. Some people feel that it produces too many arguments.

fat board A bricklayer's board for carrying mortar during *pointing*.

fat lime Lime with a high *volume yield* and good *workability*, usually *high-calcium lime*.

fat mix, rich mix A mix of mortar or plaster with a high *binder* content.

fattening, thickening An increase in the *viscosity* (*C*) of paint during storage, not sufficient to make it unusable. 'Feeding' is severe fattening.

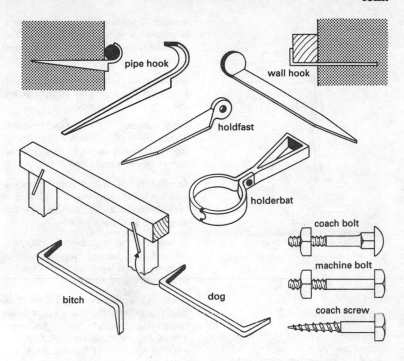

pipe hook

wall hook

holdfast

holderbat

coach bolt

machine bolt

coach screw

bitch

dog

Heavy fasteners.

fattening up, maturing Increasing the *plasticity* of *hydrated lime* by soaking in water, at least overnight.

faucet (1) (USA) A small *tap*, for example at a household sink.
(2) (Scotland) A *socket*.

fault tracing Finding the location of an electrical fault such as a broken wire (open circuit) or a short circuit.

feather A very thin piece of wood, such as a *loose tongue*.

feather-edged board *Weatherboards* tapered from 16 to 6 mm thick. See diagram, p. 166.

feather-edge rule A plasterer's metal

rule with a thin front edge and a thick, stiff back edge. It is used for *screeding* plaster, *scraped-finish* render, etc.

feathering, feather edge An edge that is thinned down to nothing (or nearly), possible with timber or *levelling compound* but not with *mortar*.

feed The supply of water to a boiler, oil to a burner, etc.

feed cistern, f. and expansion c. A *cistern* that supplies cold water to an *open-vented* boiler or cylinder, as hot water is drawn off at the taps, also allowing *expansion*.

feint In *supported sheetmetal roofing* a slight inwards fold along the edge of a

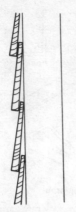

Feather-edged boards used as weatherboarding outside a building. They can also be used as fencing, with the boards upright.

capping or *flashing* to form a *capillary break*.

felt *Bitumen felt.*

felt-and-gravel roof (USA) Roofing of *bitumen felt* covered by a layer of gravel, as *ballast*, as solar protection, and to improve the *U-value*.

felt nail A *clout nail*.

feltwork Layers of *bitumen felt* bonded together for *built-up roofing*, to form *box gutters* or similar weatherproofing.

fence (1) A guide for timber on a saw bench, router, or other *woodworking machine*, to hold the cutter at a constant distance as the work piece is pushed through, when making a cut, groove, or moulding parallel to the edge.

(2) An enclosure, such as a *palisade* or an *anti-intruder chain-link fence*, or a *guard* fixed round a moving machine to prevent injuries.

fence post A main upright that supports a *fence*.

fencing *Fences* in the *external works*, or their erection.

fender A *baulk* of timber, or sections of precast concrete barrier, laid in the street to protect a *scaffold* from damage by traffic.

fender wall A *dwarf wall* carrying the hearth slab of a ground-floor fireplace and sometimes also *joists*.

fenestration The architectural arrangement of windows or any opening that admits *daylight* through the outer walls of a building.

ferrule (1) A metal band round a wooden handle to prevent it splitting.

(2) Generally a short *tubular* pipe fitting, such as a *sleeve piece*.

festooned cable A flexible electrical cable draped between insulators that slide on a steel wire, for power supply to travelling equipment.

fettle (1) To verify that a casting is free from flaws (by hanging it from chains and striking it with a hammer) and to remove roughness and *flash*.

(2) An extension of the first sense, the finishing work in any trade.

fibre (USA **fiber**) Strands of glass, steel, plastics, or wood, formerly hair or asbestos, mainly to reinforce hard, brittle materials.

fibreboard, fibre building board Many different types of *building board* made by hot-pressing a mass of wood or other vegetable fibres. Density is the main guide to their properties. They are grouped as *hardboard*, *medium board*, and *insulating board*, or the slightly different *medium-density fibreboard*, depending on how much they are compressed during manufacture. Their strength comes from the felting of woody fibres, bonded by natural wood *lignin* resins, not by added cement or adhesive. Fibreboards have in the past been the cause of rapid *flame spread*, but suitable *flame-retard-*

ant treatment can improve this property up to Class 1 (very low flame spread) to BS 476. They are made to BS 5669.

fibre cement board Usually an *asbestos-free board*, formerly *asbestos cement sheet*.

fibreglass (1) Glass *fibre*.
(2) *Glassfibre-reinforced polyester*.

fibre-reinforced concrete (FRC) Thin panels, shaped for stiffness, in concrete reinforced with fibres of *glass*, steel, or an organic material such as *polypropylene*. Generally for non-structural use, e.g. cladding panels.

fibre saturation point The *moisture content* (about 30%) above which the strength and dimensions of timber remain roughly constant. Below it the strength increases and the wood shrinks progressively as the moisture content falls.

fibrous concrete *Concrete* containing fine fibrous material such as sawdust. It is used for its lightness or for its nailability.

fibrous plaster *Plaster of Paris* shaped in-situ or by casting it in vinyl or rubber moulds, for decorative work such as *enrichments* or for dead-flat *plaster-tile* panels for suspended ceilings. The cast can be reinforced with canvas, *glassfibre*, or *lathing*. See BS 5492.

field order (USA) A *site instruction*.

field splice A *splice* in *structural steelwork* bolted on site, such as a *column splice*, rather than made in a *shop*.

field superintendent (USA) A *site engineer*.

figure The natural markings of timber, including both *grain* and colour. Figure usually adds to the beauty and value of *veneer* but rarely indicates strength. Types include: *birdseye, burr, mottle,* wavy.

figured dimension A *dimension* written on a drawing, not *scaled*.

fill Any material placed on the ground to raise its level, usually laid after topsoil has been stripped. See **backfill**. See also **hardcore** (C).

filled joint A joint completely filled with *sealant*, suitable for 'no-movement' joints. As *movement joints*, filled joints suffer overstretching and failure.

filler (1) A creamy paste spread thinly over a rough area of *substrate* for painting as part of the surface *preparation*. After hardening it is soft enough for *rubbing down*. Deep holes need *stopping* or *high-build filler*.
(2) A substance added to *plastics*, including paints, to vary their properties and sometimes also to lower their cost, resembling an *extender* in glues.
(3) A piece used to close a gap in joincry, or as a *make-up* piece.

fillet (1) A narrow strip, triangular in cross-section, at the angle between two surfaces. A timber fillet is used as a *chamfer* to strengthen *formwork* or as a shelf *cleat*, but fillets can also be formed in asphalt, cement mortar, etc.
(2) A small square wooden *moulding*.
(3) A *water check*.

fillet chisel A mason's tool for working stone to fine limits.

fillet saw A small hand saw used by a *joiner* to cut timber mouldings.

filling-in piece A timber such as a *jack rafter* or *trimmer* which is shorter than its neighbours.

filling knife A knife like a *stopping knife* but having a thinner, springier blade, used for laying on fine *knifing filler*.

fillister, sash f. (USA) A *rebate* along the side of a *glazing bar* to receive glass and putty.

film (1) The dried *paint* of one or several *coats*. Film thickness or *build* is measured as *dry-film thickness*. See **film-forming agent**.
(2) Plastics sheet stuck on outside window panes for insulation, privacy, or safety. A film made of an ultra-thin layer of metal between two polyethylene sheets reflects sunlight and has *low emissivity*. See also **solar-control glazing**.
(3) Bright-coloured *peelable* plastics used for *protection* of factory-finished surfaces, e.g. polished stainless steel.

film-faced plywood Heavy-duty plywood with a smooth coating of *phenolic resin* on the face and edges, usually deep purple-brown, used as *formply*.

film-forming agent The part of a paint *medium* that forms a tenacious, strong, continuous, flexible *film*, mainly due to its *binders*. Film-forming agents include *alkyd resins*, many *polymers* and *linseed oil*.

film glue A thin, solid sheet of *phenol-formaldehyde* resin laid between thin, costly, decorative face *veneer* and the cheap, stronger, thicker, backing veneer. Film glue is a *thermoset* and applied by *hot pressing*. Since it neither wets the veneer nor causes it to expand, it is the only glue possible with very thin veneers less than 1 mm thick.

filter A screen (the *medium*) which collects dust particles (the filtrate) from a fluid stream. The filter may be an *element*, a panel, or a *fabric roll* inside a case. As it needs to be regularly cleaned or replaced, *working space* should be provided. In air conditioning, the first filter is a fresh-air *impingement* unit, followed by a *high-efficiency particulate air filter* or an *electrostatic* filter for finer particles. Activated carbon and 'absolute' filters for specialized uses can even remove gases.

fin (1) A thin metal plate projecting from the pipe of a heating or cooling *coil* to increase the surface area and improve its efficiency as a *heat exchanger*.
(2) Concrete that has escaped into a joint between *forms* and projects from the surface after *stripping*. Fins on *direct-finish concrete* must be prevented; on other surfaces they may be removed with a *grinder*.

final account The total cost of a job, based on the *contract sum* but adjusted for the cost of *variations* in the work or *fluctuations* in costs. It is worked out by the *contractor's surveyor* and agreed with the client's representative often after building work is completed.

final certificate The *certificate* issued by the architect at *final completion*. It entitles the builder to payment of the *final account* including the second half of the *retention* sum.

final coat, finish c. The last coat of a *finishing*. Final coats of *paint* are usually *gloss*. Final coats of *plaster* are either *finish plaster* or *board-finish plaster*.

final completion The end of the *defects liability period*, usually six months after *practical completion* of a building. When all remaining *defects* have been made good by the builder, the *final certificate* can be issued.

final fixings *Second fixings*.

final inspection An inspection by the architect at *final completion*.

final sub-circuit A *branch circuit*.

financial control Checking by the contractor that a contract is not losing money, usually the work of a *contractor's surveyor*. It includes spot checks

A finger joint.

for the *costing* of isolated jobs of work against unit prices in the *estimate*, calculating the value of work completed against each *monthly instalment*, and the settlement of the *final account*.

fines The fine particles in a mix, e.g. in concrete the cement and fine *aggregate*.

fine stuff *Finish plaster*.

fine-textured wood Wood such as birch or maple, with small pores which need no *filler* before varnishing. The opposite of *coarse-textured wood*.

finger joint A strong, compact factory-made *lengthening joint* between ends machined with matching zig-zag fingers, bonded with quick-setting *radio-frequency-heated* adhesive. It is used on softwoods in *glued-laminated* beams. See diagram.

finger plate A plate fixed on the face of a door to protect it from damage and finger marks. See diagram, p. 128.

fingers A plasterer's trowel with wooden spikes for *keying* undercoats.

finial A decorative point on top of an upright, such as a *gable post* or stair *newel post*, either pointed or round a *pommel*, or carved with *enrichments*.

fining off Applying the *finishing coat* of external rendering.

finish (1) Either a *finishing* coating itself or more strictly the surface condition of a material, its appearance (colour, texture), and resistance to wear, damage, slipping, etc., which result from workmanship and surface treatment. For example: plastered surfaces may be trowelled, stainless steel is often polished, floorings should be hard-wearing, paint may be polychromatic, gloss, textured, etc.

(2) See **completion**.

finish coat A *final coat*.

finished-floor level (FFL) The top surface of a floor *screed* or *flooring*, the surface from which all levels in a room are taken, usually excluding *floor covering*.

finisher (1) A *skilled* concrete worker who produces a *finish*, usually smooth.

(2) A steel *trowel*.

finish hardware (USA) Builder's *hardware* which is seen and therefore given a good finish, such as door or window *furniture*, clothes hooks, and so on, as opposed to *rough hardware*.

finishing, finish A layer of decorative material, or any *facing* over the building *fabric*. It can be put on as a *coating*, as *paintwork*, trowelled on thick, as *plasterwork* and floor *screeds*, or fixed as individual pieces, as *tiling*, slabs, boards, blocks, etc. Finishings may need to be made from the top down (ceiling, walls, floor), then covered with *protection*.

finishing coat, fining c., setting c., skimming c., white c. A *final coat* in *solid plastering*, about 3 mm thick, usually of fine *finish plaster*.

finishings schedule A list of all rooms in a building with details of floor, wall, and ceiling finishings. It is usually a *contract document*.

finishing trades Mainly *joinery, plasterwork, tiling, paintwork*, and *flooring*. The completed work may need *protection* from damage or the weather.

finish nail (USA) A *bullet-head nail*.

finish plaster, fine stuff A lightweight *gypsum plaster* used for *finishing coats*. It can be trowelled to a smooth and fairly hard surface and includes *one-coat* and *projection plasters*.

fir A loose term which should properly be confined to the Abies family, that is *whitewood, silver fir*, and others. It is often used for *redwood* and *Douglas fir*.

fire A blaze in a building can grow quickly and be difficult to put out. Most fires in buildings occur in homes, often starting in the kitchen. Few fires spread beyond one room, but those in two-storey houses cause the most deaths, usually from *smoke* outside the room of origin, after *flashover*. Death in the room of origin is usually due to heat or burns. In Britain (apart from the unpopular Norman 'couvre feu' or curfew), laws to reduce building fires date from the London Assise of 1189. It banned thatch and required houses to be of stone and party walls to be of given height and thickness within the city. After the Great Fire, the Rebuilding Act 1667 brought in regulations on fire walls, chimneys, hearths, roofs, etc. Today the *Building Regulations*

insist on the protection of buildings and their occupants by fire-resisting construction and division into *compartments*, with *escape* routes, control of hazards, and fire fighting. See following entries.

fire alarm A bell or other *sounder* to warn people of a fire, so that they can leave the building and alert the fire brigade. Fire alarms may be set off from a *call point*, either a manual *break-glass unit* or an automatic *fire detector*. Instead of bells, computers can synthesize voice messages on fire location and recommended action. Rules for automatic fire alarms are published by the *Loss Prevention Council*. See BS 5839.

fire-alarm indicator An *indicator panel* that shows the zone where a fire is burning, or, if it has *addressable* fire detectors, the exact place. It may have a computer memory, to pinpoint problem rooms and faulty equipment.

fireback The back wall of a *fireplace*.

fire barrier, cavity b., roof screen A panel or mattress of non-combustible material (mineral fibre, brickwork, plaster) to cut off the path of smoke and flame. It is used behind *suspended ceilings* and inside the roof space above a *compartment* wall, or along the edges of a *cavity wall*. Fire barriers may make *ventilation* of the building fabric difficult. A *fire stop* is similar, but usually smaller.

fire block (USA) A *fire barrier* inside wooden floors or walls.

fire booster A *booster* for fire fighting, usually an automatic electric or diesel-driven pump supplied from a ground-level *fire reserve* to increase water pressure in a *fire riser* that goes to *sprinklers*.

firebox The space inside a heater, boiler, or furnace where fuel is burnt.

fire break A fire-resisting wall between *compartments* or an open space between buildings to reduce the risk of fire spreading to an amount acceptable to insurance companies or the law.

firebrick A *brick* made from any clay difficult to fuse, generally one with a high content of quartz. It is used in the hot parts of fireplaces and boilers for temperatures up to 1500° or 1600°C, as *refractory linings* (*C*).

fire calculations The use of mathematical analysis instead of *fire testing* when codes or regulations give no simple answer. They can assess *fire severity* and *fire resistance* in complex *fire venting* and *smoke control*.

fire cell A *compartment*.

fire cement Usually *high-alumina cement* (*C*).

fire certificate Under the UK *Fire Precautions Act* non-domestic buildings such as hotels, offices, shops, and factories must be inspected and certified by the fire brigade. The brigade informs the owner of any 'steps to be taken' to make the building comply with the Act. However, if the building use is unchanged, the steps demanded may be limited by the Building Regulations at the time of construction. The certificate specifies the means of escape, alarms, and extinguishing equipment, and any other fire precautions, such as maximum numbers of occupants, staff training, and equipment maintenance. See Guides to the Fire Precautions Act (HMSO).

fire check door A timber flush door, no longer manufactured or used as a *fire door* as it is not fully fire-resisting.

fireclay Clay free of alkalis but rich in aluminium silicate, aluminium oxide, and silica. Fired at about 1200°C, it

can make *firebrick, terracotta,* or other *ceramic*.

fire compartmentation The division of a building into *compartments*.

fire damper A fire-resisting *damper* that automatically prevents flow through an air-conditioning *duct*. It is fitted where the duct passes through a *fire wall* or floor to stop the spread of fire and smoke. Plate-type fire dampers are operated directly by *fusible link* or from *fire detectors*, but they can disturb air flow and need regular maintenance, unlike *honeycomb fire dampers*.

fire detectors Electrical devices to give early warning of a fire and set off a *fire alarm* or alert the fire brigade, close fire dampers, open fire vents, release fire doors, start *fire-extinguishing* and *smoke-control* systems, etc. They react to heat, smoke, flame, and combustion gas, and are often fixed to the ceiling, but expert choice is needed. See EN 54 or BS 5445. See also **addressable system, analogue detector**.

fire development Fire in a building usually starts from a small source, such as a cigarette or an electrical short circuit. There is a slow period of ignition, then combustion. Once flames reach the ceiling, the room heats up quickly. The speed of growth of the fire increases the hotter it gets. It can go from 100°C to *flashover* in less than a minute. Oxygen in the air can get used up and a room fill with toxic *smoke*. For this reason people should get out, stay out, and call the fire brigade.

fire door (1) **firebreak d.** A door and frame of known *fire-resistance grading* used in a *fire wall* at a *protected opening*. To be of use in fire it must be shut and so has a strong *door closer* and often a releasable *door holder* operated by *fire*

fire engineering

detectors. Fire door gradings vary with the occupancy and fire load of the building. Half-hour fire doors or FD30 doors (thirty minutes' *integrity*) look like ordinary doors, but have a special core. The same applies to one-hour fire doors and to *smoke doors.* Sheet-steel cladding is common for *fire shutters* and for longer ratings such as four hours. Fire doors are required by the *Building Regulations* as well as by insurance companies, who issue their rules through the *Loss Prevention Council.* Testing of fire doors is to BS 476, using complete *doorsets* fitted with their *ironmongery.* See BS 5588.

(2) A furnace door.

fire engineering Using measurement and calculation to provide fire safety in comfortable, efficient buildings.

fire escape An *escape stair.*

fire extinguisher, portable f.e. A container of *extinguishing medium* and a sprayer, used for first-aid fire fighting. Fire extinguishers are stored at a *fire point.*

fire-extinguishing system Fixed fire-fighting equipment connected to a large supply of *extinguishing medium* for either manual fire fighting or auto-suppression. Such systems may use water (hose reels, hydrants, sprinklers), foam, inert gas, dry powder, or halon. All are *active fire-protection* systems and require regular maintenance and inspection, usually by the fire brigade.

fire-fighting shaft A lift and stairway in a tall building to give the fire brigade safe access to a fire. It has a *fire riser* and *landing valves* in a lobby at each floor, all enclosed inside robust fire-resisting walls. The lift has a *fire switch* for direct control and a separate power sub-circuit.

fire floor (1) firebreak f. A fire-resist-ing floor that forms the top or bottom of a *compartment*, thus resembling a *fire wall.* Fire floors are tested while carrying load and may have to be re-loaded later. The test fire is from below and thus includes any protection given by a *suspended ceiling.* Concrete floors usually exceed the required performance; even steel and concrete *composite floors* may not need *fire protection* of the *soffit.* Timber floors may be able to provide satisfactory fire resistance by the use of chipboard or tongued and grooved *flooring* and a *plasterboard* ceiling. The upgrading of timber floors for fire is discussed in BRE Digest 208. Fire floor tests are to BS 476.

(2) A floor where a fire is burning.

fire grading *Fire-resistance grading.*

fire hazard Danger from burning. The hazard of materials is how much heat and *smoke* they produce, and their *flame spread.* Internal hazard arises from fuel in the contents or structure of the building, the *fire load.* It may be very high during building *refurbishment* or when using electrostatic spray painting. External hazard arises from adjoining buildings, grass, or forest. Hazards from building materials vary in severity and include combustibility, flammability, *ignitability*, heat or smoke release, and poisonous gas, e.g. from plastics. See BS 5588. Compare **fire resistance**.

fire-hose reel A reel fixed on a wall on which is wound a small fire hose that can be swiftly pulled out in any direction for first-aid fire fighting. Water is fed through the axle of the reel. It may be housed inside a cabinet.

fire hydrant An outlet from a *fire main* to which a fireman can connect his hose. It may be 19, 38, or 63 mm nominal bore and either inside the building or outside.

fire inspection The examination of a non-domestic building by *fire-prevention officers* before they issue the *fire certificate*.

fire lift The lift in a *fire-fighting shaft*.

fire-load density The amount of burnable fuel in megajoules (MJ) in the contents or structure of a building, per square metre (m²), used to predict *fire severity*. A 'low' fire load density is less than 1135 MJ/m² of floor area and this is usual for dwellings and offices. A 'moderate' one is from 1135 to 2270 MJ/m². Warehouses may have a 'high' fire-load density, 2270 to 4540 MJ/m². The 'wood equivalent' is calculated by dividing by 18.6 (MJ/kg of wood). Special requirements apply to the storage of *flammable liquids*. There is a human tendency to underestimate fire load and not allow for the worst possibility.

fire lobby A lobby leading to a *protected route* or *fire-fighting shaft*.

fire main A pipe to supply water for *fire extinguishing*, usually made of galvanized steel and *colour-coded* with red paint. It may be horizontal or a vertical *fire riser* and supply hydrants, sprinklers, or hose reels.

fireman's panel A *mimic diagram* on the wall near the entrance to a large building. It has a *plan* of each floor, controls and indicators for fire-fighting equipment, a public address microphone, etc.

fireman's switch A *fire switch*.

fire modelling The use of powerful computers to assess fire problems, such as the heating of a structure, smoke movement, time to *flashover*, roof venting, escape times, etc. A 'zone model' shows rooms as blocks of colour and gives only simple answers. A more complex 'field model' uses thousands of calculations to create a picture. Both zone and field models are interactive, and the user can re-run them to show the effect of changes in layout, fire severity, and building materials, or when sprinklers and fire detectors will operate. See BR Digest 367.

fire officer See **fire-prevention officer**.

Fire Offices' Committee (FOC) Representatives of fire insurance companies (offices) in London, now part of the insurer's *Loss Prevention Council*.

fire performance, f. behaviour A broad term used in discussing both the *fire hazard* of materials and the *fire resistance* of building elements.

fire-performance plasterboard Grades of *gypsum wallboard* or *gypsum baseboard* made for use in fire-resisting walls or ceilings.

fireplace A *hearth* for an open fire. A fireplace may also have a *hob* and a *front hearth*. See BS 1251.

fire point A place where fire-extinguishing equipment is kept.

fire precautions Measures both for the prevention of fires and for the protection of people and property. On building sites all *hot work* that makes sparks or flames should be stopped, and all fires extinguished, well before people leave the site. Accumulations of rubbish or *flammable liquids* should be avoided and duct *fire stops* always kept in place. Hazard is highest during *fit-out* and *refurbishment*. For fire precautions in completed buildings see the *Building Regulations* and BS 5588.

Fire Precautions Act 1971 (UK) Legislation applying to new and existing non-domestic buildings which requires a *fire certificate* before they can be put to use.

fire prevention

fire prevention Precautions that reduce *fire hazards* and sources of ignition.

fire-prevention officer The person in each local fire brigade who enforces the laws for preventing fires in buildings and who issues *fire certificates*. He or she should be consulted on *fire doors*, means of escape, *fire safety signs*, etc. Building plans for the officer should be prepared according to BRE Digest 271.

fire propagation index A rating of the heat output from a wall lining material (but not its *fire hazard*) when subjected to fire in a test furnace to BS 476 part 6.

fireproof A term which should not be used instead of *fire resistant*, since no practical construction can stand fire indefinitely. In the USA the term means a construction which will safely withstand the complete burnout of the contents of the building.

fire protection (1) Measures for limiting fire damage to buildings and their contents, or for protecting people, or for detection and extinction. Fire damage is limited by making the structure fire resistant and by dividing the building into *compartments*. People are protected by providing *escape routes* and by *smoke control*. The foregoing usually rely on *passive fire* protection, whereas fire detection and extinction call for *active fire protection*.
(2) The *encasement* of steelwork.

Fire Protection Association (FPA) (UK) A body which publishes advice on all aspects of fire safety and protection. It is part of the *Loss Prevention Council*.

fire rating A *fire-resistance grading*.

fire reserve Water kept for fire fighting, usually today in ground-level tanks and pressurized by a *fire booster*, although for tall buildings an *elevated gravity tank* may still be used in addition.

fire resistance The ability of a building *element* either to prevent fire from passing into another *compartment* or to continue bearing a load in a fully developed fire. Fire doors and walls need fire resistance to act as *fire breaks*. Structural elements such as columns, beams, and floors need fire resistance to prevent the collapse of the building. Compare **fire hazard**.

fire-resistance grading, fire grading, fire rating The length of time given by laboratory tests, or *fire calculations*, during which a building element has satisfactory *fire resistance*. In Britain tests are done to BS 476, which has many parts. It defines the fire-resistance grading as the time (endurance) for which the element continues to satisfy three criteria: absence of collapse (stability), flame penetration (integrity), and excessive temperature rise on the 'cool' face (insulation). If the element satisfies all three criteria for one hour, it has a one-hour fire-resistance grading, or is one-hour rated. Some of the test methods in BS 476 are: part 21, for *loadbearing* components, such as walls, floors, roofs, columns and beams, which carry a full load during the test and may have to be re-loaded later; part 22, for non-loadbearing elements, such as suspended ceilings, glazing, etc., which fail the collapse criteria if they collapse under their own weight; part 23, for the contribution of components to the fire resistance of a structure; part 24, for ducts; and part 31, for door sets and shutters. The fire resistances needed by buildings are laid down in the *Building Regulations*. Thirty minutes is often adequate for small houses, while six hours could be needed for a warehouse with a *fire-load density* of 4000 MJ/m². Fire-resistance requirements in excess of two hours cannot be justified on life safety grounds.

fire-resisting glass (USA **f. window**) Glass that will not crack and fall away from its frame when exposed to fire. Very small panes of ordinary glass have enough fire resistance for use in a half-hour *fire door*. Two hours' resistance is possible with larger panes of wire-free *borosilicate glass* in suitable frames, while *wired glass* and *copper lights* generally achieve a half to one hour. As glass allows some heat radiation to pass, the *insulation* requirement is waived, although useful fire insulation is possible with laminated *intumescent* glass. Even so, combustible materials need to be kept at a safe distance. Being non-combustible, glass is not itself a *fire hazard*.

fire retardants Chemicals that delay the burning of timber, usually applied by vacuum *impregnation*. They may be slightly acidic in high concentrations, corrode metal *fasteners*, and retard the *curing* of adhesives such as *phenol-formaldehyde* or *resorcinol-formaldehyde*. See **flame-retardant treatment**.

fire riser (USA **standpipe**) A vertical pipe in a tall building used to supply water to sprinklers, hydrants, or fire hose reels. A fire riser may be a *dry riser* or a *wet riser* and may be pressurized by a *fire booster*.

fire safety sign A sign that points out, for example, where to find fire exits (green sign with white letters), that a *fire door* must be kept closed, where to find extinguishers (red sign with white letters), etc. They are required in addition to other *safety signs*. The local *fire-prevention officer* provides information. See BS 5499.

fire severity Fire severity can be calculated from the *fire-load density* in a compartment, its size and shape, and the area of windows and ventilation openings. It indicates the *fire-resistance grading* needed by a wall, floor, door, etc. to survive the complete burnout of the building contents.

fire shutter (1) A large sliding or rolling door, like a *fire door*, but used in front of lifts, over the tops of escalators, etc.
(2) A *fire damper*.

firespray *Sprayed mineral insulation.*

fire stair An *escape stair*.

fire stop, fire stopping A narrow strip of non-combustible material that closes a gap in a *fire wall* or floor, or fills the space round pipes, cables, or ducts where they emerge from a services duct. It should allow movement, even in fire, by sliding or by its own resilience. Usual materials are compressed *mineral fibre* and *intumescent mastic*.

fire stop sleeve, pipe closer A metal collar with an *intumescent* liner, used where combustible *plastics pipes* (or pipes with plastics insulation) go through a *fire wall*, fire floor, or fire barrier. If the pipe (or insulation) melts or burns in a fire the intumescent foam closes the hole. In normal conditions it acts as an expansion *sleeve*. See diagram, p. 417.

fire-suppression system See **fire-extinguishing system**.

fire switch, f. lift priority s., fireman's s. A switch that allows the fire brigade to take over direct control of the lift in a *fire-fighting shaft*.

fire terms Caution is needed with fire terms, which change between countries and official documents. The terms used in *fire testing* are given in BS 4422.

fire testing The two main aims of testing are to estimate the *fire-resistance grading* of building elements and the *fire hazard* of building materials. Both use test procedures described in *standards* and are expensive and time consuming. In some cases testing can be replaced by *fire calculations*.

fire tower (USA) A *protected stair*.

fire tube A pipe in a *boiler* through which the smoke passes. The water to be heated surrounds the rows of fire tubes, all enclosed in a mild-steel shell.

fire vent A *damper* for *fire venting*, usually in the roof and automatically opened. It may also mean panels of easily melted plastics or aluminium roofing.

fire venting Inducing hot gas and smoke to leave a building by *fire vents* and *smoke outlets* so that firemen can see to fight the fire.

fire wall, firebreak w., division w. A fire-resisting wall from the lowest floor up to the roof, and if the roof is not fire resisting also a *parapet* 450 or 900 mm high. Solid brick walls 102.5 mm thick can have two hours' *fire-resistance grading* and 215 mm thick, 6 hours'. A fire wall forms the sides of a *compartment*, giving *passive fire protection*. Any *protected openings* in it must have *fire doors*. All pipes and ducts through fire walls (and floors) must have *fire stops* or *fire dampers*; a hole as small as 150 mm square can allow fire through it to burn out an entire building. Wall linings of high thermal conductivity, density, and specific heat can cool a fire by heat loss during the growth phase and slow its development.

firing The burning of *ceramics* in a kiln to make them partly or fully *vitrified*.

firmer chisel A carpenter's or joiner's ordinary *chisel*, originally meaning one not made to be struck with a hammer or mallet.

firring Strips of timber used as *packers* under lining materials, to straighten the support, and achieve a straight or upright surface, but also to provide a *ventilation* cavity or tapered to give a fall to roofing.

first fixings (1) *Grounds*, *pluggings*, etc., that secure *joinery*.
(2) Structural timber, *joists*, *rafters*, floors, etc.

first floor (1) In Britain, the floor which is next above the ground level and is therefore about 3 m above ground. This definition is also accepted in the USA for houses with neither *basement* nor *cellar*. In France the first floor is normally known as the 'premier étage', though it is technically described as the 'plancher haut du rez-de-chaussée', literally the high floor of the street level (ground floor), as strictly a storey is measured between *finished-floor levels*.
(2) (USA) In buildings with a basement or cellar the first floor is the first above ground level (which in Britain is the *ground floor*).

first-layer felt A *base sheet*.

first storey The space between the *first floor* and the floor above.

fish tape (USA) A *draw cable*.

fishtail fixing lug, caulk, fang, tang A split and twisted end of a metal bar, anchored by *building in* to a mortar joint, casting into concrete, etc. It should not be left projecting, as the sharp end can cause injuries. See **wall tie**.

fissured surface A flat surface with shallow irregular cracks, often used as a decorative and sound-absorbing

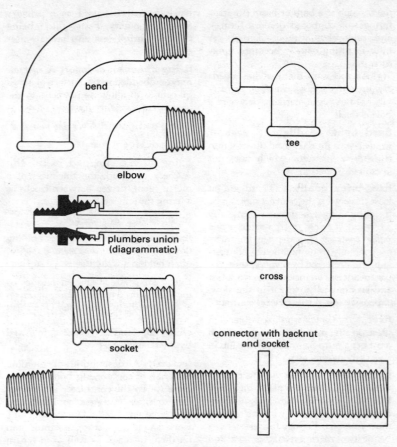

bend

elbow

plumbers union
(diagrammatic)

tee

cross

socket

connector with backnut
and socket

Fittings for screwed pipe, mainly dead mild steel tubulars and malleable iron castings.

finish, e.g. on tiles for a *suspended ceiling*.

fitment See **fitting** (2).

fit-out, fitting-out (USA **outfitting**) Work in the *finishing trades*, such as putting up suspended ceilings, fixing prefabricated joinery, laying carpets, etc. Site *fire precautions* should be observed.

fitted carpet Fixed *carpet* that covers a floor completely from wall to wall, laid loose over *underlay* and stretched tight between *gripper strips* round the edges. It may also have *threshold strips* under doors.

fitting (1) A device for joining lengths of pipe, duct, scaffold tube, cable, etc., to make end joints, corners, branches, etc. See diagram.

(2) **fitment** A unit or factory-made appliance not part of the building

fabric, such as a bath or basin (sanitary fittings) or cupboard (joinery fitting), and removable without damage. A built-in fitting usually becomes a *fixture*.
(3) An accessory that matches a general area of *finishing*, *cladding*, or *roofing*, used at edges, corners, or where a hole is made.

fixed light, deadlight, f. sash A *window* with its glass fixed directly into a *surround* or frame which does not open.

fixed-price contract There are at least three ways the *contract sum* can be fixed: by using a *lump-sum* contract, by a *schedule of prices*, or by a *measure and value contract*. Fixed-price contracts are generally recommended by consultants and preferred by owners, since the contractor has an incentive to work fast and economically, absent in the slow, expensive *cost-reimbursement* contract.

fixer A *tradesman* who fixes objects in place on site, usually *components* which have been purpose-made for the job in a workshop or factory.

fixing (1) **f. device** Any accessory which holds objects in place, including hardware, e.g. *clips, lugs, pluggings, anchors*, ties, hooks, or wire, as well as most *fasteners*. Fixings for *services* can be of two main types, *primary* and *secondary*. Lighting fittings usually have concealed fixings.
(2) Building into the works, or the labour involved in setting and securing materials in place with *adhesive*, grout, fixing devices, *shotfiring*, etc. Many of the later *trades* have work divided into *first fixings* and *second fixings*, but the term includes specialized work such as *steelfixing* or *glazing*.

fixing brick, f. block, nailing block, nog, wood brick A brick made from wood or sawdust and clay, from *diatomite*, *lightweight concrete*, or other

nailable material, used as a *primary fixing* for joinery. *Fixing fillets*, being thinner, shrink less and are a better fixing.

fixing channel, c. insert A *concrete insert* embedded in the surface of a wall or ceiling to allow adjustment in the location of *secondary fixings*.

fixing compound A *glazing material*.

fixing device A *fixing*.

fixing fillet, f. slip, pad, pallet, slip A piece of wood, the thickness of a mortar joint, driven between bricks as a fixing for joinery.

fixing gun A *cartridge tool*.

fixing strap A metal strap for holding down *supported sheetmetal roofing*, often held in a welted *seam*.

fixing strip An H-shaped plastics or metal strip used to fix *cladding* or *linings* at the joint between sheets. One sheet is put in place, the fixing strip is slipped on the edge and *secret-nailed*, followed by the second sheet.

fix only A description of building materials or components bought by the *client* (or by the contractor as a *prime-cost* item), or '*by others*', and delivered to *site* to be fixed. The contractor's work is to unload, store, assemble, and fix them in place, as well as to return the packaging.

fixture Anything *built in* and permanently attached to the *fabric* of a building. Its removal is usually prevented by law, although what are fixtures and what are *fittings* may need to be agreed between landlord and tenant.

flag, flagstone A slab of precast concrete, or cast or natural stone, used for paving footways, for landscaped *hard surfaces*, or as a floor finish.

flagging A walking area covered with *flags*, or the work of laying them.

flagman (USA) A person whose job it is to keep the public out of danger away from the area where a *crane* is working overhead.

flakeboard A *building board* made of adhesive-bonded long thin wood particles, in layers parallel to the board surface. It is made for exterior and structural use and can be *melamine-surfaced* to take paintwork.

flaking The defect of detachment of stone, paint, plaster, etc., in layers.

flame bonding *Torching-on.*

flame cleaning Removing mill scale and water from weathered structural steelwork by means of a flame. The surface is primed with *pretreatment primer* immediately afterwards.

flame detector, radiation d. A *fire detector* that reacts to flickering infrared or ultra-violet radiation, in many types of flame. Its lenses collect dirt and need regular cleaning.

flame-penetration rating *Integrity.*

flame retardant A chemical that reduces surface *flame spread*, thus affecting *fire hazard* rather than *fire resistance*. It is not a *fire retardant*.

flame-retardant paint (1) Paint of normal decorative appearance but containing added *flame retardants*. Antimony oxide and chlorinated compounds are used in oil or emulsion paints, and calcium carbonate is used in *chlorinated rubber paint*. Flame-retardant paints are used on timber, and on wood-based *building board* in an *escape route*, and should be certified by a fire test laboratory for a given thickness and *substrate*.

(2) *Intumescent* paint.

flame-retardant treatment Pressure *impregnation of timber* with *flame retardants*, which can reduce its *flame spread*

to low or very low. Some flame-retardant treatments for solid timber may reduce its durability.

flame spread, surface spread of flame A measure of one type of *fire hazard* – the way a material reacts to fire and thus the contribution it makes to the *fire load*. In Britain building materials used for wall or ceiling finishings are tested under BS 476, first for *non-combustibility*, then if they are combustible, for *ignitability* by means of several different tests to determine how quickly their surface ignites. They are grouped as follows: Class 1 (very low), 2 (low), 3 (medium), 4 (rapid) flame spread.

flame-textured granite *Granite* paving slabs with a non-slip finish.

flammable A term agreed by fire authorities in Britain and the USA, to avoid confusion over 'inflammable'.

flammable liquids Liquids such as petrol, oil, *solvents*, and *thinners*. The way they may be stored in quantity is controlled by law because of their high *fire hazard*. Use on site should be subject to a *permit to work*. Foam or dry powder are good *extinguishing media* for flammable-liquid fires.

flammable materials Materials such as plastics or fabrics with a high *fire hazard*. They often also melt.

flange (1) A flat plate on the end of a pipe for making a bolted joint.

(2) The top and bottom plates of a steel beam or column, joined by a *web* (*C*).

flank The *intrados* of an arch, near its *springings*.

flanking transmission, f. sound A type of *transmission* of *airborne noise* between neighbouring terrace houses or adjoining rooms that passes through the outside walls or windows, not through the *separating wall* as is the case with *direct transmission*. It is difficult to reduce flanking transmission below 50 to 55 dB, which similarly limits total *sound insulation*.

flanking window, wing light A window beside an external door, with its sill at *threshold* level.

flank wall, side w. A wall at one side of a building, not a front or back wall.

flap (1) One of two flat plates with screw holes that are joined by a pin to make a *hinge*.

(2) A panel in a counter top that can swing upwards on hinges.

flared joint A strong *manipulative* pipe joint made by widening the end of copper pipe to a cone shape and gripping it with a cone and nut *fitting*.

flash (1) To make a weathertight joint, called a *flashing*.

(2) **f. line, mould mark, fin** In any casting, a narrow strip of surface metal projecting along a line that was the boundary between the two halves of the mould. It is removed by *fettling*.

flashing (1) **roof f.** A strip of impervious sheet material, either *plumber's metalwork* such as lead 1.8 mm thick, or soft aluminium 1.1 mm thick or galvanized steel, or other materials (such as *bitumen felt*, which is thicker) that excludes rainwater from the junction between a roof covering and another surface. Flashings at a roof/wall *abutment* may be in two parts, with an *upstand flashing* turned up the wall and a *counter-flashing* turned down over it. Flashings at their upper end are usually wedged tightly into either a *raglet* or mortar joints raked out to receive them. The bottom edge is *dressed* down over the roofing to prevent entry of *driving rain*. No nails should pass through any part of a flashing. They are also used

over *aprons* as *damp-proof courses* in *cavities*, at roof *eaves* and *verges*, and over *patent glazing*.

(2) **pipe f., vent soaker** A preformed weathertight joint for the *penetration* through *roofing* of a vent pipe, usually in synthetic rubber such as *ethylene propylene diene rubber*.

(3) Burning bricks alternately with too much and too little air, to give them varied colours.

(4) The defect of glossier patches in the finish surface of paintwork. It happens from making a lap or join without a *live edge* or from the too rapid set of the paint.

flashing board A *layer board*.

flashover In a room fire, a semi-explosion of all combustible materials as they reach ignition temperature. At this stage of *fire development*, flames have spread across the ceiling and smoke is layering down to near the floor. Both cause unbearable heat. When the gases under the ceiling reach 600°C, 1 or 2 megawatts of energy are released, producing temperatures above 1000°C within seconds. Pressure from expanding gases may break windows or fling open lightly latched doors. The fire can become twice as hot as one outdoors if the doors are open, and drive vast quantities of deadly *smoke* to other areas. Compare **backdraught**.

flat (1) A level platform, generally a roof, particularly a lead-covered roof (lead flat).

(2) (USA **apartment**) A dwelling within a building, originally on one floor.

(3) **matt** Description of paint or varnish with low *gloss* or none.

(4) *Flush*, or a surface made smooth by *flatting-down*.

(5) A thin rectangular metal bar.

flat arch, French a., straight a.

(USA **jack a.**) A door or window arch with a level *soffit* built either of *soldiers* or of wedge-shaped *gauged brickwork* or moulded bricks which radiate from one centre point.

flat cost The cost of labour and materials only, without *overheads*.

flat cutting The *re-sawing* of timber lengthwise.

flat glass Sheets of *glass* which are not curved, e.g. *float glass*; formerly glass made by the flat-drawn process (*sheet glass*).

flat grain The grain of timber which, after *conversion*, has annual rings at less than 45° with the face of the piece, as in some *flat-sawn* planks.

flat interlocking tile A *single-lap* standard British clay roof tile normally measuring 390 × 200 × 19 mm, provided with two nail holes and no *nib*. It has a 76 mm side lap.

flat joint, flush j. A mortar joint whose surface is flush with the brickwork.

flat-joint jointed brickwork Brickwork with *flat joints* in which narrow grooves have been cut with a *jointer*.

flatness tolerance The accuracy of the surface of a finishing, e.g. one having no more than 3 mm unevenness when measured with a 2 m *straightedge*.

flat paint Paint that gives a finish with little or no *gloss*.

flat paint brush A metal-bound *brush* used for painting large areas such as walls. Its bristles are stiff enough to carry heavy paint or varnish. It is from 10 to 150 mm wide.

flat pointing The pointing of brickwork to make *flat joints*.

flat roof (Scotland **platform r.**) A roof which slopes at less than 10° to the

horizontal. The *roofing*, which is continuous, is usually *built-up* feltwork or a *membrane* with gravel protection, unless intended to be *trafficable*. An insulated flat roof should be built as a *warm deck*. Careful *workmanship* is needed to avoid high maintenance costs, and it is usual to require a long warranty from the roofing contractor. Flat roofs should have a *fall* of at least 1 in 60 towards the *rainwater outlet* to prevent *ponding*, and a fail-safe rainwater drainage system with *overflows* so that rainwater from blocked *gutters* will not damage the building *fabric*. Flat roofs should be regularly inspected and maintained, as they are not self-cleansing. See BRE Digest 312.

flat-sawn timber Timber from a log sawn with parallel cuts; a simple and economical method of *conversion*. The first and last planks have *flat grain* and the middle planks are *quarter-sawn*. See diagram, p. 182.

flatting agent A chemical added to *paint* to help it *flow* out flat.

flatting-down The rubbing of *undercoats* with fine *sandpaper* to remove minor defects and make a smooth surface for the finish coats of *paintwork*.

flat varnish Varnish that gives a finish with little or no *gloss*.

flaunching A low, wide cement mortar *fillet* surrounding the *flue terminal* on top of a *chimney stack* to throw off rainwater.

flaxboard A *building board* used for decking and *semi-solid* door cores.

fleck paint, f. finish Paintwork with blobs of one or more colours on a different colour background, giving a *stipple* finish.

Flemish bond A brickwork *bond* that shows *stretchers* alternating with *headers* along each course and alternating

Fletton brick

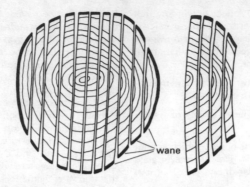

Flat-sawn timber. The log has been seasoned after sawing, and the diagram shows the effect of shrinkage on it. The surfaces are curved because the radial shrinkage is only two thirds of the tangential shrinkage. On the right is part of the same log showing the (concave-outwards) warp which the shrinkage causes. The two heart planks are fully quarter-sawn.

vertically between courses. Single Flemish bond shows on one wall face, double Flemish bond on both wall faces.

Fletton brick A low-cost pale-red brick made from Oxford clay with traces of coal that burn during firing, saving energy. It is *pressed* like many *clay bricks* and can be sandfaced, pigmented, or textured for use as *facings*.

flexible conduit Steel *conduit* with a PVC covering. It can be bent backwards and forwards, and is used for wiring to movable equipment. See BS 731.

flexible coupling (1) A rubber connection or *V-belt* drive between an electric motor and a fan or pump. It does not transmit vibration.

(2) A *flexible duct connector*.

flexible damp courses *Damp-proof courses*, in preformed rolls, made of a variety of materials such as bitumen/ethylene/propylene and fibres, pitch polymers, or polyethylene. Some need warming before unrolling in cold weather.

flexible door A plastics sheet (PVC/PVA) door on a spring-loaded arm easily pushed aside or operated automatically, used for high-traffic areas. See diagram.

flexible duct connector A short length of rubberized fabric on the outlet from a main fan or packaged air conditioner to prevent noise *transmission* to the stiff steel sheet of air-distribution *ductwork*.

flexible ductwork A round air *duct* of plastics or galvanized steel, used between a supply air duct and ceiling outlets or a *variable air volume box*. It may need to be fire-resisting and can reduce air flow if too many bends are made.

flexible glazing sealants Glazing materials used about 3 mm thick between glass and frames where high *thermal movement* or severe wind loads are possible. They can be *non-setting*, or two-part *elastomers*, and may need a primer as well as *glazing blocks* and *distance pieces*.

flexible metal conduit Spiral-wound

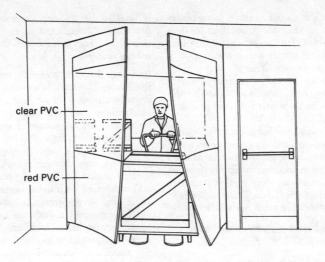

A flexible door.

galvanized steel strip conduit which allows movement in a machine or enables a *socket outlet* to be shifted without damage. A separate *earth* is needed.

flexible metal roofing *Supported sheetmetal roofing.*

flexible mounting An *anti-vibration mounting.*

flexible PVC flooring Resilient and hard-wearing tiles or sheet made with a high proportion of *polyvinyl chloride* resin and some *plasticizers.* It can be printed to imitate traditional materials and faced with a clear protective layer of PVC. Any bumps in the *subfloor* or fixing *adhesive* tend to show.

fleximer floor Expensive *jointless flooring*, which may be only 6 mm thick, containing some *elastomer* to reduce cracking and increase flexibility and adhesion. A solid sub-floor of concrete or steel is usual, and laying is done by specialists.

flier, flyer (1) A rectangular *tread*

with edges parallel to those of the other treads of a straight stair *flight* (not a *winder* or a *tapered tread*).

(2) A *flying shore.*

flight A series of *steps* between two landings of a *stair.* See diagram, p. 437.

flint-lime brick A type of *calcium silicate brick.*

flitch (1) A large timber at an intermediate stage in *conversion*, after *breaking-down* but before *re-sawing.* It may have *wane* on one or more edges.

(2) A stack of *veneers* after cutting, piled in the order in which they were in the log.

float (1) (USA **floater**) A plastering tool with a flat face and a handle, used for *floating* the surface of plaster or a *screed.* Hand floats of wood, plastics, or aluminium include the *cross-grained float* and the *Darby float.* A *power float* is used for floors.

(2) A drain pipe (copper, cast iron)

running at a shallow slope, hung under a floor *soffit* or above the floor inside a duct. It is often prefabricated and carries *soil water* or *waste water* from a *range* of sanitary fittings.

(3) A hollow ball of plastics or incorrodible metal fixed to the end of the operating lever of a *ballvalve* or a *float switch* (*C*).

(4) Spare time for completion of one or more *activities* on a *programme* using the *critical path method*, calculated from the time in addition to activity *duration* that remains between earliest start date and latest finish date. Float can be *free*, *independent*, or *total*. See **slack**.

floated coat A plaster coat compacted and smoothed by a *float*.

floated finish Plasterwork or a floor *screed* finished by *floating*.

float glass Flat window *glass* smooth on both sides made by running hot fluid glass from the furnace along a bath of molten tin (on which it floats) until the moving sheet is cool and hard enough to be lifted by rollers, flame polished and annealed. It can be from 3 to 25 mm thick and is used to make *laminated glass* and *toughened glass*. Float glass has replaced *plate glass* and *sheet glass*.

floating (1) Compacting the surface of plaster (the *floating coat*) or a floor *screed* by rubbing or scouring with a *float* held flat against the surface. This leaves a rough but dense 'floated' surface, often later smoothed by *trowelling*.

(2) **flooding** The re-arrangement or separation of *pigment* grains at the surface of a paint film, where two pigments are mixed. It can be a defect, or desired, as *leafing*, in aluminium and other metal paints.

floating coat, browning c. The second plaster *undercoat* in *three-coat work* and the first in *two-coat work*. It

is levelled by *floating*, resembles the first coat although not more than 10 mm thick, and usually has the same mix as *rendering*.

floating floor A concrete *screed*, or floorboards, on top of a structural floor, separated from it by resilient material, as *discontinuous construction* for sound insulation. Its advantage is to reduce transmission of *impact noise* downwards through a *separating floor*. A concrete screed floating floor (floating screed) should be at least 65 mm thick and reinforced with lightweight fabric. It is laid over impact sound duty (I S D) *expanded polystyrene* specially made for resilience and long life, or glass wool, which must have tight *butt joints* and an overlay of *building paper* or polyethylene sheet to keep concrete out of the joints. A floorboard floating floor may be fixed to 50 × 50 mm battens carried on *acoustic clips* or separated by a *glass-wool* quilt laid on the rough floor without nailing. See diagrams here and on p. 392.

floating screed A concrete *floating floor*.

floatvalve (mainly U S A) A *ballvalve*.

flock spraying, flocking The blowing of soft fluffy fibres on to a sticky surface, for *sprayed mineral insulation*.

flooding See **floating**.

floodlight A powerful outdoor *luminaire*, used for *security*, display, or public lighting. See **hot work**.

floor A level indoor area for walking on. A floor is one *element* of the building *structure* and usually made of concrete or timber. See **floor finish**, **flooring, storey**.

floor area ratio (U S A) The *plot ratio*.

floor blocks Hollow *blocks* of precast concrete or burnt clay, laid in rows on

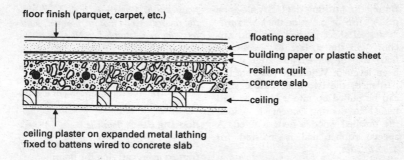

floor finish (parquet, carpet, etc.)

floating screed
building paper or plastic sheet
resilient quilt
concrete slab
ceiling

ceiling plaster on expanded metal lathing
fixed to battens wired to concrete slab

A floating screed for sound insulation between floors.

the shuttering of a concrete floor, so that the spaces between them form ribs, for either a *rib and block suspended floor* or a *voided slab* (*C*).

floorboards The most usual timber boards for flooring are 150 × 25 mm, with *tongued and grooved* edges, giving a floor that is stronger, freer of draughts and more fire-resistant than *square-edged* boards.

floor boarding, timber flooring A floor made of *floorboards* laid several boards at a time, tightened with *floor cramps*, and nailed twice into each *joist*. To make a decorative floor, nail heads are *punched*, the holes filled, and the floor then sanded and coated with *floor*

sealer. Squeaking can be difficult to cure; one possibility is to re-lay with *latex cement* in the *tongue and groove joints* and with screw fixings instead of nails.

floor centre, telescopic c. A small, extensible, folded sheet-steel beam used to support *formwork* for reinforced-concrete floor slabs. See diagram.

floor check A *door closer* with a hydraulic *check*, set into a deep hole in the floor, usually covered with a neat plate flush with the floor surface.

floor chisel A *bolster* about 50 mm wide for pulling up floorboards.

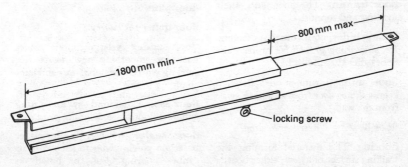

800 mm max

1800 mm min

locking screw

A floor centre.

floor clip, bulldog c. (USA **sleeper c.**) A strip of 0.9 mm thick *sherardized* steel sheet pushed into and anchored in the surface of a concrete floor slab or screed just after it has been levelled. When the concrete has hardened, the two ends of the strip can be pulled up into a U-shape and nailed to flooring *battens*, to which the finished wooden floor is fixed. A special rubber pad is inserted for *acoustic clips*.

floor coverings Thin resilient *floor finishes*, usually detachable or loose and laid on top of the *finished floor*, to improve comfort and reduce *impact noise*, e.g. carpet, linoleum, cork, and cushioned vinyl.

floor cramp, flooring c. A *cramp* for forcing *floorboards* together before they are nailed down. It is set astride a *floor joist*, which it grips.

floor drain A *gully*.

floor finish, f. finishing Material laid on top of the *sub-floor* to form the visible surface and to resist wear, *indenting*, water, and chemicals. In *wet areas* it should also be *non-slip* when wet. It can be in-situ *terrazzo* or *granolithic*, thick *floor tiles* on adhesive or mortar *bedding*, thin *floor finishes*, or one of many *floorings* or *floor coverings*.

floor framing The floor *joists*, their *strutting* and support.

floor grinder, concrete g. A machine for smoothing *granolithic* or *terrazzo* floors after they are laid.

floor guide A runner on a floor to keep a *sliding door* from swinging away from the wall.

floor heating *Underfloor heating*.

flooring The material forming the walking surface of a *floor*, either boards or panels on top of the *floor joists* or a

floor finish. Floorings include timber *floorboards*, *wood mosaic*, *parquet*, *jointless flooring*, *studded rubber*, *quarries*, *floor tiles*, etc.

flooring saw, floor s. A short saw with teeth on its curved nose, used to cut floorboards for plumbing or electrical work. See diagram, p. 388.

flooring tile A *floor tile*.

floor joist A *common joist* of a timber-framed *floor*, carrying the flooring.

floor layer's labourer A floor *tiler's* helper who unloads tiles or other materials, soaks clay tiles, mixes mortar, and brings them all to the tradesman at the work point. He or she may help in rubbing down a *terrazzo* floor and cleans up *trade rubbish* after work is completed.

floor outlet A *gully*.

floor paint Tough paint, usually based on *epoxy* or *acrylic resin*, for markings or the decorative colouring of concrete or timber floors and also capable of acting as a *floor sealer*.

floor plan A *drawing* showing the layout of all rooms, wall thicknesses, locations of columns, etc., on a floor, and the *grid* for *setting-out*. A floor plan may be repeated for *typical floors*. See diagram, p. 330.

floor plate An *anchor plate*.

floor quarry A *quarry*.

floor sander A large mobile *sander* used for smoothing new *floorboards*, starting with coarse *sandpaper* and later changing to fine.

floor saw A diamond saw for cutting concrete floors. See diagram, p. 123.

floor sealer A clear finish, either based on *polyurethane* to keep cork or timber flooring clean, or based on *epoxy resin*, for *dust-proofing* concrete.

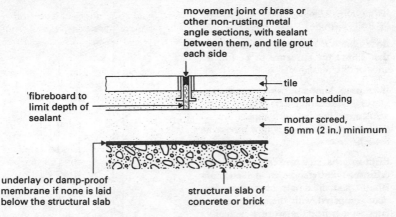

movement joint of brass or
other non-rusting metal
angle sections, with sealant
between them, and tile grout
each side

'fibreboard to
limit depth of
sealant

tile

mortar bedding

mortar screed,
50 mm (2 in.) minimum

underlay or damp-proof
membrane if none is laid
below the structural slab

structural slab of
concrete or brick

A movement joint in clay floor tiles.

floor slab A reinforced concrete *suspended floor*, or a *ground slab*.

floor spring, s. hinge A *pivot* recessed into a floor with a spring to act as a *door closer*, either single-acting or for swing doors *double-acting*. It can be a *floor check*.

floor stop A *door stop* fixed to the floor.

floor strutting *Herring-bone strutting* or *solid bridging* between floor *joists* that stiffens the floor by making the joists act together as *common joists*.

floor tile, flooring t. Thin *floor finishes* made to be *non-slip* and resistant to breakage from heavy foot traffic or small falling objects. Floor tiles for *wet areas* are usually unglazed *clay tiles* or concrete, and need to be fully bedded in adhesive or mortar. They can have square edges or edges with spacer *lugs*, and are laid with even joints a few millimetres wide, which are later *grouted*. For a more sound-absorbent, heat-insulating, decorative, or comfortable surface, tiles of *linoleum*, *cork*, rubber, *asphalt*, or *thermoplastics* are

used, or in dry locations even *carpet tiles*. *Hard surfaces* are formed by *quarries*, steel or cast-iron *anchor plates*, and *pavers*. In general *wall tiles* are too weak, soft, and slippery when wet for use as floor tiles. See diagram.

floor trap A *gully*.

floor unit A cupboard or storage unit that stands on the floor.

floor varnish Oil-based varnish, or a *floor sealer*.

floor void (1) A hole through a floor such as a *penetration* to pass services pipes or cables, or an opening for a stair or partial floor.
(2) A hollow inside a *voided slab* (*C*).

floor warming *Underfloor heating*.

flow (1) **supply** In a system with circulating fluid the movement of the water or air from the boiler or treatment point to the point of use. It comes back by *return*.
(2) **level** The property of spreading into a smooth film. Paints which flow do not show brush marks, *orange peeling*, or other defects of *application*, but

flow chart

they may *run* badly. Flow is improved by *flatting agents*.

flow chart, f. diagram A *programme* that shows the sequence of performance of different jobs.

flow pipe A pipe by which hot water leaves a boiler and enters a radiator or hot-water *cylinder*, or a pipe from a *chiller* to a *coil*, a primary flow pipe to a *calorifier*, etc.

fluctuations, rise and fall (USA **escalation**) Differences in the cost of labour and materials at the *tender* date compared with their cost at the later time when the work is actually done. A formula is usually stated in the *conditions of contract* for calculating the adjustment of payments to the contractor, based on a monthly official *price index*.

flue (1) A pipe to the outside air from a fireplace, boiler, or heater, either in a *chimney* or leading to it, to take away poisonous smoke or burnt gas. Flues are usually vertical and work by their *draught*. Flues for industrial boilers can be 600 mm dia. or larger and have special *flue linings*. The traditional domestic 230 by 230 mm clay-brick flue with pargeting has been superseded by *flue blocks* or stainless-steel flue pipes. The flue for an open fireplace may be large; that for a gas fire is much smaller. A flue can also bring in air, as in a *ducted flue* or a *balanced flue*.
(2) A ventilation *duct*.

flue block, f. brick, chimney block (1) A block with a hole for a *flueway*, but otherwise of normal size for bonding into a brickwork or blockwork wall within the wall thickness. It is made of terracotta or precast concrete with *pumice* or *diatomite* lightweight aggregate. The flueway can be 230 × 230 mm square, circular, sloping, or rectangular and the block bed faces have a

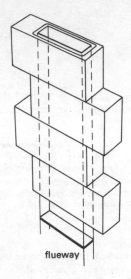

flueway

Flue blocks for a gas fire, forming part of a partition. These blocks for a straight cavity can be bonded into brickwork without difficulty. Curved flueway blocks can also be bonded in.

spigot-and-socket or *ogee* joint, which is installed socket upwards to drain *flue condensation* inwards. Flue blocks need careful handling on site to avoid damage. See diagram.
(2) A similar block, not made to be bonded, for use as a *flue lining*.

flue condensation *Condensation* can occur in a flue if the flue gases are too cool. In open fireplaces this is rarely severe and usually occurs only for a short period just after the fire is lit. In an efficient *boiler* the flue gases can be cooled too much, but the potential for extracting more heat is used in a *condensing boiler*. The main danger from flue condensate is that it contains acidic sulphur dioxide, leading to *sulphate attack* of any mortar. Flues may need an acid-resistant *flue lining* of ceramic or stainless steel. Joints have *sockets* upwards so that condensation drains inwards.

flue isolator A *damper* at the bottom of a flue, shut when the boiler is not being fired, to reduce heat loss. It should close gently, so that pressure waves do not cause *drumming*.

flue lining (1) The material on the inside face of a *flue*, which must be able to withstand high temperatures as well as *flue condensation*. Materials such as stainless steel or glazed ceramics are suitable for the severe conditions of oil-fired boilers. Domestic solid-fuel boilers may need only *flue blocks*.

(2) **relining** When a new gas-fired or oil-fired furnace is installed at an old flue, a pliable stainless-steel flue lining tube should be inserted along its full length. Because modern boilers are so efficient, they have cool gases, liable to create *flue condensation*.

flue pipe A section of flue, often an adaptor between a domestic *room-heater* and chimney flue, usually of stainless steel, but in industrial installations possibly of other materials.

flue terminal A *cowl* or cap on top of a flue to keep out rain and prevent wind gusts disturbing the *draught*. Special types are used for gas appliances.

flueway The open passage in a *flue*.

fluorescent luminaire A *luminaire* that carries single or twin *fluorescent tubes*, such as an industrial *reflector* or *air-handling luminaire*.

fluorescent tube, f. tubular lamp An efficient type of *discharge lamp* consisting of a glass tube filled with argon and low-pressure mercury vapour. It has a phosphor coating on the inside of the tube giving off light (fluorescing) when excited by an electric arc through the vapour. Each phosphor emits a particular colour and by mixing phosphors (as in tri-phosphor lamps) almost any shade of white can be produced. The light output efficiency of fluorescent lamps is around five times that of a 240 V *incandescent lamp* of similar wattage, but their reliability drops if they are not operated at the right temperature. The *starter* and other *control gear* is usually housed in a *fluorescent luminaire*, except for *instant-start* tubes or compact low-wattage fluorescent lamps with their own internal gear, which are used in a normal light socket. Some tubes can have electronic dimming.

fluoropolymers Probably the most durable and chemically resistant *polymers* available, though they are easily scratched. PVF_2 and $PTFE$ (Teflon) are fluoropolymers.

fluorspar Crystalline calcium fluoride, used as chips for *spar dash*.

flush (1) **flat** A description of a surface wholly in one plane, particularly any *finishing* or *facing*, including components which are *flush fixed*.

(2) To send a quantity of water, usually released from a *cistern*, down a drainage pipe or channel to clean it or to carry waste from a *sanitary fitting*.

flush door A flat *door leaf*, faced with *plywood* or *hardboard*, over a *core* of solid wood, chipboard, flaxboard, or cellular cardboard.

flush eaves *Eaves* with no *overhang*.

flush fixed, f. mounted, let in f. Of an accessory, *recessed* into a building element, so that the two surfaces are flush.

flushing (1) **f. up** Setting two components with their surfaces *flush*.

(2) Water released as a *flush* by operating a *flushing mechanism*.

(3) Flaking from the face of loadbearing masonry, such as *ashlar*, above and below the joints, due to pressure from bedding that is hollow, or at least not solid. It is prevented by *recessed joints* 20 mm deep which are *pointed* later during *cleaning down*.

flushing cistern A *cistern* for a *water closet* or other *sanitary fitting*.

flushing mechanism A device that delivers nine litres of water in five or six seconds, emptying the *cistern*. It used to be siphonic, but is now operated by valve and float. It may be a *dual-flush* type.

flushing trough, t. cistern A long water tank extending above a *range* of several *water closets* with peaks in demand, so that each can be flushed at short intervals. There is less plumbing required, but repairs can stop the whole system, which is not the case with individual *flushing cisterns*.

flushing valve (USA **flushometer**) A valve which supplies a precise quantity of water to flush a WC and therefore replaces the usual *cistern*. It has the advantage that there is no need to wait while the cistern refills, but, because of its direct connection with the water main, not all water authorities allow it.

flush joint A brickwork joint made by *striking off* the mortar during laying, or one that has been *flush-pointed*.

flush pipe The pipe down from a *cistern* to a WC, urinal, or other *sanitary fitting*. It should be of 32 mm internal dia. for high-level cisterns or 35 to 41 mm for low level, and has a push-on rubber or plastics *pan connector*.

flush pointing *Pointing* finished with a *flush joint*.

flush slide A small door-bolt, usually recessed into the *stile*.

flush soffit A slab without *downstand beams*, giving a flat underside.

flush valve A *flushing valve*.

flux (1) In *soldering*, *brazing*, and *welding* (C), fusible substances such as borax which cover the joint, prevent oxidation, and so help the molten metal to stick. After the joint is made, any remaining flux should be removed, as it is corrosive, particularly if water-soluble.
(2) *Luminous flux*.

flyer A *flier*.

flying form Movable *formwork* for concrete that is lifted by crane, usually wall shutters or slab decking.

flying scaffold A *scaffold* hung from *outriggers* to give access to the outside walls of a tall building. Small units may have ropes and pulleys, and a *cradle* to carry two or four men, plus tools and materials. Large units, such as those used for installing *curtain walls*, have steel cables and winches and may go round the full building, be several storeys high, and have their own *debris-collection fan* on top.

flying shore, flier, horizontal s. A temporary horizontal strut above ground level, often placed between the walls of two houses in a street when the house in between has been demolished. For spans longer than 10 m, timber flying shores long enough are difficult or impossible to find. Consequently tubular-steel scaffolding is then used, with intermediate scaffolding towers to the ground where convenient. See diagram, p. 406.

fly wire Fine mesh, of metal wire or plastics, made for *insect screens* but also used as *scrim* between the joints of *tapered-edge plasterboard*.

foam (1) *Cellular* material. See also **aerated concrete** and terms under **expanded** or **foam**.

(2) An *extinguishing medium* of fine bubbles that smother and wet a fire, even of *flammable liquids* such as petrol and oil. The bubbles are made with chemicals for hand extinguishers or mechanically generated in large systems for application by hose.

foam-backed rubber flooring *Sponge-backed rubber flooring.*

foamed plastics, foam p. *Expanded plastics*, such as *urea-formaldehyde foam*, *polyurethane foam*, and *phenolic foam*; perhaps strictly also sponge *rubber*.

foamed slag Blastfurnace *slag* sprayed with a little water as it cools and used as a reliable *lightweight aggregate* of density 320 to 880 kg/m³, giving concrete of up to 24 N/mm² strength, to BS 877.

foam glass *Cellular* glass used as thermal *insulation* in trafficable areas. It is strong and is non-combustible, although its insulation value is only about half that of common expanded plastics.

foam inlet A connection for the fire brigade *foam*-making equipment.

foil Metal which is thinner than 150 microns (0.15 mm), e.g. *aluminium*.

folded sheetmetal Metalwork fabricated by folding to shape, such as galvanized steel *ductwork*, equipment housings, and *Z-purlins*, usually thinner than *pressed-metal* sections. Panels may have *diamond break stiffening*.

folding door, sliding-f.d. A *door* with two or more shutters or doors hinged together that open or close by folding or unfolding, used in confined spaces.

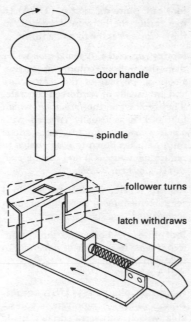

door handle

spindle

follower turns

latch withdraws

A follower.

There are many types: *bi-fold, centre-opening, multi-folding,* etc.

folding shutters *Boxing shutters.*

folding wedges, easing w., foxtail w., striking w. Wedges, often of hardwood, used in pairs to tighten up or slack off *formwork, shores, falsework* (C), and *centers* of all sorts. See diagram, p. 406.

follower A plate inside a door *lock* with a square hole. It is turned by the door handle, attached to the *spindle*, which fits in the square hole to withdraw the *return latch*. See diagram.

foot bath A *sanitary fitting*, recessed into a floor, for washing feet.

foot block An *architrave block*.

foot bolt A strong *barrel bolt* set vertically at the foot of a door.

foot cut

foot cut, plate c., seat c. (USA) In *roof cutting* the horizontal saw cut in the *birdsmouth* at the foot of a *rafter*.

footing, spread f. A *foundation* wider than the wall or column it carries. Footings are part of the *substructure* and are usually of reinforced concrete. They are cast on the *formation*, which in the UK is usually covered with *blinding concrete*. Footing excavations must be taken down to *good ground* to reduce *subsidence* (*C*) and can be *pad, raft* (*C*), or *strip footing*.

footing fabric, trench mesh Steel reinforcement *fabric* in widths to suit a *strip footing*.

foot plate (1) A horizontal timber from a *wall plate* to the foot of *ashlering*, to hold it in place.
(2) A *sole plate*.

footprints, pipe tongs (USA **combination pliers**) A pipe fitter's adjustable wrench with serrated jaws, made in different sizes to grip steel pipes from 76 mm dia. down to a few mm. It also turns nuts but burrs the flats, preventing them being turned with a spanner afterwards.

footstone A *gable springer*.

force account (USA) *Daywork*.

force cup A *sink plunger*.

forced-action mixer A *pug mill*.

forced air *Mechanical ventilation*.

forced circulation, mechanical c. The pumped circulation of fluids, or the fan circulation of air, as opposed to *gravity circulation*.

forced draught, mechanical d. Flue *draught* assisted by a *fan*, needed in efficient *boilers* which cool the smoke so much that the *flue* draws poorly, as occurs in most commercial boilers and domestic *condensing boilers*.

forced drying The *drying* of paint at or below 65°C, thus cooler than *stoving*.

force majeure, act of God, special risks In a building *contract* an event over which the builder has no control (strike, flood, etc.) and which prevents work from being carried out. It may lead to *frustration of contract*. Lists of typical force majeure events are given in most standard *conditions of contract*.

foreman An experienced *tradesman* (usually a man) in charge of other men, either the *general foreman* or a *trade foreman* under him. Foremen may work on the tools (working foreman) or merely supervise others. He must see that the *tradesmen* have all the materials and labourers they need, and check the *setting-out*, and so must be able to read the working *drawings*.

foreman bricklayer An experienced *bricklayer* in charge of the brickwork on a large site, where there may be several bricklaying *gangs*.

foreman carpenter An experienced *carpenter* who works on a building site, supervising carpentry, formwork, and joinery. Many *general foremen* have been promoted from foreman carpenter.

foreman plasterer A capable *plasterer*, usually with experience in *fibrous plaster*, solid plaster, and *dry linings*, able to work to *drawings* and the *specification*.

forend, fore-end The side of a morticed *lock* case that is visible on the edge of the door *stile*. It has holes for the *return latch*, *dead bolt*, screws, etc.

form The basic element from which *formwork* is made up. It can be any box, shutter, pan, or mould, with special types for columns, slab or beam soffits and sides, wall stop ends, etc. Forms have *draught* for easy *stripping*.

formaldehyde A strong-smelling poisonous gas which dissolves easily in water (and then sold as Formalin). It is used in producing synthetic *resins* for plastics, either polymerized into *acetal* or combined with *melamine, phenol, resorcinol, urea*, etc., to make *adhesives*, foamed *insulation*, and other products. Formaldehyde gas, if given off by *particle board* or panel products, can accumulate dangerously in poorly ventilated areas. It caused problems with some older types of *cavity fill*, now restricted in the UK and most of Europe. It is now rarely linked to *sick buildings*.

format A simple repetitive layout, e.g. of bricks or tiles.

format dimensions The face dimensions of a brick, block, or tile, plus the width of its *joints*. See **brick format, special**.

formation The surface of a completed *excavation*, usually a level bottom, ready for the following work, after *bottoming* or trimming. For footings it may need a *foundation inspection* before *blinding concrete* is laid.

formed concrete *In-situ concrete* placed in *formwork*.

former (1) A guide used in bending or shaping sheet materials, rods, or pipes.
(2) A device to create a hole in concrete, such as an *expanded polystyrene* block for making a *void* or *pocket*.

form of contract Typical *conditions of contract* and an *agreement* that can be filled out to suit a project and client, usually a *standard form*.

form of tender A one-page document including the project name, the *tender* price, the time for completion, and other details, signed by the contractor.

formply *Plywood* used for linings to *formwork*, often *film-faced*.

form tie A steel rod between two *wall forms* to keep them the right distance apart when filled with concrete. It is left in the concrete. The ends of *cone ties* or *snap ties* come off below the surface, but not *she bolts*. See **cover**.

formwork Anything that holds fresh in-situ *concrete* in place until it hardens, such as plywood *shutters*, steel *pan forms*, fibreglass moulds, or *profiled decking*, as well as its supporting *props, centering*, or *falsework (C)*, plus accessories like *wedges* and clips for tightening joints and to make *stripping* easy. Formwork must bear all loads put on it by *concreting* and be sturdy and grout-tight during *vibration*, particularly for *direct finishes*. It also should have many re-uses. Formwork is usually stripped after twenty-four hours for vertical surfaces such as columns and walls. Under *soffits* of beams or slabs, the props and forms are either left in place until the concrete has gained sufficient strength or they are removed followed by *back propping*. Usual abbreviation: 'fwk'.

formwork carpenter A *tradesman* who erects *formwork*, usually as part of a *gang*. During the *placing of concrete*, one or two formwork carpenters should be present to tighten wedges or to stop the work if a collapse appears likely.

formwork foreman A *trade foreman* in charge of *formwork* on a large project, usually an active and aggressive leader if formwork is a *critical activity*.

forward pass A calculation of the *earliest date* at which an *event* should occur, using the *critical path method*. The estimated time *durations* for each *activity* are added together from the start or *beginning event* to the *end event*.

foul air, vitiated a. Air that is no longer fit for breathing, such as that in toilet areas, inside drain pipes, or in a *sewer (C)*.

foul-air flue A ventilating duct which draws air out of a room.

foul drain A *drain* to carry *foul water* from a building to the sewers.

foul water Water that is contaminated with *soil water*, *waste water*, or industrial effluent, in a *foul drain*. This usually excludes rainwater.

foundation (1) The supporting ground underneath a building. Excavations for foundations are taken down to *good ground*. See *cracking in brick walls*.
(2) The part of a building which sits on the ground, to carry the *substructure*. Low-rise buildings and houses commonly have shallow foundations, usually *strip footings*, while medium-rise buildings may need *pad footings* or a *raft foundation* (*C*). High-rise buildings on clay are often on deep foundations such as *piles* (*C*). See **trees**.

foundation bolt A *holding-down bolt*.

foundation inspection An examination by the *building inspector* (or *clerk of works*) of the ground on which *footings* are to be built, to ensure compliance with the *Building Regulations* (and the *specification*). Once approval is given (usually in writing) either footings should be concreted immediately or *blinding concrete* laid to protect the *formation*.

four-coat system The traditional way to paint new wood. There is agreement that the *paint system* should include *primer*, *undercoat*, and *gloss* coat, but some authorities insist on two undercoats, others on two gloss coats (some gloss coats cannot be put on to another gloss coat). For internal work a good finish may be had by using two undercoats and *flatting-down* before recoating. A slight *tint* is added to one undercoat so that *skips* can be seen. For outside work, better protection should be had from two gloss coats.

four-pipe system (1) A heating system with separate *flow* and *return* pipes for central heating and for *domestic hot water*.
(2) An air-conditioning system with separate *flow* and *return* pipes to *heating coils* and *cooling coils*.

four:two:one mix (4:2:1 mix) An ordinary *prescribed mix* (*C*) of concrete, *batched* by volume, with 4 parts of coarse *aggregate*, 2 parts of fine aggregate, and 1 part of *cement*, mixed on site and used for sundry non-structural work.

foxtail wedges Wooden wedges used either in pairs as *folding wedges* to tighten formwork or in joinery for *secret wedging*.

frame (1) **framework** A loadbearing structure made up of slender *members* connected by *joints*. Building frames can be a skeleton of columns and beams which carry wall panels and floor slabs. They are usually of concrete, steel, or timber.
(2) In joinery, a rectangular surround of straight members joined at the corners, generally used for support and stronger than a *lining*. Frames are used to *hang* doors or windows, as well as round door mats or manhole covers.

frame construction Either *timber-frame construction* or a building with a frame and *bracing*, carrying infill or *cladding*. See **structural steel**.

framed door A wooden *door leaf* with a rigid frame, with at least top and bottom *rails* and side *stiles*. It can be *panelled* or *matchboarded*.

framed grounds Joinery *grounds* for high-quality work, framed together with mortice and tenon joints, to fix *linings* round door or window openings.

framed ledged and braced door,

f.b. and boarded d. A *matchboarded door*.

framed partition, stud p., stud wall A *partition* built up on its own frame of timber or steel *studs* and covered with a *lining* material.

frame house (USA) *Timber-frame construction*.

frame room A *plantroom* for a telephone *main distribution frame*.

frame saw A *gang saw* with a number of blades in one frame, used to cut several slabs of wood or stone in one operation, giving even thickness and fairly smooth faces. Frame saws with diamond-toothed blades are used to cut slates.

frame tie A *wall tie*.

framework See **frame**.

framing (1) The work of building a *frame*.
(2) Frame members, as of a door *leaf frame* or a *curtain wall*.

framing gun A large *nail gun* that drives nails 100 mm long or longer.

framing square A *steel square*.

free cooling The use of cool air from outside a building, instead of running a water *chiller*, to cope with heat from lighting and computers.

free float Spare time for an *activity* on a *programme*, left over from the early completion of previous activities. It can be used without delaying later activities.

free lime *Lime* inside set concrete, produced by chemical reactions as the *cement* sets. It may be further converted by *carbonation* or a *hardener* or, if moisture is present, cause corrosion of lead metal.

freemason In the Middle Ages and later, a skilled *mason* capable of carving

freestone. He was paid more than a rough mason.

freestanding (1) **self-supporting** Of a building element or structure, able to stay upright by itself, as do chimneys, parapets, and scaffolds.
(2) Of equipment, double-sided, e.g. a bench accessible from both sides.

freestone Building stone soft enough to be cut with steel tools and uniform enough to be carved in any direction, being free of *cleavage* – generally limestone or fine-grained sandstones with high compressive strength and good durability. Usually freestone can be laid at any angle.

free stuff *Clear timber*.

freezer A *cold store* kept below 0°C, either down to – 16°C or a *deep freezer*. The *insulation* needs a *vapour barrier* on the outside ('warm side'). Freezers on the ground need a vapour barrier over the foundations, with ventilation from its top side ('cold side'), to prevent ice forming from *rising damp* and causing *frost heave* (C).

freeze–thaw cycle A series of temperatures that rise and fall above and below freezing, which may damage saturated, porous materials. In brickwork the damage is avoided by using *frost-resistant bricks* or by excluding damp with protective overhangs, *copings*, *weatherings*, and *damp-proof courses*.

freight elevator, trunk lift (USA) A *service lift* which is used for hoisting furniture and other heavy loads in a building, but not for carrying passengers.

French casement, F. door, F. window A *casement door*.

French drag A mason's hand tool resembling a plasterer's *float*, with several metal blades set upright in the working face. Toothed drags are used for *dragged work* and plain drags are used in France today for cleaning *formwork*.

French flier A *balanced step*.

French polishing A traditional finishing for high-grade joinery of *shellac* dissolved in methylated spirits, applied in two stages, *bodying-in* and *spiriting-off*. It gives a warm appearance to timber, but is easily damaged, and can be dissolved by whisky, gin, brandy, or other spirit.

French roof A *mansard roof*.

French tile A clay *single-lap tile* occasionally imported to Britain.

Freons A trade name for vaporizing liquids used as *refrigerants*, with different grades for cool rooms and freezers. They are *chlorofluorocarbons*.

fresh-air filter The first main filter to remove dust and large particles from fresh air for *air conditioning*. It is usually a viscous *impingement filter*.

fresh-air inlet (1) The fresh-air intake for an *air-conditioning* system, with louvres to keep out rain and often a *bird screen*. It should be located well away from any *flue*, from sources of dust or pollution, and from air and spray given off by *cooling towers*, which can breed *legionnaire's disease*.

(2) A box about 150 mm square, fixed on a pipe to the *intercepting trap*, to admit fresh air to a long drain pipe, used only if no *vent pipe* connects to the manhole.

fresh-air rate The percentage of fresh air that is mixed with *supply air*, usually in a *mixing box*, replacing some of the *return air*.

fresh concrete, wet c. Mixed concrete that has not reached *initial set*. This takes one or two hours, giving time for *placing* in formwork.

fretsaw A *coping saw* with a deep U-shape frame. See diagram, p. 388.

friction catch, f. latch A small door fastener that holds with a spring or ball. It requires very accurate fitting and can go out of adjustment easily.

frit Treated and finely ground sand, glass, and flint, used for glazing bricks and other *ceramics*.

frog An indentation in one or both bed faces of a moulded or *pressed brick*. It must not exceed 20% of the brick volume. It forces the clay to spread and fill the mould during brick making. In the UK bricks are usually laid frog upwards. See diagram, p. 48.

front The *face* or side meant to be seen, not the *back*, e.g. the street side of a building or the *forend* of a lock.

frontage The length of a building, or *site* in contact with a road.

frontage line A *building line* on a street when there is no *set back*.

frontager (USA) A property adjoining a street.

front hearth The concrete floor in front of a *fireplace*, projecting into the room.

front lintel The *lintel* supporting the visible wall over an opening.

front putty *Face putty*.

frost-proofing admixture (1) An *air-entraining agent* which improves the ability of the hardened concrete to resist frost attack.

(2) An *accelerator* used to warm fresh concrete in cold weather.

frost protection (1) During *winter working*, the prevention of damage to building materials exposed to cold conditions. All *wet trades* such as concrete, brickwork, and plasterwork, need a temperature above + 5°C for cement to set

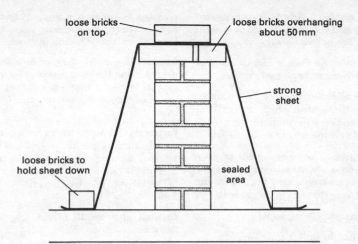

loose bricks on top

loose bricks overhanging about 50 mm

strong sheet

loose bricks to hold sheet down

sealed area

The protection of newly laid brickwork against frost. Concrete beams are treated similarly. In extreme cold the air space may need to be heated overnight. The work is warm and dry next day.

and during early *curing*. *Accelerators*, heating, or *curing blankets* may be essential. Dry trades are less affected, but paint can suffer slow or defective drying and adhesives may not set properly in frost. See diagram.

(2) Buildings can suffer severe flood damage from thawed water supply pipes, previously frozen, which have burst. There is also a risk of boiler explosion from ice blocking the cold feed to a hot-water system or the expansion pipe. Water from an open reservoir or filter beds may be not much above 0°C when delivered into the supply mains in winter. To ensure that the main does not freeze, it should be covered by at least 750 mm of earth. In Sweden 2 m is usual. Water mains should therefore rise into the house well inside the outer walls. Frost protection for roof cold-water *cisterns* includes suitable *insulation* of the cistern itself as well as its *warning pipe*.

frost-protection quilt A *curing blanket*.

frost-resistant brick The *durability* of bricks exposed to frost is classified under BS 3921. Designation F (frost-resistant) bricks are suitable for use in an exposed location and repeated *freeze–thaw cycles* even if saturated; designation M (moderate frost-resistant) bricks need protection from a *coping* or *eaves*; while designation O (not frost-resistant), e.g. *underfired* clay bricks, are only for internal walls.

froststat An automatic control, a type of *thermostat* that switches on the heat when the *circulating water* temperature falls to nearly 0°C.

frustration of contract When a contract cannot be carried out because of unforeseen events, such as *force majeure*, it is said to have been frustrated. The contract therefore comes to an end.

full, on the f. side Slightly oversized compared with the intended or *nominal size*. The opposite is *shy*.

full coat As thick a *coat* of paint or varnish as can be properly applied.

full collective control An electronic *lift-control* system which stores all car and landing calls (up/down) and sends the *lift car* to each in turn on the way.

full gloss The highest grade of *gloss*. It reflects the most light and is the easiest to keep clean, but shows the most imperfections in the *substrate*.

full height *Ceiling height*.

full-height unit A *ceiling-height unit*.

full-length bath A standard domestic *bath*, in Britain 1700 mm long by 700 mm wide.

full-length pipe A pipe as manufactured, not a shortened *cut*.

full-scale drawing A drawing the same size as the object it represents.

full size The *tight size* of an opening for glass.

full-way valve A valve which has a full-bore waterway when opened and so does not impede the flow; e.g. a *gate valve* and a quarter-turn *cock*.

fully quarter-sawn timber Boards *quarter-sawn*, with no growth ring at an angle of less than 80° to the face, which minimizes warping.

fully supported metal-sheet roofing *Supported sheetmetal roofing*.

fully vitrified ceramic A *ceramic* that has been fired at above about 1200°C and vitrified throughout its thickness. Its water *absorption* is below 0.5%.

fungicidal paint A *paint* which dis-courages *fungus* even in tropical conditions. Many disinfectants can be blended with paint for this purpose.

fungus The cause of *decay* in wood, *dry rot* being the commonest in buildings in Britain. Other types of fungus are *mould*, mildew, *wet rot*, and *blue stain*.

furnace (1) A *heater* with a fire inside, usually burning gas, oil, or solid fuel in a *firebox*, although electric furnaces are used in *fire testing*.
(2) (USA) The usual name for a *boiler*.

furnish and install (USA) *Supply and fix*.

furniture (1) (USA **finish hardware**) Decorative *hardware*. See **door furniture**.
(2) *Street furniture*.

furniture beetle (Anobium punctatum) The most common *borer*. It attacks green hardwoods or softwoods and leaves *pinholes* of about 1–2 mm dia.

furring, encrustation, scale Deposits of the minerals from *hard water* when it is heated or when the pressure drops. It occurs in *boilers* and pipes, at hot-water taps, shower heads and sprayers, in cooking pans, hot-water jugs, etc. It dissolves in vinegar.

fuse (1) A *circuit protector* with a fuse element (see below) to prevent overheating in wiring, which may cause a building fire. If the fuse is properly designed no *overcurrent*, e.g. from a *short circuit*, can last more than a fraction of a second. High-breaking-capacity *cartridge fuses* have replaced *rewirable fuses*.
(2) **f. element** A piece of thin 'fuse wire' or metal strip, to protect an electric circuit. It melts and stops current passing if an *overcurrent* occurs.

fuseboard, fusebox A *distribution board* with *fuses*, or a *consumer unit*.

fused (1) Joined by melting, as in the *heat fusing* of plastics.

(2) A circuit or device with a *fuse* for overcurrent protection.

fused switch A switch containing a built-in *fuse* that is turned during switching and makes the contact. Compare **switchfuse**.

fusible link A metal part which, until it melts (e.g. at 68°C), holds open a *fire*

door (door holder) or a *fire damper* against the force of its closer. Formerly used in fire *sprinklers*.

fusible plug A metal plug of low melting point screwed into the part of a boiler just above the furnace, under the water. If the water level drops below the fusible plug, it will melt, allowing steam and water into the fire, which puts the fire out.

fusion welding, f. joint The *heat fusing* of plastics or metals. See *C*.

G

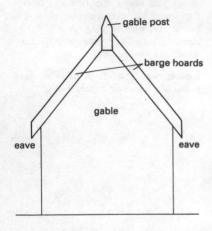

A gable.

gable, g. end The triangular part of the end wall of a building with a *pitched roof* between the *barge boards* or rafters. A gable may be of any material – masonry, *cladding* such as weatherboards, hung tiling, etc. See diagram.

gable board A *barge board*.

gable coping A coping to a *gable wall* projecting above the roof. It sits on top of the *kneelers*.

gable post A short timber post at the apex of a *gable* into which *barge boards* are *housed*. It is crowned with a *finial*.

gable roof, gabled r. A roof with *gables* at one or both ends. The edge of the roof covering is the *verge*.

gable shoulder The projection of a *gable springer* at the foot of a *gable coping*.

gable springer, footstone, skew block, skew corbel An overhanging stone, brick, or tile *corbel* at the base of a *gable*. It carries the bottom *kneeler*. See diagram.

gablet A small gable, e.g. for a *gambrel roof* or a *dormer*.

gable wall A wall crowned by a *gable*.

gaboon, g. mahogany, okoume An *African mahogany* (Aucoumea klaineana).

gain (1) An inward flow of heat which may create a cooling load on mechanical plant, usually *solar gain* or heat that warms *chilled water* pipes.
  (2) A *mortice* or notch to receive another timber or a timber *connector*.

gallet A *spall*.

gallows bracket An upright with a horizontal member projecting from its top and a *knee brace* between them, forming a triangle.

galvanized iron pipe Galvanized steel *screwed pipe* for water.

galvanizing A *coating* of zinc on steel which is quite hard and gives good protection against mild or moderate conditions of exposure, although it can be corroded by *dissimilar metal contact*. The usual process is by *hot dip* (C), given after all joints have been welded and all holes and cuts have been made. Minor damage can often be tolerated by galvanizing, which tends to 'heal' across scratches. For large areas of damage *zinc-rich paint* is used to *touch up*. Galvanizing can be further protected by *organic coatings*, but painting

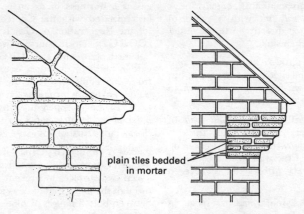

plain tiles bedded in mortar

A gable springer.

can be very difficult. Before painting, galvanizing needs cleaning with a *wash primer* or a few months' exposure to the weather, which has a similar effect, then *passivation* treatment and a *pre-treatment primer* or etch primer, followed by the top coats. See also **aluminium–zinc coatings**.

gambrel roof (1) **half-hipped roof** An end to a *pitched roof* which slopes up from the *eaves*, as does a *hipped end*, but stops part way at a vertical *gablet*. The shape resembles a horse's hind leg. See diagram, p. 376.

(2) (USA) A *mansard roof*.

gang (1) (USA **crew**) A group of operatives on a building site. A gang of *labourers* is under a *ganger*, but a gang with both craftsmen and labourers, such as a bricklaying or formwork gang, is under a *trade foreman* or *charge hand*.

(2) A description of a set of similar parts that act together.

ganger (USA **sub-foreman**) A man in charge of a *gang*.

gang form Several small panels of formwork that are re-used as a single movable *form*, not separated during *stripping*.

gang nail A *nailplate*.

gang rate The *unit price* per hour of a *gang*, which may change if the 'gang mix' changes, e.g. when more labourers are introduced to bring materials up a *scaffold*.

gang saw A reciprocating saw with several blades, e.g. a *frame saw*.

gangway (1) A path of *scaffold boards* laid for walking on or to wheel barrows along.

(2) A *circulation area*, as between machines or storage areas.

gap An *open joint*.

gap-filling glue An *adhesive* for gluing wood which is strong even with a joint 1.5 mm thick. Sawdust mixed with *close-contact glue* fills gaps also.

garage A car shelter with a roof and walls, unlike a *carport*. A lock-up garage has a door, usually of *up-and-over* or *rolling* type, often operated by electric motor with radio control from inside the car.

garbage disposal sink, s. grinder A *waste disposal unit* with a powerful grinder fitted directly on to the outlet from a kitchen sink.

Garchey sink A *waste disposal unit* that uses the sink as its entry point but with its own separate waste *stack* in addition to the normal drain pipes.

garden-wall bond A *solid wall* of brickwork 215 mm (one brick) thick. It is laid as *stretchers* showing *facework* on each side of the wall, with a course of *headers* every fifth, sixth, or seventh course.

gargoyle A roof *rainwater spout*, in the past made of stone carved as animal heads and lined with sheet lead.

garnet hinge A *cross-garnet hinge*.

garnet paper *Sandpaper* covered with grains of garnet.

gas Usually *natural gas* used for heating and cooking. Safety regulations apply to gas supply pipes, appliances, flues, and flue terminals. In high-rise buildings a *riser* pipe must be run up a *protected shaft*. Pipework usually must be installed by a registered *gas fitter*. See **bottled gas, town gas**.

gas circulator A small boiler for heating *domestic hot water* by gas in warm weather when the main *central-heating* boiler is not working. It is usually connected to a storage *cylinder*.

gas concrete The lightest, weakest, most insulating *aerated concrete*. It contains no sand and is made by foaming the mixing water with cement alone.

gas detector A *combustion gas detector*.

gas fitter A person who installs gas central heating, cookers, or other appliances, and their fittings, pipes, and *flues*. In the UK regulations require that gas fitters working with piped gas as a business or as an employee be registered with the Confederation for the Registration of Gas Installers (CORGI). Do-it-yourself work on liquefied petroleum gas (LPG) from garden tanks or cylinders is excluded from the regulations, but the law requires anyone who works on gas fittings to be competent. Any local office of the *Health and Safety Executive* can give further advice.

gasket A preformed permanently flexible layer or strip squeezed firmly between two surfaces to prevent leakage through the joint. Flat gaskets are used between bolted *flanges* of pipe fittings, and profiled rubber or plastics extrusions for *gasket glazing*.

gasket glazing, dry-g. g. Window *glazing* using *gaskets* of synthetic rubber or plastics, without the use of *putty* or *glazing material* but often relying on *compression glazing* to make the seal, particularly for *split gaskets*. Gasket glazing allows great differences in *thermal movement*, such as between glass and an aluminium surround, and is common in factory-finished window units and *curtain walls*.

gaskin Organic fibre such as tow, traditionally used for waterproofing screwed metal pipe joints or put in before the mortar jointing of *spigot-and-socket* drain pipes.

gas pliers Strong pliers with serrated concave jaws for gripping gas pipe, but made so that overtightening of fittings is difficult, to prevent damage.

gas strut A cylinder with a piston that acts as a compression spring, e.g. to open a rooftop *fire vent* once it is released by a *fusible link*.

Gastronorm pans Standard-size rectangular stainless-steel pans with detachable handles, used as both cooking and serving vessels. The largest size,

1/1, is 530 by 325 mm and there are smaller sizes, such as 2/3, down to 1/8, which fit together in a servery counter. They can be washed in a dishwasher, eliminating the traditional *pan wash*.

gate valve A *valve* with a sliding disk which opens to full pipe bore.

gathering A narrowing of a fireplace *chimney* above the *throat*, leading to the *flue* and forming the sides of the *smoke chamber*.

gauge (1) A set size, volume, or weight, or an associated measuring device.

(2) One part of the main *binder* in a mix, from which the other parts are measured. See **gauged mortar**.

(3) **batten g.** A dimension. It is the spacing between roof *battens*, as well as the distance up a roof between the bottom ends of each course of *tiles* or *slates*. It is also the slope length of the seen part of each tile or slate. For *single-lap tiles* the gauge is the tile length less the *headlap*. For *plain tiles* it is the tile length less the headlap and divided by two. See diagram, p. 378.

(4) A wooden or metal boundary strip used in asphalting to show the correct thickness of asphalt, as does a *screed rail* in plastering.

(5) The thickness of sheetmetal, or of plastics sheet, or the diameter of wire. Heavy-gauge sheetmetal is thick, light-gauge is thin. See **standard wire gauge**.

gauge board (1) A *banker* for batching and mixing mortar, plaster, etc.

(2) A *pitch board*.

gauge box A *batch box*.

gauged arch An arch of *gauged brickwork* with each brick individually tapered and splayed, often a *flat arch*.

gauged brickwork *Facework* of soft bricks (rubbers) sawn to shape or rubbed smooth on a stone or another brick and laid with very fine joints, often in pure *lime putty*, which may be 3 mm thick or even less. It was in fashion in England from 1650 to 1750 and in Colonial Virginia, then sporadically in fine Victorian buildings, but even more rarely since that period.

gauged mortar A *mortar* in which the mix proportions are given by comparison with cement as one part. See **mortar designation**.

gauged stuff, g. lime plaster Traditional *lime putty* to which either gypsum *plaster* or cement (but not both) was added to hasten the set.

gauge pot A container for pouring liquid *grout*.

gauge pressure Pressure measurement starting from atmospheric pressure as zero.

gauge rod A stick used for *setting out* distances. See **storey rod**.

gauging (1) Measuring each constituent of a *batch* of concrete, mortar, or plaster, compared with one part of *cement*, *gypsum*, or other main *binder*.

(2) Marking timber with a *mortice gauge* or *marking gauge*.

(3) Sawing and rubbing bricks to size and shape for *gauged brickwork*.

gauging board A *banker*.

gauging box A *batch box*.

gauging plaster *Plaster of Paris* or other *gypsum plaster*.

gauging trowel A triangular *trowel* for measuring plaster. See diagram, p. 332

gauging water *Mixing water*.

gaul A hollow in a *finishing coat* of plaster caused by bad trowelling.

G-cramp A steel G-shaped screw *cramp* used by *joiners* when gluing wood.

gel coat The smooth, outer protective coat of resin on *glassfibre-reinforced polyester* with no glassfibres (to project and let in damp), but with *pigments* that resist solar damage. It goes into the *mould* first, 0.4 to 0.5 mm thick, and is followed by *laying-up*. An old gel coat is difficult to remove and replace.

general areas A description of all or most of a *finishing*, without details or exceptions.

general attendance On a large project the *main contractor* provides *temporary works* such as scaffolding, power and water supply, etc., which are also used by the *sub-contractors*. Charges for their use are deducted from payments made by the main contractor to each sub-contractor, in proportion to the value of each sub-contract. It differs from *special attendance*.

general contractor A *main contractor*.

general foreman The *main contractor*'s representative on site, in charge of all labour under the *site manager*. He coordinates the work of *trade foremen*, whether employed by the main contractor or by a *sub-contractor*. He has usually been a `carpenter`, which involves contact with all trades.

geometric stair A *stair* which on plan may be circular, semi-circular, or elliptical. It has *tapered treads*, no newel posts, and often no landings between floors. The inner handrails have *wreaths*.

Georgian wired glass A *wired glass* with 13 mm square mesh.

German siding, novelty s. *Weatherboards* concavely rounded on the top edge and rebated on the inner face of the bottom edge.

German silver, nickel brass A brass alloy that contains copper, zinc, and nickel. It is strong and has good corrosion resistance.

ghosting *Core patterning*.

giant form A large piece of *formwork*, such as a wall *shutter* or *table form*, usually moved on wheels or lifted by crane.

gilding A finishing of *gold leaf*.

gimlet A small hand tool with a T-handle and a corkscrew-pointed steel shaft, for boring holes smaller than 6 mm in wood or soft materials. It was used by the ancient Greeks and in some ways resembles an *awl*.

gimlet point The threaded point of *wood screws*, *coach screws*, etc., which form at least part of their own hole as the screw is driven.

gin A tripod and *gin block* or other simple *lifting tackle* (*C*).

gin block, jinnie wheel, rubbish pulley One pulley with a fibre hoisting rope passed through it. The pulley is in a steel frame with a hook by which it hangs.

girder A *primary beam* made as a *truss* with parallel top and bottom *chords* (*C*).

girder bracket A *joist trimmer*.

girder truss A *trussed rafter*, usually of timbers thicker than standard size, that supports other roof members, such as those used to form a *hipped end*.

girt (1) **ledger, ribbon board** A small girder, a *rail* or intermediate beam in a wooden-framed building, often carrying floor *joists*.

(2) **cladding rail** A horizontal beam running between columns on the out-

side wall of a building. The main load cladding rails bear is horizontal wind load from the external *cladding*, so they are laid flat, not on edge.

girt board A timber *girt*.

girth (1) The circumference of round timber, measured by tape.

(2) The width of sheet material required to make up a folded shape, such as a roof gutter. In *bills of quantities* girth is stated to allow pricing.

(3) A *girt*.

gland (1) A seal round the shaft of a pump or between the body of a tap and its *spindle* to prevent water leaking out. Glands can be made with *packing* or rubber *O-rings*.

(2) **olive** A compressible copper or brass ring with a very shallow D-shape, which in a *non-manipulative joint* is slipped over the copper tube. When the fitting is tightened, the soft metal of the gland is compressed and deformed, making a rigid, pressure-tight joint.

(3) A *cable gland*.

gland nut, packing n. A nut round the *gland* of a water tap. It has a conical inside end that squeezes the *packing* tightly round the *spindle* as the nut is tightened. When the tap starts to leak round the spindle, the gland nut needs to be retightened. The nut is removed to wind on more packing.

glare Light that is too bright, causing either discomfort and eyestrain or disabling human vision. Glare from *daylight* can be prevented by shading or *solar-control glazing*, but it can also come from *luminaires* directly or from glossy surfaces with high *reflectance*.

glass Hard but brittle flat glass allows light, and some heat, to pass. Its main use in building is for clear *glazing*. Commonly it is made of soda-lime-silica by the *float glass* process. Excep-

tions are patterned and wired glass, made by rolling, and special glasses made to be heat- or fire-resistant.

glass block, g. brick A hollow translucent block of toughened pressed glass obscured by patterns moulded on one or both faces and with the air exhausted from the cavity. When used as *partitions*, glass blocks give a pleasant diffused light, but they have low heat insulation value. Used as *glass concrete* they can form a fire barrier, as they stay tightly in their openings and have good *fire resistance*. They also have no *fire hazard*, being non-combustible. In *pavement lights* solid glass is used.

glass concrete (1) Reinforced concrete with *glass blocks* built in, usually flush with one surface. The blocks must be clean so as to stick firmly to the concrete.

(2) *Glass-reinforced concrete.*

glass cutter A tool with a small, V-edged, hard-metal wheel for *scoring* a straight line on flat glass, so that it can be accurately broken by bending over a table edge ('score and snap'). The cracking can be initiated by light tapping. Eye-protective *goggles* should be worn.

glass door Either an *all-glass door* or a *panelled door* with flat glass in the panels, which for external doors should have *safety glass* of suitable thickness.

glassfibre, glass wool, fibreglass Fine, tough fibres made by drawing and rapidly cooling molten glass. Different types are used for heat *insulation* or as reinforcement in plastics, roofing felts, plaster, or concrete. See following entries.

glassfibre cloth Woven glassfibre *rovings*.

glassfibre mat Random chopped *strand*, used as reinforcement in many composite materials.

glassfibre-reinforced cement
Glass-reinforced concrete.

**glassfibre-reinforced gypsum
(GRG)** *Fibrous plaster* reinforced with
glass.

glassfibre-reinforced plastics Usually *glassfibre-reinforced polyester*.

**glassfibre-reinforced polyester,
fibreglass (GRP)** The commonest
reinforced plastics, suitable for making
complicated shapes such as *cladding
panels*, gargoyles, roof trim, window
sills and frames, and corrugated sheeting, and likely to have at least thirty
years' life when properly made. GRP
is usually moulded by hand *laying-up*
and surfaced with a *gel coat*. It is fairly
strong and lightweight and although
not very stiff has high impact resistance. It can be highly decorative and
does not rot or corrode, but care is
needed to ensure *compatibility* with
some foam plastics. Containing no
metal, it cannot corrode from *dissimilar
metal contact*.

glasshouse A building that uses the
greenhouse effect of glass cladding.

glasspaper Inexpensive *sandpaper*
covered with the *abrasive* powdered
glass, used mainly for *rubbing down*
wood and *flatting down* paintwork. It
clogs up or wears out fairly quickly
and is then discarded. Glasspapers are
marked with numbers to show their
grade of fineness or coarseness.

**glass-reinforced concrete,
glassfibre-r. cement (GRC)** Either
factory-made precast or spray-moulded Portland cement concrete
units, reinforced with *alkali-resistant
glassfibre* (Cem-FIL); or acrylic
polymer cement concrete units, reinforced with *borosilicate glass* fibre.
The products are only 10 to 15 mm
thick and, like other *fibre-reinforced
concretes*, include *cladding panels*,
sewer pipes and linings, cable ducts,
and street furniture. The material is
also used for *squash-court plaster*, but at
the present state of development GRC
tends to become brittle in a wet environment and is not used for structural purposes. See BRE Digest 331.

glass-reinforced plastics (GRP)
Usually *glassfibre reinforced polyester*.

glass size *Glazing size.*

glass stop (1) A device at the lower
end of a *patent glazing* bar to prevent
panes sliding down.
(2) A *glazing fillet*.

glass tile, g. slate A roof tile made of
glass and laid among other tiles to let
light into an attic. Several glass tiles
must be used together if enough light
is to pass. Apart from glass some clear
or translucent plastics are allowed: unplasticized PVC, GRP (both stabilized against *ultraviolet*), or acrylic
material. For roof coverings in towns,
fire hazard must be carefully considered, as most plastics have poor fire
resistance and high fire hazard.

glass wool, g. silk A *non-combustible*,
flexible fibre made from molten glass
and used as a heat *insulator* or sound
absorber, obtainable as a *blanket* between waterproof papers, or resin-bonded, bitumen-bonded, or loose.

glaze (1) To install *glass* or clear *plastics* sheet in a window.
(2) A glass-like waterproof protective
coating fired on to the surface of *ceramics*, such as bricks or sanitary ware,
often over an opaque *enamel*. Glazes
are made from silica *frit*, porcelain, or
inorganic base materials (formerly
from common salt).

glaze coat A nearly transparent, thin,
coloured *coat* put on paintwork to enhance the colour of a *ground coat* below.
The process is called glazing.

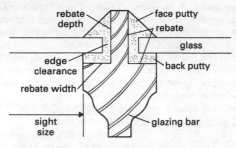

A glazing bar.

glazed door A *door* with a large or small window or a vision panel.

glazed tile An earthenware *wall tile* mainly for interior use, obtainable with a cream or white earthenware glaze, or an *enamel glaze* of many possible colours. Glazed tiles craze if heated and cooled. The only tiles that do not are unglazed tiles, e.g. *quarries*.

glazed ware, glazeware Common but outdated terms for *vitrified clayware*, a survival from the time when the best clayware was glazed by the vapour of common salt thrown into the kiln during firing.

glazier (1) A *tradesman* who cuts glass and fixes it in a window or door frame either with *putty* and *glazing sprigs* or with *glazing beads* of wood, metal, or plastics. Other processes such as *gasket glazing* are usually done in the factory.
(2) The *glazing* trade.

glazier's chisel A glazier's *putty knife* with a square end.

glazier's point A *glazing sprig*.

glazier's putty *Putty*.

glazing (1) Clear sheet materials, such as glass or plastics, with good *transmittance* of light, and the methods and materials used to fix them: *gaskets, putty, sealant*, etc. Glazing can be done in a factory or on site, and different methods are used for single or *double glazing*. In new work, glazing should have protective *manifestation*. See also **structural glazing**.
(2) The *trade* of fixing *panes* of glass or plastics into prepared openings in windows, doors, screens, partitions, etc. See BS 6262.
(3) The application of a *glaze coat*.

glazing bar (1) A T-shaped wooden divider between the small *panes* of thin glass in a traditional *sash*, often with a half-round *astragal* moulding. Its width affects the *sight size*. A horizontal glazing bar is a *lay bar*. See diagram.
(2) A vertical or sloping metal bar used for *patent glazing*, usually made of galvanized steel, extruded aluminium, or *lead-clothed* steel, with cappings of aluminium or plastics. See diagram, p. 319.

glazing bead The usual term for a *glazing fillet*.

glazing block A setting block, *distance piece*, or *location block* of fairly firm plasticized PVC or similar material, under, above, or beside a large *pane* at several points, to hold the glass centrally in the sash while the *glazing material* hardens.

glazing fillet, g. bead A wooden, metal, or plastics *quarter-round* strip with holes drilled through it for screwing to the rebate in a glazing bar. Other types clip in place over a round bead of sealant. Both systems supersede traditional *face putty* over the glass. In the best work corners are mitred. Reglazing is quick, neat, and clean because the old putty does not have to be hacked out.

glazing material, g. compound A substance that holds glass in place and makes a weathertight seal. Glazing materials include high-performance elastomeric *sealants* and *gaskets* as well as *non-setting glazing compound, metal casement putty*, and traditional *putty*.

glazing putty *Putty*.

glazing rebate An L-shaped cut into a corner for fixing *glazing*. For *putty* glazing its depth is at least 8 mm and its width the glass thickness plus 18 mm.

glazing size, glass s. The size of a *pane* of glass or clear plastics cut for glazing. The clearance between the edges and the window should be 1.5 mm all round for glass, giving a glass size of 3 mm less than the distance between the extreme edges of rebates in window bars, or *tight size*. For clear plastics greater clearance is needed. For double-glazed *sealed units* the space all round should be increased to 3 mm, so that the glass size should be 6 mm less for both height and width. Glass measurements should always be stated in the sequence length × height.

glazing sprig, glazier's point, brad A small headless nail driven into the glazing rebate of a timber sash or *glazing bar* after the *pane* has been pushed into the *back putty*. It holds the pane in place while the *face putty* is hardening.

gloss The reflection of light from *paintwork*. The stages of gloss in *finish* recognized by the British Standards Institution are:

1. flat (or matt), practically without *sheen*, even from oblique angles
2. eggshell flat (in the trade 'silk' or 'low sheen')
3. eggshell gloss ('satin' or 'low gloss')
4. semi-gloss
5. full gloss, a smooth, almost mirror-like (specular) gloss from any angle

In general the higher the gloss the harder and more *abrasion-resistant* the surface.

gloss paint Usually *alkyd paint* for *final coats*, suitable for exterior use.

glue An *adhesive*, strictly one made from animal or fish skins or connective tissue.

glue block An *angle block*.

glued-laminated timber, 'glulam' Large timber members built up of glued strips of softwood, used for *structural timber*, e.g. beams, columns, or *portal frames*. Strips can be lengthened with *finger joints* and the main limitation on size is the weight that can be handled by cranes or road transport. Durability is increased by using timber *preservatives* and *weatherproof adhesives*. See diagram.

glue line The thickness of a glued joint. Strong joints in wood have a narrow glue line made under *pressure*.

gluing clamp A *cramp*.

goat's-foot clip A folded steel plate, slit like a cloven hoof, which grips a *reinforcement bar* when hammered on. It is used to make a quick, simple clamp for concrete *formwork*. See diagram.

goggles, safety g. Eye protectors. Clear goggles should always be worn when cutting or grinding with *power tools*.

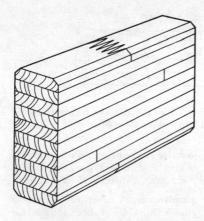

A glued-laminated beam. The laminations, 40 mm thick, are lengthened with finger joints and bonded together with synthetic resin adhesive.

going The horizontal distance between two successive *nosings* is the going of a stair *tread*. The sum of the goings of the treads is the *total going* of the *flight*, or in the USA the *run*. Going is measured along the *walking line* for curved stairs. See diagram, p. 437.

going rod A rod for *setting out* the goings of a *flight* of stairs.

gold leaf Very thin sheets of gold

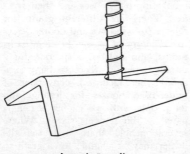

A goat's-toot clip.

used for *gilding*. They are stuck down using *gold size*.

gold plating A thin layer of electro-deposited gold used as a luxury finish to bathroom taps and fittings.

gold size Two types of gold size exist, both *oleo-resinous* varnishes. One dries quickly to the tacky state and remains tacky for some time, which helps for sticking down *gold leaf*. The other has more *driers*, hardens quickly, and is used for making *fillers*.

gondola A *cradle*.

gong A *fire-alarm* sounder worked by the flow of water to *sprinklers*.

Good Building Guides (GBGs) Leaflets issued each month by the *Building Research Establishment*, for designers and site supervisors, on aspects of building construction, such as repairing *lintels*, damp-proofing basements, where to put wall *insulation*, bracing *trussed rafters*, etc.

good ground, g. bearing g. Ground suitable for *foundations*, which may be found at the expected level or at an *extra depth*.

good practice, current trade p., site p. The accepted way of putting materials into place on site, including proper *workmanship*. See **code of practice**

goods lift A *service lift*.

gore A *lune*.

gouge A chisel with a curved cutting edge for hollowing out wood, more used by carvers and wood turners than by *joiners*.

grab rail A handle fixed to the wall near a bath or toilet, or a continuous handrail along a wall, for use by the aged, infirm, or disabled.

grade (1) Quality level. In building,

direction of grain ⟶

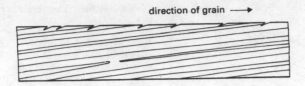

The grain of wood. The surface, rough from having been planed against the grain (chipped grain), would have been smooth if it had been planed to the right, with the grain.

grade includes the strengths of materials, especially concrete, steel, and timber. Different *mixes* of concrete are used in beams, slabs, and columns. Other types of concrete include: normal, *pumping-grade*, non-structural, and *dry lean*. For timber, see **gross feature**, **stress-graded timber**; for reinforcement steel see **grades of steel and concrete** (*C*).

(2) (USA) The *formation* level, or the ground level, or a *gradient* (*C*).

graffiti Illegal writing or drawing on a wall, sometimes discouraged by *anti-graffiti treatments*.

grain (1) The general direction, size, and arrangement of the fibres and other elements in wood. Grain is a *gross feature* and is described in relation to the length of a piece of timber, e.g. straight, sloping, cross, or end grain. A carpenter's *plane* used 'with the grain' cuts fibres that slope away from the direction in which it is pushed, leaving a smooth surface. Working 'against the grain' may cause the plane to jam or chip the wood surface. See diagram.

(2) The length of wood fibres, affecting both strength and *figure*.

graining Painting a surface to look like the *grain* of wood or marble, etc., by manipulating a wet coat of semi-transparent 'graining colour' with graining combs, brush, rags, etc.

grand master key A key that can open all door locks on a *key schedule*.

granite A hard rock, in many shades of grey or dark pink, used for *cladding* and *paving*, sometimes *flame-textured*.

granite sett A rough-hewn cube of *granite* with sides of about 120 mm for roads or 50 mm for pedestrian areas. Granite setts are decorative and non-slip and are often laid as *radial-sett paving*.

granitic finish A *face mix* resembling granite, on *cast stone*.

granolithic finish, grano. A *floor finish* made of a thin *topping* of cement, granite chippings, and sand, laid over a concrete floor slab, preferably as a monolithic *mono-grano screed*. It is hard-wearing and can be made *non-slip* by sprinkling emery or carborundum powder (1 kg/m²) over the surface before final trowelling.

grant In the UK, money provided by the government to help those who can least afford it towards the cost of renovating old houses and their *services*. There are different grants for repairs (to windows and doors, defective *damp-proof courses*, heating, bathrooms, and toilets), adaptation for a disabled person (access, bathroom, and kitchen facilities), conversion of under-used houses for letting, and for people with special needs (single parents, those over sixty). Where several people share a house, the owner may be eligible for a grant to bring it up to a basic standard of

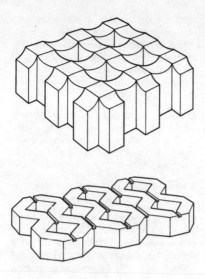

Grass concrete.

fitness by reducing *dampness* and improving light, ventilation, plumbing, and fire escapes. Minor works grants go to those on low incomes. Before work begins, the local council's grants officer should agree to it. An *architect* or *building surveyor* should supervise the work and usually the house must not be sold for three years.

granular fill Material such as *peagravel* or crushed stone which is non-cohesive, i.e. free of any sticky material like clay, and also without large stones. It is used for *bedding* under drain pipes, as well as beside and on top of them, and for other building *services* in the ground, usually · 100 mm thick.

granulated cork Small grains of *cork*, used as *loose-fill insulation*, weighing 80 to 100 kg/m³. See **cork board**.

grass concrete Precast concrete

paving slabs with holes and grooves inside which grass grows. They are used in areas of occasional car parking or emergency vehicle access. See diagram.

grated waste A *waste* that has a *grating* over its inlet.

grating, grate A series of strong bars over an opening that allows air, water, or small objects to pass. Gratings on the ground are made with different strengths, either for walking on or to carry vehicles. See **grille**.

gravel Coarse grains larger than 2 mm but smaller than 10 mm used for *bedding* under drain pipes and as *ballast* over built-up roofing. See also *C*.

gravel guard A *ballast grating*.

gravity circulation, thermosiphon A system which circulates water by *convection* from a heater to a storage *cylinder* or radiators and back again. It works because hot water is less dense than cold water and thus tends to rise in the *flow pipe* and fall in the *return pipe*. As the circulation pressure is not great, the pipes need to slope upwards from the heat source and have a large diameter, but no pump is needed.

gravity flow The movement of water in an open channel, or in a partly filled sewer or drain pipe, which goes downhill under the effect of gravity.

gravity pipe A pipe for *gravity flow*, not intended to take water under pressure.

green concrete, g. mortar Concrete or cement mortar after it has set but before it has hardened greatly, when it has a dark-green colour. It remains green for around seven days – more in cold weather, less in hot weather. See **curing**.

greenhouse effect Since a greenhouse has glass walls, the heat of the sun that enters it gets out more slowly, and the greenhouse heats up. Moisture is also kept in. The greenhouse effect is used in conservatories, *solar collectors*, and the *Trombe wall*. See **heat diode**. See also *C*.

green timber Timber soon after the tree is felled, containing its original free water. In most European species the *moisture content* is about 50%. Green timber requires *seasoning* before use in buildings or before being painted.

grey iron The original *cast iron* (*C*), far more brittle than *ductile iron*.

grid (1) A network of *grid lines* used for *planning* or *setting-out*.

(2) A distribution network for a public *service* such that supply can come from a second direction if there is a fault in one pipe or cable. The first grids were for electrical supply.

(3) The *runners* in each direction that carry the panels or tiles of a *suspended ceiling*.

grid line A line on a *grid plan*. There are two sets of parallel lines, running at right angles. One set is lettered A, B, C, etc., the other 1, 2, 3, etc., so that any location (e.g. for a column) can be given by a letter and number combination, e.g. B2, C3.

grid plan A *drawing* on which setting-out lines called *grid lines* coincide with the most important columns, walls, and other building *elements*. Steel-framed buildings are usually set out on a grid plan.

grillage A metal frame carrying tiles, slates, or shingles and replacing *battens*. The horizontal bars can be set at the *gauge* of the tiles.

grille, grill (1) A screen with holes in it, usually made of metal bars or wood,

used for decoration, security, or as a *non-vision grille*. See **grating**.

(2) A grating or screen through which air passes into (rather than from) an *air-conditioning* duct. A grille has no damper and connects to a *return-air* duct, unlike a *register*.

grinder A *power tool* with a *grinding wheel* for cutting or shaping hard, rigid materials, e.g. a hand-held *angle grinder*, a fixed *bench grinder*, a mobile *terrazzo* grinder.

grinding disc A thin, flat wheel containing *abrasive* used in *angle grinders*.

grinding wheel A rigid *abrasive* wheel for a *grinder*. There are many different sizes and types of abrasive for different grinders and materials, e.g. aluminium oxide for steel and plastics, silicon carbide for ceramics, masonry, and non-ferrous metals.

grinning, g. through Showing through, as an unwanted pattern or colour in a top coat of paint or plaster shows through from an *undercoat* or the *substrate*.

gripper strip, carpet g. A strip of wood with rows of small steel hooks on the top, fixed down to the floor round a wall and used to secure the edges of *fitted carpet*.

grit (1) Angular sand, used as an *abrasive*.

(2) The coarseness of *sandpaper* is measured as the number of grains per inch. Very coarse sandpapers have a low grit number, e.g. 12 grit, used for the first passes when *floor sanding* new hardwood, followed by 40 grit and finer. The sanding of joinery and the *flatting-down* of paintwork is done with papers from about 80 to 240 grit. Polishing sandpapers may be 600 grit or even finer.

(3) Sand removed from air by a *filter* or *grit arrestor*.

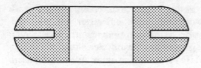

An electrical rubber grommet.

grit arrestor An *air-treatment* pre-filter, needed in climates where sand storms occur. It makes the air turn a sharp corner, flinging the sand out towards a narrow slit, through which it is expelled. Up to 10% of the air volume is also lost.

grit trap A *catchpit*.

grog, chamotte Burnt clay, crushed and mixed with fresh clay, giving it low shrinkage when fired a second time, used in *fireclay* and *refractory mortar*.

groin The curved line at which the *soffits* of two *vaults* intersect.

grommet A washer of synthetic rubber that fits inside a hole that a pipe or cable passes through, to make a seal or to prevent damage from rubbing. See diagram.

groove A long, shallow hole. There are many different types: *chases*, flutings, *raglets, throats*, etc.

gross features (USA **characteristics**) Some features in timber that can be seen, exploited in visual *stress grading*. They include *checks, knots, rate of growth, shakes, slope of grain*, and *splits*, all defined in BS 4978 (softwoods), BS 5756 (hardwoods), and ASTM D9.

ground (1) (USA) An electrical *earth*.
(2) A *substrate*. See also **grounds**.

ground breaking The start of ground-works, or a ceremonial hole dug with a new stainless steel spade at the official commencement of a building project.

ground coat An opaque coat of paint put on under a *glaze coat* or a *scumble glaze*.

ground duct, trench d. A U-shaped concrete trench in the ground along which services pipes or cables are run. It has removable cover slabs, which sit in *rebates* along the top inside edge of its walls.

grounded work *Joinery* fixed to *grounds*.

ground floor The floor which is nearest ground level. A concrete ground floor can be a *ground slab* or *suspended*. Timber ground floors are always suspended, at least 300 mm above the ground, over an *underfloor space*. In Britain the *U-value* of a floor must be better than 0.45, so in many cases thermal *insulation* is required.

ground-floor plan A *drawing* of the layout of the ground floor of a building. It shows the location of walls and columns and may show foundations, drawn as dotted lines.

ground investigations *Site investigations*.

ground level The height of ground above a level *datum*, usually sea level. Quantities of *groundworks* are calculated from the initial and final ground levels.

ground plate, sole p. The lowest horizontal member of a wooden building frame.

grounds, groundwork Timber battens fixed to a wall, enabling nails or screws to be driven to fix wallboard or *dry linings*. They can be *common grounds* or *framed grounds*.

ground shaping *Groundworks* for *landscaping*.

ground sill A *sole plate*.

ground slab (USA **slab-on-grade**) A concrete slab supported by the ground, unlike a *suspended slab*. It can be on *fill*, which should not exceed 600 mm in thickness. The *formation* may be shaped for *thickenings* and covered with *hardcore* (*C*) and sand *bedding* or a layer of *blinding concrete*. It may also need a sheet plastics *damp-proof membrane*, depending on site conditions and building *occupancy*. Large areas of ground slab are cast in *alternate bays*.

ground storey The part of a building between *ground floor* and *first floor*, usually measured between finished floor levels.

groundwork (1) **earthworks, work in the ground** The earliest *trade* in building work, mainly *excavation* to provide space for the *substructure*. It may follow or include *demolition* work or *fill* to make up levels, then later, after the substructure is completed, *backfilling* of the excavations. The work is mostly done with machines, and on large projects by a special *gang* which is changed once the job is *out of the ground*. Mistakes in *setting-out* are difficult to avoid during groundworks and may not become obvious until much later.

(2) *Grounds*.

group Concrete or mortar with particular *mix* proportions, for *blinding*, *infill*, *structural* work, etc.

group supervisory control A *lift-control* system to operate a bank of several lifts, each working on *full collective control* but usually kept within a zone of several *levels*, so that a car can quickly answer a landing call.

grout (1) A thick creamy mixture of cement, sand, and water, mainly used for fixing components in place. Neat grout has no sand. See also *C*.

(2) **jointing mortar** A cement-based product with inert fillers for *grouting* tiles. Grout should be flexible if *movement* in the floor or wall is likely.

(3) The cement and *fines* in concrete.

grouting (1) The use of *grout* for fixing or *bedding* building components, or for putting on hardened concrete to improve the bond with fresh concrete or mortar.

(2) Filling or *pointing* joints between floor or wall tiles with *grout*, which is done after the tile-laying adhesive has set.

grout loss Leakage of *grout* from *formwork* as fresh concrete is placed and vibrated. It is prevented by having tight joints between forms. Inserting foam strips between hardened concrete and the face of formwork helps to make a tight joint.

grout pocket A hole in concrete in which a component can be built in by *grouting*.

growth ring Usually the same as an *annual ring*, though it is possible in a year of exceptional weather for two growth rings to occur.

GRP *Glassfibre-reinforced polyester* or *glass-reinforced plastics*.

grub screw A short *screw* with no head and usually a *straight slot*, driven into a threaded hole in a door handle to grip the spindle tightly. It is short enough to pass completely into the hole and thus should not scratch fingers if properly fitted.

guard A security grating or solid panel over an opening, or a protective fence, or a concrete surround over a buried stopvalve beneath a *surface box*.

guard bead An *angle bead*.

guard board A *toeboard*.

guard rail, safety r. A temporary protective rail that must be put up along the edge of any floor of a building under construction, or in front of any drop of more than 2 metres, in order to prevent building workers from falling.

guide coat A very thin *coat* of loosely bonded paint applied over a *surfacer* before it is rubbed down. It is removed during rubbing, but guides the person doing the rubbing, showing the high spots.

guide rails Vertical steel rails on the sides of a *lift shaft* to keep the lift car and counterweight running straight up and down. The emergency *lift safety gear* grips the guide rails.

guide shoes, runners Attachments to a lift car and counterweight that align them with the *guide rails*.

guillotine A machine with two cutting blades, one fixed and one moving, operated either by hand or power, used to shear materials. Different types are used for cutting mouldings and sheet metal and *reinforcement bars*.

gullet The gap between the teeth of a file or saw; also the length of a saw tooth from point to root. See diagram.

gully, gulley, floor drain An outlet into a drain pipe for water from a floor in a *wet area* or from a *hard surface*. The floor or hard surface usually has a *screed to falls* towards the gully. There are many special types: bathroom, garage, kitchen, road, yard. See also *C*.

gun A hand tool with a pistol grip. There are many types, e.g. for: *caulking*, a *cartridge tool*, *hot-air stripping*, *spray painting*, etc.

gun-grade adhesive Rubbery *gap-*

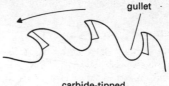

carbide-tipped
circular saw blade

A gullet.

filling glue, put on in blobs from a *caulking gun*, as a cheap and quick method of fixing either *dry linings* or trim such as *architraves* or *skirtings*, to hard, uneven *substrates*.

gun-grade sealant Building *sealant* that can be forced into a corner or gap as a uniform *fillet*, from a *caulking gun*.

gunmetal A hard, highly corrosion-resistant *bronze* containing *zinc*, used for the working parts of cast-iron stop-valves and fittings, circulating pumps, etc. It does not suffer from *dezincification*.

gunstock stile A *diminishing stile*.

gurgling The noise made from the *unsealing of traps*.

gusset piece In *supported sheetmetal roofing* a piece of metal added over a joint at a corner to make it watertight. It is stronger than a *dog ear*.

gusset plate A piece of strong *plywood* nailed or screwed, sometimes also glued, over the joint between members in a *trussed rafter* or other frame, instead of a timber *connector*. See also *C*. See diagram, p. 216.

gutter, guttering A gently sloping channel to collect water and lead it to an outlet or *drain*. Roof gutters can be run along the *eaves*, down a *valley*, behind a parapet or chimney, or let

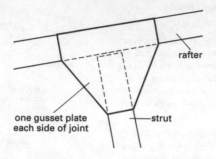

one gusset plate
each side of joint

rafter

strut

Plywood gusset plates at a joint in a timber roof truss.

into a *flat roof*. Gutters are also used for roads and *hard surfaces* or to catch *condensate*.

gutter bearer (1) A short 50 × 50 mm timber under a *box gutter*. Gutter bearers carry the *layer boards*.

(2) A timber bearer each side of a *box gutter* carrying the *snow boards*.

gutter bed *Supported sheetmetal roofing* laid over the *tilting fillet* behind an *eaves gutter* to prevent overflow from entering the wall.

gutter board A *layer board*.

gutter bolt A short *cuphead* bolt used for joining thick sheet metal.

gutter end A *stop end*.

gutter plate (1) A wall plate which supports a sheetmetal roof gutter.

(2) A *valley board*.

gymnosperms Trees with naked seeds, that is conifers (fir, pine, yew), called *softwoods*, though some may be very hard.

gypsum ($CaSO_4.2H_2O$) A solid white mineral which is heated and used as a *binder* in *gypsum plasters*; also the final stage of these plasters

after they have set. Gypsum is mainly mined, but also comes as an industrial by-product. Saturated gypsum causes *sulphate attack* on mortar and concrete.

gypsum baseboard, g. backerboard Square-edged *gypsum plasterboard*, usually 9.5 mm thick and in sheets 1350 by 900 mm, used as a base to be finished with gypsum plaster, after *scrimming*. The plasterwork which goes on gypsum baseboard is in one or two coats. Two-coat work has a first coat of *board-finish plaster* or *bonding plaster* (which must not contain lime), then a finishing coat of any *retarded hemihydrate plaster*, either neat or gauged with lime. Gypsum baseboard F has improved *fire performance*.

gypsum lath *Gypsum baseboard*, usually 1200 by 400 mm, with specially rounded edges so that joints do not have to be *scrimmed* except at corners.

gypsum plank (USA **board lath**) *Gypsum plasterboard* 19 mm thick, no wider than 600 mm and usually 3000 mm long, with a surface to take plasterwork.

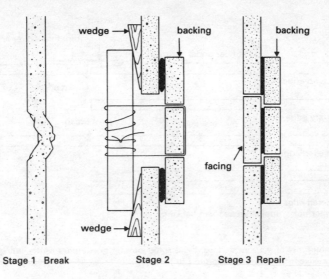

Gypsum plasterboard: repairing a break.

(1) A neat rectangle is sawn round the break.

(2) A backing piece of plaster or wood is inserted, anchored with a piece of wire or string through it. It is glued in place with plaster or glue.

(3) A facing piece is cut to the shape of the hole and glued to the backing with plaster. The edges are sandpapered and the gaps filled with plaster joint filler.

(With acknowledgements to British Gypsum Ltd.)

gypsum plaster White powdery material used as a *finishing* on interior walls and ceilings to give a smooth, warm surface suitable for paintwork. Gypsum plaster contains *hemihydrate* gypsum ($CaSO_4.\frac{1}{2}H_2O$), usually with *lightweight aggregates*, a *retarder*, and other specialist additives (but not *anhydrous* gypsum). After water is added it reverts to gypsum ($CaSO_4.2H_2O$), expands slightly, and therefore rarely cracks. Most plasters are premixed for particular purposes, such as browning, bonding, renovating, or finishing, and must never be mixed with Portland *cement*. Gypsum plaster has poor water resistance, so is unsuitable for outdoor plasterwork or *wet areas*, although waterproof gypsum plaster is marketed. Four weeks' *drying-out* time is needed before it can be decorated. Gypsum plaster improves the *fire resistance* of building *elements* as it does not heat up until all the chemically combined water is driven off; its *fire hazard* is also low, because both gypsum and the aggregates are non-combustible.

gypsum plasterboard *Building board* with a core of aerated *gypsum*, usually enclosed between two sheets of heavy paper, used as a *dry lining*. Plasterboard includes *gypsum wallboard*, which is pre-finished, and other types which take coats of paint (cream face) or plaster (grey face), e.g. *gypsum baseboard*, lath, or plank, as well as *laminated*

gypsum wallboard

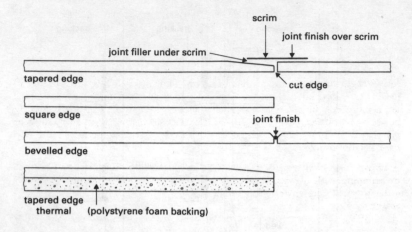

Common types of gypsum wallboard and some jointing possibilities as dry lining. (With acknowledgements to British Gypsum Ltd.)

plasterboard, fire-performance wallboard and baseboard. It is easy to cut, either by sawing or for straight lines by scoring both surfaces and snapping over a straight edge, in the same way as flat glass. Plasterboard is 'hung' on walls or ceilings by a *plasterer*. On brickwork it is held with *dabs* of adhesive or on timber studs secured with *plasterboard nails* or screws. Only a few plasters are satisfactory for finishing, and *lime* should never be added to mixes that go directly on to plasterboard. Gypsum plasterboard is *noncombustible*, or Class O under the Build-ing Regulations, and is made to BS 1230. See diagram, p. 217.

gypsum wallboard *Gypsum plasterboard* with its face pre-finished, usually with PVC, or designed to be decorated directly. Different types of jointing are used with square edged and *tapered-edge plasterboard*. Insulating 'thermal grades' are made with a layer of *expanded polystyrene* and may have an aluminium-foil backing to act as a *vapour barrier*. Gypsum wallboard F has improved *fire performance*. See diagram.

H

habitable room A room for living in, usually excluding kitchens and sculleries; a room may be taken as habitable or not by particular parts of the *Building Regulations*, or because of factors such as noise from outside.

hacking (1) *Keying* a surface, usually by chipping with a *comb hammer*, in order to make plaster or *render* stick. See **scabbling**.

(2) (USA) Laying bricks with a tilt so that the bottom edge of each course is set in from the line of the course below.

(3) A course of *rubble walling* composed of stones alternately one and two to the height of the course.

hacking knife A knife for *hacking out* old putty.

hacking out Removing old *putty* and glass from a broken window before re-glazing.

hacksaw A handsaw or mechanical saw for cutting metal, consisting of a steel blade stretched tight in a frame. The blade is replaced when it is worn out or if the teeth break off, which happens easily, as they are hard and brittle. See diagram, p. 388.

haft The handle of a small hand tool, such as a knife or an awl.

ha-ha, sunken fence A ditch used in *landscaping* to stop people or animals. It does not block the view.

hair Animal hair was used in lime-plaster undercoats (*coarse stuff*) to act as a binder and reduce cracking. The use of *gauged stuff* reduces the need for hair, but synthetic *fibres* may be used in renovation for *pricking up* on lathing.

hair cracking, hairline c. Fine, erratic, random cracks in a surface such as concrete or paintwork, usually as a result of *shrinkage*. In paint, hair cracking does not penetrate the top coat.

half bat, snap header A half brick, a full brick cut across the middle.

half bond *Stretcher bond.*

half-brick wall A *solid wall* the width of a brick, therefore today in Britain 102.5 mm thick using standard bricks.

half-hipped roof A *gambrel roof.*

half-hour fire door (FD 30) A door with a *fire-resistance grading* of thirty minutes, usually a timber *flush door* at least 44 mm thick.

half landing A platform in a stair between two floors and joined to them by *flights*. It may be a *half-space landing*. See diagram, p. 127.

half-lap joint A joint formed by *halving*.

half round (1) A half pipe, cut in two lengthways.

(2) A solid shape with one shallow curved and one flat face.

half-round bead (USA **astragal**) A high half-round *moulding*.

half-round channel A pipe cut in two lengthways, used as a channel, including *channel bends*.

half-round screw A *button-headed screw*.

half-round veneer *Veneer* cut on a lathe from a *flitch* which is roughly semi-circular. The *figure* is intermedi-

half-space landing

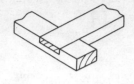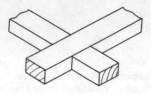

Types of halved joints.

ate in character between *sliced* and *rotary* veneer, since the cut radius can be made very large and the curvature very flat.

half-space landing A stair landing of width equal to two *flights* at which there is an angle of 180° between flights. Compare **quarter-space landing**.

hall, hallway A corridor or similar *circulating area*.

halogen lamp, tungsten-h.l. An *incandescent lamp* filled with low-pressure vapour of iodine or bromine, which improves efficiency. High-wattage straight-filament types used for portable site floodlighting should be kept level to increase their life. Their heat may be a *fire hazard*.

halons (halogenated hydrocarbons) Vaporizing liquids which act as anti-catalysts and extinguish a fire almost immediately they reach a concentration of 3–6% in air. They cause no damage to electrical equipment, but are lethal if burnt. Since halons all contain bromine, which depletes the earth's ozone layer, they are being phased out except in specialized auto-suppression *fire-extinguishing systems*.

halving, halved joint A general way of forming an *angle joint*, or occasionally a *lengthening joint*, between two timbers of the same thickness. One half of each is cut away, the cut surfaces are placed together, and the result is a joint in which the outer faces are flush. See diagram.

hammer A hand or power tool for striking, with a face for driving nails as well as for *dressing, knapping, peening, picking, scabbling*, or *spalling*. Hammer heads are commonly of steel, and handles are of plastics, wood, or steel with a rubber grip. The head may also have a peen, claw, or pick. Types include: *bricklayer's, carpenter's, bush, club*, and sledge hammers.

hammer-dressed stone, hammer-faced s. Stone which has been only roughly faced, that is with a hammer in the quarry.

hammer drill An *electric drill* with a *tungsten-tipped* bit for drilling concrete or masonry. The *chuck* is both turned and vibrated, relying on shattering, which is noisy.

hammerhead crane (USA) A *saddle-jib crane*.

hand (1) The direction of closing (clockwise, anticlockwise) is important when *hinges* and some door springs, latches, or rim locks are ordered, but mortice locks, butt hinges, parliament hinges and some others are not handed or can be double-handed. In Britain a door that to ISO 1226 is turned anti-clockwise to open is a traditional right-hand hung inward-opening door, but its rising butts or lift-off

hinges would be left-hand. American terminology is different, but in both Britain and USA a right-hand lock fits a right-hand door, a right-hand screw tightens clockwise and a right-hand stair has the handrail on the right going up.
(2) A skilled worker (charge hand, leading hand, scaffold hand).

hand drill, wheelbrace A small hand-held *drill*, usually of 8 mm boring capacity or less. It is held in one hand and turned by the other. Being lighter and more compact than the *breast drill* it is more suitable for working in confined spaces.

handed Building parts, including *joinery*, may match each other as an object matches its image in a mirror. Handed objects form a matching pair, one left-handed, the other right-handed. Their *hand* is stated when ordering.

hand float A plasterer's *float* for laying on the finishing coat. It measures about 300 × 100 × 10 mm and is of lightweight plastics or wood with its grain parallel to the length. See diagram, p. 332.

handhole An access hole in a casing or panel large enough for a hand to pass.

handle Plastics or wooden tool handles often fit on a *tang* or in an *eye*.

handmade brick A *moulded brick* made by hand.

handover, handing over When a building has reached *practical completion* it is inspected by the architect and *possession* is returned to the building owner, the *client*, usually at a meeting, sometimes at a ceremony. Handover starts the *defects liability period* and usually half the *retention sum* is paid to the contractor.

handpull A handle for opening or

closing a sliding door or window, usually fixed into a recess in its surface.

handrail A rail at hand height above floor level, or slightly higher in the case of the sloping handrail of a *stair*. Handrails can be either on top of a *balustrade*, railing, or guard rail or supported by *brackets* from a wall, and are usually made of timber, metal, or plastics extrusions on a *core rail*. Stair handrails may have *knees*, *wreaths*, or *scrolls*.

handrail bolt, joint b. A bolt threaded at both ends concealed inside a timber *handrail* at the joint between two lengths. It is tightened by turning with a handrail punch from underneath.

handrail core A *core rail*.

handrail screw Either a *dowel screw* or a *handrail bolt*.

handsaw Any joiner's, carpenter's, or mason's *saw* worked by hand, such as a *ripsaw*, *cross-cut saw*, *tenon saw*, or *block saw*. See diagram, p. 388.

hand tool, small t. A *tool* held by hand. It may be a hand *power tool*.

handwheel, wheelhead A wheel on top of the stem of a *valve* which is turned by hand to open or close it.

hang (1) To fix a door or window to its frame by hinges or other *hanging devices*.
(2) To place sheet, board, or tile finishings on a wall or ceiling.

hanger (1) A metal strap, stirrup, wire, threaded rod, etc., from which a pipe, joist, suspended ceiling, etc., is hung. It may have holes for nails, adjustment screws, attachment clips, etc.
(2) **hanging beam** A timber beam over ceiling joists and at right angles to them, held to them by *joist hangers*, to stiffen the ceiling.

hanging The work of fitting a door

hanging device

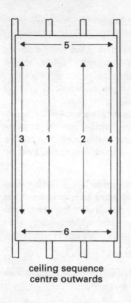

Spacing 100 mm (4in.)
along the perimeter and
200 mm (8 in.) elsewhere.

ceiling sequence
centre outwards

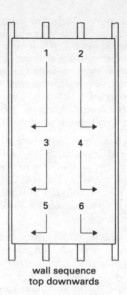

wall sequence
top downwards

Hardboard nailing sequences.

leaf (or a window *sash*) to its frame and fixing its *hanging devices*, lock, stop, etc., a job done by a *carpenter*. *Doorsets* have a leaf that is factory-fitted, reducing the amount of skilled work needed for hanging doors on site.

hanging device The *ironmongery* for hanging a door or window, such as hinges and pivots, sash balances, rollers and tracks, or tilting gear.

hanging gutter (USA) An *eaves gutter* fastened to *rafter* ends or to a *fascia*.

hanging jamb The *jamb* of a door frame to which the hinges are fixed.

hanging sash A *sash window*.

hanging space A place for hanging clothes, e.g. in a *wardrobe*.

hanging stile The *stile* carrying the hinges of a door or casement.

hardboard *Fibreboard* formed under pressure to a density of 840 to 1000 kg/m^3 is medium hardboard. Standard hardboard is denser. *Tempered hardboard* is stronger, denser (more than 1000 kg/m^3), and more resistant to water. Most hardboards, unless 'duo-faced', have one smooth and one tex-tured face. Sometimes the smooth face is pre-decorated with plastics, metal, or wood veneer, or embossed with a pattern to represent wood, leather, etc. Usually 3 mm thick, hardboards exist in thicknesses up to 12.7 mm. They are used, e.g., as facings for *flush doors*, interior linings, and as an *underlay* over rough timber floors on which *thin floor finishes* are to be laid. See diagram.

hard-body impact test A strength test for *doors, security glazing*, etc. A 50 mm dia. steel ball is dropped all over the horizontal face from increasing heights until the facing breaks.

hard burnt, well b. The condition of ceramics, such as *clay bricks*, after *firing* at a high temperature. They are fully vitrified and have high compressive strength, good durability, and low *absorption*.

hardcore See *C*.

hard dry The stage in the *drying* of a paint film at which it no longer feels rubbery and even a thumbnail will not leave a mark. It is dried throughout its depth and therefore can be safely flatted and recoated.

hardener (1) An *accelerator* for a synthetic *resin*.
(2) A material used to harden plaster casts, such as alum, dextrine, or polyvinyl acetate.
(3) **surface h.** A solution in water of sodium silicate, zinc silico-fluoride, etc., applied to a concrete floor to bind together the surface and reduce *dusting*. A hardener cannot hold together a poor concrete but may improve a good one.

hard gloss paint A popular *oil paint* like *enamel paint*. It obtains its hard glossy finish from the high proportion of *resin* in the oil *medium*, i.e. it has a short *oil length*. It is hard by comparison with *oil gloss paint*.

hardhat (USA) A safety *helmet*; also a construction worker.

hard landscaping A *hard surface*.

hard plaster, h. finish Plasters resist knocks in the following order, hardest first: cement and sand *rendering*, *squash-court plaster*, *renovation plaster*, finishing *gypsum plaster*, and *gauged stuff*.

hard plating *Chromium plating* directly on steel, without soft copper or nickel underneath. Deliberate surface *crazing* makes it stick well.

hard solder *Solder* containing copper, which therefore melts at a higher temperature than *soft solder*. It is also stronger and is used for *brazing*.

hard stopping Paint *stopping* made from powder mixed with water or oil.

hard surface, hard landscaping An outdoor area, usually for foot or car traffic, covered with stone, bricks, concrete slabs, gravel, bitumen, etc. See **grass concrete**.

hardware Originally builder's *ironmongery* and steel supplies, now also non-ferrous metal and plastics parts, including *fasteners*, *connectors*, door *locks*, *hinges*, and *furniture*. This vast subject is covered by BS 6100, part 1.3.2, in classified order, with index. The twenty-four pages of definitions include *adhesives* and *sealants*, but not parts of locks.

hard water Water containing calcium or magnesium salts in solution, which react with soap and thus prevent a lather forming easily. When hard water is heated or the pressure drops, these salts create *furring* in pipes, taps, boilers, or channels. Hard water can be softened by a *water softener*.

hardwood Wood from broad-leaved, usually *deciduous* trees, botanically *angiosperms*. Hardwoods are one of the two main groups of *timbers*, the other being *softwoods*. They are often decorative, but are not always hard, though those from European forests usually are. The *dust* from sanding hardwoods must not exceed 5 mg/m^3 in the air of a workplace.

harl (Scotland) *Roughcast*.

harmonization In the European Community the provision of equal access for all member countries to markets. This involves the use of *standards* applying equally in all the countries concerned. The creation of a uniform standard may begin with a *harmonization document* and end with a *European Standard* or a *Eurocode*.

harmonization document (HD) A *European Standard* which differs slightly between member countries for historical, technical, or legal reasons.

harsh mix A *mix* of concrete or mortar with a low proportion of *fines*, which usually results in less *workability*.

hasp and staple A fixing for doors, gates, or box lids. The slotted hasp hinges down over the upstanding *staple* and is secured by a peg or padlock.

hatch An opening in a wall through which objects can be passed by hand. It often has a shelf and a door.

haunch (1) A triangular thickening inside the angle of a *portal frame* where the post joins the rafter, to stiffen the joint, as does a *knee brace*.

(2) **hauncheon** The wide part of a *haunched tenon*, near the root.

haunched tenon A *tenon* which is narrower at the tip than the root.

haunching Concrete beside a drain pipe or behind a road *kerb* to hold it in place. It may go on top of the *bedding* and have a sloping top, but it is not a full *surround*.

hawk, mortar board A small board about 300 mm square with a handle below, for carrying plaster, filler, etc., to be applied with a trowel. See diagram, p. 332.

haydite (USA) *Expanded clay*.

hazard See **fire hazard**.

head (1) The enlarged end of a *fastener*, such as a nail, screw, or bolt, used for driving, thus coming against the surface last. It is also the part of a hammer that drives a nail, or the upper horizontal member of a door frame (a door head) or window frame, or the top of its opening, the top of a partition, roof tile, or downpipe, the end of a *cable*, etc.

(2) The height of a column of water that the pressure in a pipe, air duct, etc. could support, e.g. the minimum head at a draw-off *tap*. See **head loss**.

head board A board carried on the head of a *bricklayer's labourer* in the north of England, formerly used instead of a *hod* for carrying bricks or mortar.

head casing (USA) The part of the *architrave* outside and over a door. It may be topped by a *weathering*.

header (1) A *brick* or *block* laid across a wall to *bond* together its two sides, and by extension the exposed end of a brick (103 mm wide × 65 mm high).

(2) **h. joist** (USA) A *trimmer joist* in floors or walls of *timber-frame construction*. For the difference between American and British senses, see **trimmer**.

(3) A *rainwater header*, but see also *C*.

head flashing A *flashing* resembling a small gutter round the edge of a *penetration* through a roof, such as a skylight.

head guard A *damp-proof course* over the head of a door or window frame.

heading bond Brickwork *bond* of *headers* only, used for footings and curved walls.

heading course A course of *headers*.

heading joint A joinery *lengthening joint*.

head joint, cross-j. (USA) A brickwork *perpend*.

headlap The *lap* between the top of one *roof tile* and the bottom edge of a tile in the next course above, which partly covers it. See diagram, p. 378.

head lining A *lining* to the head of a door opening.

head loss A drop in air or water pressure from partly blocked flow. In *continuous mechanical ventilation* each inlet has only a small opening, so that the low air pressure inside the large-diameter duct is shared fairly equally.

head member A *door head*.

head of water Water pressure, measured in metres of water.

headroom, headway The height of a ceiling or roof beams above a floor (or stair nosing). Stairs need two metres vertical headroom, as well as *stair clearance*.

head tank An elevated water storage tank which pressurizes water pipes.

headtree A T-top *prop* used to carry a beam *soffit form*; or a *bolster*.

headworks Fittings to supply and blend hot and cold water for a *bath*.

Health and Safety Executive (HSE) A UK government department which advises on health, safety, and welfare standards in industry under the Health and Safety at Work Act 1974 and ensures its enforcement. The HSE has three regional headquarters (for policy making, research, and publications) and twenty-one local offices. The local office must be informed of building work (including maintenance and demolition) that will take more than six weeks and of site accidents or ill-health.

heart The centre of a log. It may have *heart centre* and *heartwood*.

heart bond A *bond* for walls which are too thick for *through stones*. Two *headers* meet within the wall and their joint is covered by another header.

heart centre, pith The core of a tree,

containing weak *parenchyma* and most of the log's defects. It is contained in *boxed heart* or a *centre plank*.

hearth The floor of a *fireplace*. It may have *hobs*.

heart plank A plank containing the *heart centre* of a *hardwood* log.

heartshake A radial *shake* originating at the heart (BS 6100).

heartwood, duramen Wood that has finished growing, usually the best timber in a log, darker and more durable than *sapwood* except for the *heart centre*.

heat bridge A *cold bridge*.

heat capacity *Thermal capacity*.

heat detector A *fire detector* that reacts to a given high temperature, e.g. 65°C, or to a *rate of rise* of temperature. Point heat detectors usually have a *thermistor* (*C*) sensor (formerly a bi-metal strip (*C*)). Line (or linear) heat detectors more often have a taut wire, capillary tube, or fusible link.

heat diode A device with a selective flow of heat, having less resistance in one direction than in the other. Examples are *heat pipes*, the *greenhouse*, and *low-emissivity coatings*.

heated floor, h. screed *Underfloor heating*.

heater An appliance used for *heating*, whether by convection, radiation, or warm air, itself warmed by burning a fuel, by electric *heating elements*, or by hot water from a *boiler*.

heat exchanger A device that transfers heat from one fluid to another, for instance air/air, water/air, or water/water, usually working between pipes or ducts and with special tubes to carry the hotter and colder fluid. Examples include: *calorifiers, coils, economizers*

(*C*), *heat pipes, heat-recovery wheels,* and *plate exchangers.*

heat fusing, thermofusion welding, fusion w., polyfusing Making joints in *plastics* by heating the two ends to be joined until they are soft and stick together, with or without a filler rod of the same plastics. The method is suitable for most *thermoplastics* and some synthetic rubbers, and is widely used for joining plastics pipes (except PVC pressure pipes). The heater can be an electric coil embedded in the plastic, a *heat gun*, or a blowtorch.

heat gun A compact electric hand tool with an air heater and a fan that blows out very hot air through a nozzle. It is light enough to be held in one hand and has many uses, e.g. *heat fusing* plastics, *hot-air stripping* of old paint, etc.

heating Keeping a building warm in winter (*space heating*) and the production of *domestic hot water*. Buildings can have *central heating*, supplying radiators, convectors, concealed *coils*, or warm-air *unit heaters*, unless they have *air conditioning*. See **comfort index**.

heating and ventilation engineer A *mechanical engineer*, often a member of the *Chartered Institute of Building Services Engineers*, concerned with all building *services*, including *heating, air conditioning, sprinkler systems, centralized vacuum cleaning* plant, etc.

heating battery A *heat exchanger* which warms ducted air from piped hot water.

heating coil (1) A *heating battery*.
(2) Pipes embedded in walls or floors for *coil heating* or *underfloor heating*.

heating element A coil of wire heated by an electric current passing through it, used in *heaters*, hot-water *cylinders*, etc.

heating medium A *heat-transfer fluid.*

heating panel See **panel heating**.

heating, ventilation, and air conditioning (HVAC) Mechanical services which depend on each other and so are often grouped as one *trade.*

heat insulation, thermal i. See **insulation**.

heat load The total amount of heat that has to be transferred by an *air-conditioning* unit. It comes from *solar gain*, lighting, office equipment, etc.

heat loss In a building in winter, the flow of heat outwards through the envelope (walls, roof). The building's boiler is usually chosen according to the maximum heat loss, which can be reduced by *insulation.*

heat of hydration The heat produced in concrete, plaster, mortar, etc., from the chemical reactions of setting. It can be used for *self-curing* in winter, to give concrete some *frost protection.*

heat pipe A sloping tube containing *refrigerant* and a metal wick. Heat causes evaporation at the top, and vapour returns to the bottom and condenses. It is a simple device, without moving parts that wear out, used in sets for *heat recovery*. See also *C.*

heat pump A machine that uses the *vapour compression cycle* (usual in refrigerators) for *heating*. The evaporator draws in a large quantity of low-temperature heat, from a river, a buried coil, outdoor air, an unglazed *solar collector*, etc. Condenser panels are used as *radiators*. Running its compressor costs about a third the value of its heating. The French radio and TV (ORTF) building in Paris is heated from the river Seine by a heat pump.

heat recovery The use of a *heat exchanger* to transfer heat as an energy-saving measure, e.g. from warm *exhaust* air to cold winter fresh air.

heat-recovery wheel, thermal w. An efficient *heat exchanger* with a large wheel that rotates between the incoming and outgoing air *ducts* of an air-conditioning system. Some types can partly recover both *sensible heat* and *latent heat*, with efficiencies of about 75%. In summer they can cool incoming air.

heat-resistant paint Usually *enamel paint* dried by *stoving* at high temperatures. It is used on *radiators* or similar equipment. Types containing *silicone* resin binders have working temperatures of 350 to 500°C. Coatings for higher temperatures include vitreous *enamel*.

heat-resisting glass A glass with a low *coefficient of expansion* (*C*), such as borosilicate glass, used in the windows of *room-heaters* and as wire-free *fire-resisting glass*.

heat-transfer fluid, coolant, heating medium Usually water, with *corrosion treatment*, used in *central-heating* systems.

heat-transmission value The *k-value*.

heat welding gun A *heat gun*.

heavy-bodied paint Either a viscous paint or one that makes a strong *film*.

heavy-body impact test One of the tests for *door strength* in which a 30 kg bag of sand is swung as a pendulum three times on each face, level with the lock, from increasing standard heights.

heavy protection Heavy material, such as a mortar *loading coat* or gravel *ballast*, laid over *built-up roofing* to prevent uplift and give *solar protection*.

heel (1) The part of a beam or *rafter* resting on a support.
(2) The rear end of a *plane*.
(3) The lower end of a *stile* of a door.

heel strap A U-shaped steel strap bolted to the *tie beam* of a timber *truss* near the *wall plate*. The strap passes over the back of the *principal rafter*, joining it to the tie beam, and transmits the rafter thrust to the tie beam.

helical hinge A double-acting *spring hinge* with two coil springs used for lightweight *swing doors*, hung from the *door frame*.

helical stair The correct but not the usual name for a *spiral stair*.

helicopter A *power float*.

helmet, safety h. (USA **hardhat**) A strong hat for protection against falling objects. Wearing of helmets is compulsory on building sites in most countries and notices that helmets must be worn are required.

helper (USA) A *mate*.

helve The handle of an axe, pick, sledge hammer, or similar heavy tool that is wielded with both hands. Helves are made of plastics, ash, or hickory.

hemihydrate plaster The basic material of all *gypsum plaster*, obtained by gently heating *gypsum* ($CaSO_4.2H_2O$) at about 150 to 170°C so that it loses three quarters of its chemically combined water and becomes half-hydrated ($CaSO_4.\frac{1}{2}H_2O$, also written $2CaSO_4.H_2O$). It is used pure as *plaster of Paris* or more often as *retarded hemihydrate*.

hemlock, hem-fir A non-resinous straight-grained brownish-white *softwood* which is mainly from western Canada and is similar to *Douglas fir*. It dries slowly, has good *workability*, and is used for structural *carcassing*, e.g. roof framing.

hepa filter A *high-efficiency particulate air filter*.

herring-bone strutting (USA **cross-bridging**) The stiffening of a floor on *common joists* by *bridging* between them with pairs of light struts forming crosses, fixed from the bottom of each joist to the top of its neighbour. There is either one row at mid-span or *double bridging*. See diagram, p. 65.

hessian (USA **burlap**) Open-weave brown fabric of jute or hemp fibres used to make sacks, wrap electric power cables, reinforce *fibrous plaster*, etc.

hew To shape with an axe or hatchet in squaring timbers or dressing stones.

H-hinge A *parliament hinge*.

hickey (USA) A *bending tool*.

hiding power, opacity The ability of a *paint* etc. to obscure a colour beneath it.

high bay lighting Lighting for industrial areas, usually with efficient *discharge lamps* in batwing *reflector* fittings. To give even *illuminance* the lamps need to be closely spaced, at a ratio of 0.25 to 0.75 times the bay height.

high-breaking-capacity (HBC) **fuse** A *cartridge fuse* with shaped wires surrounded by insulating material inside a glass body. They protect circuits safely and reliably against *overcurrent*.

high-build coating A *coating* that can be put on thicker than normal yet still harden properly. It may also need to be heavy and viscous to prevent flow down walls. Examples include bituminous *waterproofing* and high-solids *masonry paint*.

high-build filler *Two-part* polyester *filler* for deep holes in timber or steel.

It is generally suitable for interior or exterior use.

high-calcium lime, fat l., non-hydraulic l., rich l., white chalk l., white l. A pure lime (mainly CaO) giving a very plastic *lime putty*. It does not *set* but slowly hardens by *carbonation*. It can safely be mixed with Portland *cement*.

high-efficiency particulate air filter, hepa f. In *air conditioning*, a filter with a *medium* of cellulose fibres (like blotting paper) or high-quality cotton, usually folded zig-zag to increase the surface area.

highlighting Emphasizing the relief of a surface by painting certain parts of it paler than the general colour.

high-pressure hot-water (HPHW) system A method of *central heating* with small pipes and rapid circulation of water at above 200°C and 16 bars pressure. It is used in industrial and commercial systems, having largely replaced the *low-pressure hot-water* systems which required awkwardly large pipes. In domestic heating a *small-bore system* is commoner.

high-pressure system A *high-pressure hot-water* or a *high-velocity system*.

high-rise building A tall building, usually with more than eight *storeys* and therefore at least 28 m from street level to the roof top. Common since the invention of *lifts*, structural steel, and concrete in the nineteenth century, they make *fire protection* difficult.

high spot A part of a *substrate* that is higher than the general area.

high-velocity system A space-saving air-conditioning system with small round ducts through which the *supply air* is forced by powerful *fans*, usually with unducted return of air along corridors. Each *re-heat unit* needs a pressure-reducing valve, which creates

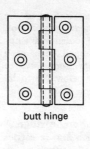

butt hinge

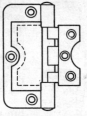

non-mortice butt

parliament hinge

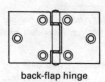

back-flap hinge

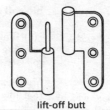

lift-off butt

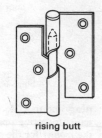

rising butt

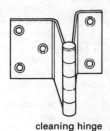

cleaning hinge

Hinges.

noise, and an *attenuator*. Systems can have *dual ducts* or a single duct going to *variable air volume boxes*, or the air passes straight out through *induction units*. The fans use more energy than those of a *low-velocity system*.

hinge Two flaps joined by a pin through their knuckles, used for *hanging* a door leaf from its frame. Hinges are usually made of steel (ironmongery) and fixed with screws. The commonest hinges in building are *butt hinges*, *non-*

mortice hinges, *parliament hinges*, and *back-flap hinges*, which are symmetrical and not *handed*. All the other types of hinge are handed, including *lift-off hinges* and *rising-butt hinges*, for which the hand must be stated. For heavy doors or *swing doors*, other *hanging devices* may be preferred. See diagram.

hinge bolt, dog b. A projecting steel lug on the hinge *stile*, which enters a hole in the *jamb* as the door closes to prevent the door being forced.

hinge-bound door

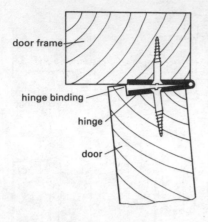

A hinge-bound door.

hinge-bound door A door which tends to spring open slightly because the *rebates* for the hinges are too deep and the leaf comes into contact with the frame before it is fully closed. See diagram.

hip, h. ridge (Scotland **piend**) The outstanding edge of a *hipped roof* formed by the meeting of two roof slopes. Rainwater flows away from a hip and towards a *valley*.

hip capping The uppermost strip of *bitumen felt* or other *trim* over a hip.

hip hook, h. iron (Scotland **piend strap**) A visible bar nailed to the lower end of a *hip rafter* to hold the lowest *hip tile*.

hip knob A *finial* to a *ridge* where it meets a *hip* or *gable*.

hipped end, hip The sloping triangular end of a *hipped roof*. In traditional timber framing it has short *jack rafters*.

hipped-gable roof A *jerkin-head roof*.

hipped roof, hip r. A *pitched roof*

which has four slopes instead of the two slopes of an ordinary gabled roof. The shorter sides are roofed with sloping triangles, the hipped ends, each bounded by two *hips* from *eaves* to *ridge*, and by the eave below. Normally the eaves are at the same level all round. See diagram, p. 376.

hip rafter, angle r., angle ridge A roof timber forming a *hip*. It carries the top end of each *jack rafter*.

hip roll (1) **ridge r.** A *wood roll* with a V-cut beneath to cover a *hip*.
(2) The *supported sheetmetal roofing* covering which fits over a wooden hip roll (see above).

hip tiles Clay or concrete tiles over the roof tiles that meet at a *hip*. Angular, round, and *bonnet* hip tiles are standardized in the UK, and several manufacturers offer *dry hip* systems. Traditional hip tiles should be nailed at the top end and bedded in mortar at their lower edges. The mortar may need *movement joints* on a long hip. The lowest hip tile is held by a galvanized steel *hip hook* and the top one is usually covered by a *ridge tile*.

hip truss A *trussed rafter* that forms a *hip*.

hire of plant When particular *contractor's plant* (*C*) is required for only a short time it is usually hired, at a far lower cost than buying. Plant hire firms often specialize, e.g. in excavators, power tools, scaffolding, etc.

hit and miss fencing Fencing with boards placed side by side but with wide gaps between them.

hoarding A tall close-boarded fence round the boundary of a building site to keep out the public and children (for whom more care is required) and safeguard against theft. Its cost is a *preliminary* item.

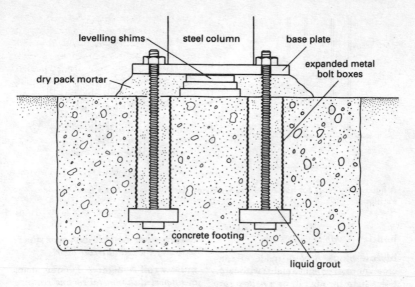

levelling shims   steel column   base plate

dry pack mortar

expanded metal
bolt boxes

concrete footing

liquid grout

Holding-down bolts.

hob A raised flat area each side of a fireplace, shower tray, etc.

hob unit, cooktop A cooker panel with gas burners or electric hot plates that can be built into a kitchen work top (six-burner hob, electric hob, etc.).

hod A three-sided plastics tray about 400 × 230 × 230 mm, with a straight wooden handle projecting underneath, forming a Y-shape. It is used to carry bricks or mortar, resting on the *hod carrier*'s shoulder and steadied by one hand, and so can be taken up a ladder.

hod carrier, hodman A *bricklayer's labourer* who carries a *hod*.

hoist A machine for lifting building materials, which are carried on a platform or in a skip. Passengers are not allowed. Compare **builder's lift**.

hoisting Raising or lowering the load

on a *crane*, one of the three basic *movements*.

hoistway (USA) A *lift shaft*.

holderbat A traditional steel *fixing* with a long dovetail end for building into brickwork, usually with a ring on the outer end to support a pipe.

holding-down bolt, foundation b. One of a set of mild-steel *bolts* cast into a concrete footing for anchoring a *baseplate*. During concreting they are held by a *template* and *bolt boxes*. See diagram.

holding-down clip A *tingle* shaped to fit a *capping* to anchor and join adjacent lengths.

hole saw, annular bit, tubular s. A tool, usually driven by a powerful *electric drill*, which cuts a ring-shaped sinking and can cut out a complete cylinder of wood, plastics, etc. leaving a circular hole.

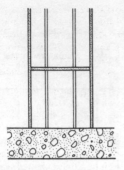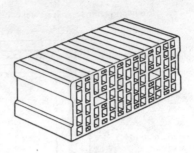

A hollow clay block for house walls in Germany. Grooves in the block faces provide a key for plaster.

holiday A *skip* in paintwork.

hollow block A *block* made of concrete (normal or lightweight) with large holes right through it; or a *hollow clay block*. It is used for building walls and may require *shell bedding*. Even if the holes are filled with *insulation* it is still a hollow block.

hollow brick Any *brick* with more than 25% volume of through holes, or exceeding the other limits of a *perforated brick*. They are not made in the UK but are used with *hollow clay blocks*.

hollow clay block (USA **structural c. tile**) Burnt-clay *blocks* with large perforations through their middle, laid to form vertical airways. For external walls they are *shell-bedded* and as warm and dry as a *cavity wall*. Their *fire resistance* is lower than that of a solid brick wall, and *fire stops* may be needed. They are used all over the west European Continent, but not in the UK. See diagram.

hollow clay tile, h. pot Tiles used for *drained cavities* or for *hollow-tile floors* (C).

hollow concrete block See **block**.

hollow-core door A *flush door* superseded by the *cellular-core door*.

hollow roll A method of joining strips of *supported sheetmetal roofing* running from the ridge to the eaves of a sloping roof, using only simple tools. The edges of adjacent sheets are laid together, lifted up, and bent round to form a cylindrical roll without a *wood roll*, but sometimes with a metal fixing strip, a *tingle*.

hollow wall A *cavity wall*.

hone, oilstone A block of fine-grained quartz stone with very flat faces, used to give the final, durable polished finish (honing) to an *edge tool* after grinding to shape. Hones are also used to give a matt polished surface to stone or *terrazzo*.

honeycomb bond The bond used in a *half-brick wall* built of *stretchers* with gaps between; the bricks are held only by bed joints at their ends. The bond is used in *sleeper walls*, which allow *ventilation* under the wooden ground floor of a house with no basement.

honeycomb-core door A *cellular-core door*.

honeycomb fire damper A *fire*

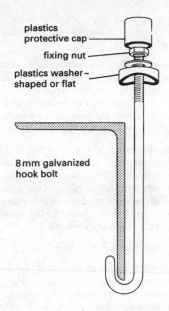

plastics
protective cap

fixing nut

plastics washer –
shaped or flat

8 mm galvanized
hook bolt

A hook bolt holding profiled roofing sheet to a steel angle purlin or rafter. (With acknowledgements to Alan Everett, *Finishes*, Batsford, 1979.)

damper with an open honeycomb containing *intumescent* material. It is maintenance-free, and used in air-conditioning *ducts* that are inaccessible, for air velocities up to 10 m/sec.

honeycombing (USA **rock pocket**) Unwanted gaps in concrete from a lack of cement and *fines* between coarse *aggregate*. It can be due to *segregation*, low *workability*, or insufficient *vibration*. Care must be taken to avoid honeycombing in *direct-finish concrete*, as *patching* is usually not allowed.

hood A large stainless-steel canopy hanging from the ceiling above kitchen equipment, connected to an *extract fan* to take away vapour and fumes.

hood moulding, drip cap A projecting moulding over the outside of a door or window opening, to throw off rain. In France hood mouldings are also used over upstand *flashings*. A stone hood moulding is a *dripstone*.

hook bolt A galvanized-steel *bolt* with its unthreaded end bent to a J-shape, which hooks behind a roof *purlin* or wall *girt*, for fixing profiled sheeting. The long, threaded end goes through a hole in the roofing or cladding and is secured with a nut and shaped washer. See diagram.

hook-bolt lock A lock for a *sliding door* which *butts* against its frame. It is morticed into the *stile* and its bolt swings into a hole in the *jamb*.

hook curtailment A *reinforcement bar* with its end bent back on itself round a fat pin.

hook height The maximum lifting height of a crane.

hook intake, cable sleeve, turndown A pipe with a turned-down end up which a cable is passed to enter a building without causing a leak through the roof or walls.

hook joint A joint made by the *meeting stiles* of a double *casement window*, which are shaped to mesh together. A hook joint is usually shaped to keep out rain and weather and can be held closed with a *Cremona bolt*.

hook terminal The end of an *expansion pipe* or *vent*, turned down to keep flying objects from entering and polluting the water.

hook time In planning the use of *cranes*, the number of hours per day given to each *trade* (e.g. concreting, moving formwork, erecting cladding panels).

hopper A large squarish funnel shape that narrows from top to bottom.

hopper head A *rainwater header* in the shape of a *hopper*.

hopper light A window with a *sash* hinged inwards at the bottom (bottom hung) and with draught-preventing side panels, which are often *glazed*.

hopper window A window with one or more *hopper lights*, usually one above the other.

horizontal shore A *flying shore*.

horizontal work Asphalt work done by pouring the mix on to a sub-base such as a concrete slab. More work can be done per hour than in *vertical work*.

horn A door *jamb* that extends beyond the *head* of the frame to avoid weakening in the *mortices* for the head during transport and handling on site. The horns are sawn off once the frame is finally in place, and preservative is painted on the *end grain*. See diagram, p. 360.

horsed mould A *run moulding* in plasterwork.

hose reel A *fire-hose reel*.

hose union tap, h. cock (USA **sill c.**) A garden tap with a thread on its outlet to take a hose fitting.

hospital sink A sturdy rectangular bowl made of *fireclay* or stainless steel for rinsing bed sheets and mackintosh aprons, often also used as a *slop hopper*. It has a P or S trap like a *water closet* and hot and cold water sprayers.

hot-air stripping A safer method of *paint removal* than *burning-off*. It uses a *heat gun* that blows out hot air at 600°C (compared with 1900°C for a blowtorch).

hot bonding compound (USA **hot stuff**) *Bonding compound*.

hot coil A *heating battery*.

hot cupboard A *servery* where food is kept warm, usually at about 75 to 85°C, either under a counter or in a full-height cabinet.

hot-melt glue A thermoplastic *adhesive* that softens when heated to about 160 to 200°C in an electric glue gun. It is then squeezed on to the parts to be joined, which are kept under *pressure* until the glue cools, when they stick.

hot pressing Gluing *plywood* by *platen-pressing*, usual for *thermoset* resins such as *film glue*, and common for other *adhesives* which set more quickly when heated. On site, veneers can be stuck to chipboard by hot pressing with an electric clothes iron.

hot surface A *hungry* substrate.

hot-water system Building *services* to supply *domestic hot water*, e.g. a storage *cylinder*, a *calorifier*, an *instantaneous water heater*, etc. Cold water comes from a *feed cistern* in open-vented low-pressure systems, or straight from the mains supply in *unvented systems*. See diagram.

hot work Work using a naked flame or heat-producing equipment, including arc welding, bitumen boilers, blowlamps, brazing, grinders, high-intensity lighting, etc.

hot-work permit A *permit to work* issued to a person before doing *hot work*, as a precaution against fires on site.

house breaking The *demolition* of houses.

housed joint, let-in A joint in which a shallow sinking in the face of one board encloses (or houses) the end of another timber, in the way that steps are housed into a *close string*.

housed string A *close string*.

house service cutout An *intake unit*.

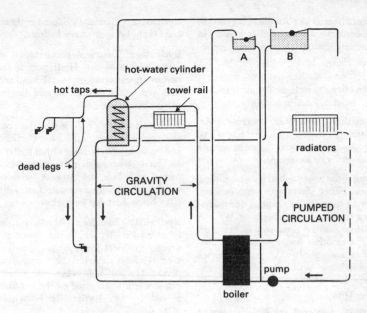

A hot-water cylinder heated by gravity circulation with a few radiators heated on the one-pipe system. The towel rail, not on the heating circuit but on the primary flow-and-return, is an exception, enabling it to heat the bathroom even when the central heating is turned off. A = expansion tank; B = cold cistern. (With acknowledgements to *Central Heating*, Consumers' Association, 1991.)

housewrapping Enclosing a house under construction with *breather membrane*.

housing (1) The sinking made for a *housed joint*.
(2) A dwelling, or a casing to protect equipment.

HSE The *Health and Safety Executive*.

hub (USA) The enlarged end (bell or *socket*) used to make a pipe joint.

hubless pipe (USA) An *unsocketed pipe*.

hue The property of redness, yellow-ness, blueness, etc. The hues of the spectrum, seen in a rainbow, are the most intense. Eight standard hues are used to describe *colours*.

humidifier An *air-treatment* unit which makes air wetter. Cold outdoor air contains very little water vapour and thus, even if originally damp, has a low *relative humidity* after heating to room temperature. The extra heat needed to evaporate the water can come from a preheater, or the temperature can be made up later by a reheater. Humidifiers that use steam are the least likely to cause *humidifier fever*.

humidifier fever An allergic reaction in people to a fungus that breeds in low-temperature *humidifiers*, sometimes linked to the *sick-building* syndrome.

humidity Dampness; for air it is usually stated as *relative humidity*.

humidity controller, hygrostat An *automatic control* which reacts to changes of *humidity* in the air to operate a *humidifier* or *dehumidifier*.

hung A fixture attached to the building fabric, either directly or on a *bracket*, may be described as 'hung', for example a *wall-hung* washbasin or WC. A window *sash* or door *leaf* is hung on hinges or other *hanging devices*. Casements are usually side-hung, sashes double-hung, and ventlights bottom-hung. Lift-off butts make doors easy to unhang.

hungry, starved (1) **hot** A hungry *substrate* is one which is too absorptive for the amount or kind of paint put on it. The paint film appears to be thin and patchy.
(2) A hungry plastering mix is one of low *plasticity*.

hurdle A low wooden frame made of two pegs driven into the ground and a board between them, used as a setting-out *profile*. It is easily damaged, and if repaired should be checked from a *recovery peg* (*C*) by the *site engineer*.

hush latch, mechanical striking plate A *striking plate* that opens under gentle pressure, allowing a door to close without forcing back the latch bolt.

HVAC *Heating, ventilation, and air conditioning.*

hydrant An outlet from a water main, usually a *fire hydrant*.

hydrant bend A shallow S-bend used to connect a *hydrant* to a water main.

hydrated lime Slaked lime, Ca(OH)₂, bought as *dry hydrate*.

hydration Chemical combination of water with a *binder*. Hydration of the *cement* in concrete gives off heat and uses up water, which should be replaced by *curing*.

hydraulic binder A *binder* that can set under water, such as Portland *cement* or *hydraulic lime*. Other materials have hydraulic properties when mixed with *lime*, for instance *pozzolan* or *fly ash* (*C*) (pulverized fuel ash). Clay brick dust can be hydraulic.

hydraulic breaker A demolition *breaker* with a hydraulic ram that compresses an inert gas 'spring', which is then released to drive the hammering piston. Hydraulic breakers are usually massive units mounted on the boom of a *backhoe*. See **hydraulic hammer** (*C*).

hydraulic cement A *cement* which hardens under water. See **hydraulic binder**.

hydraulic door closer A *door closer* with a hydraulic *check*.

hydraulic equipment Equipment operated by a watery oil, which can be pumped under pressure to open or close *rams* or turn motors. Powerful, easily controlled, reliable, and efficient, it is usual on new bulldozers and excavators.

hydraulic lift A *lift* moved by an electric pump driving a hydraulic ram, either direct-acting or indirect-acting. A direct-acting ram is below the car, in a borehole as deep as the full *travel* height. An indirect-acting ram is mounted beside the car, which it raises and lowers through ropes and pulleys. The pump is usually on the lowest floor, not above the top floor. See diagram, p. 267.

hydraulic lime *Lime* burnt with up to 22% clay, first made by Vicat around 1818, shortly before the invention of *Portland cement*. Natural hydraulic limes were known since 1760. Although not made in the UK, natural and artificial hydraulic lime are imported from France. The stronger artificial hydraulic lime, made from Portland cement clinker (thus strictly a cement) and limestone fillers, is used for plastering.

hydraulic test A test for drains before they are buried. See (*C*).

hydrophobic cement A cement protected against *warehouse set* by packaging it as small balls covered with wax, which is worn away during mixing.

hygroscopic salts Salts which suck in moisture, commonly found in the ground. These salts travel up a wall in *rising damp* to be left behind and con-centrate in the wall, as the water that carried them then evaporates in a warm room. The salts always hold some water, so that the bricks and plaster are never dry, even in persistent drought. Several possible treatments exist after the rising damp has been overcome. All are expensive but some have been successful. Hygroscopic salts can also enter timber from its treatment with some inorganic *preservatives*.

hygrostat A *humidity controller*.

hypalon (cspe) A synthetic *rubber* used for *single-ply roof* membranes.

hyperbolic paraboloid A sweeping geometric shape that results from straight lines running between the sides of a rectangle folded across its diagonal. Although it has *double curvature*, it can be made with cables, or tapered planks, or as a thin but stiff reinforced-concrete shell.

I

I (intensity) The symbol for electric current, measured in amperes.

IEE *Institution of Electrical Engineers.*

ignitability The ability of sheet or slab materials to be ignited by a small flame, one of the *fire hazards*. Ignitability tests are done to BS 476.

illuminance (E) The amount of light falling on a surface (the *luminous flux* divided by area, or *lux*), used in the *lumen method* for predicting *daylight* and *artificial lighting*. (E is for the French 'éclairement'.)

illuminated push A *push button* with a tiny red light visible in the dark.

illumination *Artificial lighting.*

imbrex In *Italian tiling* the over-tile, which is semi-circular and fits over the tegula or under-tile.

imbricated Overlapped, usually resulting in a decorative pattern.

immediate demould A method of making precast concrete *blocks* and bricks using an open-bottom mould box into which the *mix* is dropped, then powerfully compacted with vibrating feet. The mould is pulled upwards over the stopped feet, leaving the blocks behind. They are strong enough to be picked up carefully and taken away for *curing*. Many different shapes can be made and the method gives close *tolerances* on size.

immersion heater An electric heater inside a case submerged in a hot-water *cylinder*.

impact noise The effect of footsteps on an upper floor, which make the whole floor vibrate, disturbing the occupants below. Hard floor surfaces (and hard shoes) create the most impact noise, but resilient *floor coverings* such as carpet, cork, or foam-backed rubber reduce it. A *floating floor* prevents the *transmission* of impact noise to the building fabric.

imperishable Imperishable materials are those which do not decay or corrode, such as *ceramics*, *stone*, *glass*, *plastics*, well-made *concrete*, and some metals.

impingement filter, viscous i.f. A *fabric roll filter* coated with a sticky wax to catch dust in intake fresh air for air conditioning.

impregnation of timber The thorough penetration into timber of a *preservative* or chemical *fire retardants* by *double vacuum* or *vacuum-pressure* treatment. It requires special plant and is best done by the timber wholesaler.

improved nail A *ring-shanked nail*.

improvement line A *building line* planned for the future.

inactive leaf The leaf of a *double door* which is not normally used. It has lock bolts top and bottom and is *rebated* to mate with the *active leaf*.

incandescent lamp A tungsten filament inside a glass envelope under vacuum, or filled with inert gas, so it can be electrically heated without burning out. Efficiency is fairly low but better in a *halogen* lamp. It gets very hot and needs a heat-resisting lamp holder, *luminaire*, and cable.

incentive scheme A wage system in which operatives receive more pay if they do more work, for example by *bonus* or piece rates. However if, as in Britain, the extra pay makes a large increase in the amount of tax paid, the incentive may not be a reason to work steadily. It may encourage an operative to work hard for a short time and then stay away from work so that some tax is refunded.

incise To cut or carve stone, glass, wood, etc.

inclement weather, unfavourable w. Rain, snow, or high winds that slow down or stop building work, particularly before the *enclosed stage*, may give the contractor the right to an *extension of time*.

inclinator An inclined hoist *stairlift* installed in his home by Groucho Marx, who also invented the word.

inclined shore A *raking shore*.

incombustible building material A *non-combustible* building material.

incorporated into the works Materials are said to be incorporated into the works when they are fixed in place.

increase or decrease in prices *Fluctuations*.

increaser A *taper* pipe increasing in diameter in the direction of flow.

incrustation (1) A layer of corrosive material which collects on stone or brick in an industrial district. It should be removed periodically to prevent the wall wearing away.
(2) *Furring*.

indent A notch made into a surface, such as the gaps at the end of a brickwork wall, between every other course for *toothing*.

indented joint A carpentry joint cut with mating notches (which may be wedged), to strengthen it.

indenting (1) Deep marks made in a floor finishing, roofing, etc., by heavy point loads, from foot or wheel traffic, etc.
(2) *Toothing*. See also *C*.

indenture A legal agreement used, for example, between an *apprentice* and a master.

independent float Spare time for an *activity* on a *programme*, calculated on the basis that all previous activities are completed as late as possible and all subsequent activities are started as early as possible.

independent scaffolding A *scaffolding* with its own uprights and braces. It has ties to the building for sideways support.

indicating bolt A privacy latch for a WC door that shows on the outside when it is locked or unlocked, e.g. by colour (red or green) or wording (vacant or engaged).

indicator, i. light A small light that gives a signal, such as red for danger, green for safe.

indicator panel, annunciator A panel with several *indicators*.

indirect cylinder A storage hot-water *cylinder* in which is a coil of pipe forming a *heat exchanger* that separates cylinder water from boiler water. *Direct cylinders* fur up quickly in *hard-water* districts. To prevent or reduce *furring*, the internal coil is connected to the boiler by *primary flow and return pipes* forming a sealed unit in which the water is never changed. These pipes, at least 19 mm in diameter, use *gravity circulation*.

indirect lighting Room lighting with light reflected from a hidden lamp by a large area, e.g. a white ceiling. The light is more restful than direct lighting, being diffuse with much less glare. Examples are *uplighters* and *cove lighting*.

indoor pool A *swimming pool* inside a building, for use in winter. The materials in an indoor pool area should be able to withstand a 26°C air temperature and 100% *relative humidity*.

induced siphonage *Unsealing* of the *trap* in one sanitary fitting due to air pressure changes in the drain pipes caused by flow from another fitting.

induction unit An *air terminal unit* placed under a window and blowing upwards that takes its primary air from a *high-velocity* duct and releases it through rows of nozzles, which cause the secondary room air to circulate through the unit. The casing can have a heating or cooling *coil* and *filter* and needs only regular cleaning, as there are no moving parts to wear out and break down.

industrial building A utilitarian building that covers a large area, usually with high bays framed in *structural steelwork*, a concrete *ground slab*, and sheet roofing and cladding.

industrialized building methods Building methods involving a high degree of prefabrication, often of the structural framing, roof, and cladding, so as to reduce skilled site work to the minimum. This involves careful planning and the maximum of standardization. The amount of factory work on the main building *elements* is deliberately increased so as to reduce the cost and improve the quality and speed of construction. Examples are *system build-ing*, *kit homes*, *trussed rafters*, and prefabricated services for a *services core*.

inert gas extinction Smothering a fire with inert gas, usually *carbon dioxide*, less often nitrogen, to displace air and cut off oxygen to the fire.

infill concrete Mass concrete under foundations, to replace poor ground.

infilling (1) Material placed within a building frame or *partition* for fire resistance, insulation, weather protection, or stiffness. Brickwork is the usual infilling material in Britain, but *insulation* is also used.
(2) *Infill concrete.*

infill wall A *non-loadbearing wall* made of *infilling*.

infiltration The leakage of air through cracks, round doors or windows, or at joints in *cladding* or *roofing*, causing draughts. The loss of heat in winter from infiltration can account for 40% of heating costs, leading to the use of *weatherstripping*. Rates of infiltration for windows are measured at a pressure difference of 600 pascals (0.6% of atmospheric pressure) and given in m³/h. See also *C*.

inflammable See **flammable, non-flammable**.

inflated structure An *air-inflated structure*.

Information Paper (IP) One of the publications of the *Building Research Establishment* outlining the performance in service of particular building products or systems, warning about typical defects, advising how to investigate and diagnose failures, and suggesting remedies.

infra-red drying *Stoving*.

infrastructure The public roads and other *services*, such as water, electricity,

An injection damp course.

telecommunications, gas, and sewers, which society uses to function properly.

ingo, ingoing (Scotland) A *reveal* to a door, window, etc.

ingo plate (Scotland) A *reveal lining.*

inhibiting pigment A pigment used as a *corrosion inhibitor* in paint, particularly *metal primer*, usually to slow down or stop the rusting of steel.

initial ground levels *Natural ground* levels.

initial rate of absorption The *absorption rate.*

initial set The first hardening of concrete, mortar, or render, before which all *placing* should be done. Unless *accelerators* or *retarders* are used, it occurs one or two hours after water is first added to the *cement* during mixing. During that time there may be *setting shrinkage.*

initial shrinkage, i. drying s. The main part of *drying shrinkage*, which

occurs in the early stages of the setting of concrete, mortar, render, etc.

injection damp course, chemical i. In walls of old buildings, built without a *damp-proof course* and where a *sawn damp course* is impracticable, the injection of a *water-repellent* silicone may provide a good remedy for *rising damp*, even in a rubble wall. Holes are drilled into the bricks or stones (not mortar) at about 150 mm centres and are connected by polyethylene tubes to the waterproofing liquid, which can be injected at any desired pressure. Water under pressure cannot be held back. Specialist systems of injection may have an *Agrément Certificate.* For details such as wall replastering, see BS 6576. See also diagram.

inlaid parquet *Parquet* flooring glued in blocks about 600 mm square on to plywood, then fixed to floor boards.

inlay A *finishing* with a decorative design made by cutting away a surface (or leaving a hole) and filling it with a different material.

inodorous felt A brown *bitumen felt* made with flax fibres, used as an *underlay* to prevent corrosion of *supported sheetmetal roofing*.

input system A system of *mechanical ventilation* in which air is blown into rooms with a fan. A filter, a silencer and an air heater may be included. Disturbing noise from the fan is difficult to eliminate, unlike that from an *extract system*.

insect screen, fly wire s. A panel of fine *fly wire* in front of a window or door to keep insects out of a building.

insert (1) A patch of *veneer* which fills a hole in one of the veneers of a sheet of *plywood*.
(2) A *concrete insert*.

inside glazing *External glazing* placed from within the building. It provides more security than *outside glazing*.

in-situ concrete (USA **cast-in-place c.**) Concrete placed where it is required while fresh and allowed to harden, usual in *structural* work. It may be supported by *formwork*, or by the ground, as are *trench-fill foundations*. Horizontal surfaces are usually levelled with a *screed board*, followed by *floating* and *trowelling*, or given one of many types of finish. Decorative vertical surfaces may have a *direct finish* from the formwork.

inspection A check on work by a qualified person, to ensure that it has been done properly. Inspections are made in stages, before work is covered up, in *foundations* or wall *cavities*. Some inspections are made by the *building inspector*, others by a *clerk of works*, if there is one. The *architect* also makes monthly and final inspections before issuing each *certificate*. Inspection is limited to visual assessment and differs from *supervision*.

inspection certificate A note from a *local authority* certifying that drains are satisfactory and that a building is ready for occupation.

inspection chamber A *pipe inspection chamber*, but see also *C*.

inspection cover A metal plate in a fixed frame over an *inspection chamber* (*C*). It may have a single or double seal to make it *airtight*.

inspection door A panel or trap that can be removed from a wall, ceiling, or duct to reach *service* pipes, ducts, and cables and their fittings.

inspection fitting, i. eye An *access cover*.

inspection junction A *rodding point*, or a *lamphole* (*C*).

installation (1) The equipment and system of pipes, cables, or ducts for building *services*.
(2) **internal i.** The parts on the consumer's side of an intake cock or switch, or the *sanitary pipework* and buried drains before they reach the sewer connection, including those for temporary use on site.
(3) The supply and fixing of components; incorporation into the works.

installed load The total possible demand for power from an electrical installation if everything were to be switched on together, without *diversity*.

instantaneous water heater A gas or electrical *water heater* usually able to supply enough hot water for a shower and rated at about 7 kW.

instant-start tube A *fluorescent tube* that strikes an arc and lights up immediately without delay. It may also restrike quickly.

Institution of Electrical Engineers (IEE) The professional body for electrical engineers in Britain (see next entry).

Institution of Electrical Engineers Regulations for Electrical Installations, IEE wiring regs (UK) A *code of practice* published by the Institution of Electrical Engineers, applying to building *wiring*. It is harmonized with European regulations, working through the *European Committee for Standardization*. The IEE wiring regs have no statutory force under the Electricity Acts, but the electrical contractor has to certify that a new installation complies with them before a local electricity company will connect the power.

instructions to tenderers (USA **notice to bidders**) An early page of *bills of quantities* that gives guidance on pricing, such as the need to enter a price for each *item*, to ensure that prices include all labour and materials and other costs, etc.

insulate To reduce *transmission* of heat, sound, or electricity, usually by lowering conductivity, using either *insulation* or *discontinuous construction*.

insulating board, softboard Lightweight *fibreboard*, with a density usually less than 350 kg/m³ and a *k-value* better than 0.058 W/m°C, used for thermal *insulation* or to make *acoustic tiles*.

insulating material, insulant Material used for *insulation*.

insulating plaster Special plaster with a high proportion of *lightweight aggregate*. If too much cementing material is used it becomes less porous and less effective. It has low *thermal capacity* and warms up quickly, which can reduce *condensation* on walls.

insulating plasterboard Thermal-grade *gypsum wallboard*.

insulation (1) **thermal i.** Material that slows the heat loss, used to reduce the energy needed to keep a building warm or a refrigerator cold, or to provide *fire protection*. It can often control *condensation* if correctly installed. The most common types are cellular and reflective insulation.

Cellular insulation (foam, expanded materials) has many tiny air or gas bubbles, which reduce *conductivity* and usually make it lightweight and not very strong. It is therefore difficult to fix securely, requires *mechanical protection*, and has low *thermal capacity*. It comes in the form of *insulating boards, loose fill, moulded insulation* or *flock spraying* and will usually save money in the long run. The saving can be worked out from its *economic thickness*. Cellular insulation may need protection from *interstitial condensation* by having a *vapour barrier* on its 'warm' side and *ventilation* of its 'cold' side.

Reflective insulation, such as *aluminium foil* and *low-emissivity coatings*, reduces heat loss or gain by radiation.

Building insulation can be installed on top of the ceiling, under or over the roof, in a wall cavity, outside a wall, under a ground floor, or as window *double glazing*. See **U-value**.

(2) **acoustic i.** See **sound insulation**.

(3) **electrical i.** A non-conducting covering over cables or inside *power tools* to prevent short circuits and electric shocks. Usual materials are plastics such as *polyvinyl chloride* and *cross-linked polyethylene*, synthetic *resins*, or rubber. See **mineral-insulated cable**.

insulator A material used for *insulation*.

insurance A contract in which one person (the insurer) agrees to bear a financial risk on behalf of another (the insured) in return for a payment (the premium). Conditions of this contract are stated in a written agreement (the policy). The insured must act in the 'utmost of good faith'. Insurance against injury to workpeople is required by law in the UK. Proof of contractor's all-risk insurance is usually required before a building contract is signed. Insurance against construction defects is voluntary in Britain but has been required by law in France since 1805.

intake unit, house service cutout A box for connecting the incoming electricity supply *service cable* through a *fuse* to cables that go to the electricity meter.

integral waterproofing The *waterproofing* of concrete by including an *admixture* with the mixing water or cement. Probably only good-quality concrete can be successfully waterproofed in this way.

integrity In *fire-resistance grading* the absence of cracks through a door or wall under test that would allow flames to pass to the cooler side.

intelligent building, smart b. A building in which computer predictions are used in the *building management system*. The computer may be able to reprogram itself for changes in season or patterns of use.

intelligent fire detector, smart d. A *fire detector* with a built-in memory holding computer data about usual room conditions and changes. It is used in an *addressable system*. Apart from giving a fire alarm signal it may also give 'fault' and 'pre-warning' signals.

intercepting trap, interceptor (1) A *disconnecting trap*.
(2) A *trap* that prevents unwanted waste from getting into a drain, e.g. a *grease trap* (C) or a *petrol interceptor*.

intercom A *communications* system between different parts of a building, e.g. for an *entry phone*.

interference *Clashing*; also electrical interference or *cross-talk*.

intergrown knot A *live knot*.

interim certificate Usually a *monthly certificate*.

interim payment Usually a *monthly instalment*.

interior adhesive (INT) A low-durability *adhesive* with some resistance to cold water but not reliable against insects or fungus.

interior paint, internal p. The most usual finishing for internal walls is *emulsion paint*. It can be used on most types of plaster without the separate primer needed on wood or metal. Interior paints may not be suited to *wet areas*.

interior plywood *Plywood* made with *interior adhesives*.

interlock (1) A joint with an edge that holds another similar edge, such as a loose tongue joint between sheets of *particle board* or metal sheet folded round metal sheet in the slats of a *rolling shutter*.
(2) A safety *lockout*.

interlocked grain, interlocking g., twisted fibres A *grain* of timber in which the fibres slope one way in one series of rings, then slowly reverse their slope in the next growth rings, and so on. *Spiral grain* is one example. Wood with this grain is difficult to cut or split.

interlocking joint A joint in *ashlar* in which a projection on one stone beds in a groove on the next one.

interlocking paver, interpaver A *paver* with zig-zag edges that does not slide along others when laid with tight joints, although the paving can flex.

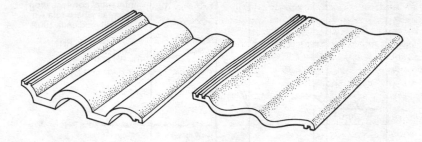

Interlocking tiles.

interlocking tile A concrete or clay *single-lap* roofing tile shaped to fit other similar tiles. The *sidelaps* are weatherproofed by matching grooves under each right edge and on top of each left edge. The *head* may also have a turned-up edge, to go under the *tail* of a tile in the course above, in which case the tile is said to interlock on four edges. The main surface of the tile can be of any shape, *double Roman* or *flat interlocking*. In general, interlocking tiles make a lighter roof than *slates* or *plain tiles*, but their *gauge* must be set out accurately because little adjustment is possible. See diagram.

intermediate joist A full-length *common joist* from wall to wall.

intermediate landing A *landing* between floors.

intermediate rafter A *common rafter* that runs the full length from *wall plate* to *ridge*, not a *jack rafter*.

intermediate rail A horizontal *rail* of a door joining the *hanging stile* to the *shutting stile*, e.g. a *lock rail*.

intermediate sheet A thick, heavy *bitumen felt* with a smooth face, bonded on to the *base sheet* of *built-up roofing*.

internal angle The corner of a room seen from the inside.

internal diameter (ID), bore The usual description of pipe sizes is to state their *nominal* inside size, plus the wall thickness. For plastics, the outside nominal diameter is stated.

internal door A *door* for communicating between rooms, made for *sound insulation* or *fire resistance*, but not for security and exposure, as are external doors.

internal dormer A vertical window in a sloping roof within the general line of the roof, with a flat surface or deck outside requiring an *apron*.

internal hazard An internal *fire hazard*.

internal leaf The inside leaf of a *cavity wall*.

internally reflected component In the *daylight factor method* the light reflected from objects inside a room.

internal pipework, i. plumbing Building *services* which supply hot and cold water, plus *sanitary pipework*.

International SfB A classification for building information subjects. The system originated in Sweden but is now sponsored and administered by

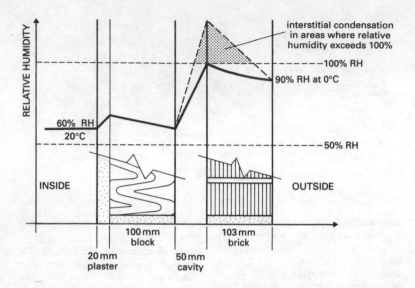

Interstitial condensation. This takes place when the warm surface is not impervious. The air passes through it and drops its water as dew wherever it is cooled to its dewpoint.

CIB (International Council for Building Research Studies and Documentation). The CI/SfB English-language version is the responsibility of the RIBA.

International Standards *Standards* issued by *ISO*.

interpaver An *interlocking paver*.

interstitial condensation *Condensation* inside the walls or *insulation* of a well-heated building, thus usually invisible. It wets the material of the wall, or the insulation, thus providing a path for heat loss. In severe cases it can freeze and cause damage, rot and corrosion. If a *vapour barrier* is placed on the warm face of the wall and there is adequate *ventilation* of the cold face, interstitial condensation should never occur. See diagram.

interstitial level A *plant level* between

two normal floors of a tall building, usually with the central *air-conditioning* plant for about six levels above and below it, to limit the length and size of *ducts*.

intertie An intermediate horizontal member of a *framed partition* which stiffens it at a door head or elsewhere between floor levels.

in-the-white joinery Unprimed joinery, such as doors from a factory.

intrados The curved *soffit* forming the underside of an arch or *vault*.

intruder alarm system, burglar a. A *security* system with detectors activated by an intruder and with an alarm panel or sounder. Detectors can work with invisible microwaves, ultrasonics, or infra-red, or be simple magnetic door or window contacts. Expert advice is needed to avoid false alarms.

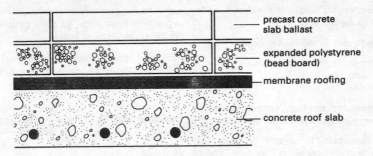

precast concrete
slab ballast

expanded polystyrene
(bead board)

membrane roofing

concrete roof slab

An inverted roof.

intumescence Rapid expansion on heating. In a fire intumescent paints make an insulating foam that delays *flame spread* and may in thick layers improve *fire resistance*. At 100°C their contained water boils to make a foamed *thermoset* resin. They can act as steelwork *encasement* for action in fire, but the full thickness of coating must be present, so in an exposed place an extra protective paint coating may also be needed. Intumescent strips and sealants are used for *fire stops*, e.g. to upgrade the sides of existing timber doors by preventing flame penetration. Intumescent materials are also used in *honeycomb fire dampers* and in laminated *fire-resisting glass*, which turns opaque when heated, thus providing some insulation.

inverted roof, protected membrane roofing (USA **i.r.m. assembly**) Roof covering with *external insulation* on top of *membrane roofing* rather than under it (the traditional arrangement), made possible by the development of plastics-foam insulation boards such as extruded *expanded polystyrene*, which resist moisture and insulate well when wet. The advantages are that this forms a warm deck, or *warm roof*, so that the membrane acts as a *vapour barrier* for the insulation outside it, and the insulation in its turn

provides *solar protection* for the membrane. To stop the wind blowing it away, an inverted roof needs to be weighed down with pre-loaded insulation boards, concrete slabs, or, for sheltered sites, gravel *ballast*. See diagram.

invertor A device that converts power from a battery into alternating current (usually at mains voltage), used in *uninterruptible power supplies*.

invited bidder (USA) A *selected tenderer*.

ionization chamber smoke detector A fast and sensitive *fire detector* that reacts to both visible *smoke* and earlier, clear *combustion gases*.

ironfighter (USA) A *steelfixer*.

ironmongery *Hardware* usually for doors or windows, such as locks, hinges, and closers, mostly made of steel but also of aluminium, nylon, brass, or bronze.

iron paving tile A triangular floor tile made of *cast iron* (*C*), with 305 mm sides and feet at each corner. The feet project through the mortar *bedding*, the tiles are laid in and rest on the concrete subfloor. This heavy-duty industrial flooring withstands high temperatures and grease.

ironwork Decorative *metalwork* made of *wrought iron* (*C*).

irregular bond *Broken bond.*

irregular coursed rubble *Rubble walling* built to *courses* of varying depths.

irregular paving Natural stones of varying shapes laid to fit together.

irrigation The supply of water for growing plants or for *leaching*. Drip, spray, and trickle methods are used for *landscaping* round buildings.

ISO (International Standards Organization) The body in Geneva that coordinates international standards, working with national standards bodies such as the *British Standards Institution* or the *American National Standards Institute*, as well as the *European Committee for Standardization*.

isocyanurate Poisonous material in some *two-part* polyurethane paints and thermal insulation that can be dangerous if breathed in as a 'spray mist', vapour, or fumes. For spraying, breathing apparatus is required as well as one-piece overalls and gloves. Otherwise good mechanical ventilation may be enough. In the UK it comes under regulations administered by the *Health and Safety Executive*.

isocyanurate foam *Polyisocyanurate foam.*

isolated ceiling A *false ceiling* carried by ceiling *joists*, not by floor joists, used as *discontinuous construction* to improve sound insulation.

isolated column A *column* some dis-tance from other columns and *measured separately* from them to allow pricing of the extra *labour* to build it.

isolated excavations *Groundworks* for foundations and services trenches. They are dug after the *bulk excavations*.

isolated footing A *foundation* to carry a point load, such as a *base* or *pad footing* under a column.

isolating membrane An *underlay*, usually of *building paper* or *sheathing felt*, between *asphalt* roofing and its supporting roof deck, to allow the asphalt and the deck their different *thermal movements*. The friction should be enough for the asphalt neither to contract in cold weather nor to slide easily.

isolating strip An *expansion strip.*

isolating valve A *stopvalve* to close off the water supply to part of an installation but not to regulate flow.

isolator A *coupler.*

Italian tiling (USA **pan and roll tiles**) *Single-lap tiles* which form a roof covering with two different sorts of tile, the curved over-tile or imbrex and the flat tray-shaped under-tile or tegula. Compare **Spanish tile**.

item A numbered line or paragraph in a *bill of quantities* stating the *description*, *quantity*, and unit of measure (m, m², m³, kg, etc.) of one part of the building work. When *tendering*, the contractor writes a price in the money column opposite the item.

J

jack arch A *Welsh arch* or a *flat arch*; but see also *C*.

jacket An outer casing over the insulation of a hot-water *cylinder*.

jacket ladder A *safety ladder*.

jackhammer A *breaker* used to dig rock, demolish concrete, etc. Safety boots should be worn and feet kept clear of the pick (except when starting) in case it breaks.

jack-leg cabin A *volumetric building*, with steel legs which can be adjusted to suit uneven ground. Lifted into place by crane, the cabins can be stacked on top of each other and are used for *site accommodation*. Fire safety should be carefully considered – they have only modest *fire resistance*.

jack plane A hand-held *bench plane* for cleaning saw marks from timber or for preliminary rough *dressing*. Electric *planers* are used for similar work.

jack rafter (1) A short roof *rafter* from *eave* to *hip* or from *valley* to *ridge*.
(2) **j. truss** A shortened *trussed rafter*, one end of which may be carried on a *girder truss*, although details vary between manufacturers.

jack shore A *back shore*.

jamb The vertical flank of a wall opening, to the full thickness of the wall; often also the joinery *lining* on it, including the *reveal*, or the upright side members of a door or window frame. For a single door or *casement* window, one is the *hanging* and the other the *shutting* jamb.

jamb form A three-sided box the thickness of a wall, placed between two shutters to form a door opening in an *in-situ concrete* wall.

jamb lining A timber facing covering a *jamb*.

jemmy A *pinch bar* about 380 mm long.

jenny A *gin block*.

jerkin-head roof, hipped gable, shread head (USA **clipped gable**) A pitched *roof* with a *hipped end* until halfway down to the *eaves* and gabled from there down, unlike a *gambrel roof*. See diagram, p. 376.

jet A nozzle or sprayer through which air or a liquid is forced, emerging at high speed, thrown far, and mixing with the surrounding air or liquid.

Jetfreezing A method of *pipe freezing*.

jib The movable arm of a *crane* from which the hoisting rope hangs.

jib door, gib d. A door whose face is flush with the wall and decorated to be inconspicuous.

jig A clamp or *template* for holding work or guiding a tool so that repetitive jobs can be accurately worked without repeating the marking out.

jig saw A portable *power tool* with a reciprocating, projecting saw blade. It rests on top of the work and is guided by one hand, to cut straight lines or sharp curves. One weighing about 3 kg can cut mild steel 6 mm thick, aluminium 15 mm thick, or softwood 60 mm thick. See diagram, p. 346.

jinnie wheel A *gin block*.

jobber, builder's handyman A

semi-skilled worker who can do any house repair such as bricklaying, plastering, painting, plumbing, joinery, roofing, etc.

job site (USA) A building *site*.

job specification (USA) The *specification* for a particular building project. It may have to be read along with a standard specification.

joggle An up-and-down shape, with square or sloping sides and a flat top, e.g. a double crank, a *stub tenon* joint, or a *tongued and grooved joint*. See next entry.

joggle joint, stop end and key A *construction joint* in a concrete slab with a bevelled groove made by a trapezoidal former fixed inside the *stop end*. The groove forms a tongue along the edge of the next bay of concrete, making a joint that can transfer shear forces.

joiner A craftsman who makes *joinery* working mainly at the bench on wood that has been shaped by the machinists. The work is finer than that of a *carpenter*, much of it being done in the good conditions of a joinery shop, where it is not exposed to the weather.

joiner's gauge A *marking gauge.*

joiner's hammer, Warrington h. A *hammer* with a *cross-peen* head, weighing from 110 to 570 grams.

joiner's labourer, carpenter's mate A helper to a *carpenter* or *joiner.*

joinery (1) (USA) **finish carpentry** Making and fixing the wooden finishings and trim to a building, such as doors, *skirting boards*, *architraves*, linings, windows, and picture rails, which are usually painted or varnished.
(2) (USA) The joints in *carpentry*, joinery, or cabinet making.

joineryfittings (USA **millwork**) Prefabricated *joinery* components, such as cupboard and bench units, storage fitments, benches, reception desks, etc., manufactured *off-site* in a machine shop (USA millwork plant, planing mill), using *shop drawings*. The *dimensions* of the building *fabric* must be checked on site before joinery is made, and it is usual to allow for a *scribed joint.*

joint A connection between two components, which are shaped to fit each other, or fastened or sealed by another material such as *adhesive* or a *weld* (*C*). In brickwork and blockwork a joint is the mortar (usually 10 mm thick) between adjacent bricks or blocks, the *bed joints*, *perpends*, and *wall joints*, or the method of surfacing the mortar by *jointing*. In carpentry and joinery, many intricate shapes have been used to form corners (angle joints) and for lengthening or widening a piece of timber (see BS 1186). Pipes and fittings mostly have some form of *spigot-and-socket joint*, made strong or kept from leaking by *joint rings*, *solvent welding*, or *capillary jointing*, or with a screwed *union*. Metal sheets can be joined by folding the edges to make a *seam.*

joint backing A *backup strip.*

joint bolt A *handrail bolt.*

joint cement, j. filler, j. finish, jointing compound Proprietary materials used to make joints between sheets of *dry lining.*

Joint Contracts Tribunal (JCT) The body which publishes several different *standard forms of contract* for building works, such as JCT 80 for large projects, with or without *quantities*, JCT Minor Works 88, etc. JCT members come from organizations representing architects, surveyors, contractors, local authorities, etc.

joint cover Trim over a joint, such as the capping strip of a *wood roll.*

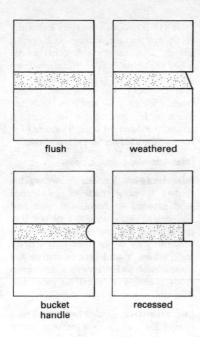

flush   weathered

bucket handle   recessed

Types of pointing.

jointer (1) A bricklayer's tool used after bricklaying for *jointing*.

(2) **jointing plane** A long hand *plane* for straightening edges of wood to be fixed with *adhesive*. Steel jointers are about 600 mm long.

jointer saw A coarse-toothed steel hand saw for cutting stone.

jointing (1) Filling the joints of brickwork or blockwork with wet mortar, performed as a single operation as *bedding and jointing*; also any forming of the *face joint* done while the mortar is still wet, e.g. striking off flush for *pointing*. Jointing can be *bucket-handle*, *weatherstruck*, *flat-joint*, or *keyed*. See **tooling**.

(2) Sealing a pipe joint with compound, tape, sealant, caulked lead, etc.

(3) Making or assembling a woodwork joint.

jointing compound (1) Putty-like material inserted into a pipe joint to make it leak-proof, largely superseded by *jointing tape* or *joint rings*.
(2) *Joint cement*.

jointing fluid Liquid used to make a *solvent-welded joint*.

jointing mortar Tiling *grout*.

jointing rule A long *straightedge* used by bricklayers to keep a *jointer* on the right line.

jointing strip Preformed foam and mastic used to make joints.

jointing tape Thin *polytetrafluoroethylene* tape used for *jointing* pipe *taper threads*.

jointing tool A 13 mm dia. steel bar with turned-up ends and a central handle, used to make a *bucket-handle joint*.

jointless flooring A floor surface formed in-situ rather than laid as tiles. The simplest is a concrete slab, finished by a power float and *hardeners*. The term includes toppings such as *granolithic* and *terrazzo, composition flooring, asphalt* or *pitch mastic, epoxy resin, polyester* resin, and *polyurethane*. Despite the name, many jointless floors, particularly those containing cement, are best laid with joints about 2.50 m apart to allow for *drying shrinkage*.

joint mould, section m. A *template* used in plastering to shape a *run moulding*.

joint reinforcement Steel mesh in the *bed joints* of *reinforced masonry*.

joint ring, rubber r., sealing r. A ring of rubber or plastics used to make a *push-fit* joint in a drain pipe. Many types exist, for concrete, vitrified clay, cast iron, or plastics pipes, such as *O-rings, D-rings*, and *lip seals*. See diagram, p. 252.

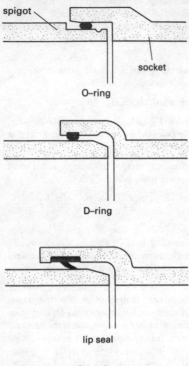

spigot

socket

O–ring

D–ring

lip seal

Joint rings.

joint runner, pouring rope Asbestos rope or similar material placed round the outside of a cast-iron pipe while pouring a *lead-caulked joint*.

joint tape (1) Rolls of *scrim*, usually 50 mm wide, either self-adhesive or stuck down with *joint cement* over the joints between *gypsum wallboard*, before the joint is finished with more joint cement.
  (2) Woven tape used to join rolls of *carpet*. It is either ironed on or heated by passing an electric current through an element in the tape.

joint venture A temporary union of two or more companies for one large project.

joist (1) A wooden or steel beam directly supporting *flooring* or a ceiling lining, in common with other joists. They are either full-length *intermediate joists*, *trimmer joists*, or *trimmed joists*.
  (2) In the USA rectangular *lumber* from 2 in. to 5 in. thick and 4 in. or more wide, graded for its bending strength loaded on edge. If graded for strength loaded on face it is a *plank*.

joist anchor A *wall anchor*.

joist hangers Various *connectors* for timber framing made of galvanized steel pressed to shape, with punched nail holes. One common type is a box that carries the end of a floor joist so that the joist is not built into a brick wall, where it could rot or reduce *fire resistance*. Another type is a steel strap from a *hanger* to a ceiling joist. See diagram.

joist trimmer, girder bracket A *joist hanger* for nailing to a *girder* to carry one end of a *joist* at right angles to it. See diagram.

journeyman A *tradesman* who practises his trade with full responsibility.

judas A small hole in a door at eye

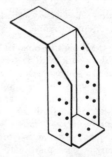

A joist hanger of galvanized steel for building into brickwork.

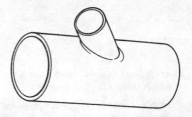

A junction.

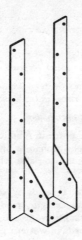

A joist trimmer of galvanized steel, nailed or screwed to a joist so that another can hang at right angles to it.

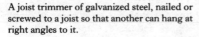

level through which visitors can be seen.

jumbo brick (USA) A *brick* larger than usual, intentionally or by mistake.

jumper (1) A *stretcher* brick which covers more than one *perpend*.

(2) A brass or nylon, mushroom-shaped part in a *screw-down valve* which carries the *washer* on the lower face of the mushroom disc. The stalk of the mushroom points upwards into a hollow guide. The jumper is forced down on to the valve seat as the handle is turned. The jumper is lifted by water pressure in the pipe after the tap is opened, but it does not allow *back-flow*.

(3) **jump lead** A cable used to make a temporary electrical connection.

jump form A climbing form made of individual panels that are taken off the bottom and moved to the top, for the form to take the next *lift* of concrete.

jumping jack A *consolidating rammer*.

junction A *special* drain pipe made to join a *branch* to a main run. It has a short tail off the main run at a 45° angle, or can be a *wye*. See diagram.

junction box A box which covers the joints between the ends of electrical conductors in *wiring* or underground cables.

K

kauri A New Zealand conifer the wood of which is used for fine joinery.

Keene's cement Hard-burnt *anhydrous* gypsum with an alum *accelerator*, no longer made, formerly used as a hard-finishing *plaster* that could be trowelled smooth.

keeper (1) A *striking plate* on a door jamb in which a lock bolt engages.
(2) A metal loop over the fall bar of a *thumb latch* to limit its travel.

keeping the gauge Maintaining the spacing of brick *courses*, usually four per 300 mm height for a metric English brick of 65 mm thick and 10 mm joints.

keeping the perpends Laying *bricks*, stones, slates, or *tiles* accurately so that the cross-joints (*perpends*) or the visible edges of slates or tiles in alternate courses are in the same vertical line.

keratin A *retarder* added to gypsum *hemihydrate plaster* to make *retarded hemihydrate*, obtained from the horns, hoofs, nails, or scales of animals.

kerb, curb (1) A low fender or buffer to separate a road from a *walkway*, usually of precast concrete supported by a *haunching* of *dry lean concrete*.
(2) A low *upstand* above the general line of a roof, as at *fire vents*, *extract fans*, or *expansion joints*, against which a *flashing* is made.

kerf The slot cut into wood, stone, etc., by a *saw*.

kerfed beam A beam which is bent by *kerfing*.

kerfing Making saw cuts in one side of a piece of wood and bending it

towards that side, a convenient way of curving the risers of bullnose steps.

kettle A *bitumen boiler*.

kettleman, potman A building *operative* who melts asphalt or bitumen and carries the hot material to where it is to be laid.

Keuper Marl brick A strong *clay brick*, usually *wirecut*, made of fine-grained siltstone from ancient wind-blown deposits in western Europe, including Britain.

key (1) Keys for locks come in many shapes and sizes, some types having large numbers of *differs*.
(2) An operating handle, e.g. for a stopvalve under a *surface box*.
(3) A *loose tongue*.
(4) The roughness of a surface, the texture which enables *plaster*, *mortar*, *adhesive*, etc., to grip with a *mechanical bond* (*C*).

key changes, k. combinations *Differs*.

key cut The metal removed from a key blank to give it a profile to operate a particular lock or suite of locks.

key drop, keyhole cover A flap that covers the keyhole of an *escutcheon*.

keyed (1) Held or locked by a *key*.
(2) Said of a *dowel* which is grooved lengthwise to allow air and excess *adhesive* to be driven out.

keyed beam A timber beam with a *lap joint* into which slots have been cut in each member. Hardwood or metal rectangular wedges (keys), driven into these holes, increase the bending strength of the joint.

keyed construction joint A *joggle joint*.

keyed joints Raked-out brickwork *joints* for keying plasterwork.

key event A term from planning and scheduling, corresponding to a *stage* of work progress.

keyhole saw A *compass saw*.

keying Roughening a surface to enable later work to stick, e.g. hacking walls to take *plaster*, scabbling slabs to take a *screed*, or scratching plaster undercoats for the finishing coats.

keying-in *Bonding* a brick wall to another already built.

keying mix A *bonding treatment*.

key-in-knob set A *bore lock*.

key plan A *location plan*.

key plate An *escutcheon*.

key schedule A chart that lists all doors and their keys, organized into *key suites* for *master keying*. It is a *contract document*.

keystone A central wedge-shaped arch stone at the crown of an *arch*, put in last. It is often large and carved, but no more important than the other arch stones, though its insertion completes the arch.

key suite A set of keys listed on a *key schedule*.

keyway miller A *brickwork chaser*.

khaya The most usual *African mahogany*.

kick A change in slope of a roof. A *mansard roof* has a 'kick in', a steep roof with a shallow veranda has a 'kick out'.

kicker The first 70 mm of a concrete *column* or wall, identical to the remainder, but cast before *shutters* are stood

up, to hold them in place and plug their bottom end. Kicker *setting-out* lines should be marked on the floor slab in pencil, where the carpenters actually have to put the *kicker frame*.

kicker frame The *formwork* for a column *kicker*, usually a timber rectangle. It is 70 mm high and held in position by being cleated to the column *starter* bars.

kick plate, kicking p. *Door furniture* fixed across the bottom of a door leaf to protect it against kicks, usually a stainless steel plate 150 mm high.

killed spirits A *flux* for soldering galvanized steel, made by dissolving enough zinc in hydrochloric acid for it to become a solution of zinc chloride.

kiln A furnace in which cement, brick, or lime is burnt, or a chamber through which warm air passes to dry timber.

kiln-dried timber, kilned t., ovendried t. Kiln drying is quicker than *natural seasoning* and the high temperature kills *fungus* and *borers*, but the pores of the timber are forced open so that it absorbs water quickly if exposed to the weather without being painted. Some timbers, such as *eucalyptus*, may suffer *collapse* and need *reconditioning*.

king closer A *special* brick with a long triangle cut off one corner, vertically with the brick flat, to leave a half face, and a half end, to *format* sizes, respectively 103 and 46 mm long. It may be called a three-quarter brick, as three quarters of the original faces remain, plus a splay.

king-post truss A traditional timber roof *truss* with a vertical post from the apex to the centre of the bottom *tie beam*, suitable for spans up to about 11 m, but like the *queen-post truss* not now used for new work in Britain.

kiss mark A mark on a *brick* from having touched another in the kiln.

kitemark The trademark of the *British Standards Institution*, which can be used by an approved product manufacturer as a *conformity mark*.

kite winder Of three *winders* in the corner of a stair, this is the central one, so called because on plan it looks like a kite.

kit home A prefabricated house, usually timber-framed, assembled on a builder's foundation either by a builder or an owner. Kit homes are part of a trend towards *industrialized building methods* and can consist of the frame only, a complete *superstructure*, or a full house with cladding and even *external works*. Kits may include fixing instructions and data needed for the *building permit*.

kit joinery Factory-finished standard doors, windows, cupboards, shelving, etc., made for easy assembly on site and usually supplied complete with all metalwork, fixing screws, glass, etc.

knapping hammer A *hammer* for shaping building stones.

knee (1) **elbow** A sharp right-angled pipe *bend*.
(2) An intersection between the horizontal part of a stair *handrail* and the downward sloping part, shaped like a human knee. Timber knees are *mitred*.
(3) **crook** A naturally curved short timber.

knee brace, corner b. A short strut across the angle between a post and a beam (or roof rafter), forming a triangle that stiffens the joint against wind.

kneeler, skew, s. table The sloping-topped, level-bedded stones in a *gable coping*, above the *gable springer*.

knifing filler A *filler* fine-grained and soft enough to be put on with a *filling knife*.

knob (1) A *door knob* or similar *furniture* on a cupboard or window.
(2) A rounded end-piece or button, e.g. a *pommel* on a stair *newel post*.

knob set Usually a *bore lock* with *door knobs* on both sides.

knocked down Of building components, having been completely cut and shaped for a job. They are delivered for assembly on site.

knocking up *Retempering*.

knockings Stone chips smaller than *spalls*.

knockout A *popout*.

knot A place in a tree trunk from which a branch has grown out, a *gross feature*. Knots which reduce the strength of wood are *dead, loose*, or *unsound* knots. Harmless knots are *live*, pin, *sound*, or *tight*.

knot brush A thick paint *brush* with its bristles or fibres bunched in round or oval shapes, for painting walls (one-, two-, or three-knot brushes).

knotting *Shellac* dissolved in methylated spirit, used as a local *sealer* over *knots* in new softwood. It enables knots to be primed without danger of resin dissolving the paint and bleeding through. Other quick-drying compositions are used, as well as *aluminium primer*.

knotting, priming, and stopping The three main steps in *preparation* of new wood for painting.

knuckle The holes in a *hinge* through which the pin passes.

knuckle bend A pipe *bend* which turns as sharply as possible.

knuckle joint A sharp joint between the two slopes of a *mansard roof*.

kraft paper Strong brown paper which is put to many uses, e.g. as the containing sheet of insulation *blankets*, in *building paper*, etc. (Swedish for strong.)

k-value, thermal conductivity, heat-transmission value The amount of heat that passes through unit area (m^2) of a material of unit thickness (m) for unit temperature difference (°C). Low-density cellular plastics used for *insulation* have a k-value of about 0.025 to 0.035 W/m°C; bricks have a value of around 0.3 to 0.7 W/m°C.

L

laboratory receiver A *diluting receiver*.

labour constant The cost of labour per unit, used in *estimating* from bills of quantities, e.g. for 1m² of brickwork 215 mm thick.

labourer A building worker who does physical work but who has learnt no trade and is therefore paid less than a *tradesman*. A labourer usually works as part of a *gang* and can be unskilled, *semi-skilled*, or *skilled*.

labours Work, especially that done on a material for which there is an item in *bills of quantities*, but needing extra attention, e.g. a construction detail. The cost of labours is added to the price of the existing item. A labour may be a separate item in a bill, but one item may contain several labours.

laced valley A roof *valley* formed of *tiles* or *slates* without a *valley gutter*. Slates or tiles of one and a half times the normal width are laid on a *valley board*. The two slopes intersect sharply and do not blend into each other as in a *swept valley*.

lack of cover Lack of *cover* by concrete over *reinforcement bars* gives less protection against corrosion.

lacquer A paint which dries rapidly by *solvent release* from the *medium*. It is usually sprayed, to prevent it re-dissolving an undercoat. It is prone to *blushing* and *orange peeling*. Lacquers are used on metals which they 'wet', giving good adhesion, and even directly on brass (transparent lacquer). Unlike *varnish* and *enamel paint*, most lacquers are based on cellulose compounds.

ladder (1) Ladders have two *stiles* spaced apart by horizontal *rungs*. Fibreglass and aluminium ladders are much lighter than wooden ones, but the latter conduct electricity. Ladders should be at least 1.05 m longer than the roof to be reached and used strictly to the maker's instructions to prevent falling accidents. The ideal angle is 75° (1 horizontal for 4 vertical). Top ends can slide sideways and must be tied. Bottom ends can slide outwards under the thrust of a person's weight and the bottom should be secured if it is resting on a smooth surface. Short ladders should be held by a second person (footing). A ladder placed on top of a scaffold tower can overturn it. Nothing heavy should be carried up a ladder, nor anything left on it. Usual types are the *builder's ladder*, pole (wooden) ladder, *extension ladder, steps*, and fixed *safety ladders*. See *Health and Safety Executive* guidance notes, BS 1129, and BS 2037.

(2) **cable l**. A metal framework like a ladder to support runs of cable, which are secured to the rungs with *cleats* or ties.

lag bolt, l. screw (USA) A *coach screw*.

lagging (1) Thermal *insulation* on the outside of pipes or tanks to reduce heat loss from warm surfaces, or on cold surfaces to reduce heat *gain* and *condensation*. Air-conditioning pipes and ducts can have wrapped or *moulded insulation*, usually applied by a specialist contractor.

(2) Sheet material used for *earthwork support*, carried by *soldiers* (*C*).

258

laitance A weak milky skin on top of concrete, composed of cement and *fines* and caused by excess water in the mix or by light hand *trowelling* too early, before *initial set*. After *power floating* it is rare. Laitance on untrowelled structural slabs is removed by *hacking* or *scabbling* before a *screed* is laid.

lake A *pigment* made by dyeing an insoluble base, such as aluminium hydroxide.

lake asphalt A natural *asphalt* containing bitumen, clay, and water, found in low-lying land. The deposits in Trinidad are being lowered by constant extraction and no longer re-fill overnight.

laminate (1) **decorative l., laminated plastics** Thin but tough sheeting made of paper or fabric bonded with melamine or phenolic *resins*, forming a hard-wearing decorative finishing for work tops. It can also be made into bars, cylinders, and other sections. Tests are given in the 'dual-numbered' *standard* BS EN 438.
(2) To bond layers of material together in the form of a sheet, beam, etc., called *laying-up* in the case of *glass-reinforced plastics*.

laminated-fibre wallboard *Hardboard* used for panelling walls, ceilings, etc., with a surface which is smooth or pebbled, painted or prepared for painting. Preformed boards made to curves from 150 to 600 mm dia. can be used for forming concrete columns.

laminated glass Flat glass sheets joined by interlayers of tough plastics, such as *polyvinyl butyral*. Different types and thicknesses are used for security, for fire resistance, for *sound insulation*, and to control *ultraviolet*.

laminated joint A *combed joint*.

laminated plasterboard A *gypsum*

plasterboard which is used in *double-leaf separating walls*. It may consist of two layers of 20 mm plasterboard, three layers of 13 mm, or four layers of 10 mm, at least partly glued together in the factory and there enclosed in a timber frame. The last layer is often fixed on site with its joints in the opposite direction to those made in the factory. The 40 mm thickness of plasterboard weighs only 25 kg/m².

laminated plastics See **laminate**.

laminated wood Strips of wood glued or mechanically fastened together with their grain parallel. See **glued-laminated timber**.

laminate veneered board A wood-based *building board* with plastics *laminate* bonded to its faces, usually produced in a factory. Cutting the laminate without chipping is done with a shallow-set fine-toothed *circular saw* or by scoring a line before hand sawing. With care it can be *edge-shot*.

lamination (1) See **laminate**.
(2) A plane of weakness through a material along which it may crack.

laminboard A fine-quality wooden *building board* built up of core strips not more than 7 mm wide, glued between two or more outer plies. The finished board can be from 12 to 50 mm thick and is similar to *blockboard*, which is cheaper.

lamp (1) The light source of a *luminaire*, usually replaceable. There are many types, including the familiar *incandescent* light bulb as well as *discharge lamps*, e.g. the *fluorescent tube* and high-intensity mercury or sodium-vapour lamps. All lamps produce heat, but high-efficiency lamps give more light per watt.
(2) A *luminaire*.

lamp base, 1. cap The end of a *lamp* by which it is held and connected.

lamp holder A fitting which carries a *lamp*. Different types are made for *bayonet caps*, *Edison screw caps*, and other types of lamp.

landing A platform at the top or bottom of a stair, or between two *flights* (an intermediate landing), e.g. a *half-landing*, *half-space landing*, or *quarter-space landing*. Sometimes it is reached by a *lift*.

landing button A button on the wall of a *landing* used to call a *lift* (landing call). Large lifts usually have two landing buttons, one for 'up' and one for 'down'.

landing valve An outlet for connecting hoses to a *fire riser*, usually at each floor of a tall building, in the lobby of a *fire-fighting shaft*.

landscaped office An open area of floor divided into work stations separated only by furniture or screens which can be moved round to suit changing needs.

landscaping The land shaping, planting, and roads round a building. It is part of the *external works* and divided into *soft landscaping* and *hard surfaces*.

lantern (1) A square, octagonal, circular, etc., upstand above a roof, containing windows and its own roof, to admit light and allow ventilation through a ceiling. See **rooflight**.

(2) A *luminaire*.

lap, overlap, lapping (1) The distance one material passes over another.

(2) A weathertight joint between the sides or ends of lengths of *profiled sheeting*, or the sides and heads of *roof tiles*, slates, etc.

(3) The length that *reinforcement bars* run together where they are joined, to provide a *bond length*.

(4) To apply one coat of paint or varnish beside another and over its edge; or the joint so formed, which can be invisible if there is a *live edge*.

(5) A *passing*.

lap joint, lapped j. A carpentry *joint* made by laying the ends of two timbers one over the other and securing them by *adhesive*, bolting, or a *connector*. See diagram.

larch (USA **tamarack**) (Larix) A hard, strong, resinous *softwood* grown in the UK, not easy to convert owing to knots and resin and therefore used in the round for fences and in *carpentry*. Vitruvius, the Roman architect, noted that it is very heavy and resists decay and borers 'by the great bitterness of its sap' and that it 'cannot be kindled by fire'. It is one of the few deciduous *conifers*.

large-panel system A building *system* with precast concrete *cladding panels* that run the full width between columns (e.g. 7.20 m) and from window head to the sill above. Panels are lifted by crane, set on *shims* (*C*), and bolted to the *structure*. They usually have *open-drained joints*. Many large-panel systems have failed and needed repair or are enclosed with other cladding. See BRE reports BR 93, BR 154, and BR 185.

larmier A *drip*.

larrying, 1. up (USA) A method of bricklaying using fluid *mortar* in which the *bricks* are slid into position, after which mortar is poured in to fill the vertical joints.

latch (1) **l. set** A device with a bevelled tongue kept pushed out by a spring and a handle which withdraws it. Latches are fixed to the edge of a door leaf, so that the tongue engages in the door frame to hold the door closed. As anybody can turn the handle, it does

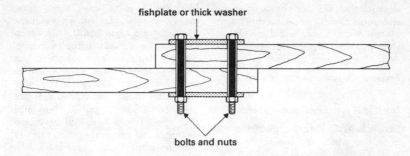

fishplate or thick washer

bolts and nuts

A bolted lap joint.

not provide the security of a *lock*. There are many types of latch, for doors, cupboards, and WC cubicles (privacy latches).

(2) The tongue of a latch, usually a *return latch*.

latchet A *tingle*.

latent heat Heat given off during *condensation* and taken in during *evaporation*. A different but smaller amount of latent heat is given off during freezing and taken in during thawing. Water vapour in the air contains latent heat which is increased by the sweat from people, who lose it and thus cool themselves. Latent heat affects the *comfort index* but does not show on a dry-bulb thermometer as *sensible heat* does. In refrigeration the *vapour compression cycle* uses the flow of latent heat.

latest date, l. event time In the *critical path method* a programme date before which an *activity* needs to be started or finished in order to prevent *slippage* of dates. It is calculated by a *backward pass*.

latex Originally the sap of the rubber tree, but now more likely to be *styrene/butadiene copolymer*. It is used in *emulsion paints*, *latex cement*, and foam rubber.

latex cement A *contact adhesive* used to stick down *floor coverings*, such as linoleum, carpet, or *wood-block flooring*, and as a *levelling compound*. It gives off ammonia, which can damage *copper pipes* embedded in a floor slab.

lath Split wooden slats, the original *lathing* material.

lath and plaster A traditional *finishing* made by nailing *laths*, with gaps between, to *studs* or ceiling *joists*, then trowelling on *plaster*. Some plaster, squeezing through the gaps, forms a neck which can break, allowing plaster to drop off. Movements in the laths may also cause cracking.

lathing Any base for plaster, including specially made *plasterboard* (gypsum lath), *metal lathing*, *woodwool*, etc., or traditional *laths*.

lathing plaster, metal l.p. A lightweight plaster mix used for undercoats on *metal lathing*. It is easy to apply and reduces rusting.

lath, plaster and set *Two-coat work* – a *floating coat* and a *finishing coat* – often used on backings such as *plasterboard* or *insulating board*.

laundry A place for washing soiled

linen, clothes, etc., and drying, ironing, and storing clean linen, sometimes with a *linen chute*. Large laundries with a continual discharge of hot water may need heat-resisting drain pipes.

lavatory A trough *ablution fitting* or a *water closet*.

lay bar A horizontal *glazing bar*.

layer board, lear b., gutter b. A board on which the lowest sheet, or sole, of a metal *box gutter* is laid.

laying-off The final, gentle *brushing* of a wet paint surface with a 'dry' brush to eliminate brush marks or roller stipple marks and help *levelling*.

laying-on trowel A rectangular steel trowel about 260 × 110 mm with which *plaster* is laid on a surface.

laying trowel A *brick trowel*.

laying-up, laminating The main stage in moulding *glass-reinforced plastics*, in which layers of *resin* then *glass-fibre* are placed over the *gel coat* and rolled until consolidated.

laylight A window fixed horizontally in a ceiling.

layout The location of each part of a building, its *fabric, finishings*, and *services*. Layouts are shown on a *drawing*, usually a plan, used on site for *setting-out*.

leaching The removal of substances by dissolving them in water. Alkalis leached from brickwork can etch marks in window glass. In arid-region *irrigation*, heavy watering of the soil leaches polluting salts and carries them down below the plant roots. See **leach, leachate** (*C*).

[In all the following terms down to 'lead wool' (except where otherwise indicated) the syllable 'lead' is pronounced 'led' and refers to the metal.]

lead (1) A soft, heavy metal used mainly in buildings for its excellent durability and high *ductility* (*C*), which allow it to be *dressed* to complex shapes. The Romans used it for plumbing, but Vitruvius warned that water stored in lead vessels was unhealthy; we now know that most lead compounds are poisonous. Many public buildings from the Middle Ages still have lead roofing. Lead rainwater pipes, often of very beautiful design, were used until the nineteenth century. Today, lead sheet is mainly used for roof *flashings*, but also for its low sound *transmission*. Lead corrodes aluminium, copper, steel, etc., from *dissimilar metal contact*. The risk to workers in most building work is liable to be 'not significant', but hands should be washed before eating or smoking (see the Control of Lead at Work Regulations).
(2) A lead *came* of a *leaded light*.

lead (*pronounced 'leed'*) (1) (USA) A *conductor*.
(2) **corner l.** (USA) A section of brickwork, plumbed exactly and built up ahead of the remainder by steps in the courses, called *racking-back*.
(3) *Lead time*.

leadburning The welding of lead using lead as the filler metal, instead of tin/lead *solder*. A high temperature is needed (about 327°C for pure lead as against 180°C for solders). The work is difficult, though less so if the lead contains antimony (regulus metal). Joints can be made that are strong, malleable, and sometimes inexpensive. Lead pipe can be joined to brass by leadburning. An oxy-acetylene or oxypropane flame, to provide the intense local heat needed, is best. Some toxic gases are emitted during leadburning.

lead-caulked joint A pressure-tight joint between cast-iron pipes made by filling the space between the *socket* and *spigot* with molten lead poured in behind a *joint runner* or *pipe-jointing clip*. After the lead has cooled and hardened it is trimmed, then hammered home with a *caulking chisel*. A similar joint can be made with *lead wool*, but both are now less used than *push-fit joints*.

lead-clothed glazing bar A *glazing bar* with a steel core and a *lead sheath*. It also has wings and a drainage channel, and ends sealed with lead.

lead dot The usual name for a *soldered dot*.

lead driers Lead paint *driers*, such as lead linoleate, lead resinate, and other organic salts of lead, but rarely the more toxic *litharge*. They are no longer used in paints for retail sale.

leaded light, lead glazing A window in which small pieces of glass are held between lead *cames*, common with decorative stained glass.

leader head (*pronounced* '*leeder*') (USA) A *rainwater header*.

lead flat A *flat roof* covered by *milled sheet lead*, laid as *supported sheetmetal roofing* with *rolls* and *drips* at suitable intervals and secured with *soldered dots*. Lead flats are best laid in areas smaller than 2.2 m² and shorter than 3 m, falling at 1.2% (12 mm per m) or more. The lead should be placed on a waterproof building-paper *underlay* to isolate it from its substrate and allow free *thermal movement*. The substrate can be wooden *decking* of planed boards or plywood, but *oak* should never touch lead as its tannic acid attacks the metal. Loose sheet lead is not suitable for steep slopes, which can be clad with *lead laminate*. The thickness of sheet lead is specified according to its *substance*, the lightest being Code 3 (1.32 mm thick), used for roofs with no traffic, and the heaviest Code 8 (3.55 mm thick), used if *bossing* is required.

lead-free paint Paint without lead compounds, even safer than *lead-restricted paints*, used in food packing.

lead-free solder A *soft solder* mainly of tin, used in *drinking-water* pipe joints. Tin/copper solders have 0.45 to 0.85% copper, tin/silver solders have 1.8 to 3.7% silver. Both can have traces of bismuth, cadmium, lead (0.1%), zinc, etc.

lead glazing A *leaded light*.

leading hand (*pronounced* '*leeding*') A *charge hand*.

lead joint A *lead-caulked joint*.

lead laminate A durable *composite* material made of thin sheet lead bonded to plywood, chipboard, or steel sheet, usually produced by specialists. Panels can be made to decorative shapes and are used for *cladding*, sound insulation, and radiation protection.

lead monoxide *Litharge*.

lead paint A poisonous paint containing *red lead*, or more *white lead* than allowed for *lead-restricted paint*. There is opposition to the use of lead paints, although they make very good *primers*, and they are not allowed on public buildings under EC law. If they have to be removed during renovation, dry sanding and hot-air stripping are both dangerous. Safer methods include using chemical *paint remover*, wet rubbing, *water blasting*, and enclosed *sand blasting* with dust extraction. In all cases the toxic waste must be collected, sealed in bags, and properly disposed of, but never burnt.

lead plug A small cylinder of lead inserted into a hole drilled in a wall, used as a *fixing* for a screw.

lead-restricted paint Paint which contains less than 5% *litharge* calculated according to the British *lead paint*

lead roof

regulations of 1927. Such paints are sometimes regarded as *lead-free*.

lead roof A *lead flat*.

lead safe (USA also **dip sink**) A lead-sheet floor *tray* with low edges over timber fillets.

lead sheath An enclosure of lead round a steel core, combining strength and durability, a technique used on iron by the ancient Greeks and continued in the *lead-clothed glazing bar*.

lead slate A *flashing* to waterproof the hole in a roof where a pipe comes through it. It is shaped like a top hat with its crown cut off.

lead time (*pronounced 'leed'*) Time taken before work can begin, e.g. for delivery of materials, for ordering and making special equipment, for putting up *temporary works*, for the *curing* of concrete, etc. On a *programme* it goes in front of an *event*.

lead wool Lead made into thin strands and used to make a *lead-caulked joint* simply by driving it in – it does not have to be melted.

leaf (1) A *door leaf*.
(2) (USA) **withe, tier** One half of a *cavity wall*, generally a *half brick* thick. Inner leaves are usually in blockwork or timber framing and outer (external) leaves in facing brickwork. The two leaves are joined by *wall ties*.
(3) *Gold leaf* or metal *foil*.

leaf frame, l. structure, framing The structural parts of a *door leaf*, such as the *stiles* and top and bottom *rails*, often acting together with the *facings*.

leafing The *floating* of paints containing small flakes, which protect the paint film by lying flat in it. Examples are *aluminium* paint and paints containing mica. They are often used as primers or *sealers*, e.g. over bitumen, tar, etc.

lean concrete Usually *dry lean concrete*.

lean lime *Lime* that contains impurities, such as silica and magnesium compounds.

lean mix A *concrete* with a low *cement content*, or any *mix* with little *binder*.

lean mortar A *lean mix* of mortar. It is harsh and difficult to spread, unlike a *fat mix*.

lean-to roof, half-span r. A *pitched roof* sloping one way only, with its top edge meeting a wall higher than the roof, forming an *abutment*.

lear board A *layer board*.

Leca (Light Expanded Clay Aggregate) A *lightweight aggregate* used for making insulating screeds and light but strong building *blocks*.

ledge (1) The narrow shelf on top of the wall inside and below a *window*.
(2) One of the two or three horizontal timbers on the back of a *matchboarded door*.

ledged and braced door, batten d. A *matchboarded door*.

ledgement (USA) A *string course*.

ledger A horizontal framing member, either a *ribbon board* or a tube in a *scaffold* to carry *putlogs*.

legionnaire's disease A severe pneumonia, caused by the bacteria legionella, spread through the air inside sprayed droplets of water. Legionella bacteria occur in small numbers in soil and water supplies. To thrive, they need warm water (20 to 50°C) that is slightly starved of oxygen, has iron present, and contains the right nutrients (algae, protozoa). These conditions usually occur only in large building complexes, rarely in the home. To spread from a source, water contaminated with bacteria must be sprayed into the air, forming an aerosol inside which the legionella can survive. These droplets can be blown by the wind for distances of 500 m, or carried through air-conditioning *ductwork*. If inhaled by a susceptible person (usually male smokers over fifty who are unwell) the disease takes ten days to incubate and can be fatal. The main danger comes from incorrectly installed or operated *cooling towers*. If an outbreak occurs the cooling tower has to be stopped and flushed with caustic detergent and chlorine, by people wearing protective clothing and breathing equipment. Once the air-conditioning fans are stopped the danger is much reduced, but full precautions include evacuating the building and disinfecting all ductwork. Other sources of infection are evaporative condensers, whirlpool spa baths and hot-water taps or showers with a *secondary circulation* system. The disease was first recorded in Philadelphia in 1976 after a conference of the American Legion (ex-servicemen). See **sick building**.

legislation The control of building in the U K is provided by the *Building Act*, under which the *Building Regulations* are made, but other legislation, such as the Health and Safety at Work Act, the Factories Act, and the Electricity Supply Act (which gives the *IEE* regulations 'deemed to comply' status), affect work on site. The latest Act on the relevant subject should be referred to. Acts of the European parliament touch on subjects as wide apart as *lifts* and free market competition.

Leighton Buzzard sand The *standard sand* for cement tests in Britain.

lengthening joint, heading j. A *joint* made between the ends of two pieces of timber, such as a *butt joint*, a *finger joint*, or a *half-lap joint*.

let in *Housed* or *flush fixed*.

letter plate An item of *door furniture* to receive letters, consisting of a slotted plate fixed outside the door, as trim to its slot. The best types have a *door tidy* inside.

level (1) A distance above the ground, or a *datum*, e.g. of a *floor*, the top of a steel beam, or a water surface. Levels may also describe intensity of sound or lighting.
(2) **spirit l., mechanic's l.** A tool with bubbles or *level tubes* (*C*), usually in an aluminium *straightedge*, from 230 to 900 mm or more long, used for setting out vertical or horizontal lines.
(3) A group of spaces in a building at the same height above ground, such as the *storeys*, but also including *sub-levels*, roof level, *interstitial levels*, *mezzanines*, etc.
(4) To spread material to a smooth, flat surface. See also *C*.

levelled finish A *textured finish* that has been flattened with a trowel or *float* while still wet in order to squash down spiky high spots.

levelling (1) Working a material to flatten its surface. Paintwork is levelled by its *flow* or by *laying-off*.
(2) Setting a component to a given height.
(3) Placing objects exactly horizontal.

levelling compound, dressing c., synthetic screed Fine-grained *mortar* or similar substances used for smoothing a rough surface to be covered by a *thin floor finish*. It can be safely placed on a *sub-floor* in a thin layer of 5 mm or even less (which is impossible for cement *screeds*, since screeds break up, even at 12 mm thickness). It can also be used to make repairs or to fill holes.

levelling rule

It is spread with a trowel, but the trowel marks disappear after a minute or two. Some are relatively cheap (three times as expensive as cement mortar) but non-wearing; others are expensive *resin mortar* with epoxy, polyester, or polyurethane binders, plus curing agents and fillers. Both types can be used for smoothing a floor to receive *thermoplastic tiles*, sheet rubber, or *polyvinyl chloride* sheets. The resin mortars are hard-wearing and chemical-resistant; they can be laid on concrete, timber, or metal and can be used for *feathering*. See **self-levelling screed**.

levelling rule A *straightedge* about 3 m long, used with a spirit *level* for bringing *dots and screeds* to a uniform surface.

lever boards Adjustable wooden *louvres*.

lever handle A *door handle* with a short usually horizontal lever parallel to the door face, easier to use than a *door knob*. It may have a turn-back end to prevent clothes being snagged.

lever lock A good-quality traditional *lock* in which the key must move several levers to release the bolt when it is shot or withdrawn.

lever mixer, l.-operated thermostatic mixing valve A convenient *mixer* for hot and cold water operated by a single lever handle, which is lifted up to give more water or turned left and right to make it hotter or colder.

liability Legal *responsibility for construction* (*C*), which in a building project may fall upon the contractor, architect, local authority, or building owner, according to *legislation* or the *conditions of contract*. Often *insurance* can be taken out against the cost of liability.

lid A *cover*, or a concrete slab on top of a *lift shaft*.

lift (1) (USA **elevator**) A machine with a *lift car* driven vertically, used for carrying passengers and goods from one level to another. In buildings, lifts usually run in a *lift shaft*, but 'scenic lifts' go up and down an outside wall, in an *atrium*, etc. Lifts, which have come into use gradually since the 1850s, have helped make today's tall buildings possible. Early types, raised by high-pressure water from a 'hydraulic main', have given way to those using electric motors, which wind wire ropes or pump an oil-filled hydraulic ram. See entries below, **builder's lift**, and BS 5655. See also diagram.

(2) A single stage of work divided in its height, usually to allow *compaction* by such means as the vibration of concrete, the rolling of *fill*, or the ramming of *backfill*.

lift car The moving cabin of a *lift*, usually of standard size and finish. Office lifts have automatic centre-opening doors to save time in loading and unloading passengers. They run at high speeds.

lift controls (USA **elevonics**) Automatic devices for operating one or more *lifts* as directed by passengers pressing buttons on a landing (landing call) or in the lift car (car call). Calls are stored and answered according to the number of lifts: one (simplex operation), two (*duplex*), three (triplex), or more (multiplex), and the type of lift control system: *push-button, down-collective* or *full-collective*, and *group supervisory*. In a *building management system* lifts can react to control signals by other services. Manual control is mainly used for *fire-fighting*.

lift drive The machine that moves a *lift*. Large, fast lifts are held on steel-

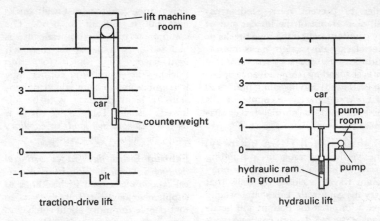

lift machine
room

4
3
2 — car
1
0
−1 — pit

counterweight

traction-drive lift

4
3
2 — car
1
0

pump
room

hydraulic ram
in ground

pump

hydraulic lift

The two main types of lift.

wire ropes over a traction *sheave* (*C*)
and have a counterweight. Smaller lifts
have a drum and no counterweight.
Hydraulic drives have a pump and
ram.

lifting beam (1) A *strongback* hung
from a crane hook. It has several hooks
to carry the weights from long loads,
such as a beam, column, or pile. Lifting
beams are usually specially made for a
particular job and are justified by savings in materials and the assurance that
no damage will be caused by lifting. A
lifting frame is similar.

(2) A steel beam fixed to the building
structure for moving heavy equipment,
as in a *lift machine room*.

lifting equipment Plant such as
cranes and *hoists* and their *lifting tackle*
(*C*), or building installations such as
lifts, all of which must be built to strict
safety rules, have overload protection,
and be load-tested before use.

lifting eye A ring to which a crane
hook or other lifting device can be attached.

lifting frame A *lifting beam* made for
long, wide loads such as precast slabs.
Lifting frames often have wire ropes
and pulleys to equalize lifting forces.

**lift machine room, l. motor r.
(LMR)** The *plantroom* for the motors,
winding gear, and controls for raising
and lowering a *lift*, usually built on top
of the *lift shaft* and reached from the
roof or by a *loft ladder* (except in the
case of *hydraulic lifts*). The LMR
should have a *lifting beam*.

lift-off hinge, loose-butt h. A *hinge*
that can be taken apart by lifting one
flap from the other. A door on lift-off
hinges can be unhung simply by lifting,
but only when it is fully open. See
diagram, p. 229.

lift pit, runby p. A deepening of the
bottom of a *lift shaft* to enable the car
or counterweight to over-run the
normal *travel*. Deep lift pits are needed
for fast lifts or those with heavy *duty
loads*.

lift safety gear Hinged metal teeth
fixed to a *lift car* or frame that close
and jam tight on to the *guide rails* in

order to prevent over-speed, free-fall, or over-run of the lift car and of any counterweight if the rope breaks or goes slack. Other safety measures include keeping doors closed until the car is at a landing (some advance opening is allowed) and installing buffers in the pit to stop the car or counterweight going beyond its normal travel distance.

lift shaft, l. well (USA **hoistway**) The vertical opening in a building through which a lift car and counterweight travel, enclosed by walls that carry the *guide rails*. Lift shafts usually have a *lift pit* at the bottom and headroom at the top. In-situ concrete work for lift shafts is often a *critical activity*, but see next entry.

lift shaft module A prefabricated length of *lift shaft*, usually one storey high, that fits a hole left out of each floor slab. The shaft can thus be built quickly, advancing *programme* dates and allowing the builder to use the lift.

ligature A *stirrup*.

light A *window*. It can be fixed or opening, and may be separated from other lights by *mullions* or *transoms*. A light can have one or several *panes* held on *glazing bars*.

light alloys *Aluminium* or magnesium alloys.

light fitting, l. fixture A *luminaire*.

light-gauge copper tube Copper tube with an outside diameter of from 6 to 159 mm and corresponding wall thicknesses of from 0.6 to 7.6 mm, described by BS 2871 in a dozen tables of tube dimensions. Tubes may be supplied in one of several different conditions, the main three being 'annealed' (soft), 'half hard', and 'hard-drawn'. Tube bought in a coil is invariably annealed, which makes it easy to straighten or to bend. Hard-drawn tube has the least wall thickness but is difficult to bend. The commonest tube for domestic use is half hard, 15 mm outside diameter, light-gauge, thin walled (0.7 mm thick). When specifying copper tube it is important to state which dimensions table in BS 2871 is to be followed. Usually Table X, Y, or Z is used for *compression joints* or *capillary joints*. See also **copper pipe**.

lighting Either *daylight* or *artificial lighting*. Interior lighting from electrical *luminaires* is used in buildings to supplement daylight from *windows* so that people can move about safely and carry out precision tasks. *Deep-plan buildings* and *emergency exits* also need lighting. Exterior lighting is mainly used for public areas, the floodlighting of façades, and air-navigation beacons.

lighting bollard A low post with a *luminaire* that directs light on to the ground, used for public lighting and in landscaping.

lighting column A tall post used for the public lighting of large open spaces, usually with an *outreach arm*, *discharge lamp*, and downlighting reflector.

lighting fitting A *luminaire*.

lighting installation The electrical supply services for *lighting*, run from a *consumer unit* or *main switchboard* and separated from the *power* supply.

lighting panel A *distribution board* for lighting circuits.

lighting point The place where a *luminaire* is to be fixed. Wiring is run by the electrician and may be left as a *cable tail*.

light-loss factor (LLF) (USA **dirt-depreciation f.**) The loss in light output from a *luminaire* due to dirt on the lamp, luminaire, or room surfaces, used in calculating *illuminance*. It replaces the former maintenance factor.

lightness, value, tone The amount of light reflected by a *colour*, regardless of hue.

lightning arrestor A device for protecting high-voltage power lines against *overvoltage* from lightning, by diverting the surge to an *earth*.

lightning-protection system, l. conductor, l. rod The parts of the electrical installations that protect a building against lightning. They have an *air termination network*, a thick copper *down conductor*, and an *earth electrode*. See BS 6651.

lightning shake The compression failure of wood, seen as a cross-break.

light switch A hand *switch* to control a *lighting* circuit, usually fixed at about 1.20 m above a finished floor, in an *architrave* or on a wall, unless it is a ceiling-mounted *pull-cord* type. It can be a *one-way switch*, a push button for a *timer*, or a *two-way switch*.

lightweight aggregate Low-density aggregates such as natural pumice or *diatomite*, but more often industrial products such as expanded clay, *foamed slag*, or sintered pulverized fuel ash, used for making *lightweight concrete*. See BS 3797. See also **ultra-lightweight aggregate**.

lightweight concrete *Concrete* which is lighter than 1900 kg/m³. There are two main types: (a) *aerated concretes*, weighing around 800 kg/m³ or less, which are highly insulating and not very strong (used for the *encasement* of steel beams or columns in non-traffic areas), and (b) concretes made from *lightweight aggregate*. These can be used for structural purposes (columns, beams, and slabs) and are less good

insulators, though better than dense *concrete*.

lightweight concrete block, l. masonry unit A building *block* made from *aerated concrete*. The block is said to be solid rather than hollow, as it has no large internal holes, although the material is *cellular*. See BS 6073.

lightweight plaster One of several different *plasters*, usually *premixed*, containing *lightweight aggregates*, including exfoliated vermiculite or expanded perlite, with *retarded hemihydrate* gypsum or cement as their *binder*. They have only low shrinkage, and being both thermally *insulating* and *non-combustible* can be used for fire *encasement*. Types include *browning*, *lathing plaster*, *bonding plaster*, and *finish plaster*.

light well, air shaft An unroofed space within a *deep-plan building* that provides some light and ventilation for windows facing it. The walls need finishings with high *reflectance*, such as white paintwork or sand-lime bricks. Since fire can spread up a light well, *fire-resisting glass* may be required in windows. An existing light well is often converted into an *atrium*.

Lignacite A building *block* made of sawdust, sand, and cement. The material is nearly three times as insulating as dense concrete, and easy to nail and saw. The block may be hollowed out to save weight and improve insulation.

lignin A natural resin in wood which (together with cellulose) cements the wood fibres together. It is the binder in *fibreboard*.

lime Chalk or limestone burnt in a kiln, or a by-product in making acetylene. The anhydrous product is *quicklime* (CaO), which is slaked with water to make *lime putty* or *dry hydrate* (Ca(OH)₂), a white powder sold in bags. As a mineral it varies in composition (dolomitic, magnesian, high calcium, lean limes). It can be *hydraulic*. Lime is a slow-setting *binder* used in *coarse stuff* and *composition mortar* for plastering and bricklaying, and can react with *pozzolans*. Lime also results from the setting of cement in concrete or mortar, giving *free lime*.

lime concrete A mixture of gravel, sand, and *lime* used since Roman times until Portland *cement* was made. It does not really set, but hardens very slowly by *carbonation*. Compare **Roman cement**.

lime plaster Neat *lime* or a mixture of lime and sand, formerly used for plastering.

lime putty Wet *hydrated lime* which has been soaked overnight or longer to give it *plasticity*.

lime:sand mortar, coarse stuff A *ready-mixed mortar*.

limit of works The point where work by one contractor, or work in one *trade*, begins or ends, such as a *service* connection, an electrical *cable tail*, or a *substrate* for paint.

limit switch A switch that stops a drive motor at the end of travel.

limpet washer A conical *washer* (*C*) fixed under the nut of a *hook bolt* to hold down a corrugated sheet. It is shaped to fit the top of the corrugation. A *diamond washer* has the same function.

line (1) **bricklayer's l., builder's l., stringline** A length of fine cord stretched taut between two points for *setting out* building work. It can be held at the ends by nails in a *profile*, or by bricklayer's *line pins* or *corner blocks*, or used as a *chalk line*. Lines are quick to use, simple, and reliable compared with a *theodolite* (*C*), but are easily

blown sideways by the wind and sag over long distances. This is easily seen and corrected by a *tingle*. A line is never knotted, but attached by winding.

(2) A pipe for the distribution (or collection) of water (pipeline, drainline); or an electrical services cable (power, telecom line).

line and level The *setting-out* of buildings and their parts so that they are in the right position and at the right height, often as two separate operations, with different equipment.

lineal measure, run Length.

linear cover The width of a roof *tile* along a *course*, less its *sidelap*.

linear diffuser, slot d., strip d. One or more long narrow slots through which *supply air* emerges, to mix with room air. It can be mounted on the wall or on the ceiling.

lined eaves *Eaves* with a *soffit board*.

line detector See **heat detector**.

line level A small spirit *level* which, suspended at the middle of a taut string *line*, can be used to level to within about 1 mm per metre.

linen chute, laundry c. A vertical duct to a basement *laundry* from a bathroom or service room into which soiled sheets, towels, etc., are dropped.

line pins Flat steel pegs about 80 mm long inserted into the mortar joints at the ends of a wall and used for holding the bricklayer's *line*.

line-tapped connection A branch in electrical wiring made without cutting a main cable run, usually in a *tap-off unit*. The alternative is *looping-in*.

lining (1) An inside covering or coating, e.g. *gypsum plasterboard* on a wall (dry lining), mortar inside a water pipe, or plywood facing on concrete form-work. Compare **cladding**.

(2) A timber *surround* to a door or window, put in the wall opening to cover the *reveal*. It is thinner than a loadbearing *door frame*.

lining paper Paper pasted on to a wall as a base for wallpaper or paint.

lining plate In *supported sheetmetal roofing* a *tingle* at *eaves* or *verge*.

link (1) One of the connecting parts of a chain, or an electrical *coupler*, or a *fusible link*.

(2) A narrow reinforcement *stirrup* between one top and one bottom bar.

(3) **point of articulation** A *key event* on an *arrow diagram*, where the lines representing several *activities* come together, showing that they all have to be completed before the work *programme* can progress.

linoleum, lino A detachable, loose *floor covering* made from *linseed oil* and hessian, with fillers of flax, chalk, or softwood, which is *calendered* to a smooth surface. Lino comes in many bright colours and is soft, resilient, and easy to maintain, but can suffer from *indenting*. It slowly gives off linseed oil vapour, which kills bacteria without being toxic to humans, and it tends to oxidize and harden with age. Concrete is an ideal sub-floor for lino and *underfloor heating* can be used up to about 27°C.

linseed oil A *drying oil* used for making *oil paints*, *linoleum*, and *cork carpet*, as well as for preserving wooden tool handles and reducing the water absorption of *quarries*. It can be either raw linseed oil, *boiled oil*, *blown oil*, or *stand oil*.

lintel, lintol A small beam over a door or window *head*, usually carrying a wall load only. Lintels for a *cavity wall* are usually made of *powder-coated*

pressed steel, or more expensive *stainless steel*, or as a precast concrete *boot* lintel. Lintels should have a *drip* so that alkaline water from masonry falls clear of glass or anodizing. The repair or replacement of lintels is discussed in the BRE Good Building Guide 1. See **combined lintel**. See also diagram, p. 272.

lintel damp-course A *combined lintel*.

lipping, banding, edging strip A strip of solid wood along the edge of a door *leaf*, etc., to protect a *cellular core* or soft-timber core.

lip-seal joint A *push-fit* pipe joint with a sliding *joint ring* round the inside of the *socket* that has an inward-projecting thin blade of rubbery plastics. When a plain *spigot* pipe end is inserted the lip seal makes a watertight joint, acting like a car windscreen wiper blade. See diagram, p. 252.

liquidated damages A sum of money agreed in advance as a fair and resonable amount of *damages for delay*.

listed building See **ancient monument**.

listing A slope given to a *slate* or *tile* to throw off rainwater.

litharge, lead monoxide (PbO) A poisonous, yellow to brown *pigment*. See **lead paint**.

litigation Taking a matter under *dispute* before a court of law. The advantages are that evidence is given under oath and the decision is binding, but law courts are slower and more costly than *arbitration*.

little joiners Small pieces of wood which are used to hide and fill holes in wood, for example *pellets* and *inlays*.

live (1) The red wire in electrical wiring for *low-voltage* power or lighting. Fuses and switches are on the live

A lintel.

wire, which makes a circuit with the *neutral*.

(2) **alive** Said of a cable connected to a source of power; of a gas or water pipe under pressure; of a drain carrying water, etc.

live edge, wet e. An edge of a fresh coat of paint that has not dried enough to prevent picking-up as the work is continued. A lap made against a live edge does not produce *flashing*.

live knot, intergrown k. A *knot* that has fibres intergrown with the wood. It is a *gross feature* allowable in structural timber up to certain sizes of knot.

live main A gas or water pipe under pressure (the opposite is a dead main).

live tapping The connection of a water pipe to a *live main* using a *saddle*.

load (1) The weight carried by a structure. See *C*.

(2) The demand for current from a power circuit, either the *installed load*

or the load reduced by *diversity*.

(3) The weight lifted by a *crane*, a *lift*, or a *hoist*. See **crane safe working load** (*C*), **duty load**.

(4) *Heat load* or *fire load*.

loadbearing wall A wall that supports the *structure* above it, including the walls, floors, roof, and their loads, such as wind, snow, earthquakes, etc.

loading (1) The addition of a *filler* or *pigment* to a paint, adhesive, etc.

(2) Filling a pipe with hard-packed sand to prevent it distorting during bending if an appropriate *bending spring* is not available.

(3) Weighing down with *ballast* to prevent uplift.

(4) See **load**.

loading coat, l. slab A concrete *screed* or slab laid over asphalt *tanking* or on top of *roofing* to resist water pressure from below, wind uplift, etc.

load schedule A list of electrical circuits for power distribution that shows all the *installed loads*.

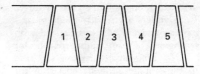

pipe cut at 7¹/₂° angle to
make six joints

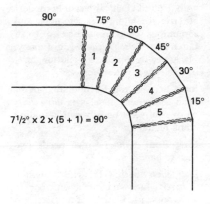

$7^1/_2° \times 2 \times (5 + 1) = 90°$

A five-part lobster-back bend.

load shedding, automatic l. limitation Switching off power to some normal circuits so that an emergency generator can supply the *priority circuits*. As power is restored the circuits are *reloaded*.

load smoothing *Resource smoothing*.

lobby, vestibule A short passage with a door at each end, one of which is normally closed to act as an *air lock* or for *security*. It usually has *door closers* to shut one door before a person can reach the other. Lobbies are mainly used at entries and for *fire protection* or *sound insulation*.

lobster-back bend, cut and shut A *bend* that is fabricated by cutting a pipe or duct into angled segments to form several close-spaced *mitred joints*. In light-gauge steel sheet they can be made by folding. See diagram.

local area network (LAN) A network of cables to carry *data* between computers or other equipment such as printers. Connections are made at a LAN socket or data socket.

local authority, l. council In the UK, a city, town, borough, burgh, district, island, or county council or similar body which among other things is responsible for *building control* by enforcing the *Building Regulations*. It is not always the planning authority but is usually in the same building. Before work is started on building, altering, or demolishing a building, an application must be made for a *building permit*. The foundation ground must be inspected before the *footings* are concreted. After a completed building is inspected, an *inspection certificate* is issued. In the UK, councils own houses and flats that are rented to those on low incomes. They can usually be bought by tenants.

local control Usually a *thermostat* for the automatic control of *air conditioning* according to the ambient temperature in one particular room.

location block A resilient non-absorbent block of material, such as plasticized PVC, that prevents the movement of *glazing* in its frame. It should be at least 25 mm long and positioned near each quarter point of the frame side.

location plan, key p. A *drawing* which shows the placing and dimensions of a building site and the proposed building.

location schedule A list of rooms with their *finishings* and equipment.

lock (1) door l. A fastening for a door operated by a key which shoots a bolt.

lock block

The lock can be *morticed* or *rim-*mounted, and have a *cylinder* or *lever* action. Compare **latch**.

(2) To *throw* a *dead bolt* or engage a *snib*.

(3) To secure firmly, e.g. by dressing a *welt* or padlocking a *lockout*.

lock block A wooden block to contain the mortice for a lock in a *cellular core door*.

locking stile A *shutting stile*.

lock joint A *seam* in *supported sheet-metal roofing*.

lockout A safety device that prevents mechanical or electrical equipment from operating dangerously. It can be automatic or a padlock.

lock rail An intermediate rail of a *framed door* to carry a lock.

locksaw A *compass saw*.

lock set Usually a *bore lock*.

lockshield valve A *discharge valve* with no handle on its spindle, so that its setting cannot be altered by the normal user. It is used on *radiators* as a *balancing valve* and on drinking fountains and automatic showers.

lock stile A *shutting stile*.

lock-up stage The *stage* of construction at which a building can be locked up to prevent theft after working hours. Fittings can then be installed.

loft (1) A store room in the *roof space*, usually with access through a *trap*. A loft can be converted into an *attic*.

(2) (USA) An entire upper floor of a commercial building, let bare to a tenant.

loft insulation *Ceiling insulation*.

loft ladder, disappearing stair (UK) A folding or telescopic ladder which is normally concealed above the door of a *trap* in a ceiling. The trapdoor is hinged to open downwards when pulled by a cord or a *long*

arm which hooks into an eye or catch. Loft ladders are mostly made of *aluminium* and are counterbalanced with springs. They are indispensable for transforming an inaccessible loft into usable storage or living space, or for gaining access to a *lift machine room*.

London stock brick A yellowish *clay brick* hand-made of London clay mixed with ground chalk. It has only moderate strength, high water *absorption*, and sometimes good resistance to frost. Often it is not *self-cleansing*, but many, blackened by former air pollution, have been cleaned by *wet blasting*.

long arm A rod for operating overhead equipment, such as a *rolling shutter* or roof vent, usually with a folding crank handle.

long float A *float* which needs two men to handle it.

longhorn beetle (Hylotrupes bajulus) A *borer* which attacks the sapwood of softwoods, usually in roof timbers, and against which treatment with *preservative* is required by the *Building Regulations* in parts of England. It leaves an oval bore hole 6–10 mm wide.

long oil An *oil paint* or varnish with a high ratio of oil to resin which tends to dry to a soft and flexible *film*. It is long in *oil length*.

longscrew The parallel threads of a pipe *connector*.

longstrip roofing A method of laying *supported sheetmetal roofing* in full-length strips, thus eliminating welted cross joints. The strips run from eaves to ridge and their edges are joined with double-lock *standing seams*, made in-situ with a hand-guided power *seamer*. The holding-down *clips* are folded into the seam. Aluminium, copper, or stainless-steel sheet is used, usually of a heavier gauge than for hand-made seams, at least 0.9 mm thick.

long sweep bend A *slow bend.*

loop, ring A *circuit* or *main* that can provide two paths for power, water, etc.

looped-pile carpet Woven *carpet* with long parallel rows of yarn that are looped through the *backing* but not cut.

looping-in, loop-in method A way of eliminating T-joints from wiring by running only uncut lengths of cable between the *terminals* in each *fitting.* For *luminaires* a separate pair of wires is run to the light switch. More wire is needed than usual as the conductor must double back on itself, but there are fewer *tap-off units.*

loop vent (USA) A *branch vent* that joins the first and last *trap* in a branch, then runs to the *vent pipe.*

loose-butt hinge A *lift-off hinge.*

loose-fill insulation Thermal insulating materials such as *granulated cork,* loose *expanded clay,* or other *lightweight aggregate,* or *mineral wool,* used as *ceiling insulation* or *cavity fill.*

loose key (1) A *loose tongue* in joinery.
(2) A T-shaped bar with a square recessed end, inserted through a *surface box* to operate a buried stopvalve. See **turn cock.**

loose knot A *knot* which is not held tight and may drop out. In grading timber according to *gross features,* it is a *defect.*

loose material Scale, rust, or flaking paint on a *substrate* for painting. It is removed during surface *preparation* by wire brushing, blast cleaning, scraping, etc.

loose-pin butt, pin hinge A butt *hinge* with a withdrawable hinge pin which enables a door to be unhung merely by removing the pins from the hinges, without the labour of unscrewing the butts.

loose ring A *joint ring* that is not attached to the *socket,* e.g. a rolling *O-ring* for a drain pipe or the *gland* of a *non-manipulative* pipe fitting.

loose side, slack s. The side of timber *veneer* which was in contact with the cutting blade. It was bent outwards and slightly broken, being the inner side of the veneer as it was sliced or peeled from the log. The outside is the *tight side.*

loose tongue, key, slip t., slip feather, spline A strip of wood or plastics that fits into two grooves, one on each side of a joint, to keep the two sides aligned, such as between sheets of flooring *chipboard.*

Loss Prevention Council (LPC) A body funded by British insurers and Lloyds involved in all aspects of loss prevention, including fire safety. It has a technical centre at Borehamwood for fire testing and certification, which publishes rules for *fire alarms, sprinkler systems,* and *fire extinguishers,* but is also concerned with *intruder alarms,* and provides information and advice through the *Fire Protection Association.*

lost-head nail A round wire *nail* with a very small head which can be concealed by driving it slightly below the surface with a *punch.*

lot (USA) A building *site* in a town.

loudness The volume of sound or noise heard by the human ear, as judged by a normal observer compared with the quietest sounds that can be heard. The two measurement systems used in building acoustics are *dBA* and *Noise Ratings.*

louvre (USA **louver**) A ventilator originally with horizontal wooden *slats* to keep out rain, now often of glass strips horizontally pivoted in a window, or vertical metal blades in a *cell ceiling*. In a medieval hall, it was a combined chimney and window.

low-angle light Light that falls on a flat surface from a shallow angle and shows up any imperfections.

low bid (USA) The *lowest tender*.

low block A *podium*.

low-build coating Liquid material that can be applied only as a thin *film*.

low-emissivity coatings (low-E c.) Coatings that reduce the rate of heat loss from a surface. On glass they can be an invisible thin layer of silver which reduces heat losses outwards, while doubling the reflection of heat from the sun, without reducing daylight. They are used in double glazing *sealed units* and as *solar-control glazing*. See **solar collector**.

lowering wedges *Wedges* that are *eased* before formwork *stripping*.

lowest tender (USA **low bid**) The lowest price offered by one of several contractors to do a particular job, usually the *tender* that is accepted.

low-level suite A *water-closet* pan with a matching *close-coupled cistern*.

low-pressure hot water (LPHW) system A central-heating system with water as a *heat-transfer fluid* and a *flow* temperature usually below 80°C, used for both large commercial installations and domestic *small bore* or *microbore*.

low-rise building A building with from one to eight storeys, not a *high-rise building*.

low-velocity system Quiet and economical *air conditioning* with large *ducts* from the central plantroom and often back to it. The ducts are usually rectangular and branch off to increasingly smaller ducts at the end of each zone. The ducts are larger than for a *high-velocity system*.

low voltage (1) Alternating current at less than 250 volts, but higher than *extra-low voltage*.

(2) The *electrical services* for power and lighting at 240 volts, supplied direct from the *mains*.

luffing, derricking Altering the radius of a crane *jib* or a *derrick* (C), by lowering or raising it. Compare **trolleying**.

lug (1) **ear** A projecting metal *fastener* with a hole in its end for a nail, screw, or other *fixing*, or a *fishtail* for *building-in*.

(2) A *spacer* on the edge or back of a floor or wall tile (spacer-lug tiles), used as a guide in tile laying to give uniform, narrow joints.

(3) A piece of metal attached to a wire which makes an electrical connection, e.g. to a *terminal*.

lug end The end of a *lug sill*.

lugged pipe bracket A *fixing* for attaching a pipe to a wall, with a *lug* and usually a *two-piece cleat*.

lug sill A *sill* with its ends built into the window *jambs*, being longer than the *opening*. Usually the middle section has a *weathering* between the *stools* of the *lug ends*. Compare **slip sill**.

lumber (1) (US) term for converted logs which have been sawn and sometimes *re-sawn* into *dimension lumber*. Boards of lumber may be up to 24 ft (7.3 m) long, while *timber* may be 60 ft (18 m) long. Lumber is measured by the *board foot*.

(2) Imported square-edged sawn hardwood of random width (BS 6100).

lumber core (USA) *Blockboard* or *laminboard*.

lumen (lm) The unit of *luminous flux* used to measure the amount of light given off by a *luminaire* or falling on a surface (*candelas*/steradian).

lumen method A method of finding out what *luminaire* will give a particular level of lighting in a room. The required lumen output (RLO) is:

$$\frac{\text{illuminance (E)} \times \text{room area (A)}}{\text{utilization factor (UF)} \times \text{light-loss-factor (LLF)}}$$

luminaire, lighting fitting A lighting fitting which controls the pattern of light emitted (using reflectors, diffusers, etc.) and does not overheat. Ceiling luminaires may be *air-handling luminaires*. The maintenance schedule of an office building may include regular cleaning and *relamping* of luminaires once they reach a given *light-loss factor*.

luminance (L) (French, luminance) The brightness of a surface in a given direction, usually measured in *candelas* per square metre (cd/m²). Luminance figures are used in calculations for *lighting* and *colour*.

luminous ceiling A translucent *suspended ceiling*, or a *cell ceiling*, containing lamps to light the room below.

luminous flux (F) The flow of light energy from a source, or reflected from a surface, standardized for the human eye and measured in *lumens*. It is used to calculate *illuminance*.

luminous intensity (I) The amount of energy in a cone of light from a source, in *candelas*.

lump hammer A *club hammer*.

lump-sum contract A *fixed-price contract* in which the *client* agrees a *contract sum* before work is commenced, plus the cost of *variations* and *fluctuations*. The work to be done is shown on the contract *drawings*.

lune, gore A figure enclosed by two arcs of circles. Sheet materials can be cut to a lune, then halved and fitted together to cover a dome. Such a series of half lunes follow a dome's *double curvature* fairly closely.

luting Sealing with stiff material, such as clay, cement, or asphalt, e.g. into a flange round a *rainwater outlet*.

lux The unit of *illuminance*, measured in *lumens* per square metre. Normal values are: bright sunlight, 100 000 lux; worktop near window, 3000 lux; precision task lighting, 1000 lux; office, bench top, reading, 300 lux; corridor, storage, 100 lux; full moon, 0.25 lux.

luxmeter A photoelectric cell that reacts to light by producing an electric current, which is then read from a meter to show *illuminance* in *lux*.

lyctus beetle *Powderpost beetle*.

M

machine base A large block of concrete, as a foundation for mechanical equipment. It may be separated from the floor and usually projects above it as a *plinth*.

machine float, m. trowel A *power float*.

machine grading The *stress grading* of timber by means of a *stress-grading machine*.

made ground Ground which has been raised by *fill*.

magnesian lime *Lime* with less magnesium oxide (5% to about 40%) than *dolomitic lime*.

magnetic catch A cupboard *catch* with a small magnet that holds the door closed until it is pulled open. This is convenient, but offers no security.

mahogany Tropical *hardwoods*. They are mostly either *African* or *Brazilian*.

main The supply service's pipe or cable, run in the street.

main bar The thickest *reinforcement bar* in a beam, column, etc.

main contractor, general c., prime c. The *contractor* who has the head *contract* with the *client* and is directly responsible for all work on site, including the work of *sub-contractors*, as well as their *coordination* and providing *attendance*, usually acting through a *site manager*.

main distribution frame, building d.f. An open frame with *terminals* for connecting *telephone* lines and exchange equipment, housed in a *frame room*.

main services duct A tunnel for pipes, usually running underground between two buildings and large enough to walk through.

mains pressure system, m. connected s. An *unvented system*.

main stopvalve The *valve* which controls all water supply to a building.

mains voltage *Low-voltage* power, in the UK at 240 volts.

main switchboard, main low-voltage switchpanel In a large installation the *switchboard* for low-voltage *outgoings* to each of the *distribution boards*.

maintenance The work of keeping things in good condition. Materials used in the *fabric*, *finishings*, and *services* of buildings, and even their *external works*, are increasingly either 'maintenance-free' or made to reduce maintenance cost. Large buildings should have a maintenance schedule for 'planned preventive maintenance' of major equipment, such as air conditioning, lifts, lighting, and fire-protection equipment. In general, good *workmanship* helps to reduce the maintenance cost. *Contractor's plant* (C), like all machinery, also needs regular servicing.

maintenance factor (1) A correction for dirt on glass in the *daylight factor method*, from 0.9 to 0.5, depending on the air pollution and the slope of the glazing.

(2) A former term for the *light-loss factor* of a luminaire.

maintenance painting *Repainting*.

maintenance period (1) The *defects liability period*.

(2) For mechanical plant and other building *services*, the time after *handover* for which the equipment installer agrees to keep it in proper working condition. In practice it may be difficult to separate from *commissioning*.

maisonette (USA **duplex apartment**) A self-contained dwelling on two levels having its own internal stairs. Large blocks of flats are sometimes built in this way to obtain quiet for the occupants, avoiding the costly methods of *discontinuous construction*.

make good To repair as new, usually after cutting away for access.

make up To add extra material so as to bring something to the required level or quality.

makore (Mimusops heckelii) An *African mahogany*.

mall, maul A heavy *mallet* or *beetle*.

mallet A tool like a hammer, with a wooden, plastics, or rubber head of any shape, used for striking thin, brittle, or weak things. See diagram, p. 281.

management contract (USA **construction m.**) A contract in which the *main contractor*'s chief function is not as a builder but as a manager of work packages done by *sub-contractors* and as an advisor to the architect, structural engineer, and other consultants. A management contractor should be appointed early and is usually concerned with programming and *coordination*, having no say in the choice of sub-contractors nor the right to do the work directly. The

system is well known in the USA and in France (as 'pilotage') and is aimed at taking care of the client's interests for large fast projects with a high proportion of *mechanical and electrical* equipment that is difficult to price before work starts. It allows *fast-track procedures* and has been claimed to promote intelligent cooperation. The client can expect greater security than with a *design-build contract*, although the system has a reputation for creating confrontation between members of the building team.

managing contractor The *main contractor* of a *management contract*.

man door A *wicket door*.

manganese drier Manganese dioxide (MnO_2), and other organic and inorganic salts of manganese, used as *driers* in linseed oil paints.

manhole A hole big enough for a man to pass. See also *C*.

manifestation, white blob A removable mark put on clean new *glazing*. It prevents damage or injury by showing that the glazing is there.

manipulative joint A pipework joint made by opening out the end of the tube with special tools to form a *socket* or a *flare*, usually after softening by heat. Socket types are used for *capillary joints*, thus saving *fittings*, and flare types are used for *compression joints*, but both make a strong seal.

man-made fibre *Synthetic fibre*.

mansard roof, curb r. (USA **gambrel r.**) A *pitched roof* which has a break in each slope, with a shallow top part and a steeper lower part, usually with *dormers* for the attic space. See diagram, p. 376.

manufactured gas Gas piped to customers after processing, originally made from coal (town gas) or oil. It has been almost entirely replaced by *natural gas*.

marble A metamorphic rock derived from the natural alteration of limestone or dolomite under heat and pressure, found in many parts of Europe, e.g. Carrara in Italy, and quarried as solid blocks or slabs. Broken marble chippings from quarrying are used in *mosaic* and *terrazzo*, or *reconstituted marble*.

marble cladding An exterior facing of *non-loadbearing* marble slabs fixed to walls or columns with *dowels* or *cramps*, helped by *dabs* of weak mortar or *lime putty*. The slabs should be 20 mm thick for small areas and up to 40 mm thick for larger units. Their back should be coated with *shellac* or paint to prevent staining by moisture absorbed from the building *fabric*.

marbling Copying the appearance of marble by paint or other artificial means.

margin (1) A narrow strip of finishing, such as a border round a general area, a *drafted margin* to stonework, or the edge of a *stile* or *rail*.
(2) The visible part of a lapped roof tile; its *gauge*.
(3) The projection of a *close string* above the line of stair *nosings*.
(4) Money added to an *estimate* to allow for risk and profit in the *tender* price. The margin is decided during *adjudication*.

margin template A *pitch board* with an edge strip equal to the width of the *margin* from the top of the *string* to the *nosing line*.

margin trowel A narrow rectangular *trowel* used for working in a narrow width.

marker tape *Tracer*.

marking drawing A *drawing* showing

where prefabricated building elements are to go, e.g. for *structural steelwork*, it gives the letter and number of each beam, column, joist, etc., and its location on a *grid plan*.

marking gauge, joiner's g. A beechwood bar with a steel point projecting at right angles from it near one end and a hardwood block sliding along it which can be locked at any point along the bar. It is used for marking lines parallel to the edge of the wood which the block travels along. See diagram, p. 67.

marking-out *Setting-out*, or showing where to make a cut, fold, etc.

marouflage To glue a sheet of canvas over cracks or joints, as on a wall which is to be covered by a mural painting. The canvas is put on by rollers and glued with paste (often *gold size*), forming a strong, matt *substrate* for the painter.

mash hammer A *club hammer* (French 'masse', or sledge hammer).

mashrabiyya An openwork screen of interlocking wood elements used in Arab countries to allow cool air to enter without people seeing in.

masking Tape or paper covering an area beside which paintwork is to be applied, to make a neat edge. Usually put on after surface *preparation*, it is carefully peeled off when the paint is *dry to handle* but not yet *hard dry*.

mason A *tradesman* who cuts stone or sets stonework. In Scotland or the USA a *bricklayer* is usually also a mason. He is generally in England either a *fixer* or a *walling mason*. See also **freemason**.

masonry A *loadbearing* structure built of individual *blocks* of precast concrete, burnt clay, or stone, or any type of *brick*, usually laid in mortar. These

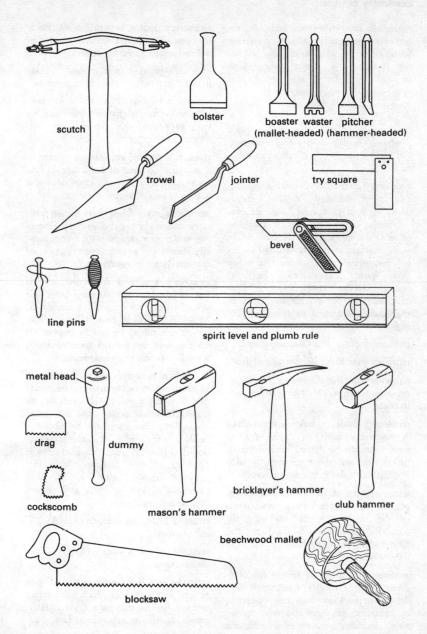

scutch

bolster

boaster waster pitcher
(mallet-headed) (hammer-headed)

trowel   jointer   try square

bevel

line pins

spirit level and plumb rule

metal head

drag   dummy

cockscomb

mason's hammer

bricklayer's hammer

club hammer

beechwood mallet

blocksaw

Mason's and bricklayer's tools.

materials are strong in compression but weak in tension. Masonry walling (of stone or brick) is no longer used for tall buildings, its great thickness at ground level being too costly and space-consuming. Today in the UK 'masonry' usually has the more limited meaning of work with cut stones. For the painting trade, masonry includes *rendering*. Both masonry and rendering tend to be damp and alkaline, so difficult to paint.

masonry cement A *cement* consisting mainly of ordinary *Portland cement*, often with a *plasticizer*, e.g. an *air-entraining agent*, and clay, *whiting*, or other mineral fillers. Mortars made with masonry cement have good working properties and behave like other mixes containing ordinary Portland cement, but have lower strength. Their main uses are for bricklaying, plastering undercoats, and external rendering.

masonry cleaning *Cleaning-down.*

masonry drill A *tungsten-tipped drill.*

masonry fixing *Cramps, dowels*, or *expansion bolts* (*C*) for fixing stonework in place.

masonry nail (USA **concrete n.**) A twisted square nail made of hard steel that can be driven into ordinary bricks, as are *drive screws*, or into pre-drilled holes in harder materials.

masonry paint, organic rendering, plastic p., stone p. Exterior wall paint, up to 2 mm thick and usually emulsion-based, much used on the Continent. Its *high build* is suitable for *textured finishes*.

masonry primer A *primer* for concrete, cement render, brickwork, etc. These surfaces are often damp, porous, and alkaline. If *oil paint* is to be put on new masonry, an alkali-resistant primer should be used.

masonry unit A building *block, brick*, or *stone*. The term is used in official definitions to avoid repetition.

mason's joint A mortar *joint* consisting of a projecting triangle of mortar.

mason's labourer A skilled *labourer* who helps in the mason's yard and on site, supplying the mason *fixer* with mortar and grout.

mason's mitre, m. stop A corner in a wall formed out of solid stone. The actual joint is not a *mitre* but usually a *butt joint* away from the corner.

mason's putty *Lime putty* mixed with stone dust and Portland *cement* for jointing *ashlar* stonework, with a mix usually between $1:1:6$ and $1:4:15$ of cement:lime:stone dust.

mason's scaffold A *scaffold* which stands free of the wall being built and is supported on two rows of *standards*. Mason's scaffolds cannot (like bricklayer's scaffolds) be partly supported on the wall by *putlogs* because these would disfigure the stonework.

mass law A simple, approximate rule about the main factor in sound *transmission* through a solid wall or floor, its mass. The law states that the *sound insulation* value increases by about 5 *decibels* each time the thickness is doubled (i.e. its mass per unit area is doubled). Other factors which improve insulation value include using materials of low stiffness and *discontinuous construction*.

master antenna system (USA) *TV distribution.*

master clock system (USA) *Time distribution.*

master key A door *key* made to open several locks, each of which has its own *servant key*. It can be a floor master, grand master, or sub-master key.

master keying An arrangement of

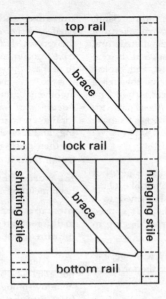

top rail

brace

lock rail

shutting stile

brace

hanging stile

bottom rail

A matchboarded door.

locks to enable *master keys* to be used. The families of suites and sub-suites are shown on a *key schedule*. Master keying makes access easier for cleaning but reduces *security*.

mastic Any permanently plastic, usually sticky waterproofing material that can be used as a *sealant* round timber door or window frames and panels where movement is relatively small.

mastic asphalt A mixture of *bitumen* and mineral fillers, or natural *lake asphalt*, that can be spread by a hand float once it is heated. *Asphalt work* includes roofing, waterproofing, damp-proof courses, flooring, and paving.

matchboard, matched board, match lining Timber board with edges that are *tongued and grooved* on opposite sides and often shaped with

mouldings (*vee joint, quirk*) so that one edge fits (matches) the groove on the other. Matchboards are fixed side by side for flooring, wall lining, and so on.

matchboarded door, framed ledged and braced d., batten d. An external door faced with vertical *matchboards* on a simple frame, usually with horizontal *ledges* and diagonal *braces*. See diagram.

matched floor *Matchboard* flooring.

matching (1) Decorative finishings such as timber *veneer* or *marble*, arranged so that their colour, *figure*, and other markings form a pleasant pattern. *Sliced veneer* often shows *book matching*.

(2) *Matchboard*.

mate (USA **helper**) A *labourer* who assists a *tradesman*.

materials, building m. Sand, cement, and timber, or manufactured products such as fabricated structural steelwork, readymix concrete, bricks, tiles, carpet, pipes, paint, and cables. Materials are often ordered through a builder's *purchasing officer*. A builder's prices may include *supply and fix*, with on-site labour using materials to make building *elements*.

matt *Flat*.

mattock A hand tool like a *pick* but with one end broadened out like a hoe, the other end being sharpened to a point, chisel, or axe blade. It is used for cutting tree roots and digging in stiff ground.

maturing Improvement with age, e.g. the *curing* of concrete or the *fattening-up* of lime.

mat well, m. sinking A shallow hole for a *doormat*.

mat well frame A *doormat frame*.

maul, mall A heavy *mallet* or *beetle*.

maximum demand The *after-diversity maximum demand*.

mean radiant temperature (MRT) The overall effect of heat radiated from surfaces in a room, measured according to their area and temperature, used to calculate *comfort indexes*. A typical MRT figure would include a large area of wall at 20°C, a small area of radiator at 70°C, a window at 10°C, etc. An MRT close to air temperature is comfortable, one slightly above air temperature gives a fresh environment.

measure (1) The quantity of a material, given in metres (lineal, superficial, cubic), number, mass (kilograms), or *substance*.
(2) A method of measuring, such as *board measure* or heaped capacity.

measure and value contract A *fixed-price contract* in which the contractor receives *bills of quantities* with the *tender* documents. Where there is time to prepare the bills as well as drawings, this sort of *contract* is preferred by architects or engineers, as it simplifies *assessment of tender responses*. Contractors have precise information for tendering, and less work is needed to make up their prices than with unbilled contracts.

measured item An *item* which has been included in a *bill of quantities*.

measured separately (ms) (1) *Billed* in different categories, e.g. *general areas* measured separately from *narrow widths*.
(2) Mentioned in a *description*, but not part of the particular *item*, e.g. 'window lining in brickwork opening (ms)' measures only the lining.

measurement The *quantity surveyor*'s duties of *taking off* from drawings, the quantities of materials required for a job, or any later remeasurement on site of the work done and to be paid for. The rules are given in a *Standard Method of Measurement*.

measuring frame A *batch box*.

measuring tape A *tape* measure.

mechanical and electrical (M and E) Both *mechanical services* and *electrical services*, which overlap and may be grouped together.

mechanical barrow A *motorized barrow*.

mechanical circulation *Forced circulation*.

mechanical core A *services core*.

mechanical draught *Forced draught*.

mechanical engineer A person qualified to design *mechanical services*, in

Britain usually a university graduate and member of the Institution of Mechanical Engineers.

mechanical floor, m. level A *plant level*.

mechanical plastering *Projection plastering*.

mechanical protection The protection given by strong materials covering weaker ones, preventing damage from impact or abrasion, e.g. steel *conduits* or cable covers, concrete pipe *surrounds*, door *kick plates*, *lippings*, etc.

mechanical room (USA) A *plantroom*.

mechanical services, m. engineering, m. installations Building *services* such as *heating, ventilation, and air conditioning* and gas supply.

mechanical striking plate A *hush latch*.

mechanical ventilation Basic *air handling* which moves fresh air into rooms and removes odours and humidity using electric *fans*. Domestic systems often work by *extract* alone, either directly or through ducts with *continuous mechanical ventilation*. Commercial systems can have *combined extract and input* and lack only the air-treatment equipment of full *air conditioning*.

mechanic's lien (USA) The rights of a supplier against real property if payment is not made. A supplier who has provided materials, labour, or professional services and has not been paid can prevent a building's ownership title being transferred until the debt is settled.

medical gases Carbon dioxide, oxygen, nitrogen, compressed air, and vacuum supplied by service pipes to points in a hospital from a central storage area.

medium (1) **vehicle** The liquid part of *paint*, its main *film-forming agent*. Some writers say that 'vehicle' includes *thinner* but that 'medium' does not. (2) The filtration material of a *filter*. It may be a replaceable *element* that fits into a filter unit. See also **sand filter** (*C*).

medium board A *fibreboard* of 350 to 800 kg/m^3 density, with good resistance to *fungus* and *borers*. Low-density types are used indoors for pinboards. High-density types may be factory-decorated and suited to external finishes. No additional *binders* are used, as they are in *medium-density fibreboard*.

medium-density fibreboard (MDF) Strong, dense *fibreboard* (often over 800 kg/m^3) upgraded by adding resin *binders* so that it performs like joinery hardwood. It can be machined to a smooth surface ready for painting.

medium-voltage installation Cables and *builder's work* for power supplied direct to industry at 33 kV or 11 kV, usually for loads of one megawatt or greater.

meeting Building projects have two main types of meeting, *site meetings* and *coordination meetings*. Their agenda usually includes a progress report, followed by discussion and agreement about future action, which are recorded in the *minutes*.

meeting rails, check r. The horizontal rails of a *sash window* which touch when the window is closed.

meeting stile A middle *stile* (or shutting stile or closing stile) of a *double door*, double *casement window*, or *folding door*. They are always in pairs, one shaped to stop, or to hold, the other.

megger, megohmmeter An electrical meter with a small generator that is rotated by hand to give a high voltage. It is used to measure very high resistances of millions of ohms (megohms), such as the resistance to earth of electrical wiring.

melamine-formaldehyde (mf) A synthetic *resin* used to make *adhesives*, *laminated plastics*, and *stoved* finishes, but rarely used unmixed with other resins. It has good resistance to wear and to heat, and keeps its colour well.

melamine-surfaced chipboard Resin-bonded wood *chipboard* with a thin coating of factory-applied *melamine-formaldehyde*, usually on both faces. It is cheap, stable, and fairly strong.

melamine-urea formaldehyde (muf) A *moisture-resistant adhesive* made by adding *melamine-formalde-hyde* to *urea-formaldehyde* to form a *copolymer* with better properties. It is used for making *chipboard*.

member A part that is joined with others to make a *frame*, such as a building *column* or *beam*, or truss *struts* and *ties*, or door and window *rails*, *stiles*, *sills*, etc.

membrane A skin, usually waterproof, such as a *curing membrane*, a *damp-proof membrane*, or *membrane roofing*. Membranes are also used for *air houses*.

membrane roofing, m. waterproofing A waterproof roof covering, usually one that has sealed joints and can be used on a *flat roof*. It can be *built-up* bitumen feltwork, or asphalt, although the term most often means high-performance *bitumen-polymer* sheeting, used in a *single ply*.

mensuration The *measurement* of lengths and the calculation of lengths, areas, and volumes; or the methods used in these operations.

merchantable timber Clean, sound, and properly *seasoned* timber, without any combination of *defects* that would make it unsuitable for *carpentry*.

mesh Reinforcement *fabric*, or any metal grid, grille, or sieve, of wire or *expanded metal* (*C*); or the spacing of wires in fabric.

metal-casement putty A traditional *glazing material* made for *steel windows*, as steel has more *thermal movement* than glass. It is also used for hardwood or concrete windows. Galvanized-steel casements need to be *degreased* and given a coat of *pre-treatment primer*, and other surfaces should be sealed to prevent absorption of the oil. The putty should be overcoated with paint within twenty-eight days to prevent it becoming brittle in a few years. The maker's instructions should be followed; there is no BS.

metal coating A thin film or films of nickel, copper, cadmium, chromium, aluminium, or zinc laid over a surface, e.g. a metal or plastics. On metals like steel the coating protects either by the *barrier effect* or by *sacrificial protection* (*C*), or a combination of both. *Galvanizing*, *sherardizing*, and *chromium*, *cadmium*, or nickel plating are usually applied before a part is built in. Zinc, aluminium, or cadmium coatings can, however, be put on by *metal spraying*. Compared with paintwork or *organic coatings* the cost of metal coatings is high, but they are much harder, scratch-resistant, and not affected by *ultraviolet*.

metal conduit Thin steel (or copper) tube *conduit*, which costs more than *plastics* but provides better *mechanical protection* and an *earth* path. Steel conduit built into walls should be treated against corrosion.

metal decking *Profiled sheeting*.

metal door frame A *pressed-metal* door frame.

metal flooring Anchor plates, or *open metal flooring*.

metal lathing *Expanded metal* (C) not less than o.56 mm thick, weighing 1.6 kg/m² at least, used as a base for plaster or exterior *render*. Steel lathing should be galvanized or painted. It is stronger, more durable, and more fire-resistant than wood *laths*. Expanded metal lathing is made to be quickly fixed. Some types have stiffening ribs, but even then it is not a rigid *background* and the shrinkage of cement-lime-sand plaster distorts the mesh if the mix is too strong or insufficient time is allowed between coats. *Lathing plaster* may be the most suitable plaster. Cut edges that expose bare steel should be painted with bitumen to prevent rusting.

metal lathing plaster *Lathing plaster*.

metallic paint Paints containing tiny flakes of metal, e.g. *aluminium paint, aluminium primer*, and *zinc-rich paint*.

metallic parts, earthy metalwork Building components which are connected together by the earth *bonding* system, e.g. metal door frames, cable trays, switchboard cabinets and doors (joined with braided cable), steel *reinforcement bars* and structural steelwork.

metal primer The first coat of paint on bare metal, or the coat put on over a *pretreatment primer*. Steel primers often have an oil *base* that 'wets' the surface and sticks well. They reduce or stop corrosion, either by the *barrier effect* or by the action of an *inhibiting pigment*. Metal primers which contain compounds of *zinc* or are *zinc-rich* have largely replaced *lead paints*. Most metal primers should be *overcoated* before

they are exposed to the weather, as they are porous and need the gloss top coats to provide full corrosion protection.

metal-sheathed mineral-insulated cable *Mineral-insulated cable*.

metal spraying, metallization The spraying of a molten *metal coating*, usually aluminium or zinc on steel, at high temperature. The process can be done on site and is suitable for large bulky objects, giving a coating that will withstand severe conditions. Since it is porous, a *sealer* should be applied within forty-eight hours.

metal trim Pieces of shaped metal, e.g. *angle beads, stop beads, architrave beads*, etc., fixed along the edge of an area of plastering to guide the final thickness, to form a neat break between materials, or to allow for *movement*.

metal valley A *valley gutter* lined with lead, zinc, copper, or aluminium.

metalwork The *trade* of purpose-made non-structural work using metals, usually carried out in a workshop off site and requiring skill to install on site. Metalwork can be architectural (decorative), builder's (utilitarian), or *plumber's metalwork*. Examples are duct covers, handrails, and doormat frames, but metalwork normally does not include standard metal items like lathing, lintels, or fasteners.

meter bypass A *bypass* pipe with a *stopvalve* to allow a full flow of fire-fighting water, for which no payment is made, past a *water meter*.

methane A flammable gas, lighter than air. It can be emitted from an old rubbish tip. To prevent the risk of fire and explosion, new houses with a *ground slab* in such areas must have a methane barrier (often plastics sheet) under them which satisfies the *Building Regulations*.

methyl methacrylate (polymethyl methacrylate) *Acrylic resin.*

Methods of Assessment and Testing (MOAT) The performance conditions which a building system or product must fulfil for issue of an *Agrément Certificate.*

method specification The most usual sort of *specification*, one which describes in detail how to do a job of work.

mezzanine A floor *level* within the height of a tall *storey*, covering less than the full area. The mezzanine as well as the floor below may have less than normal headroom.

micaceous iron oxide paint A traditional paint with a *pigment* which tends to keep out water, used on steelwork.

microbore pipework, minibore p. *Central heating* using copper or nylon circulation pipes smaller than those used for *small-bore systems*, usually of 6, 8, 10, or 12 mm outside diameter. The microbore pipes start from larger copper main flow and return pipes from the *boiler*. They can be run inside walls or timber floors. The larger pipes form a *manifold* (*C*) for connections of microbore pipe, serving up to nine radiators from one point. The tubes are easy to bend and install, are less conspicuous than larger pipes, and lose less heat. High-pressure *circulating pumps* are needed to drive the water through the small tubes.

micron (μm) One micro-metre (one millionth of a metre), equal to one thousandth of a millimetre. Household aluminium foil is about 17 microns thick. A normal *dry-film thickness* for coats of paint is 50 to 100 microns.

milestone A key *event* on a *programme* using the *critical path method.*

milkiness The defect of a whitish appearance in a *varnish* film.

milled sheet lead *Lead* metal rolled into sheets from cast ingots.

mill-finish aluminium Extruded *aluminium* untreated after it leaves the *die* (*C*). It becomes pitted and unsightly in most environments, but is used for unseen work in sheltered locations.

millwork (USA) A *joinery fitting* prefabricated in a planing mill.

mimic diagram A panel with diagrams of a building, usually floor plans divided into zones, with indicator lights and lines to represent equipment, showing quickly its location and how the system works.

mineral-fibre tiles Convenient lightweight *acoustic tiles* that form the panels of a *suspended ceiling*, usually with interlocking edges made to fit (or conceal) the supporting *runners* and join adjacent tiles. Their surface is often *fissured* for sound *absorption*, but they have little effect on sound *transmission* or *fire resistance*. Care is needed in handling, removing, or replacing mineral-fibre tiles – they are easily dirtied, or damaged along the edges.

mineral-insulated cable, m.-i. metal-sheathed c., m.-i. copper-covered (MICC) c. An electrical cable with a single inner wire surrounded by magnesium oxide, an insulating material unaffected by heat or age (it should never need rewiring), compressed inside a metal outer sheath. The inner wire is a solid strand (copper, aluminium, sometimes a heating element) and the outer ꜱheath is copper or aluminium. Moisture must be excluded with *glands* that seal the cable ends. They can have a PVC outer *serving* and PVC shrouds over the glands to give

---

mitre saw

protection against damage from steel trowels or from corrosion due to *dissimilar metal contact* with galvanized *cable tray*. Mineral-insulated cables are used to connect *fire detectors*, for running *earthed concentric wiring*, or with a heating-element inner wire for *underfloor heating*.

mineral spirits *White spirit*.

mineral-surfaced bitumen felt A *bitumen felt* coated with a *self-protective* dressing of coloured mineral particles on its top surface, plus talc or sand on the other side, used as a single layer on sloping roofs or as a *cap sheet* for *built-up roofing*. It is made 1 m wide in rolls 10 m long.

mineral wool Fine fibres of rock, slag, or glass made up into *blankets* or panels for *fire stops*, used as *sprayed mineral insulation*, or laid loose for heat *insulation* and sound *absorption* inside a closed space. Mineral wools absorb water easily, but are not damaged by it, and are non-combustible.

miniature circuit breaker (mcb) A small *circuit breaker* that fits into a modular *distribution board*, often in a *consumer unit*. It is used to protect a single *pole* for domestic power or lighting loads and is tamper-proof.

minibore pipework *Microbore pipework*.

mini kitchen unit An assembly of *sink*, *hob unit*, and refrigerator.

minus 60 mm hardcore Granular material containing all sizes of stone up to 60 mm maximum.

minutes A record of what was said at a *meeting*, as well as action to be taken and who is to take it. Minutes are written after the meeting, usually by whoever took the chair, and sent to everybody present. Minutes of the pre-

Mitred joints.

vious meeting are read early in the next meeting and discussed.

mist coat A very thin coat of paint, usually highly diluted and sprayed on. It can be used as a *sealer* to stabilize a loose *substrate*.

mitre (USA **miter**) One of two matching ends, each cut off at half the bend angle to make a *mitre joint*. (French 'mitre', from 'mi-tour', mid-turn.)

mitre box A square U-shaped box of plastics or wood with saw cuts at 45° to guide a *handsaw* while cutting a *mitre*. An electric *mitre saw* is more accurate.

mitred joint, mitre j. A neat *joint* made with two *mitres* that form a line at a corner, usually between members of similar cross-section. The most common mitre joint is at 90° and requires two mitres at 45°, e.g. at the corner of an *architrave*. Wood mitre joints can be glued and butted or reinforced with a *loose tongue*. See diagram.

mitred knee A *knee* between the horizontal and sloping parts of a timber *handrail*, cut to a *mitre* so that its shape matches end-to-end.

mitre saw A fast, accurate electric *circular saw* mounted on an arm fixed to a bench. The arm can be turned and locked at an angle, e.g. 45° for most *mitres*. See diagram, p. 346.

mitre square A *bevel square*.

mitring machine A *trimming machine*.

mix The proportions of each constituent in a batch of *concrete, mortar, plaster*, etc. See also *C*.

mixed construction A *composite*.

mixer (1) **mixing valve, combination tap** A *tap* supplied with hot and cold water that lets out *blended water* through a single *spout*. It can have manual, thermostatic, or *touchless controls*. Different types are made for sinks, basins, bath/showers, etc., often in a *monobloc* unit with single *lever* operation.
(2) A *concrete mixer*.

mixing box An *air-handling unit* with *dampers* to vary the flow from two different ducts, such as the hot and cold air in a *dual-duct system*, or fresh and recycled air. In a *high-velocity* dual-duct system the mixing box also contains a constant-volume regulator and an *attenuator*. It usually has *automatic controls*, including a local thermostat.

mixing valve A *mixer*.

mixing water, gauging w. The water used in making concrete or mortar, which should be free of impurities. In general *drinking water* is suitable, but not sea water.

mobile crane A *crane* which can move around on tyres or crawler tracks. Most have a *luffing* jib. Road-going types with tyres have *stabilizers* and a telescopic jib which can be extended once it is set up.

mobilization costs (USA) *Preliminaries*.

mock-up A full-size model used for tests or to show how a room will look.

Model Water Byelaws The regulations which control water supply in Britain. Since 1986 they have allowed *unvented* hot-water cylinders.

modification (USA) A *variation*.

modular brick, metric m.b., standard m.b. A *brick* 190 × 90 × 65 mm, which when laid has *format dimensions* of 200 × 75 mm. See BS 6649.

modular coordination *Dimensional coordination* using a *modular system*.

modular masonry unit (USA) A *brick* or building *block* which measures, laid up with its mortar joints, a multiple of 4 in. (101.6 mm) both in plan and vertically. There are several systems; some do not *bond* easily and may require expensive fair cutting with a *diamond saw*.

modular sizes The sizes of building *components* or *elements* which fit the basic *module* or *multi-modules*, such as 200 mm square wall tiles and metric *modular bricks*. Non-modular sizes may be metric measure for old imperial sizes, e.g. $4\frac{1}{2}$ in. (114.3 mm) or 6 in. (152.4 mm).

modular system The system of planning buildings and their components to fit a *planning grid* related to a *module*, allowing *dimensional coordination*.

modulated controls Gradual automatic controls which can change a mechanical or electrical system continuously, increasing or decreasing supply to match demand. They are smoother and more responsive than *on/off control* and (most important) result in less *equipment noise*.

module (1) **building m., planning m.** A standard unit of length which is repeated many times and controls the sizes of *components* and the layout of a building. A module can be small, like the present standard (or fundamental basic) module of 100 mm. Other modules have been used, e.g. 4 in. (101.6 mm), 900 mm, 1 m, and 40 in. Japanese buildings have a standard module, used since long before European influence, of the *tatami* floor mat (6 × 3 Japanese ft), with room sizes described as a number of mats. When a module is used, few dimensions are needed on the *drawings*, as all sizes are indicated by the grid lines. Building components such as bricks, blocks, and tiles are made to *modular sizes*, and so is some construction equipment, such as *panel forms*.

(2) Any prefabricated assembly, such as a *bathroom pod, lift shaft module, packaged plant*, or storey-height lengths of *stack pipe*.

moisture barrier A *vapour barrier, damp-proof course*, etc.

moisture content In wood, the amount of water expressed as a percentage of its *oven-dry* weight, often exceeding 100% for freshly felled timber. Before use in building the moisture content is reduced by *seasoning*, to improve strength and reduce *decay*. The reduction of moisture content from the green state to about 25% has little effect on the strength, but a reduction from 25% to 12% may double the compressive strength. With a moisture content above 18%, wood should neither be painted nor primed. At the time of installation in a building, timber should be at its *equilibrium moisture content*. The smallest moisture content at which *fungus* can grow in wood is 20%. Below 25% the moisture content can be measured to an accuracy of 2% or slightly closer by electrical moisture meters, which measure the electrical resistance of the wood between two sharp electrodes driven into it with a special hammer. In the conditions of building sites it is unrealistic to insist

on moisture content being closer than ±2% to the required figure (BRE Digest 287). To take *shrinkage* into account, BS 4471 gives cross-sectional sizes for softwoods at 20% moisture content, and figures for adjusting sizes for higher or lower percentages. Many thermally insulating materials lose *insulation* value when wetted. See also **moisture movement**.

moisture expansion The permanent increase in size of *ceramics*, such as clay bricks or tiles, starting from the time of *firing*. Both dry and saturated ceramics expand by the same amount – very little water is needed. Bricks expand up to 1 mm per metre (1 in 1000) in the first eight years, half of it in the first week, depending on clay type and firing temperature. The brickwork containing them expands less, about half (0.6 mm/m), and mostly within two years. Walls should therefore be built with bricks at least three weeks old (never kiln-fresh). In addition, bricks have reversible *thermal movement*, so that in order to prevent *cracking* brickwork should always be built with horizontal and vertical *movement joints*. Ceramic *tiles* can lift off a floor or wall if they are laid too soon after firing or with tight joints and over-strong *grouting*.

moisture movement Concrete and timber expand when their moisture content increases and shrink as they dry out, often merely in response to a change in the weather, without getting wet at all. Metals, glass, etc., have no moisture movement; granites, plastics, and some hard plasters have very little, and ceramics (clay bricks, tiles) have only *moisture expansion*. The change in size of concrete and cement-based materials from moisture movement is usually less than that from their initial *drying shrinkage*. Buildings are made so that moisture movement can occur freely at *movement joints*, in finishings, and in timberwork, both *carpentry* and *joinery*. See BRE Digests 227, 228, and 229.

moisture-resistant adhesive (MR) A glue with good resistance to insects, fungus, and cold water, and to hot water for short periods.

molding (USA) *Moulding*.

monitor, m. roof, console A raised area of roof with glazed sides which form a continuous *lantern* light.

monitoring Watching, often including the use of *sensors*, to operate an *alarm*, prevent an unsafe condition, control *security*, run a *building management system*, etc.

monkey tail A downward scroll at the end of a *handrail*.

monkey-tail bolt An *extension bolt*.

monobloc mixer, single-hole m. A *mixer* for a single *taphole* sink. It has one spout, two taps, and two copper *pipe tails* all in one body.

mono-grano screed A *monolithic screed* of *granolithic*.

monolithic screed, m. finish A wearing *topping* of *granolithic*, *terrazzo*, or other mix laid on a fresh concrete slab while it is still plastic (within three hours), thus becoming part of the slab and undergoing the same *drying shrinkage*. It is usually only 10 mm thick, but can be accepted as a structural part of the slab up to 20 mm thick. A monolithic screed will never separate from the slab, and cracking is usually limited – common faults with other *screeds*. See diagram, p. 392.

monopitch roof A *pitched roof* that slopes in only one direction, either with no walls above and capped with mono-pitch *fittings*, or a *lean-to roof*.

monthly certificate, interim c., progress c. A *certificate* which states the amount the *client* owes the *contractor* for work done during a particular month. It is issued by the *architect* and usually based on the contractor's *monthly statement* with a deduction for *retention*. Usually the client may not interfere with the architect's decision.

monthly instalment, interim payment, progress p. The sum of money paid by the *client* to the *contractor*, according to the *monthly certificate*.

monthly statement, claim A valuation of work completed in a *contract*, usually based on the amount of work completed in each *trade*. It is prepared by the contractor's *quantity surveyor*, usually as part of the overall *financial control* of a project.

mopboard (USA) A *skirting*.

mordant wash A *wash primer*.

mortar A mixture of *sand*, *cement*, and water that sets hard within a few hours to form a thick, rigid, semi-porous mass. It is usual to make mortar with Portland cement, or as a *composition mortar* with lime, although other binders such as *masonry cement* or *resin* can be used. Sand for site-mixed mortar is usually *sharp* and must be clean and suitably stored. In addition to the basic constituents the mix can also contain *plasticizers* or *polymer* modifiers. Site batching and mixing is done in a *cement mixer* or on a *banker* board. Alternatively, use is made of *ready-mixed mortar*. The main uses of mortar are for the bedding and jointing of *brickwork*, for laying floor tiles, and as *render* in plasterwork, as well as for general jobs such as *building-in* or sealing holes and as *dry pack* under baseplates, all of which are *wet trades*. Stiff mixes give the best bond with non-absorptive materials and reduce rain penetration.

The size of a batch of mortar mixed at one time should be kept small, so that it does not set before the work is completed. If that happens the set mortar should be discarded, not *retempered*. See BRE Digest 362. See also **mortar designation**.

mortar board A *spot board*, or a *hawk*.

mortar cube test A compressive strength test for cement, which is mixed in a standard way with *standard sand*. See BS 4551.

mortar designation A classification of mortar for *bricklaying* according to mix proportions, which is the main factor affecting its properties. There are five groups, designated 1 to 5, from strongest to weakest. Designation 1 mortar is always a cement:lime:sand mix, in the proportions 1 :$\frac{1}{4}$: 3. The other designations can be cement:lime: sand (giving the best bond and resistance to rain penetration), or *masonry cement* and sand, or cement:sand with *plasticizer*. Designation 1 mortar (10 N/mm²) gains strength quickly, has high shrinkage, and may cause cracking in brickwork, which can be reduced by *curing*. It is normally used only in the ground or in the bottom *damp-proof course*. Designation 3 mortar (2.5 N/mm²) is adequate for external walls in most low-rise buildings, allowing *movement* by being weak and flexible. Designation 5 mortar is used for internal walls.

mortar mill A forced-action *pug mill* for mixing stiff, dry mortar.

mortice, mortise A slot cut in wood or stone, in which a *tenon* from another member is glued or pinned, or a *lock* is placed. Its *cheeks* should not be wider than one third the thickness in wood, less in stone.

mortice-and-tenon joint A *joint* between members at right angles to each other, such as a door *rail* tenoned into its *stile*. It is tightened against its *shoulders* by *drawboring* and *dowel pins*, or with *folding wedges*, or simply glued and cramped.

mortice chisel A thick, strong steel *chisel* with parallel sides, used to cut *mortices* in wood, usually to the full width of its blade.

mortice gauge, counter g. A *marking gauge* with two marking points, one fixed and movable, which mark out an accurate *mortice* to be filled exactly by its *tenon*.

mortice joint A *mortice-and-tenon joint*.

mortice lock A *lock set* in a *mortice* (within the door thickness). The lock is hidden and the joinery is of better quality than that for a *rim lock*.

morticing machine A *woodworking machine* which cuts *mortices* in timber. It may be one of two general types, the square chisel, or the chain morticer. The square chisel morticer has an *auger bit* rotating in a square steel shell with holes cut into the sides through which the chippings are thrown out. The chain morticer has a projecting jib and chain with cutting teeth, and is used for cutting large mortices.

mosaic A decorative wall or floor finishing made of small cubes (*tesserae*) of marble, glass, burnt clay, etc. They can be laid as plain unicoloured sheets, as a pattern, or made into a picture. Mosaic has excellent water and frost resistance and its frequent joints improve slip resistance.

mosaic parquet panels *Wood mosaic*.

mother-in-law test The *heavy-body impact test*.

motive power circuits Circuits in the *electrical services* able to supply the high starting current demand from *electric motors*, typically seven times the current in normal service, used only for large motors, e.g. in *lifts*.

motorized barrow, mechanical b. (USA **buggy, concrete cart**) A power-driven barrow with two wheels, mostly used for carrying fresh *concrete* across a hard, level surface.

motorized valve A *discharge valve* operated by an electric motor, used in large automatic *central-heating* systems and giving *modulated control*.

mottle A *figure* which greatly increases the value of *veneer* for cabinet work.

mottler A thick flat paint *brush* for *graining* and *marbling*.

mould (USA **mold**) (1) A pattern used to make a *moulding*. In plasterwork it is a sheet of stiff material cut to form a profile, usually fixed to a wooden handle.
(2) A *fungus* that stains materials but does not rot wood. Before repainting its spores should be deactivated with preservative (BS 8000).
(3) See *C*.

moulded On which a *moulding* has been cut or cast.

moulded brick A *clay brick* shaped out of soft mud, in a mould, usually with a *frog*. Since the clay is not compacted, the bricks distort and crack in drying and firing, and are weak and easily chipped, but they often have a creased texture prized for *facing brick*, as well as marks from being stacked to dry or from sand used in the process. The Romans made moulded bricks in England, but the art was lost, to be revived in the early Middle Ages. In Tudor times both soft moulded and *stock bricks* were made by hand and usually covered with plaster to imitate stonework, except in chimney stacks and *finials*, which were often elaborate *specials*. The plain bricks or stucco of Victorian England brought a return to fashion of moulded and multi-coloured brickwork, led by Ruskin in 1852. Handmade as well as machine-made moulded bricks are used today, both for new work and for *refurbishment*.

moulded-case circuit breaker A large *circuit breaker*, used as a main switch or to protect *three-phase* circuits.

moulded insulation, sectional i. Thermal *insulation* shaped to fit round steam, hot water, or refrigerator pipes and *fittings*; a ready-made *lagging*.

moulding (USA **molding**) (1) A continuous projection or groove on an outside wall or column, used as decoration to throw shadow, as for *dressings* and *enrichments*, or sometimes to throw water away from a wall. It may be in stone, brick, plaster, joinery, aluminium, plastics, and so on. Joinery mouldings are either formed on the solid (*stuck*) or fixed with adhesive or nails (*planted*).
(2) The work of cutting mouldings with a *moulding machine* or a hand *plane*.

moulding machine A *woodworking machine* for cutting *mouldings* on wood, for example a *spindle moulder*.

mounting The supports under a machine, such as springs or rubber pads, which are resilient and *anti-vibration*; the term also includes wheels or skids.

movable partition A *partition* made up of separate sections that can be moved, rearranged, or stored, usually by the building occupants.

movement (1) Changes in the length,

area, or volume of materials from expansion and contraction. In building materials the most usual types are *thermal movement* from heating and cooling, and *moisture movement* from wetting and drying, which are both types of reversible *cyclic movement*. Some types of movement are irreversible, notably the *drying shrinkage* of concrete and the *moisture expansion* of bricks. Movement can generate forces strong enough to break materials unless *movement joints* are provided. Materials that undergo little movement have good *dimensional stability* (BRE Digest 343).

(2) Each of the actions of a *crane*: hoisting, luffing (or trolleying), and slewing, or shifting the whole crane by travelling.

movement detector A wall-mounted box which gives off a harmless beam of low-power microwaves, picks up their reflection, and analyses it for very slight changes in frequency (the Doppler effect) caused, e.g., by people moving towards it or away. It can be used to control an automatic *door opener* or outdoor lights, or as an *intruder alarm*.

movement joint, control gap An open or sliding joint to allow *movement* and prevent *cracking* in a large panel of rigid material (concrete floor, brick wall, ceramic tiles, metal roofing, etc.). For exterior *brickwork*, exposed to rain, heat, and cold, it is about 20 mm wide. Maximum spacings between vertical movement joints are 15 m for clay bricks, 9 m for *calcium silicate bricks*, and 6 m for concrete bricks or blocks. They should be kept at least half these distances from wall corners or angles, which are usually fixed points; a short *return* may cause *cracking in brickwork*. Brickwork horizontal movement joints are made every third storey (maximum 9 m), underneath a *shelf angle* or *lintel*. Vertical movement joints in *cladding* are usually *open-drained joints*. Movement joints may be made weathertight with a strip of *sealant*, formed in-situ against a *backup strip* to limit its depth and prevent the making of a *filled joint*. In sheltered locations preformed *sealing strips* are suitable for movement up to about 40% of joint width. See diagram.

mudsill A *sole plate*.

mullion A vertical dividing member of a window frame that separates the

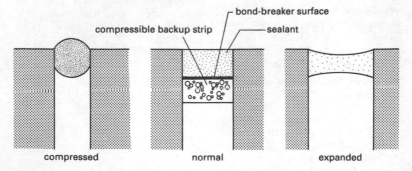

bond-breaker surface
compressible backup strip — sealant

compressed   normal   expanded

Movement joints. Sealant is inserted at normal gap width (avoiding extremes of temperature) against a flexible backup strip with a bond-breaker surface.

multi-folding door

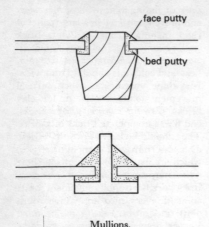

face putty

bed putty

Mullions.

lights from each other. Compare **transom**. See diagram.

multi-folding door A *folding door* with several leaves, thus very compact.

multi-module A *module* that is a multiple of 100 mm.

multi-ply *Plywood* with more than three plies, which should be *balanced*.

multi-point lock A door *lock* with three or more latching points, to give extra *security* against forced entry.

multi-point water heater An *instantaneous* (non-storage) gas water heater that supplies hot water to several taps. To reduce its running cost it should be installed next to the tap most fre-

quently used. A typical multi-point unit heats 7 litres of water per minute through 45°C, whereas a single-point heater (instantaneous sink heater) heats only 3 litres per minute through 45°C. The *Building Regulations* allow only *room-sealed* gas heaters in bathrooms.

multi-storey building A building with several *storeys* above ground.

multi-unit wall (USA) A wall built of two or more thicknesses brick or block, called *withes*. A British *cavity wall*.

Munsell reference system A system in which any *colour* is fully described by its hue, value (lightness), and chroma (saturation).

muntin (1) A subsidiary vertical joinery framing member in a *panelled door*, framed into the *rails*, separating the panels and usually of the same width as the *stiles*.
(2) (USA) A *glazing bar* or a *mullion*.

murder clause (USA) A *contract* clause that is worded unfairly, usually with the aim of shifting responsibility to someone who should not carry it.

mushroom-headed push button, emergency shutdown A *push button* with a large round top, usually coloured red, used to stop machinery in an emergency.

mute (USA) A *slam buffer*.

N

nail A straight metal *fastener*, usually with a head, hammered into position to secure one component to another (BS 6100). Nails can easily split timber if driven too close to an end, particularly with seasoned *hardwoods*, although splitting can be reduced by pre-boring a *pilot hole*, by *nail blunting*, or by clamping. Skill and care are needed to avoid bruising surfaces with the hammer, and some jobs need a *nail punch*. The commonest nails are cold-forged from bright round-steel wire of diameter between 1.2 mm and 6 mm and have a *bullet* or flat head. The largest nails (over 130 mm long) are known as spikes, the smallest are pins, tacks, oval brads (not to be confused with flooring brads), and glazing sprigs. *Clout* nails are large-headed galvanized wire nails for fixing *bitumen felt*, *corrugated sheeting*, *slates*, etc. Other wire nails include *ring-shanked nails* and *drive screws*. Those nails which are not made from wire are usually of black steel and of rectangular cross-section. They are either sheared from steel plate (*cut nails*) or have forged heads, as have rose-head nails (wrought nails). *Galvanized* nails are made for nailing *plasterboard*. Nails are usually sold by weight.

nailable Able to hold *nails* and have nails driven into it.

nail blunting Placing a nail head-first where it is to be driven into a heavy timber and striking its point firmly with a hammer, thereby blunting the point. The nail is then reversed and driven. Its point cuts the wood fibres, reducing splitting.

nail float A *devil float*.

nail gun, nailer A *power tool* for driving *nails*, which are fed in on a disposable cardboard belt or from a magazine as in a machine gun. They are usually operated by compressed air supplied through a lightweight plastics hose. Small nail guns are called *bradders*, large ones *framing guns*. See diagram, p. 346.

nailing ground A *common ground*.

nailing machine A *nail gun*.

nailing strip, n. block (USA **nailer**) A strip of *nailable* material, such as a *fixing brick*, built into brickwork or blockwork.

nailplate, gang nail, truss plate A timber *connector* made of galvanized-steel sheet punched to form many rows of nails. They come in different sizes and are mostly used as pairs of *gusset plates*, one on each side of a joint in a *trussed rafter*. In different types the nails are all driven together by hydraulic press or separately driven with a carpenter's hammer.

nail puller, pry bar A tool used for withdrawing *nails*, more delicate than a *claw hammer* and shaped something like a screwdriver with a curved, forked claw.

nail punch, n. set (USA also **brad p.**) A short blunt steel *punch*, tapered at one end to the diameter of a small *nail*, used to drive nail heads below the surface of *joinery*. Its point is concave, to hold the nail head.

nail sickness The weakening of wood

round nails rusted by moisture in the wood and its natural acetic acid, and by copper-based *preservatives*. The wood is attacked by the corrosion products.

nap Raised fibres on the surface of felted *carpets*.

narrow-ringed timber, close-grained wood, fine-grained w., fine-grown w. Wood which has grown slowly, with narrow, inconspicuous *annual rings*. It is therefore usually stronger than *wide-ringed timber*, which has a high *rate of growth*.

narrow widths In quantity surveying, narrow widths are paintwork, plasterwork, etc. less than 150 mm wide, *measured separately* to allow pricing of *labours* compared with work in *general areas*.

National Building Specification (NBS) A six-volume (for small jobs two-volume) loose-leaf publication updated quarterly. It is obtainable on subscription from NBS Services Ltd, Newcastle upon Tyne, a subsidiary of the RIBA. It is a 'library' of clauses for *specifications* from which architects and other consultants can selectively copy when writing specifications or *preambles* to *bills of quantities*. The clauses are simple, short, and, unlike most specifications, easy to understand. Ample references are given to *British Standards* and to other authorities. The information can be supplied in at least four possible forms: (1) conventionally printed on loose-leaf sheets, (2) on magnetic disk suitable for most common word-processing or computer systems, (3) by computer-to-computer transfer over the telephone, (4) by IBM magnetic card for a few word-processor or computer systems. NBS is also a word-processing bureau for building specifications based on its publications.

National Fire Protection Agency

(NFPA) (USA) A body with a far-reaching reputation for its publications on all aspects of *fire protection*.

National House Building Council (NHBC) A body which encourages better house building. Its services include the voluntary registration of reputable house builders, conciliation between builders and purchasers, the NHBC warranty, ten-year insurance on new houses, and advice on correct site practices, including erection techniques, distance from *trees*, the installation of important items such as *fire barriers*, and standards for the *preservation* of structural timbers.

National Vocational Qualifications (NVQ) (UK) Qualifications offered to trained candidates, who undergo skills testing, under a government system aimed at improving the economic performance of British industry. In the building industry they are awarded by the *City and Guilds* and the *Construction Industry Training Board*.

natural bed A stone is laid on its natural bed when its *bedding planes* are horizontal. This is advisable for load-bearing stones, especially out of doors. Sometimes the natural bed must be marked at the quarry, since, particularly with igneous rocks, it is not always obvious. Rarely are stones *face-bedded*.

natural cement A low-strength *cement*, like *hydraulic lime* but burnt at a lower temperature, sometimes used for exterior plasterwork.

natural draught (USA **n. draft**) *Draught* in a *flue* produced only by the *stack effect*, not by a fan, as in *forced draught*.

natural gas Any fuel *gas* that flows from the ground. Most often it consists largely of methane with other paraffins and olefins.

natural ground level, initial g. l. (USA **n. grade**) The ground *levels* that exist on a building site before *groundworks* are started, often recorded as the *agreed levels*.

natural seasoning The drying of timber by stacking it so that it is exposed to the air all round but sheltered from sun and rain, a process which usually takes years and is therefore costly. This old method may be preferred to *kiln drying*.

natural stone Stone which has been quarried and cut, not *cast stone*.

natural ventilation *Ventilation* obtained by allowing air to move by itself, using *flues* or *vents*, without the fan needed in *mechanical ventilation*.

naturbetong Concrete made by placing coarse aggregate in *formwork*, then filling the voids with liquid cement *grout*. After *stripping*, the surface is *sand-blasted*.

navvy pick A heavy double-pointed hand *pick*, or one with a point and a chisel edge, generally with a *helve* 900 mm long.

neat cement grout *Grout* made with cement and water only, without sand.

neat size Timber *dressed dimensions*.

needle (1) **n. beam** A stout beam of wood or steel placed through a hole in a wall to transfer the weight of the wall above to *dead shores* under its ends. See diagram, p. 406.
(2) A block that forms a *key* between a brick wall and the vertical *wall piece* of a *flying* or *raking shore* to transfer the weight of the wall to the wall piece. It is let into a hole made by removing a brick from the wall. See diagram, p. 406.

needle gun A rugged power tool with a bundle of steel rods of about 2.5 mm

dia. which are forcefully vibrated forwards, to bounce off a work surface. It is used for cleaning hard materials such as steel or concrete.

needle-punch carpet A lightweight *carpet* made by punching a batt of fibres into a foam-rubber *backing* with needles. It is laid fully stuck down with adhesive. As it is thin, its *substrate* should be smoothed with *levelling compound* or by fixing down *hardboard*. It can be obtained with a *pile*.

needle scaffold A *scaffold* carried on *needles* driven into a wall.

needling The insertion of a *needle* beam.

negative pressure Air pressure below atmospheric may be produced by a *fan* or by wind. In a vertical drain pipe of any sort, such as a *stack*, falling water creates a negative pressure behind, which may unseal a trap.

negotiated contract A building *contract* in which the *contract sum* is discussed rather than arrived at by *tender*. An *estimate* is needed, but there is no waste from unsuccessful tendering.

negotiation stage Usually *tendering*, but see the previous entry.

Neoprene A trade name for *polychloroprene*, a synthetic rubber which is flexible, oil-resistant, and has the other excellent properties of non-flammability and durability when exposed to *ultraviolet* rays or ozone. It is usually black and used for *gasket glazing*, foam rubber, roofing washers, etc.

nest of saws Several saw blades which can be used at different times in the same handle.

net A *safety net*.

network (1) An *arrow diagram*.
(2) A grid of lines, pipes or cables, or an *air termination* or *earth electrode*.

network analysis The methods and rules used in the *planning and scheduling* of large complex projects, for example the *critical path method* and *PERT*.

neutral The black or blue wire in a *low-voltage* power or lighting circuit, which is connected to *earth* by one means or another. The other wire is the *live*.

neutral brought out A wire connected to the windings of a three-phase transformer to allow a *neutral* tapping.

newel The inner side of a *stair* that turns. It can be a *solid newel*, usually with a *newel post*, or an *open newel*. (From the French, 'noyau'.)

newel cap A wooden top to a *newel post*, such as a *finial* or *pommel*.

newel drop A downward decorative projection of a *newel post* through a *soffit*.

newel end The end of an *escalator* balustrade.

newel post The main upright carrying the *strings* of a timber *stair* or forming the inside of a *circular stair*.

nib (1) A small projection from a flat surface, or a small *return*, *upstand*, or *kerb* at the edge of a wall or slab.

(2) **cog** The downward-projecting lug at the head of a *roof tile* for hooking over the tiling *batten*. Continuous nibs can be obtained that for example enable an inverted tile with the nib downhill to be used as a *drip* for a window sill or *coping*.

(3) The part of the top edge of *asphalt work* which fits into a *chase*, securing it to the wall it protects.

(4) The defect in paintwork of a small particle that projects above the surface of the *film*. A film with nibs is called *bitty*.

nibbler A hand tool with jaws for biting holes in thin sheet metal. It can be worked by handles, like a pair of pliers, or by an electric *power tool*.

nickel brass *German silver*.

nickel sulphide A compound occurring in some glassmaking processes, trapped inside glass as tiny inclusions. They swell slowly, eventually breaking the glass.

nicker A mason's broad *chisel* used for grooving stone before splitting it.

nidged ashlar, nigged a. Stone, particularly granite, dressed roughly with a pointed hammer. The hardest granite can be dressed in this way.

night lock, n. latch A *cylinder lock*.

night vent, n. light A *ventlight*.

ninety-degree bend, 90° bend A *quarter bend*.

nippers, steelfixer's nips, tower pincers A pair of wire cutters with cross-jaws and long handles, used by *steelfixers* to tighten *binding wire* round *reinforcement bars*, twist it until secure, and snip off waste ends.

nipple (1) A short pipe threaded outside at each end, usually with a *taper thread*, used as a *tubular* pipe fitting for joining two *couplings* or internally threaded pipes. Types include: *barrel*, *close*, hexagonal, parallel. See **socket**.

(2) A small valve at the high points of a hot-water system, used as an *air-release valve* to prevent or clear air locks.

(3) **grease n.** A rounded steel stud with a one-way valve, to which a grease gun can be coupled, used to pump in grease and lubricate a mechanical bearing.

node (1) A meeting point of lines on a *programme*, showing *activities* leading up to, and then starting from, the one *event*.

(2) A meeting point of the members of a *truss* (*C*).

no-fines concrete Concrete made with only coarse *aggregate* and cement, without sand. It contains large pores which prevent *capillary action* letting in *driving rain* or *rising damp*. It has been used in Britain for building house walls (as at Crawley New Town) in roughly the same thickness as the *cavity wall* it replaces. A typical no-fines aggregate consists of 95% of material between 10 and 20 mm size. Since it is not *nailable*, *chases* are formed during casting and a *foamed slag* nailing strip is plastered into the chase after the shuttering has been stripped. No-fines concrete is not vibrated or rammed, but lightly punned. Only lightweight formwork is needed. It is usually not reinforced except for a few diagonal *trimmer* bars across the corners of openings. It has good mechanical *key* for plaster and low *suction*, and must be rendered outside to strengthen it and seal it against wind. Often it is regarded as a *lightweight concrete*, especially when made of *lightweight aggregate*.

nog A *fixing brick*, or a *nogging*.

nogging (1) **nog** Short horizontal timbers cut to fit between vertical *studs* of a *framed partition* to stiffen the studs and make them act in common.

(2) *Bricknogging* or blockwork between columns.

noise Unwanted, disturbing, painful, or ear-damaging sound. Inside buildings noise is pollution when it comes from outdoor sources such as traffic or aircraft, or a nuisance when the source is indoors, from people (talking, music, work, *impact noise*) and *equipment noise*. The best way of reducing noise inside buildings is by planning to keep noisy activities together and separated from quieter ones. This applies also to *zoning*

outdoors, e.g. dwellings built as a *maisonette* or offices with a *services core*. Noise reduction by means other than planning, e.g. by *sound insulation*, is usually less effective and many times more expensive than noise reduction incorporated into the design of a building, though planned noise reduction may also be expensive (e.g. a tunnel for a motorway under a housing estate).

At work, employers are under a general duty to reduce the risk to workers from noise. If noise exceeds 85 dBA, *ear protectors* must be provided. Above 90 dBA and for brief, loud noises, e.g. from *cartridge tools*, ear protectors are compulsory under the Noise at Work Regulations. These ear protection zones must be marked with a *safety sign*. Details can be obtained from *Health and Safety Executive* information centres.

noise absorption Reduction of noise and reverberation within a room by *absorption* in non-echoing materials.

noise control on site Before work is started on a building or demolition site in Britain, a statutory notice is issued by the *local authority*, under the Control of Pollution Act, stating the measures to be taken to limit noise and vibration. Compliance with the notice protects the contractor from legal proceedings over noise. See BS 5228. See also **noise**.

Noise Criteria (NC) See **Noise Rating**.

noise insulation See **sound insulation**.

Noise Rating (NR) A series of curves used in Europe to adjust the readings from a *sound-level meter* to give a single index figure for the *loudness* of continuous sounds such as *equipment noise* from building services. They are standardized through *ISO* and similar to American Noise Criteria curves.

noise reduction coefficient A measure of *sound insulation* value.

nominal dimension, n. size The *dimension* used in naming something. Its actual size may be larger (full) or smaller (shy) by an allowed *tolerance*. Sawn timber usually measures less than its nominal dimension, and is further reduced, by another 3 or 4 mm, by planing or otherwise working it to its *dressed dimension*. A 100 by 50 mm nominal-size timber may actually be only 87 by 42 mm. Pipes and tubes are named by their diameter, e.g. a 50 mm outside-diameter drain pipe, and this is usually very close to their actual size.

nominated sub-contractor A company chosen by the *client*, not the *main contractor*. The system is used mainly in the U K. The work is described in a *bill of quantities*, and during *tendering* the main contractor allows for it as a *prime cost* sum, plus the cost of *attendance*. Nominated sub-contractors have often designed their work and are directly responsible to the client (for delay or *default*), not to the main contractor. This can cause problems on site.

nominated supplier A supplier selected by the *client* or his representative. Care must be taken not to infringe laws on free competition.

non-bearing wall A *non-loadbearing wall*.

non-combustibility test A test to BS 476 part 4 in which materials are heated by a small furnace at 750°C. If they do not raise the temperature more than 80°C or give off flame for more than ten seconds, they are considered to be non-combustible. Non-combustible materials include ceramics, concrete, glass, steel, and plaster. See **Class 0, combustible**.

non-concussive action The slow closing of an automatic valve or tap to prevent *water hammer*.

non-drip paint, drip-free p. A *paint* that does not drip from the brush, being not a liquid but a jelly, with *false body*. It is none the less easy to apply and flows well.

non-flammable Having a low *fire hazard*.

non-hydraulic lime *High-calcium lime*.

non-loadbearing wall, non-bearing w. A wall which carries its own weight, plus wind loads if it is *cladding*, but not the load of walls or floors above.

non-manipulative joint A joint in pipes that can be made with simple tools and little skill, requiring only that the ends of the pipe be cut off cleanly. A sleeve is inserted into a soft tube. Special *compression joints* are needed, with different types for *light-gauge copper tube* and *plastics pipe* (poly pipe), which have nuts that force *glands* or *loose rings* into close contact with the outside of the pipe to make a watertight or gas-tight joint. They can be easily unscrewed, for example to change a pipe layout.

non-mortice hinge, surface-fixed h. A *butt hinge* which when closed is only as thick as the metal in one of its *flaps*. The flaps may be cut to fit one inside the other. See diagram, p. 229.

non-performance When the *contractor* does not do the work contracted, he or she is in *default* of the contract *obligations*. This may entitle the *client* to *determination* of the contract.

non-return valve A plumbing *fitting* inserted into a water supply pipe to prevent *backflow*. The *jumper* in a domestic *tap* or in a *screw-down valve* has the same function.

non-rising spindle A tap or valve *spindle* which does not move up or down when it is turned. It moves a threaded piston up or down to control the flow.

non-setting glazing compounds Traditional *glazing materials* of many different types. Support is given by *distance pieces* and *glazing blocks*. If the thickness is at least 3 mm the core should stay flexible even if the surface hardens.

non-slip floors, treads or **nosings** *Floor finishes* that do not become slippery when wet, waxed, or greasy. Methods of making floors non-slip include: using *inlays* of abrasive cubes of carborundum; sprinkling a *granolithic* floor with abrasive powder; forming a patterned finish or using raised studs as on *studded rubber flooring*; filling joints with abrasive powder and cement, etc. Some floorings are naturally non-slip, e.g. *mosaic*.

non-splittable activity An *activity* on a *programme* that has to be fully completed before the next activity can start.

non-tainting materials Materials suitable for contact with drinking water.

non-trafficable roof A *roof* intended only for light maintenance foot traffic, not made to resist heavy wear or indenting.

non-vision glass *Obscured glass*.

non-vision grille A *grille* with overlapping metal chevrons or blades.

northlight roof, sawtooth r. A factory roof with rows of steel *lattice* (*C*) girders running across the building from one side to the other, carrying

sections of sloping roofing on triangular half-trusses from the top of one girder to the bottom of the next. The vertical sides of the girders have *patent glazing*. The roofing, which slopes parallel to the length of the building, discharges rainwater into crossways *valley gutters* under the patent glazing. They are slow and expensive to build, but give uniform, gentle natural lighting. In the southern hemisphere they are built facing south. See diagram, p. 377.

nose (1) Any blunt overhang; a *nosing*.
(2) The lower end of a *shutting stile* of a door or casement.

nosing, nose A half-round overhanging edge to a stair tread, flat roof, window sill, etc., in concrete, stone, or timber. See diagram, p. 437.

nosing line, pitch l. The slope of a stair *flight*, measured along the nosings. The *margin* of a *close string* is measured along it. See diagram, p. 437.

notch, gain A groove in timber to receive another timber.

notch board (1) A *cut string*.
(2) (USA) A *close string*.

notched joint A woodwork *joint* made by cutting out part of one or both timbers in the joint.

notched trowel A rectangular steel trowel with square teeth at regular spacings, used to apply *adhesives* or spread thin *bedding* materials evenly.

notice to bidders (USA) *Instructions to tenderers*.

notice to proceed An instruction from a *client* for the *contractor* to start work.

novelty siding *German siding*.

nylon (polyamide 11) One of the stronger *plastics*. Although it has only one tenth the strength of steel, has

nylon

high *thermal movement*, and is *combustible*, nylon is so easy to mould, so wear-resistant, corrosion-resistant, and cheap that it is used to make hardware such as barrel bolts for cupboard doors,

wood screws, and gear wheels. It can be heat-fused on to steel as a *powder coating*. *Carpets* can be made of electrically conducting, antistatic nylon fibres which refract light and so look like silk.

O

oak (Quercus robur, Q. petraea, etc.) *Hardwoods* which grow in temperate climates throughout the world. English oak is hard and has twisted grain, making it difficult to work, but for long it has been a valued structural timber, although it is not generally as stiff or as strong as *beech* or *ash*. Its *durability* is good, except for attack by *death-watch beetle*. For finishings it is decorative, particularly when *quarter-sawn*. All imported oak is straighter-grained than English oak and therefore has better *workability*.

oakum Fibres from hemp or un-twisted rope, tarred or oiled and used as *caulking*, or to make watertight the threads of a pipe joint.

obligation In a building *contract*, what-ever the *contractor* and the client or *employer* agree on, mainly to do the work or make payment for it. A full list of the obligations of each is given in the *conditions of contract*.

obscured glass, non-vision g., trans-lucent g., vision-proof g. *Glass* with a roughened surface which light can pass through, although objects cannot be distinguished. It may be sand-blasted, acid-embossed after manufacture, or patterned during cast-ing.

occupancy The number of people and the activities for which a building or room is intended. The *Building Regu-lations* give room sizes, facilities, *serv-ices*, and *fire precautions* for different types of occupancy, which require ap-proval from and inspection by the *local authority*.

ochre An *earth colour*, hydrated iron oxide used as a paint *pigment*. It is yellow or yellowish brown, paler than *umber* or *sienna*.

offcut Any small piece left over when materials are cut to size. In *bills of quantities* offcuts are *measured* as *cutting waste*.

offer The first step in forming a *con-tract*, which comes into being when there is *acceptance*. A *tender* is one type of offer. Bargaining for a lower price may be taken as the formal rejection of an offer.

offer up To place one thing against another, e.g. to check that a tenon will enter, to mark a *scribed joint*, or to drill through and insert a fixing.

off-form concrete *Direct-finish con-crete*.

offset (1) A point at a given distance from a line, used in *setting-out*, particu-larly when columns block the line of sight along the *grid* line. It is common to use two 600 mm offsets, one at each end of the building, but they should always be marked to prevent them being mistaken for the grid itself.
(2) A ledge in a wall where it changes thickness, or a step caused by *formwork* in the face of concrete. Offsets are usually not allowed on the face of *direct-finish concrete*.
(3) **swan neck** A pipe fitting forming a long flat S-shape, used to join two parallel lengths of pipe. Offsets allow *thermal movement* but should not be too short at the top of a *downpipe*, causing a rainwater outlet to tilt.
(4) A *crank*.

off-site Located outside the building *site*, e.g. a joinery workshop or a public tip.

off-tamp finish, tamped f. A rough, non-skid finish to a concrete slab for a road, car park, or loading area, made by lifting and dropping the *tamper* at close intervals after *screeding*.

off-white *Broken white* with a grey or yellow tint.

ogee joint, O G j., rebated j. A *spigot-and-socket joint* made within the wall thickness of a concrete pipe, *flue block*, etc., by matching *rebates*, one on the inside and the other on the outside, usually so that each rebate is about half the pipe wall thickness.

oil gloss paint An *oil paint* made for use as a *gloss* finishing coat.

oil length The oil/resin ratio in *oil paint*, either *long oil* or *short oil*.

oil-modified alkyd paint The most usual type of *gloss paint*.

oil paint High-quality paint containing a large percentage of *drying oil*, such as *linseed*, soya, or tung oils, and a synthetic *resin*, such as *alkyd*. It dries by the evaporation of any *solvent*, then the oxidation of the oil, which usually has *driers*. The solvent vapours from large areas of fresh indoor oil paint can be hazardous to health. Breathing apparatus may be required, or safer, waterborne *emulsion paints*. Oil paints are used for *final coats*, usually in *gloss*, *undercoats*, and some *primers*. However, if applied on alkaline fresh concrete, they may fail from *saponification*. Traditional oil paints with simple linseed oil and *white lead* driers are no longer made. Although oil paints usually contain some alkyd resin, they give a coating which is tough and durable, with a very flexible *film* more permeable than *alkyd paint*. This allows ex-terior woodwork to breathe. Thinning and brush washing is with *white spirit*.

oil paste, colours in oil A concentrated paste of *pigment* and oil used for *tinting* oil paints or for making paint by adding oil or *thinners* and *driers*.

oil stain A solution or suspension of dye or other colouring matter in oil used as a stain on wood. It may be fairly permeable to *vapour* and allow the wood to breathe.

oilstone A *hone*.

okoume *Gaboon*.

oleo-resin Mixed oil and resin from the sap of trees, e.g. pitch pine.

oleo-resinous paint Originally a long-lasting, glossy *paint* or varnish containing *oleo-resin*, today replaced by *oil-modified alkyd paints*.

oncosts *Overheads*.

one-and-a-half-brick wall A *brick-and-a-half wall*.

one-brick wall A *solid wall* that is as thick as the length of a brick, formerly 9 in. Today in Britain a one-brick wall is a 215 mm solid wall.

one-coat plaster, single-c. p. A pre-mixed *retarded hemihydrate* gypsum plaster, usually with other additives and *lightweight aggregate*, made to be applied by hand to the full coat thickness in one operation and trowelled to a smooth and hard finish. It may be as thin as 2 mm and can be used on all normal backgrounds. *Projection plaster* is also applied in a single coat.

on edge A term describing a *brick-on-edge* or a stone which is *face-bedded*.

one-hole basin A *basin* with one *tap-hole*.

one-line diagram An electrical *single-line diagram*.

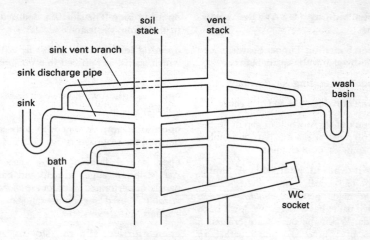

A one-pipe system of drainage.

one-part products, one-pack p. Materials that require no mixing before use, particularly building sealants, paints, and adhesives, which may also exist in *two-part* versions. One-part sealants are usually gun-grade and can be either *curing* type, which have a curing agent activated by traces of moisture from the atmosphere, or non-curing type, which set by *solvent release*. For moisture-curing types the outer layer of cured sealant may prevent further moisture entering, so that a thick mass can remain uncured in the middle. This happens more with *polysulphide* than with *silicone* or *polyurethane*. Other non-curing solvent-release sealants such as *acrylic* and *butyl* tend to remain soft and often shrink.

one-pipe system, single-p. s. (1) A system of *sanitary pipework* for taking away *soil water* and *waste water* in a combined *stack*, with a separate *vent stack* for the *branch vents*. In a modified one-pipe system some vents are omitted. See diagram.

(2) A *central-heating* circuit in which the radiators are connected 'in series',

with one pipe leading from each radiator into the next, unlike in the *two-pipe system*. The furthest radiator is therefore much cooler than the radiator on the *flow* pipe nearest to the boiler. See diagram, p. 235.

one-way switch A simple *switch* with only one circuit path, used in single-phase *low-voltage* wiring.

on/off controls Cheap and sturdy *automatic controls* that either start or stop mechanical or electrical equipment. They can be abrupt and cause more *equipment noise* than *modulated controls*, as the motors re-start often and run at full speed.

on-site Located within the site *boundary*, not *off-site*.

opacity, opaqueness *Hiding power*.

opaque glass Glass that does not let light through, used in *curtain walls*. It can get hot from the sun, as does *solar-control glazing*.

open assembly time The maximum delay between applying an *adhesive* and bringing the two parts together.

open bidding (USA) *Open tendering.*

open boarding Timber boards placed side-by-side with gaps in between.

open-cell ceiling A *cell ceiling*.

open cornice (USA) *Open eaves.*

open defect Any hole or gap in timber or ply, such as *checks*, splits, knotholes, wormholes, or open joints.

open-drained joint, d.j. A simple type of vertical *drained and ventilated joint* made between *cladding panels*. Each panel has a V-groove down both ends. When the panels are erected a pair of V-grooves come together to make a square vertical slot. A slat of durable plastics, such as *polychloroprene*, is slid down the slot to form a *baffle*. The joint allows a little *driving rain* to enter, but breaks its force so that the water falls out of a weephole in the bottom of the panel. A bead of *sealant* is used to weatherproof the interior side of the joint. Open-drained joints are used for *patent glazing* and *large-panel-system* cladding.

open eaves (USA **o. cornice**) *Eaves* without a *soffit* lining, leaving visible the *rafter* ends and underside of the roofing or roof *boarding*.

open floor A *floor* in which the *joists* are exposed underneath.

open-grain timber *Coarse-textured timber*.

open-grained timber *Wide-ringed timber*.

opening (1) A hole in a wall for a door or window, or a similar *penetration* in the structure.
(2) The clear horizontal width between wall *jambs*, available for installing a frame, usually about 6 mm wider than the frame.

opening face The side of a door *leaf* furthest from the frame.

opening leaf A leaf of a *folding door* which opens, as opposed to a *standing leaf*.

opening light A window which opens, not a *fixed light*.

open joint, gap A joint with a space between the two parts.

open metal flooring Close-spaced steel strips on edge, or strip and bar panels, or perforated sheet, or *expanded metal* (C) used for industrial stairs, gratings, duct covers, etc.

open mortice, slip m., slot m. A *mortice* in the end of a timber member, open on three edges. It can be cut with a *circular saw* and is used to make an *angle joint*.

open newel stair A *stair* without *newel posts*, e.g. a *geometric stair*.

open plan The arrangement of a large floor area which is divided only by furniture or screens, without full-height walls or partitions, mostly used in *deep-plan* offices with many work stations in rows. The *services* can be run through an *access floor*.

open rise The gap between *treads* of a *skeleton stair*, which has no *risers*.

open roof A roof in which truss members can be seen from below, since it has no *ceiling*.

open stair (USA) A *stair* which is open on one or both sides.

open string A *cut string*.

open tendering (USA **o. bidding**) *Tendering* for a building contract which is publicly advertised and a tender taken from all contractors, although *prequalification* may be required.

open valley A *valley* in which the *valley gutter* is exposed. The roof tiles, slates, or other covering overlap the valley gutter, but do not cover it.

open-vented system A pipe circuit for *central heating* or *domestic hot water*, with a feed and expansion *cistern* at the top, not a *sealed* or *unvented* system.

open-well stair A *stair* with a gap outside the *flight*, between it and the wall.

operative A worker on a building site, usually meaning a machine operator or a *labourer* rather than a *tradesman*; but see **craftsman**.

optical smoke detector A *fire detector* with a beam of light and a photoelectric cell that reacts to light scattered by the solid particles in smoke. It works best with smouldering fires, not flames.

orange peeling, orangepeel The defect of unwanted roughness in a paint finish due to insufficient *flow* during application, usual only in sprayed *lacquer*. It is caused by insufficient *solvent*, lack of coat thickness, low air pressure, etc. It can resemble deliberate *stippling*.

orbital sander A *sander* with a flat plate driven in a circular rubbing action. It is useful for *flatting down* undercoats, where any circular sanding marks are covered. Sanders often take a one-third sheet of sandpaper. See diagram, p. 346.

ordinary Portland cement, OP cement A *hydraulic cement* made entirely from *Portland cement* clinker. Portland cement can also be *blended*.

organic coatings, colour c., plastics c. High-performance finishes factory-applied to metals, often at very high temperatures. They are made of *polymers*, as in polyester and nylon *powder*

coating, or PVC *plastisols*, or stoved acrylic, or PVF_2. Steel *substrates* need previous galvanizing and passivation, and aluminium is usually *anodized*. Many organic coatings have excellent resistance to exposure, but because they are easily scratched or abraded (rather like Teflon frying pans) and are difficult to *touch up*, they must be carefully handled and protected during delivery and installation.

organic landscaping *Soft landscaping*.

organic rendering *Masonry paint*.

organosol A polyvinyl *organic coating* applied in the steelworks to the inner face of galvanized steel *profiled sheet*, about 50 *microns* thick.

oriel window An upper-storey overhanging window. Unlike the *bow window* it is carried on *corbels*.

orientation The direction in which a building, or one of its outer walls (or *elevations*), faces (north, east, etc.).

oriented strand board (OSB) A fairly strong *building board* used for structural panels, wall linings, *flooring*, etc. It is not suitable for *fascias* or *joinery*.

original ground levels The shape of a site before *groundworks*. They are recorded to work out *quantities*, or for reinstatement after excavations are closed.

O-ring joint, snap-ring j. A flexible *push-fit joint* between *drain pipes* made with a round rubber *joint ring* which is rolled into place to make a watertight seal. After pushing fully home, the pipe is pulled back a short distance to open the joint slightly and allow for movement. O-rings have many other uses, e.g. as *glands* for tap spindles. See diagram, p. 252.

outer string The *string* of a framed *stair* furthest from a wall, as opposed to the *wall string*.

outfitting (USA) *Fitting-out*.

out for tender The time in which contractors prepare their *tenders*.

outgo The pipe *tail* from a *sanitary fitting*, gully, etc., connected to the *drain*.

outgoing circuit The wiring from a *switchboard* to the use points, including its associated control gear, *protection gear*, and enclosures.

outlet (1) A *rainwater outlet*.
(2) An *air terminal unit* for supply air, or a *fire vent* to let out smoke.
(3) A connector or *socket* for gas, power, telephone, data, etc.

out of plumb Not vertical.

out of the ground The stage at which the *substructure* works in the ground are completed.

outreach arm, bracket a., carrying a. An arm that carries the *luminaire* of a *lighting column*.

outrigger A beam projecting from a building, usually wedged against a ceiling within it, to carry a *debris-collection fan* or *flying scaffold*. See also *C*.

outside glazing *External glazing* placed from outside a building.

outstanding works *Defects*.

oval-wire brad A *wire nail* formed from oval wire in lengths from 19 mm to 100 mm. It is driven with its flat side parallel to the *grain*, to reduce splitting.

oven-dried timber *Kiln-dried timber*.

oven-dry timber Timber dried in a ventilated oven at 103°C ± 2°C until it loses no more moisture.

overcladding Outside wall *external insulation*, added after construction.

overcloak In *supported sheetmetal roofing* the bottom edge and downstand of an upper sheet that laps over the *undercloak* at a *drip*, *roll*, or *seam*.

overcoating The application of additional *coats* of paint over the *primer* or *undercoat* to form the full *paint system*. Overcoating is done after the *recoating time*.

overcurrent, excess current Electrical current that is greater than the maximum a circuit is intended to carry because of a *short circuit* or other overload. Damage from overcurrent is prevented by *protection gear*.

over-fascia vent An *eaves vent* along the top of a *fascia*.

overflow An outlet from the highest water level of a storage tank, cistern, bath, *box gutter*, etc. The overflow is connected to a *waste* or to a *warning pipe*.

overflowing rainwater gutters Blocked roof *gutters* are probably the commonest cause of *dampness* in walls and the *fabric* of buildings. To prevent blockages, leaves should be cleaned out every autumn after the leaf fall. Flat roofs and *box gutters* should have a *scupper* to make them fail safe.

overhand work Facing bricks laid from inside a building by a bricklayer standing on the floor or a scaffold inside. Overhand laying does not need an exterior scaffolding, but the bricklayer cannot see properly and needs to be highly skilled.

overhang A projection of a roof edge (eave, verge), or floor *nosing*, or other horizontal part beyond the wall which carries it.

overhanging eaves Rafters that project beyond the outside face of a wall. They give more protection against rain or heat from the sun than do *flush eaves*.

overhead door An *up-and-over door*.

overheads, administrative charges, establishment c., oncosts. The costs of electric light, roads, supervision, accounting, director's fees, etc., which cannot fairly be charged to one job and must therefore be distributed as a percentage over the *flat cost* of all *items* in a contract. See *preliminaries*.

overlap A *lap*.

overlay A *blanket* of thermal *insulation* spread over the top of a ceiling.

overpainting Either *repainting*, or painting over a material such as *preservative* or *sealant*, which must be *compatible* with the paint used.

overpanel An opaque panel above a door, often replacing a *fanlight*.

override A switch to allow manual operation of an automatic control.

oversail To overhang, as does brick or stone *corbelling*.

oversite work Work to the full area of a building *site*, or the *solum*, such as excavations for basements, *oversite concrete*, a *damp-proof membrane*, etc.

oversite concrete A layer of *dry lean concrete* about 100 mm thick, with a trowel or spade finish, required by the *Building Regulations* to seal the earth under the *ground floor* of a house, particularly if the house is made of timber. See **blinding concrete, solum**.

oversize Larger than *nominal dimension*.

overslung hanger A *hanger* which is at least partly above its support, often with a spring.

overspray Paint and solvent from *spray painting* that does not land on the work. Surrounding surfaces may need *masking* against overspray, although it is often blown away and wasted.

over-tile The *imbrex* of *Italian* or *Spanish tiling*.

overtime Additional wages paid for time worked above the normal number of hours, first at *time and a half*, then at *double time*.

overvibration The *vibration* of fresh concrete that is continued for longer than the time required for full compaction. It can cause *segregation*.

overvoltage, excess voltage Abnormally high voltage in power mains, from lightning or other surges. Damage to equipment from overvoltage is prevented by *protection gear*, which for computers may need ultra-fast-acting electronic devices.

owner, building o., client The owner of a proposed building and usually also of the land under it. During construction the owner gives site *possession* to the builder. In contracts the owner is usually called the *employer*.

oxidation The *curing* of paint *drying oils* as they absorb oxygen from the air, which hardens them to form a durable *film*. See **solvent release**.

oxter piece An ashler piece, the vertical in *ashlering*.

oxy-acetylene flame A flame obtained from a torch that mixes oxygen and acetylene supplied through high-pressure hoses from steel storage cylinders. The hottest flame in common use, it can melt steel for welding or *oxy-cutting*, as well as other metals for *soldering* or *brazing*.

oxy-cutting The cutting of steel with

oxy-cutting

a blowpipe. First the steel is melted using an *oxy-acetylene flame*, then a lever is pressed to supply a large flow of oxygen, which reacts with the hot steel, producing even more heat. As the burning steel melts it is blown away.

P

package deal A *design-build contract*.

packaged plant Mechanical plant made up in a factory as a *module* containing all equipment in the one housing, delivered to site as a *unit* and there connected to pipes, ducts, wiring, etc.

packer A small piece of hard material, placed under something or between two things, to separate them or to bring a top surface to a required level; it can be of timber, plastics, or metal. A metal packer is thicker than a *shim* (*C*).

packing (1) A strip or sheet *packer*, or a timber *firring*.
(2) **stuffing** Tarred hemp, or *polytetrafluorethylene* tape, used to seal a *packing gland*.

packing gland A seal round the stem of a tap. It is filled with *packing* and made watertight by screwing down the *gland nut*

packing nut A *gland nut*.

pad (1) **padstone, cushion** A block or stone placed under a heavy load to spread the load over the wall below.
(2) A flexible or *anti-vibration mounting*.
(3) A *tool pad*, usable as a *padsaw*.
(4) A *fixing fillet*.

padbolt A *barrel bolt* that can be padlocked.

pad footing, base An *isolated footing*, usually square and of reinforced concrete, cast directly in an *excavation*. It may have *starters* for columns, ground beams, or basement walls, or *holding-down bolts* for steelwork.

padsaw A *saw* blade in a *tool pad*.

paint (1) A coloured opaque liquid applied to building materials in coats which dry hard within a few hours. Paints consist of *pigment* and a *medium*. The most used building paints are *emulsion* or *alkyd* based, with different types for *interior* and *exterior* use, and for *primer, undercoat*, and *final coat*. Paints have different *gloss* and vary in physical and chemical properties according to whether they are applied to wood, metal, or masonry *substrates*, with different *spreading rates* and *hiding power*. With the development of *self-finished* products, often with excellent *durability*, paint is less often applied on site than it was in the past. Many paints need to be stirred before use; a test for sediment on the bottom of an unopened can may be made by turning it over and tapping lightly, then doing the same with the top. A duller sound may indicate sediment.
(2) *Paintwork*.

painter, house p. A craftsman who does the work of *painting* buildings. See diagram, p. 314.

painter's caulk A fluid *filler* supplied in soft plastics tubes about 200 mm long and 40 mm diameter. It is expelled by cutting off the top corner and turning a wire handle from the bottom, like a *caulking gun*. It is easy to use, but some types have considerable shrinkage.

painter's labourer A skilled labourer who helps a *painter* by preparing surfaces, stripping old wallpaper, washing ceilings, masking, brushing, etc.

stripping knife

stopping knife

chisel knife

two-knot distemper brush

flat brush for varnish or paint

palette knife

glazier's hacking knife

lining fitch or lining tool

sable writer or pencil

ground brush

shave hooks

Painter's tools.

painter's putty Glazier's *putty* used as a *filler*.

painting The skilled *trade* of putting *paint* on a surface to make a *finishing*, for decoration or protection, or both. First the *substrate* is given surface *preparation*. This is followed by the *application* of the coats, with time for *drying* between each coat, then *sanding* and recoating until the complete *paint system* is built up to the required *dry-film thickness*. The painting of woodwork is discussed in BRE Digest 354 and the painting of buildings in BS 6150.

paint removal Old paintwork suffering from poor adhesion, flaking, peeling, blistering, cracking, crazing, severe chalking, or powdering is partly or completely removed before *repainting*. Methods include *hot-air stripping, sanding*, the use of a *scraper* or *paint remover*, and *blast cleaning*. Care should be taken with *dust*, and special precautions are needed when removing *lead paints*.

paint remover, p. stripper A liquid which softens a paint or varnish film for easy removal. Paint removers are of two types: (1) Alkaline solutions, made up from caustic soda or bought as sugar soap, used mainly for *oil paint*. They may darken some hardwoods and attack anodizing. Since they injure the skin and eyes, care is needed and rubber gloves and goggles should be worn. (2) Those containing *solvents*, which are less danger to the substrate and the hands. Several applications may be needed and time allowed for them to soak in. Both types require thorough rinsing and cleaning of cracks, to remove all residues before re-painting.

paint scraper A *scraper*.

paint stripping *Paint removal*.

paint system A succession of *coats* which build up a total *film* thickness adequate to protect a surface and give a decorative finish. It is preceded by the *preparation* of the *substrate*, which may include the use of a *pre-treatment primer*. Paint systems include the *primer*, *undercoats*, and *final coats*, which must be *compatible*. Each play their individual part in the overall result. The usual paint system for wood is the *four-coat system*.

paintwork (1) A *coating* of *paint* on a surface, usually applied on site at normal temperatures, unlike *organic coatings* or *stoving*. Paintwork *defects* can be due to poor *workmanship*, but

also to product unsuitability or lack of *compatibility*.
(2) The job of *painting*.

pair (1) Two oppositely *handed* but otherwise similar objects.
(2) Two wires for a telecom line.

pales, paling Upright wooden or metal stakes or boards in a *palisade*.

palisade *Fencing* of wooden or metal *pales* driven into the ground.

pallet A thin *slip* of wood used as a *fixing fillet*. See also *C*.

pan (1) A shallow tray for catching liquids such as *condensate*.
(2) **bowl** The main part of a *water closet*, a deep funnel with water in the bottom leading to a trapped outgo.

pan connector, sanitary c., WC c. A flexible coupling that fits on the *spigot* outlet of a *water closet* pan to make a joint with the *soil pipe*. It is usually made of plastics, with rubber sealing rings. There are different types for *P-traps* and *S-traps*.

pane A sheet of glass or plastics cut to *glazing size*. A 'full-pane' window has no *glazing bars*. Traditional *sashes* may have 'small panes' or 'squares'.

panel (1) An infilling of wood, glass, or other sheet material let into grooves or rebates in the members of a timber *frame*, leaving the panel and frame free to move. See **panelled door**.
(2) Brickwork or blockwork infilling in a *panel wall*.
(3) A prefabricated *cladding panel* for a *curtain wall*.
(4) An electrical *switchboard* or *distribution board*.
(5) An access *trap*.
(6) A single span or *bay* of a concrete slab, a truss, etc.

panel form A small standard-size *shutter*, joined with other similar panels to

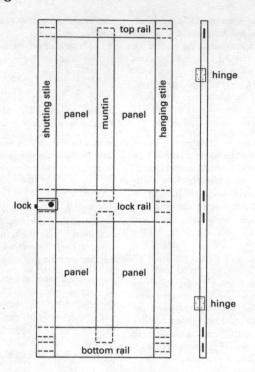

A panelled door. (*Right*) Edge view showing seen tenons. Mouldings round the panel edges are not shown.

make up a large piece of modular *form-work*.

panel heating (1) Unobtrusive, safe electric radiant *heating* by flat panels with a surface temperature of about 38°C flush with the wall, or by pressed-steel *radiators* about 50 mm clear of the wall at a higher temperature.

(2) *Coil heating*.

panelled door, panel d. A timber *door leaf* with a visible *leaf frame*. The framing members (*stiles, rails, muntins*) have grooved edges to hold the tongued edges of *panels* of thinner material.

Traditional timber panelled doors had panels of flat timber, with a thin tongue shaped to match *mouldings* on the framing members. They were the best doors in existence before *plywood* was made, and wereo built to allow *moisture movement* without stopping the door working properly. However, they have low *fire resistance* owing to the poor *integrity* of the thin tongue. They may be difficult to upgrade and still retain their attractive appearance, which is often imitated on *flush doors*. Today the panels are of plywood, or they can be glazed. See BS 4787. See also diagram.

panelling A decorative *finishing* arranged in *panels* that are usually identical and prefabricated.

panel pin A thin round wire *nail* between 1 and 1.6 mm thick and from 10 to 50 mm long. It has a small head, and in *joinery* may be nearly invisible when driven below a surface.

panel planer A *thicknessing machine.*

panel products Large panels for lining walls and ceilings, or for flooring. They include *building boards, strawboard, woodwool slab* and *oriented strand board.* See **wood-based panel products.**

panel saw A fine-toothed handsaw about 560 mm long. The teeth are at about 2.5 mm spacing (10 points per inch) and are sharpened and set for a *cross-cut.* See diagram, p. 388.

panel wall A *non-loadbearing wall* between the columns in *timber-frame construction,* usually in brickwork or blockwork. Carried by the floor and restrained by *wall ties,* it has a *movement joint* along its top edge.

pan form, waffle form A *coffer* used as *formwork* for a *waffle slab.*

panic bolt *Panic hardware* to operate the *Cremona bolt* of a double door.

panic hardware *Door hardware* for opening an *exit door* from the inside. It has a push bar across the inside face at waist height, which releases the door latch or bolt and is easily worked in darkness or by a pressing crowd. Studies on the behaviour of people in fire show that 'panic' is a misused term.

panic latch *Panic hardware* to operate the *latch* of a single door.

pantile A *single-lap* interlocking roof tile shaped like a shallow S made of clay or concrete. The standard British small clay pantile measures 355 × 240

mm. It has a *sidelap* of 40 mm and a *headlap* of 65 mm. About 200 tiles are needed per 10 m² and the weight of the pantiles is about 50 kg/m². The minimum roof *pitch* is 22½°. Pantiles generally have a nail hole and a *nib* at the head. Larger, heavier pantiles, up to 469 × 276 mm, exist, but they weigh less per 10 m² than the smaller ones and have larger laps.

pan wash sink (1) A kitchen *sanitary fitting* for washing cooking pans. It is not needed for *Gastronorm pans.*
(2) A *bedpan sink.*

parallel coping A *coping,* not *weathered* but of uniform thickness, for covering the sloping top surface of a *gable* parapet.

parallel gutter A parallel-sided *box gutter* behind a *parapet.*

parallel thread A screw *thread* of uniform diameter, such as on bolts and pipe *connectors.* Pipe *unions,* however, have *taper threads.*

parallel tread A stair *flier.*

parapet A low wall standing above a roof, on top of either an external wall or a separating *fire wall.* Since it is exposed on its face, back, and top to the weather, a parapet needs either frost-resistant bricks or a wide *coping* and a good *damp-proof course.* At roof level the damp-proof course should be sanded on both sides to give a good grip on the mortar and *movement joints* provided every 6 m. To avoid *sulphate attack* low-sulphate bricks should be used wherever possible, preferably laid in sulphate-resisting cement mortar. Similar recommendations apply to free-standing walls.

parapet gutter A parallel or tapered *box gutter* behind a *parapet.*

parenchyma The soft and weak wood tissue of the *rays* of timber, seen in the *silver grain* of oak; also the main part of *heart centre*.

parge A mixture of *coarse stuff* with *hair* and cow dung, used for *pargeting* flues. Today, cement mortar is used.

pargeting, parging (1) *Rendering* the inside of a brick *flue*, superseded by *flue blocks* and *linings*.
(2) Decorative plastering with lime plaster on the outside walls of seventeenth-century houses. Repetitive patterns, sometimes very beautiful, modelled in the lime plaster before it had hardened, can be seen in East Anglia.

paring chisel A long, thin, beveledged wood *chisel*, used only by hand, not struck with a *mallet*.

Paris white *Whiting*; chalk powder.

parliament hinge, H-hinge, shutter h. A *hinge* with two lengthened T-shaped *flaps*, joined to form an H. The knuckle (at the middle of the cross-bar of the H) projects beyond the face of the closed door or shutter, allowing the door to clear *architraves* and lie flat against the wall when opened. It is used for outside shutters. See diagram, p. 229.

parquet composite, laminated overlay Factory-made panels of *parquet floor* 19 to 25 mm thick, tongued and grooved on four edges. They can be nailed directly to *floor joists* or glued to flooring grade *chipboard* with a cushion underlay as a *floating floor*.

parquet floor Decorative *wood-block flooring* of *hardwood* from different types of tree, arranged in geometrical patterns. Traditional parquet in panels is *secret-nailed* to thin battens on a wood *sub-floor* and polished after laying. Parquet is also available as *plywood parquet*, *parquet composite*, and *parquet strip*.

parquet strip, overlay flooring A *parquet floor* consisting of tongued and grooved *hardwood* boards 9.5 mm thick, in random lengths which are *secret-nailed* and glued to a wooden *sub-floor*.

particle board Wood-based *building boards* (or panel products), such as chipboard, oriented strand board, or fibreboard. European Standard tests for *formaldehyde* in particle boards are listed in EN 120.

partition A wall between rooms, nonloadbearing and generally one storey high. Partitions can be built in an enormous variety of ways, the commonest being plastered brick or *blockwork*, but *dry construction* is possible, which is usual for *demountable partitions*. They can also be *framed* or made of *woodwool* or *strawboard*, with various *coverings* and *infillings*. They are mainly used in office buildings to make group or cellular workspaces. See BS 5234.

partition block A *hollow clay block*.

party floor A *separating floor*.

party wall A *separating wall*.

pass door A *wicket door*.

passing The *lap* of one layer of *supported sheetmetal roofing* over the next in *flashings*, ridge coverings, *gutters*, etc.

passivation, passivating (1) A *pretreatment* of metals that improves the resistance to corrosion of the *paint system* or of *organic coating*.
(2) The treatment of paints to prevent their oils becoming acidic.

passive fire protection The construction of a building so that its *structure* has adequate *fire resistance*; the inside walls and floors form *compartments*, and the outside walls and roof prevent fire spreading to or from other buildings.

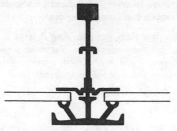

Extruded aluminium glazing
bar with lead wings

(Pillar Patent Glazing)

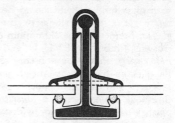

Glazing bar with galvanized
steel core and pvc cap

(Crittall)

Patent glazing.

passive solar heating Solar heating obtained by manipulating building *elements* rather than by using the pumps or fans of *active solar* systems. The orientation of buildings to face the sun (within 30°) and the minimizing of overshadowing can give solar benefits, usually at no additional cost. Extra *insulation* to walls, in the *ceiling* or roof, and under the ground floor, or the more expensive *double glazing*, can save on heating costs. Correct *ventilation* is still needed to control *condensation*, although *weatherstripping* may improve comfort. *Solar collectors* include using the building *fabric* as a heat store, as in the *Trombe wall*.

patching Minor repairs to a concrete surface after removing *formwork* in order to fill *blowholes* or *honeycombing*. Since patching with *mortar* tends to look darker unless *silver sand* is used, and the texture is usually different, it may not be permitted for *direct-finish concrete*.

patent axe A *comb hammer* with four to ten thin toothed blades, used for *scabbling* concrete.

patent glazing Glass or similar materi-als used as translucent external *cladding*, usually for industrial buildings such as factories and warehouses. Most systems have strips of glass supported on two edges by metal *glazing bars* up to several metres long, which are set at 600 mm spacings. The glazing bars have wings and a clip-on capping with *cushions* for the sides of each strip of glass, or can be made for *gasket glazing*. Joints between glazing bars and glass are not sealed, and the glazing bars have hidden grooves and channels to collect and carry away rainwater. The overall effect is less decorative than a *curtain wall* but suited to large areas of external walling, sloping roofs, or *northlights*. Makers of patent glazing usually supply and fix their own work. See BS 5516. See diagram.

patina A thin, stable, adherent film which forms on some metals exposed to the weather, particularly the green coating on copper and copper alloys (verdigris) or the dark grey on lead. It protects against corrosion and usually takes many years to form naturally, but can be made artificially more quickly by pre-patination solutions.

patio door An external *sliding door* with glass infill (which should be of *safety glass*), either single or *double glazed*. Frames can be aluminium extrusions, timber, or PVC-U, and all *ironmongery* should be non-corrosive. Patio doors have an overhead guide rail and run on floor tracks, with threshold drainage and *weatherstripping*, enabling them to withstand severe *exposure* to wind and weather.

patterned glass An *obscured glass* with a decorative pattern on one surface, formed as the glass is rolled.

pattern staining The discoloration of plasterwork caused by different heat conductances of backings. Where *plasterboard* is fixed over ceiling *joists* the part in contact with the joists becomes darker than the part with no backing. This part has a higher temperature difference from the air in the room than the surrounding plaster. The air therefore circulates over it more freely and drops more dust on to it. It is different from *core patterning*.

pavement lens, p. prism A *glass block* fitted in a *pavement light*.

pavement light, vault l. A window formed of solid *glass blocks* cast into a concrete slab in the pavement over a basement.

paver, paving brick, pavior A type of *brick* made for outdoor *hard surfaces*, either brick-shaped or *interlocking*. Pavers should be hard-wearing, non-slip, and frost-resistant. Clay pavers can be type PA, for pedestrian areas and car parks, or the stronger type PB, for heavy vehicle areas, but clay *commons* are usually too soft. Concrete pavers are suitable, as are some *calcium silicate bricks*. See BS 6677.

paving An outdoor *hard surface* of *engineering bricks*, *flags*, *floor tiles*, *pavers*, *quarries*, *radial setts*, *slabs*, etc., usually laid on sand *bedding*.

paving slab, precast flag A large slab of precast concrete or *cast stone* sometimes made with hard granite chippings, suitable for pedestrian areas.

pavior A skilled worker who lays *paving*.

peagravel Coarse rounded gravel mainly 6 to 8 mm size, used as-dug for *bedding* and as a *surround* to buried drain pipes.

pebbledash A *dry-dash finish* to the outside of a wall in which clean washed natural (or artificial) pebbles are pushed into the wet *render* and left exposed.

pedestal basin A *bathroom basin* on top of a pedestal which stands on the floor. The pedestal conceals the hot and cold water pipes, and the *trap* and *waste branch*, which is tidy but makes maintenance awkward. The basin and pedestal can be made as one unit or two matching units bolted together.

pedestal WC A *water closet* with a *pan* and pedestal made as one unit.

peelable, strippable Made to come off easily, such as a protective *film* or finishings, e.g. *wallpaper* and some fully stuck-down floor coverings.

peeling (1) The defect of dislodgement of paint or plaster from a surface due to lack of adhesion or a weak backing. (2) The *rotary cutting* of veneer.

peen A blunt wedge or ball-shaped end to a *hammer* head, opposite the striking *face*. Wedge peens can be *cross*, straight, or *claw*.

peen hammer A *hammer* with one or two *peens*.

peening The working of metals by striking with a *peen hammer*.

peg (1) **tile p.** An *oak* or galvanized-steel rod 6 mm dia., or a short fat nail, passed through the hole in a *peg tile* to hold it in place.
(2) A *dowel*.
(3) A metal pin which secures glass to a metal window frame.

peg stay A *casement stay* which holds a window in place by means of a peg through one of its holes.

peg tile A small plain clay roof tile with a hole near the top for a *peg*, instead of a *nib*.

pellet, stud A circular disc of wood, cut to match the grain of wood into which it is inlaid, to cover the head of a *countersunk* screw.

pelmet A board or very short curtain along the head of a window, hiding the curtain rail, blind fittings, etc.

penalty In the UK, a sum of money greater than the real loss. For late *completion* of work a penalty is illegal, not being fair *damages for delay*.

penalty clause (USA) See **damages for delay** and preceding entry.

pencil arris, p. round An *arris* rounded to a radius of about 3 mm.

pencil bar A *reinforcement bar* of 6 or 8 mm diameter.

pencilling Painting the mortar joints of brickwork with white paint so that the joints contrast with the brickwork.

pendant fitting, droplight A *luminaire* hung from the ceiling that distributes light in all directions, although many types cast a shadow above. Its cable down from the *ceiling outlet* should be heat-resisting if high-temperature *incandescent lamps* are used.

penetration (1) A hole made in the structure or envelope of a building (walls, floors, roof), either formed as work proceeds or cut out later. The *trade* requiring the penetration must give details to the main contractor well in advance of the work, to allow *coordination* between trades and the forming of the penetration as work proceeds, or an *extra* may be charged for cutting-out afterwards, which is more expensive.
(2) **water p.** The entry of *driving rain* through a wall or roof, or of damp by *capillary action*.

penthouse A building on top of a flat roof, with walking space round it, such as an apartment or a *lift machine room*.

penthouse roof A roof sloping in one direction only. It covers the top wall, and thus differs from a *lean-to roof*.

perforated brick A *wirecut* clay brick, defined by BRE Digest 273 as one having any hole right through between the two bed faces. BS 3921 limits the total volume of holes to less than 25% of the brick's volume, or any one hole to 10%, and the solid material in any cross-section to at least 30%. Perforated bricks have higher thermal *insulation* than solid bricks and equal resistance to the penetration of *driving rain* if properly laid. They require fairly careful handling and although easily cut often break in the wrong place. Partly completed brickwork should be covered overnight to prevent rain filling the holes in the bricks. See diagram, p. 48.

performance A formal term used in *conditions of contract* for doing building work, the builder's main contract *obligation*.

performance bond, completion b. A surety put up by the *contractor*, usually in the form of a *bank guarantee*, paid to the *client* if there is *default* in the performance of the contract.

performance specification A *specification* which describes materials to be used and results to be achieved, unlike the more common *method specification*.

perimeter angle Trim round the edge of a *suspended ceiling*, where it meets the wall, to conceal the edges of the tiles, panels, etc.

perimeter beam One of a row of beams along the top of an outside wall of a *structural steelwork* frame, between the heads of the columns.

perimeter diffuser In *warm-air heating*, an air outlet along the inside of an outer wall, for example under the windows, to neutralize the downdraught.

perlite A volcanic glass found in the USA, heated to make *expanded perlite*.

perlite plaster Gypsum plaster which contains only *expanded perlite* aggregate, no sand. It is a good insulator and is easy for the plasterer and labourer to work because it is light in weight.

permanent formwork, p. shuttering (USA **sacrificial form, absorptive f.**) *Formwork* left in place after concrete has set. It is used in *void formers* (C) or any enclosed space from which it would not be worth recovering. Permanent formwork for slab soffits includes *permanent metal decking formwork*, or *composite decking*. Other examples are *expanded metal* (C) mesh to form *construction joints*, and *woodwool slabs*. Permanent formwork should always be of rotproof materials.

permanent metal decking formwork, non-participating d. Galvanized-steel troughed sheets used as *formwork* for a slab *soffit*, which does not bond to the concrete. It may need some *props* to support the weight of the fresh concrete and loads from construction. Compare **composite decking**.

permeability The rate of movement of a liquid or gas through a solid. The permeability of wood affects its rate of drying, the amount of preservative it can absorb, and any paint put on it. In *hardwoods*, heartwood is usually denser and less permeable than sapwood. Complete *paint systems* may be chosen for high permeability, to allow masonry to dry out, or timber to 'breathe'. Cladding systems, including windows, are usually made to have low air permeability. Building *sealants* vary in their rates of water-vapour permeability. The very low rates of *damp-proof courses* or *vapour barriers* are measured in meganewton.seconds/gram (MNs/g) of water. Gases with small molecules (e.g. methane) can pass through most plastics, but not through *aluminium foil*. Others with large molecules (e.g. *radon*) are more easily stopped.

permissible deviation *Tolerance*.

permit See **building permit**.

permit to work Written permission to do hazardous work, to control access to a building site or part of it, or to services (e.g. telecom rooms). Permits should state who issued them and their position; the work involved and the person responsible for it; dates and times; associated risks; safety precautions and protective clothing needed. The permit should record the completion of the work, with the return of keys, badges, etc. See **hot-work permit**.

perpend (1) **p. joint** (USA **head joint**), **cross-j.** The vertical mortar *joint* between bricks or blocks in the same *course* of brickwork or blockwork, which shows as an upright *face joint*. Perpends are the main weakness

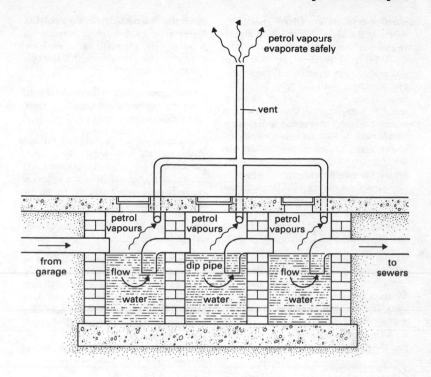

petrol vapours evaporate safely

vent

petrol vapours

petrol vapours

petrol vapours

from garage

flow

dip pipe

flow

to sewers

water

water

water

A petrol interceptor.

in *brickwork*, being more difficult to fill and usually weaker than *bed joints*, which are compacted by the weight of courses above. Good workmanship includes accurately *keeping the perpends* and making sure that *buttering* gives a joint well-filled with mortar, to resist penetration by *driving rain*.

(2) The corner of a brick wall erected first and carefully plumbed to serve as a guide for the wall between. This requires *toothing* or *racking back*.

(3) The sloping (so-called vertical) joint between adjacent slates or tiles.

(4) (USA) A *perpend stone*.

perpend stone, parpend A *bonder* which passes through a wall and is seen on both faces. (French 'parpaing', block.)

personnel door A *wicket door*.

Perspex, Plexiglas Proprietary brands of clear *acrylic sheet*.

PERT (Project Evaluation and Review Technique) An early type of *network analysis* developed for the 1957–8 US Polaris missile programme.

pet cock An *air-release valve*.

petrol bend A cast-iron *dip pipe* for a *petrol interceptor*.

petrol interceptor, p. intercepting trap A *trap* through which the surface water from a car park or filling station must pass to prevent petrol (gasoline) entering the *sewer* (*C*), where it could release explosive *petroleum vapours* (*C*). It is built as required by the *local authority*. Large types have three separate chambers, with *dip pipes* (or baffle boards) which allow water to pass but not the petrol or oil floating on the water. Each chamber is 900 mm square inside, can hold a 914 mm depth of water, and has a *vent* taken to a point well above human reach, through which the volatile and explosive part of any petrol can evaporate. The chamber walls can be in *engineering bricks*, which are not *rendered*, or the complete petrol interceptor can be of precast concrete. See BS 8301. See also diagram, p. 323.

phase (1) Part of a large project undertaken as a separate *contract*. The work done in Phase 1 is followed later by Phase 2, etc.

(2) The *live* wire in single-phase *low-voltage* distribution, or the red, yellow, or blue wire in *three-phase supply*, but not the *neutral*.

phenol-formaldehyde (pf) The most common *phenolic resin*, dark brown in colour, used in adhesives for plywood and particle board.

phenolic foam Foamed *phenolic resin* insulating boards, weighing only about 48 kg/m³, which have the advantage over many other *expanded plastics* foam boards that they burn only with considerable difficulty, with very little smoke or toxic gas, and are usable at temperatures up to 150°C. The boards are not strong enough to carry the feet of a ladder, but covered with quarry tiles their performance is greatly improved.

phenolic resins Synthetic resins from which solid high-pressure moulded *plastics* are made, originally *bakelite*, but now also as *phenolic foam* and *film glue*, or as liquid products such as *weatherproof adhesives* and *phenolic varnish*.

phenolic varnish, spar v., yacht v. An exterior gloss *varnish* based on a *phenolic resin*. Phenolic varnishes have good durability for a varnish, although less than paint.

phenol-resorcinol formaldehyde (prf) An assembly *adhesive* for severe environments.

Phillips head screw An early version of the *recessed-head screw*, still in use, particularly for *self-tapping screws*. Its main disadvantage is that the recess is easily burred, after which the screw cannot be turned.

phon A unit of *loudness* no longer used in building acoustics.

phosphating The protection of a metal surface with hot phosphoric acid, forming a thin, adherent film of iron phosphate. Like *pickling* (*C*) it is a *pretreatment* and *corrosion-inhibiting* coat rather than a finished surface. Phosphating should therefore be followed by *overcoating* so that it has the protection of a full *paint system*.

phosphogypsum A by-product *gypsum* from phosphate making, unsuitable for use in building as it gives off *radon* gas.

photographs A usual way of recording progress, photographs of building work may be required each month by the *client*. They are often also used by the contractor as evidence in support of *variations* or to avoid disputes.

piano hinge (USA **continuous h.**) A long *butt hinge* made in lengths which can be cut as required.

piano wire Thin, strong steel wire, used for *setting-out* or *plumbing* over long distances. It is less affected by the wind than a string *line*.

pick A digging tool with a sharp point,

used for breaking up loose ground or digging stiff gravel or clay. It can be a *navvy pick*.

pick and ⌣ip, New England method (USA) See **shoved joints**.

pick axe A *navvy pick*.

picked stock facings *Facework* in selected and blended *stock bricks*.

pick hammer A slater's tool used for holing *slates* and driving or drawing nails. See diagram, p. 332.

picking, stugging (Scotland **clouring**) Surfacing a stone in *rubble walling* with a steel point struck at right angles to the surface to make many small closely spaced pits. Picking can be rough or fair.

picking up Joining *live edges* of paint.

pictogram A simple diagram used as a direction sign, e.g. the figure of a man or a woman on a toilet door.

piecing-in A repair to a damaged surface made by cutting a large regular hole and fitting in a matching piece made to exact size.

piend (Scotland) A *hip*.

pier (1) The loadbearing brickwork in a wall between openings.
(2) A short *buttress* (*C*) on one or both sides of a wall, bonded to it for stability. It can be a *pilaster*.

pig lug A *dog ear*.

pigment An insoluble, finely ground coloured powder, the *stainer* or main colouring agent in *paint*. It may have *hiding power* and rust-*inhibiting* qualities, as well as colouring concrete, *plastics*, etc. Paint pigments are generally ground in a *medium*, such as oil. They are so fine that only ¼% is allowed to be larger than 0.06 mm, giving a smooth paint with high *spreading rate*. Their water *absorption* is below ½%.

Most pigments are metal oxides and except for lakes have better *hiding power* than extenders. White pigments are not considered to have *colour*, black pigments are. Pigments are therefore classified as white or coloured.

pilaster A rectangular *pier*, sometimes fluted, projecting from the face of a wall, having a cap, shaft, and base. It is similar to an *attached column*.

pile The fibres that project from a woven *carpet* and some *needle-punch* carpets. For woven carpets they are either *looped* or *cut*. See also *C*.

piling The *defect* of thick and uneven paintwork, caused by the lack of flow of quick-drying paints that become sticky and cling to the brush.

pillar A *column* or *pier* of stone or cast iron.

pillar tap A water tap with a vertical body fed upwards from below through a threaded *tail* that passes through a *taphole* in the edge of a basin or bath. It can have a high or low neck.

pilot hole A small-diameter hole pre-bored before a larger hole is drilled, or into which a screw or nail is driven without splitting the wood.

pilot nail A temporary nail, partly driven to hold timber while the main nails are being driven.

pin (1) A slender wire nail, such as a *panel pin*.
(2) The pin of a *loose-pin butt*, or a *pintle*.
(3) A wooden *dowel*, peg, or *treenail*.
(4) The electrical contact of a *fluorescent tube*.
(5) See **pinning up**.

pinch bar, case opener, claw b., jemmy, wrecking b. A steel hexagonal or round bar 13 to 19 mm dia., about 300 mm to 1 m long, with one end bent and flattened to a rough chisel shape and the other end hooked and split to a *claw* shape. It is light enough to use in one hand, unlike a *crowbar* (*C*).

pinch rod A rod used like a *storey rod* for checking the width of a gap, such as a door or window opening or the gap between floor slabs.

pine (Pinus) Many softwoods are called pines. The only one mentioned in this book is *redwood*.

pin hinge A *loose-pin butt*.

pinhole A *worm hole* of about 1.5 mm dia., without bore dust and usually dark stained, made by a *furniture beetle*. Pinholes are an allowable *defect* in timber to be painted, provided they are filled with *stopping*, but not for clear finishes like *varnish*.

pinholing Tiny holes in a dried *film* of paint or varnish, a *defect* arising from moisture or air being trapped during spraying or brushing, or from lack of *compatibility*.

pink primer A *wood primer*, originally containing mainly white and red lead *pigments*, but now of very vague composition.

pinning (1) Fixing joinery by nailing with *panel pins*.
(2) Securing a woodwork *joint* by driving in *dowels*.

pinnings (Scotland) Stones of different colour and texture set in a *rubble wall* to give a chequered effect.

pinning up Ramming *dry-pack mortar* into the gap between *underpinning* and an existing foundation.

pintle The fixed pin of a *lift-off hinge*, over which the eye is passed. The pintle is usually attached to the *jamb* and points upwards.

pin tumbler lock A *cylinder lock* with spring-loaded pins and drivers over the keyway, which is a long radial slot in the cylinder plug. Pin tumbler locks have good security, with a practically infinite number of *differs*.

pipe Pipes for most purposes are in plastics or metal. Drain pipes, *overflows*, and *vents* are in *vitrified clay*, concrete, plastics, and many other materials. Pipes are made in lengths to standard diameters and with plain ends or *spigots and sockets*.

pipe bracket A support for a pipe, fixed to a wall or ceiling.

pipe clip A *two-piece cleat*, or a *saddle clip*.

pipe closer A *fire-stop sleeve*.

pipe cutter A tool for cutting metal pipes. It has a hard-metal cutting disc which forms a V-groove as the tool is passed round and round the pipe and tightened on to it by two rollers. Some of the metal is forced inwards as the pipe is cut, reducing the pipe diameter. The pipe bore is recovered with a *burring reamer*.

pipe duct A *cable duct*.

pipe fitter A *tradesman* who installs *pipework* for water, steam, gas, oil, etc.

pipe fitting A *fitting* for joining pipes, making corners and connecting branches. Pipe fittings include *bends*, *couplings*, *elbows*, *access covers*, etc. Many are *tubulars* and each pipe material has a different type of *joint*. See diagram, p. 177.

pipe flashing A roof *flashing* at the point where a pipe or vent passes through the *roofing*, to make a weathertight joint which allows *thermal movement*, often made of synthetic *rubber*, e.g. *ethylene propylene diene rubber*. See **ridge terminal**.

pipe freezing There are several methods of temporarily freezing a central heating or water supply pipe, enabling the pipe to be opened without draining it. One of these uses liquid nitrogen at − 196°C, others use liquid carbon dioxide, which changes into the solid 'dry ice' as soon as it leaves the cylinder. A closely fitting jacket round the pipe, injected with the cold gas, enables pipes up to 100 mm dia. to be frozen, the smaller diameters usually within fifteen minutes. The jacket must have a vent hose to carry excess gas outdoors. The main danger is of escaping cold gas collecting in the bottom of a confined space, so that the workers involved cannot breathe.

pipe hook A spike *fastener* driven into a brickwork *joint* or a timber. It has a curved end for holding a pipe. See diagram, p. 165.

pipe inspection chamber A box-shaped cast-iron pipe *fitting* with a bolted cover, used to join several branch *soil pipes* to the main drain; to BS 437.

pipe-jointing clip An asbestos and metal ring put round a cast-iron pipe to hold molten lead when making a *lead-caulked joint*.

pipe layer A *drainlayer*.

pipeline Many lengths of pipe joined together.

pipe ring A *two-piece cleat*.

pipe sleeve A pipe through a wall or floor, as an expansion *sleeve*.

pipe tail A short length of pipe, usually from a *fitting* such as a *waste* or *mixer*, for connecting to a branch pipe.

pipe tongs *Footprints*.

pipework The pipes of the plumbing systems inside a building above ground level. They carry hot and cold water, soil and waste water, or heating and chilled water for air conditioning, but the term excludes underground *drainage*. Pipework needs to be supported, usually with clips or cleats at intervals. It should be installed so that *thermal movement* can occur without causing damage or making noise, in general by separating short runs with 90° bends, *offsets*, etc., and by avoiding straight pipelines between two fixed points, or by having *push-fit joints*, expansion *sleeves*, etc.

pipewrap *Wrapping tape.*

pipe wrench, cylinder w., Stillson A heavy plumber's wrench with serrated jaws for gripping steel *screwed pipe*, used to tighten or undo joints.

pisé, p. de terre (USA **rammed-earth construction**) Walling for *earth buildings* made with dry gravelly loam without too much clay, sometimes mixed with cement, and rammed into *formwork*, usually resulting in flat and massive shapes. The word 'pisé' comes from the Lyon region in south-east France.

pit A hole in the ground or a floor. See **lift pit** and *C*.

pitch (1) The angle that a sloping roof, stair, etc., makes with a horizontal line, or the ratio of height to part span (eaves to ridge). The pitch of tiled roofs is the slope of the *rafters*, not the slightly lower slope of each tile. For example a plain tiled roof should have rafters pitched at not less than 35°, which usually gives an effective tile pitch of about 22°.

(2) The distance between parallel objects at uniform spacing, such as nails in wood, reinforcement bars in concrete, bolts in steel, the threads of a *screw*, or the distance from one stair *nosing* to the next.

(3) **rake** The slope of the *face* of a *saw*

tooth measured from the perpendicular to the line of the points.

(4) A hard, black, sticky, waterproof residue left from the heating of coal, petroleum tar, or pine resin.

pitch board, gauge b. A triangular *template* for a stair, cut with one side equal to the *rise*, the second to the *going*, and the third the hypotenuse, equal to the distance from one *nosing* to the next. It is used for setting out the lines to which the *strings* should be cut or *housed*. It can be used with a *margin template* or other templates for marking out the thickness of treads and risers on the string.

pitched roof The commonest roof, usually one with two slopes at more than 20° to the horizontal, meeting at a central *ridge*. It may have *gables* or *hips* and the space inside usually makes a *cold roof*. Compare **monopitch roof**. See diagram, p. 376.

pitch-fibre pipe (USA **bituminized fiber p.**) Inexpensive pipe made from 25% wood or asbestos fibre and 75% refined coal-tar pitch, superseded by *plastics pipe*.

pitching piece A horizontal timber fixed into the wall near the *landing* of a *stair* to bear the *carriages* and strings of the *flight* and the floor *joists* of the landing.

pitch line The stair *nosing line*.

pitch mastic A *jointless floor* made from *pitch* with limestone or silica sand *aggregate*, fluid when hot, spread to a thickness of 16 to 25 mm. It is less affected by oils and fats than *asphalt* and can have a polished or matt finish.

pitch pocket A *resin pocket*.

pitch-polymer Waterproof sheet made of *polymer*-modified pitch, a very flexible and elastic material for *damp-proof courses* and *membrane roofing*. It contains coal-tar pitch with PVC,

fillers, and plasticizers, plus synthetic fibre reinforcement, and does not squeeze out under pressure in hot weather.

pith The *heart centre* of timber.

pit-run gravel Gravel straight from the quarry, not screened or crushed.

pitting The *blowing* of plasterwork.

pivot A door *hanging device* for heavy *swing doors*, with rotating parts fixed to the top and bottom *rails* of the door leaf which join matching parts in the floor and *transom* of the door frame. Pivots are stronger but more costly than hinges. The top pivot can be *cocked* to make the door self-closing, but usually they do not have springs, as *door closers* have. They require a shallow *mortice* and accurate fitting, although some adjustment is usually possible.

placing of concrete, pouring Moving *fresh concrete* from the handling equipment (skip, concrete pump) into its final location, in the ground or in *formwork*. Placing is done by *concrete workers*, who must make sure that the formwork is clean, check *workability*, and use *vibrators* to drive out air and compact the mix. During a large pour it is usual for one or two *formwork carpenters* and a steelfixer to be present.

plain ashlar Surfaced stones, smoothed with a *drag* or other smoothing tool.

plain bar Steel *reinforcement* with a smooth surface, not a *deformed bar* (C).

plain concrete Concrete without *reinforcement*. See *C*.

plain-sawn timber *Flat-sawn timber*.

plain tile The traditional flat *roof tile* of concrete or burnt clay, which in reality has a slight spherical camber, convex above. Its size is standardized

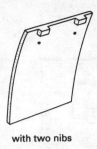

with two nibs

with continuous nib

Plain tiles.

in Britain at 265 × 165 mm and 10 to 15 mm thick, with at least two *nibs* at the head and two nail holes of 6 mm dia. They are laid so that each tile has a *headlap* over the two *courses* below it, and for this reason plain tiling is heavy, about 590 being needed to cover 10 m² of roof at 100 mm *gauge*, and weighing 78 kg/m². *Single-lap tiles* are lighter per unit area of roof, though individually heavier. Plain tiles should not be laid on a roof with a *pitch* of less than 35°, preferably 40°. Although made for roofing, plain tiles are put to other uses, such as tile *creasings, gable springers*, etc. See BS 402 and BS 5534. See also diagram.

plan A *drawing* showing a layout in a horizontal plane, of a *site, floor*, reflected ceiling, etc., usually drawn to *scale* and with a *grid* or *dimensions* for setting-out. See diagram, p. 330.

plancier piece (USA) A *soffit board*.

plane A hand tool for smoothing or shaping the surface of wood. In the past it was used for cutting *mouldings*, work which is now done with hand *power tools* or *woodworking machines*. Large jobs of flat planing may be done with *planers*, although *bench planes* are still much used. Wood that has been planed is referred to as 'wrot' or 'dressed'. A plane must be used 'with the *grain*'. See diagram, p. 67.

planed timber Wood with its surface dressed with a *plane*.

planer, rotary p. A portable electric *power tool*, held and handled like a *bench plane*, except that far less force is required and the cutting is done by the adzing action of a *cutter block*. Planing is therefore quick and relatively effortless. A 1 hp motor is needed for a cutter 63 mm wide.

planing machine, planer A heavy, stationary machine for surfacing wood, metal, or stone. There are many woodworking types, the most common being the *surface planer* and the *thicknessing machine*, but other types are made for bevelling, chamfering, or moulding. They stand in a joinery shop or planing mill, with mechanical feed, dust extraction, and computer controls.

planing mill (USA) A sawmill where timber is also planed and moulded into matchboards or floorboards, or other shapes for *millwork*.

planing-mill product (USA) A mass-run item such as floorboards, ceiling boards, and *weatherboards*, but not assembled *millwork*.

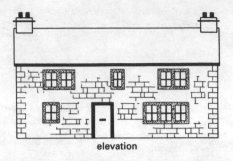

elevation

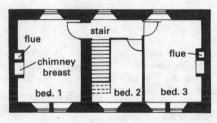

The plan of a stone house and its elevation.

plank (1) A piece of *square-sawn* timber 47 to 100 mm thick and 275 mm or more wide. In hardwoods it is sometimes *waney*, depending on the country of origin.

(2) (USA) Lumber thicker than 1 in. (25 mm) laid with its face horizontal, as is a floorboard, not on edge, as is a *joist*.

planking and strutting Vertical planks, *walings* (*C*), and struts used in an excavation for *earthwork support*.

plank-on-edge floor (UK **solid wood f.**) A floor used in the USA for its fire resistance and solidity. The floor *joists* are laid touching each other; no rough floor is laid, the finished floor being fixed to the joists.

planner The person who does the *planning and scheduling* for a large project; or a *town planner*.

planning (1) In building design, plan-

ning is the organization of the layout details of rooms and circulating areas and is best done by an *architect*, though many houses are planned by the owner or builder.

(2) Decisions by the *local authority* on land use, or the construction, alteration, and demolition of buildings, usually done by *town planners* or municipal or civil engineers with an architectural bias, or architects with an engineering bias.

(3) *Planning and scheduling*.

planning and scheduling The organization of the sequence in which work is done so that a project is completed efficiently. The work is divided into *activities*, then logical connections are added, followed by *network analysis*. The result is shown on a *programme* for use on site.

planning grid A network of perpendicular lines usually one *module* apart, used by *architects* to help them arrive at a building layout. It is not a *grid plan*.

plant and equipment Mechanical equipment for building *services*, or *contractors' plant* (*C*).

planted (1) Of a *moulding* of wood, fibrous plaster, etc., fixed to the surface it decorates with adhesive, screws, or nails, and not cut or moulded out of the solid material, as is a *stuck moulding*. *Architraves* and similar trim are planted.

(2) Surface-mounted, e.g. a *rim latch*.

planter A container for garden plants, used in *landscaping*.

plant level, mechanical floor, m. level A complete floor of a tall building taken up by mechanical equipment such as central *air-conditioning* plant. It can be a basement, the top floor, or an *interstitial level*.

plantroom (USA **mechanical room**) A room in a building for mechanical plant and equipment, e.g. central *air-conditioning* plant, *boilers*, *lift machinery*, etc., usually of solid utilitarian construction in concrete and blockwork, left bare or with plain *finishings*.

plaster Material used for *plastering*, creating a smooth *finishing* on a wall or ceiling. Plaster is put on wet in *coats* and allowed to harden. Interior *plasterwork* is usually based on *gypsum plaster*, exterior work on cement *render*.

plaster base A *background* for plasterwork.

plaster bead An *angle bead*.

plasterboard The usual name for *gypsum plasterboard*.

plasterboard trowel A *plasterer's trowel* with a thin, flexible blade that springs straight as it smooths over joints in *tapered-edge plasterboard*.

plasterboard nail A *sherardized* or otherwise zinc-coated nail with a shank at least 2.5 mm thick and a flat head at least 7 mm wide, used for fixing *plasterboard*. Its length is usually at least three times the thickness of the plasterboard, or as given in BS 8212.

plasterboard screw, drywall s. A thin trumpet-head screw, driven by an electric screwdriver, for fixing *plasterboard*.

plaster dabs Liquid *dabs* of plaster or gypsum-based adhesive used to fix *angle beads* and similar metal trim or for *bedding* plasterboard. For angle beads they are at 600 mm intervals. For plasterboard on walling they are spaced at about 300 mm apart horizontally and 450 mm vertically. If the walling is irregular the plaster dabs may be raised on solid *dots*.

plasterer A *tradesman* who usually works with *plasterboard* dry linings, but may still do some *fibrous plaster* or *solid plaster* work. See diagram, p. 332.

plasterer's float A *float*.

plasterer's labourer A helper for the *tradesman*, the *plasterer*. He or she does unskilled work such as surface *preparation*, mixing and bringing plaster, and cleaning up rubbish.

plasterer's small tool A small steel tool with a flat, spoon-shaped spatula on one end and a flat blade on the other, used for working in corners and shaping *enrichments*.

plasterer's trowel A steel *laying-on trowel*, also used for smoothing.

plastering Putting coats of *plaster* on walls or ceilings in order to cover up unevenness in the *background* and provide a smooth, crack-free surface for other *finishings* such as paint or tiles. See BS 5492. See also **plasterwork**.

plastering machine

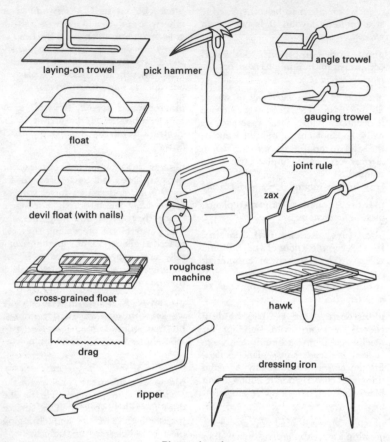

laying-on trowel

pick hammer

angle trowel

float

gauging trowel

joint rule

devil float (with nails)

zax

cross-grained float

roughcast machine

hawk

drag

dressing iron

ripper

Plasterer's tools.

plastering machine (1) A machine for *projection plastering*. It has a *twin-shaft paddle mixer* with a feed hopper above and a pump under. The pump forces the mix through a hose to a spray outlet. The hopper is either fed with dry bagged plaster or supplied through another pipe from a delivery silo at ground level. See diagram opposite.

(2) A *roughcast machine*.

plaster of Paris, hemihydrate p.,

Class A p. (to BS 1191) Pure *hemihydrate* gypsum which is mixed with water and sets in about ten minutes, expanding slightly and warming. Because of the expansion it is an excellent casting plaster, but in building it is mainly used in *fibrous plaster* work, or for quick small repairs. See **retarded hemihydrate plaster**.

plaster stop A *stop bead*.

plaster tile A dead-flat *fibrous plaster*

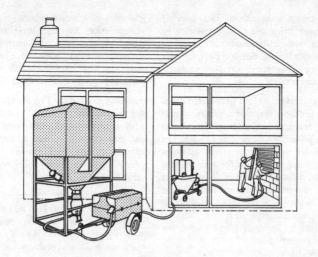

A plastering machine.

tile for *suspended ceilings*.

plasterwork A wall or ceiling *finishing* of plaster, including the *trade* of plastering and similar work, such as *rendering*, dense floor *screeds*, and plasterboard *dry linings*. Plasterwork in *solid plaster* includes the *preparation* of the background and the *application* of the coats of plaster by hand or machine. As in other *wet trades*, care has to be taken to provide *frost protection*. Fresh plasters containing alkaline cement or lime should be kept off glass and anodizing, which are both attacked by alkalis.

plastic (1) *Plastics*, the commonest synthetic materials, such as PVC.

(2) Having the property of yielding at a given pressure and continuing to yield under that amount of pressure. See *plasticity* and *C*.

plastic emulsion *Emulsion paint*.

plastic glue *Adhesive* containing synthetic *resins*.

plasticity, workability, fatness The property in a plaster or mortar of being smooth and easy to work with, obtained with cement mortars by adding a *plasticizer* or *lime*, or by using *soft sand*. Mixes containing *masonry cement*, *lime putty*, and *gypsum* are naturally plastic.

plasticizer (1) **water reducer** An *admixture* in *mortar* or *concrete* which can increase the *plasticity* of a mix so that it will flow without *segregation*, even with a low water content, thus improving strength and decreasing shrinkage. See **superplasticizer** and *C*.

(2) A non-volatile substance mixed with the *medium* of a paint, varnish, or lacquer to improve the flexibility of the hardened *film*.

(3) During plastics manufacture, plasticizers are added to make 'plasticized plastics', which are soft, flexible, and easy to process, although they are not as strong as the unplasticized product, e.g. *unplasticized polyvinyl chloride*.

plasticizer migration

plasticizer migration The loss of *plasticizer* from plasticized plastics in contact with other materials, making the plastics brittle. Some types of plasticizer can migrate from *thermoplastic tiles* into concrete.

plastic paint (1) Thick *high-build* paint, used on walls or ceilings, which can be manipulated after application to give a patterned or *textured finish*. On conscientiously scrimmed plasterboard, there is no need for even a skim coat of plaster under this paint. It comes as a powder or as a ready-mixed *emulsion paint*. Some types containing *gypsum* are suitable only for indoors, while others contain *alkyd resins*, for use as *masonry paint*.
(2) *Emulsion paint*.

plastics Materials containing natural or synthetic *polymers*, usually divided into two main classes, *thermoplastics* and *thermosets*, which require different jointing methods. Their use in building is still expanding as cheaper, improved products are developed by manufacturers. They tend to be watertight, chemically inert (therefore do not corrode metals), lightweight, tough, and fairly strong. Many plastics have a limited life when exposed to *ultraviolet* from the sun, although some, such as *fluoropolymers* and *polyester*, perform well. In general they are soft enough to cut with woodworking tools, lack stiffness, and have high *thermal movement*. Most plastics can be burnt and, although they may not produce much heat, are a *fire hazard* from smoke containing toxic or lethal *combustion gases*. None are *non-combustible*, although PVC-U is self-extinguishing and others can be made so. In a fire, thermoplastics usually melt easily.
Plastics can be formed into sheet, pipes, fibres, and mouldings, or applied to other materials as coatings, or expanded to make foam, laid up to make

glassfibre-reinforced *plastics*, etc. An abbreviation for the long names of many plastics is often printed on the material, for easy identification. Guidance on the hundreds of tests for plastics is given in Part O of BS 2782.

plastics coatings The forerunners of *organic coatings*; enamels, lacquers, varnishes, paints, and even *jointless floorings* are made of plastics, particularly *alkyd*, amide, polyamide (nylon), phenolic, silicone, or polyvinyl resins, and *chlorinated rubber*.

plastics conduit Electrical *conduit* made of PVC tube. It is less expensive than *metal conduit* but requires an *earth* wire. Plastics conduit is easily *cast in* to concrete walls and slabs, which protect it against fire. Joints and bends are made with plastics fittings and joined by *solvent welding*. Wiring in areas where corrosive chemicals may be present is usually run in plastics conduit.

plastic sheet Usually *polyethylene film*, but see *plastics sheet glazing*.

plastics pipes, poly p. Plastics pipes all have a smooth bore which discourages any deposit of scale and makes for low friction losses and good flow. They weigh only about one sixth as much as iron or steel pipes but are not so strong, have very low electrical conductivity, and cannot be used for earthing an electrical circuit. They are mainly used for drain pipes (but not where there are continuous discharges of hot water) and for cold-water supply. *Polybutylene* can be used for discontinuous hot-water supply and nylon pipe is used in *microbore pipework*. Plastics yield under load and this severely limits the pressure that pipes can stand over a long time. The layout of plastics pipes must allow for their high rate of *thermal movement*, which is five to twelve times that of steel. Plastics pipes are combustible and if they pass through a fire *compartment* floor or wall they need a *fire-stop sleeve*.

The four main types of plastics pipes are *unplasticized polyvinyl chloride* (PVC-U), *polyethylene* (polythene), *polypropylene*, and ABS (*acrylonitrile butadiene styrene*). PVC-U pipes are commonest for *sanitary pipework* and rainwater drainage above ground, and can be joined by *solvent welding* (which polyethylene pipes cannot be). Compression joints and push-fit joints are also available for most plastics pipes.

plastics sheet glazing, glass substitute Transparent plastics such as *polycarbonate*, *acrylic*, *polystyrene*, and vinyl polymers, used as flat or curved sheet for the *glazing* of windows or corrugated for *rooflight sheet*. They all break less easily than glass, both from impact and *thermal stress*, and weigh less than half as much, which may allow savings on their frames and supports. Low-melting-point vinyl plastics sheet is used in *fire vents*. Since plastics all expand about ten times more than glass when heated, they need more clearance and more flexible glazing methods. Glazing must also be done with chemically compatible materials. See BS 6262.

plastics-surfaced board *Building board* faced with factory-applied *polyvinyl chloride* or *melamine-formaldehyde*.

plastisol An *organic coating* of *polyvinyl chloride* put on to galvanized steel as a powder dissolved in a hot *plasticizer*, usually textured and 0.2 mm thick. It has long life expectancy and little need for maintenance.

plate (1) A horizontal timber, about 100 × 50 mm, laid flat and supported throughout its length, particularly one along the top of a wall on which the ends of roof timbers are laid (*wall plate*). In *timber-frame construction* it is the corresponding member capping the *studs*.

(2) (Scotland) A broad thin board, planed on one or both sides.

plate cut (USA) A *foot cut*.

plate exchanger A *heat exchanger* consisting of a stack of metal plates with spaces between, arranged for *heat recovery* between two different air streams flowing in odd or even spaces. The unit needs *working space* for regular cleaning.

plate glass Thick cast and polished sheet *glass*, which was expensive to make, now superseded by *float glass*, which is less costly and just as good.

platen-pressing The manufacture of *chipboard* or *plywood* by pressing between heated flat steel plates, which are held closed during *curing* of the *thermosetting* adhesive.

platform floor An *access floor* with a deep void under the deck panels, usually supported on adjustable steel jacks.

platform frame Traditional *timber-frame construction* in which the floor is built first, then used as a work platform to assemble and erect the wall panels.

platform roof (Scotland) A *flat roof*.

plenum An air chamber in an air-conditioning system, kept at a pressure slightly above atmospheric, usually formed by the space above a *suspended ceiling* or built as a *duct*.

plenum system A system of air conditioning with a single *plenum* which supplies outlets, often through short *ducts*. Air is allowed to escape through gaps in doors and windows, and return unducted through corridors and stairways.

Plexiglas Clear *acrylic sheet*.

pliable conduit Corrugated *conduit* which can be easily bent to shape and is then usually built in. It does not need any allowance for expansion and is made of self-extinguishing plastics to BS 4607.

plinth (1) Extra thickness at the base of a column or wall to give it more strength against buckling and to spread its load over its *footing*.
(2) A *machine base*.

plinth block An *architrave block*.

plinth stretcher A *standard special brick* with a *splay* on its long edge.

plot plan A *drawing* showing land for a building.

plot ratio (USA **floor area r.**) The maximum ratio between the area of a *site* and the total above-ground floor area of the buildings on it that is allowed by *town planning* laws. See **air right**.

plough (USA **plow**) A *plane* which makes grooves, invented in the sixteenth century together with the *panelled* construction for which it was an essential tool, now superseded by electric *routers* and circular saws.

ploughed and tongued joint A joint made with a *loose tongue*.

plug (1) A small cylinder of durable fibre, plastics, or wood pushed into a hole drilled in a wall, for a *screw*, nail, or other *fastener* to be fixed into it.
(2) A *bag plug* or *screw plug* for testing drains.
(3) **socket inlet** An electrical connector to fit into a *socket outlet*. In the UK it is usually a *three-prong plug* with a 13 amp fuse, and a cord that runs to an appliance. Many different types of plugs are used worldwide.
(4) A pipe *fitting*, threaded outside, which closes the end of a pipe by screwing into it.

plug cock, p. tap A *cock*.

plugging A *fixing* made by filling a hole with a *plug* of material to hold screws or nails; or a *plug nail*.

plugging chisel, p. drill (USA **star d.**) A short steel bar with a cross-shaped tip, held in one hand and struck with a *hammer* to make a hole from 3 to 20 mm dia. in hard materials such as concrete, brick, or masonry.

plug-in connector, flex cock, plug-and-socket gas c. A method of connecting a short, flexible gas pipe by plugging in to a supply as conveniently as with an electric plug. The plug is pushed into a *socket outlet* and, when turned, allows gas to pass to the appliance. Removal of the plug automatically closes the gasway.

plug-in switchgear, withdrawable s. Electrical *switchgear* that can be removed from a *switchboard* more easily than fixed equipment.

plugmold (USA) Usually *skirting trunking*.

plug nail A convenient all-plastics *fastener* which acts as both a plug and a nail. It fits into a hole drilled through the component to be fixed, and into the wall, and is hammered home. A cross-slotted head allows it to be unscrewed.

plug tenon, spur t. A *stub tenon* shouldered on all four faces, as for a corner post (BS 6100).

plumb To place something exactly upright or exactly above something else.

plumb bob, plummet A weight hanging on a string (the *plumb line*) to show the direction of the vertical.

plumb cut (USA) In *roof cutting*, the vertical cut in the *birdsmouth* at

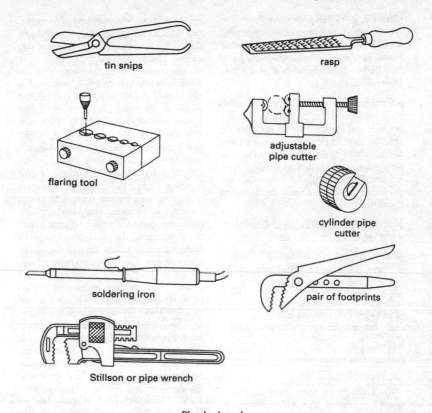

tin snips

rasp

flaring tool

adjustable
pipe cutter

cylinder pipe
cutter

soldering iron

pair of footprints

Stillson or pipe wrench

Plumber's tools.

the foot of a *rafter* where it fits over
the wall plate; also the vertical cut at
the *ridge* (top cut).

plumber A *tradesman* who works
mainly with pipes, for the *sanitary
pipework* and other piped building *services*, although a plumber may also do
rainwater plumbing, *roofing*, and *drain-
laying*. The word originally meant
somebody who worked with lead, but
a plumber may work with other metals
such as copper, steel, and aluminium,
and with *plastics*. In Britain a plumber
may hold a *City and Guilds* certificate,

or the highest qualification, that of a
Registered Plumber. See diagram.

plumber's labourer, p. mate A
skilled *labourer* who helps a plumber
by bringing tools and materials, cutting
pipes, tidying rubbish, etc.

plumber's metalwork Work in sheet
metal for *rainwater plumbing*, including
installing *supported sheetmetal roofing*,
fixing and dressing down *flashings* and
soakers, fixing rainwater gutters and
downpipes, etc.

plumber's union A *union* pipe fitting.

plumbing

plumbing (1) See **plumber**.

(2) Setting something exactly upright, such as *structural steelwork* or a *door frame*, or transferring a *setting-out* point to a higher level, as with a *plumb bob*, *plumb level*, etc.

plumbing unit A prefabricated bathroom/toilet or kitchen made as a *module* and placed by crane during multi-storey construction, mostly used for hotels, where a high standard of finishings is needed, and to save time on site. Pipe, duct, and cable *tails* are provided for connection to building *services*.

plumb level A spirit *level* fitted with a bubble for setting out vertical work.

plumb line A *line* on which a weight is hung for vertical *setting-out*. It can be of braided string, fishing line, or *piano wire*.

plummet (1) A *plumb bob*.

(2) An *optical plummet* (*C*) or *laser* (*C*).

plunger See **sink plunger**.

ply (1) *Plywood*.

(2) A *veneer* or thin sheet of wood, used for making *plywood* or *laminated wood*.

(3) One sheet of *built-up roofing* or other material used in several layers.

plymetal *Plywood* faced on one or both sides with a sheet of metal, which may be galvanized steel, aluminium, Monel metal, etc.

ply-web beam A composite beam with solid wood flanges and a *plywood* web, either box or H-shape, combining light weight and stiffness.

plywood A *building board* with more uniform strength, less tendency to split and better *dimensional stability* than plain wood. It is made of *veneers* bonded together with *adhesive*. There

are usually an odd number of veneers, laid so that neighbouring sheets have their grain at right angles to each other, unlike *laminated wood*. Three-ply is the commonest and cheapest. It easily gives *balanced construction*, but multiply is also much used, as are related products like *film-faced plywood* for formwork and marine ply. Plywood was used by the joiners of Europe, particularly in France, in furniture-making some centuries ago and probably also by the ancient Romans and Egyptians. Plywood, as we know it now, was used by the American piano industry in 1830 for the planks which held the pins to which piano cords were attached. It was then made of *sawn veneer*. At this time, obviously, piano makers understood its superiority over solid wood in strength and stability in varying conditions of dampness. *Blockboard* desk tops were made in 1883 and plywood *panels* for doors in 1890, flush doors following much later. Plywood was known in the USA as 'veneered stock' until 1919 when, to avoid the widespread prejudice against veneer, the old Veneer Association changed its name to the Plywood Manufacturers' Association of the USA. Plywood in Britain is mostly imported.

plywood parquet A form of *parquet floor* suitable for light traffic and made of strips of *plywood* with a top *veneer* 3 mm thick of *oak* or *mahogany* bonded to them. The strips, which are about 6.5 mm thick overall and 914 mm long × 70 mm wide, are laid to a *basketweave* or herring-bone pattern. They have interlocking edges and are *secret-nailed* through the plywood if there is a wooden *sub-floor* (and glued in the best work), or stuck down to any smooth surface with *latex cement*. The nail holes are filled and the floor sanded after laying. If the sub-floor is rough it should be made smooth by a plywood or hardboard underlay at least 5 mm thick all over it.

pneumatic structure An *air house*.

pneumatic tools, air t. *Power tools* driven with *compressed air*. There is little to choose between them and electric tools, except that they can be used in wet areas without risk of shock and have a slightly better power/weight ratio. But compressors are expensive and ordinary air hoses heavy. Pneumatic *breakers* are rugged and have to be effectively silenced. Hand drills, grinders, and *nail guns* are worked off a lightweight air hose.

pocket, box-out (USA **blockout**) A hole in concrete formed during concreting rather than cut out later. The *former* used to make a pocket can be of any material, such as *expanded polystyrene*, usually removed by dissolving with petrol (gasoline), or normal timber boxing, metal *bolt boxes*, etc. Pockets are used for *building in* components or to make *penetrations* in the structure for pipes, ducts, or cables.

pod A *volumetric building*, such as a *bathroom pod*, or a *plumbing unit*.

podium, low block A wide, low building, above which a tower block rises. It is often one or two storeys high and can cover an entire city site.

pod urinal A *bowl urinal*.

point (1) A sharp end, like that of a *saw tooth*. A fine saw has many points per inch (25.4 mm), a coarse saw few points.

(2) A lampholder, *socket outlet*, or other terminal at which current can be taken for electric *power* or *lighting*.

point detector A *fire detector* at a particular point, not along a line, as are detectors with a wire, tube, light beam, etc.

pointed vault The twelfth-century structural innovation of using stone *ribs* built on *centering*, then infilling with pointed stone vaults, which could be built without further centering. The daring architecture it contributed towards was later described as Gothic, after people from west Sweden, although the pointed vault was first used in Assyria, occupied by the first Crusade.

pointing (1) Raking out brickwork *joints* 20 mm deep, then filling the hole with a pointing *mortar*, usually coloured, which is pushed in off a *fat board*. The joint face can be left *flush*, projecting (tuck pointing), or finished by tooling in the same way as *jointing*. The bedding mortar is disturbed and may not bond to the pointing mortar. See **repointing**.

(2) Thin infilling at the face of a joint with *sealant*, grout, mortar, etc., such as the *grouting* of wall and floor tiles, the insertion of mortar under the edges of roof *ridge tiles*, the application of sealant round window frames, etc.

point of articulation A *link*.

pole (1) A long piece of round timber, often placed upright.

(2) A terminal on an electric circuit. Three-phase *circuit breakers* may have four poles, including the *neutral*.

pole-frame construction A house construction method in which the house is suspended between preservative-treated timber poles from ground to roof, often embedded directly in the ground, used on steep or difficult sites, usually in tropical climates. It saves on the cost of foundations and allows ventilation under the ground floor.

polished work, glassed surface, p. face Building stone of crystalline texture (limestone, marble, granite) which has been smoothed with *abrasives* and buffing to form a glass-like surface.

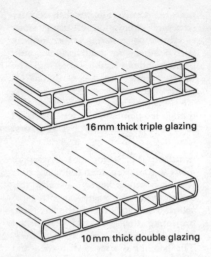

16 mm thick triple glazing

10 mm thick double glazing

Extruded transparent polycarbonate sheet glazing.

poll (1) A striking face of a *hammer*. (2) To split (knap) flints.

polyamide (PA) *Nylon*.

polybutylene A strong but expensive *plastics*, used for cold-water pipes, or for hot-water pipes without *secondary circulation*. It is usually grey.

polycarbonate (PC) Transparent, extremely tough *plastics* sheet, used for security glazing. It may also be an insulating light panel, double walled or triple walled, of 10 or 16 mm thick. It is not a *fire hazard*, as it has low *ignitability* and low *flame spread*, and releases little heat and little smoke if burnt. It can be coated to resist damage from *ultraviolet*. See diagram.

polychloroprene The chemical name for *Neoprene* synthetic rubber.

polychromatic finish Paintwork with several different colours.

polyester, unsaturated p. (UP) A synthetic *resin* with many uses, cheaper than *epoxy resins* and used to produce *alkyds* for paint. One of the toughest fibres, it is used in *bitumen felts*. Solid polyester, supplied as a *two-part product*, is used in *levelling compounds*, to bond *terrazzo*, and in *glassfibre-reinforced polyester*, but should not be made too thick or its high shrinkage can cause cracking. Polyester is probably the best *powder coating* for exterior use. Work with polyester should be done in a well-ventilated place, and its *dust* and vapours avoided.

polyethylene (PE), polythene A chemically inert, electrically insulating, waterproof synthetic *rubber* with many uses. It comes in two main forms, high-density and low-density, but is also *cross-linked* for electrical insulation. The transparent type lasts only three years in shade or one year in sunlight because it is destroyed by *ultraviolet*. More than ten years' life can be expected, however, even in bright sunlight, from the variety with 2.5% carbon black *filler*, which stops the ultraviolet penetrating and doing damage. Some of its many uses are as bottles and pipes for chemicals, as plastic sheet, and as (lubricated) substitute ice for skating. See the two entries below.

polyethylene film Waterproof plastic sheet which has many uses on site – as a protective cover over timber or bricks, as peelable protection on finishes, for *underlays, vapour barriers, damp-proof courses*, etc.

polyethylene pipe Black pipe for cold-water supply. It can be bent cold to a fairly large radius or given a permanent bend using boiling water. It should be laid with some excess length in hot weather by 'snaking' in the trench. This is not necessary in deep

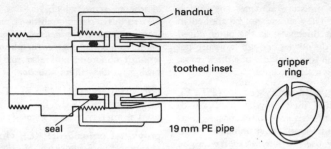

handnut

toothed inset

gripper
ring

seal

19 mm PE pipe

A non-manipulative fitting with a polyethylene body, for joining polyethylene cold-water pipe.

frost, as the pipe will expand in warmer weather. Pipes can be joined with threaded *non-manipulative* polyethylene compression fittings or by *heat fusing*. Blue medium-density polyethylene is used for main service pipes. See diagram.

polyfusing *Heat fusing* of plastics.

polyisocyanurate foam, isocyanurate f. A foam resembling *polyurethane foam*.

polymeric binders Mostly synthetic *resins*, used in paints and adhesives.

polymeric sheet Usually *bitumen-polymer* or *pitch-polymer* sheet.

polymerization The *curing* of a monomer *resin* to become a *polymer*.

polymer-modified bitumen, elastomeric b. *Bitumen-polymer*.

polymers Synthetic organic materials including *plastics*, *resins*, *rubbers*, *elastomers*, and *sealants*. They have large molecules combining many hundreds of smaller molecules of the monomer. Polymers are a large family, with overlapping properties. They can also modify the properties of other materials such as *bitumen* and *concrete*.

polymethyl   methacrylate (PMMA) *Acrylic resin*.

poly pipe See **plastics pipes.**

polypropylene (PP) One of the toughest *plastics*. It is rigid and unaffected by the oils, greases, and bleaches found in waste water. It is used in solid form for pipes and as fibres for *fibre-reinforced concrete* or as foam. It has good resistance to sunlight, but lasts longer if fully protected.

polystyrene (PS) A crystal-clear, rather brittle plastics. High-impact grades are tougher but not as clear. See the two entries below.

polystyrene bead aggregate concrete (PBAC) A *lightweight concrete* containing beads of *expanded polystyrene*. It bonds well to other types of concrete and is used for *insulation* in composite *cladding panels*.

polystyrene foam *Expanded polystyrene*.

polysulphide   sealants (USA **polysulfide s.**) High-performance gun-grade *sealants* with good adhesion to most surfaces, unless porous, when a *primer* is needed. The *two-part* types have rapid and complete *curing*, remaining tough and flexible even in severe conditions. The newer *one-part* types cure slowly, as atmospheric moisture activates the curing agent. They should not be applied in a great thickness, as the outer cured layers do not let enough moisture get through to the middle of a thick mass, making it slow to cure.

polytetrafluorethylene (PTFE), Teflon A fluoropolymer *plastics* with excellent resistance to the most aggressive liquids or environments. It is mainly used as *jointing tape* wrapped round a screw thread and supersedes the use of tow, hair, or *jointing compound*. Although of low fire hazard, under intense heating it gives off supertoxic fumes.

polythene See **polyethylene**.

polyurethane (PU) Resilient synthetic *resins* with excellent stability to light, whether clear or pigmented. The liquid forms are elastic and not subject to hair cracks. They are used in paints and varnish, particularly *floor paint*, as well as in *sealants* for outdoor movement joints. Both *one-part* and *two-part* types are used, but some give off dangerous *isocyanurate* fumes. See next entry.

polyurethane foam *Expanded plastics* with excellent properties as *insulation*. It can be made flexible, as in bituminized *sealing strips*, or into rigid board, as in *flush door* cores, or foamed in-situ. It is used in sprayed roof renovation, in which the raw materials expand and set several centimetres thick. Complex roof shapes can be followed and although PU foam is durable it is usually coated with a reflective weather-resistant finishing. In enclosed spaces PU foam is a *fire hazard*; in burning it gives off large amounts of lethal hydrogen cyanide gas. It is no longer allowed in new furniture.

polyvinyl acetate (PVA) Rubbery

synthetic *resin* which is not fully waterproof in damp conditions, used in interior wood *adhesives, emulsion paint*, as a *bonding agent* for plaster on smooth concrete, and also for other backgrounds with high *suction*.

polyvinyl butyral (PVB) Tough, tear-resistant, clear flexible sheet, used as interlayers in *laminated glass*.

polyvinyl chloride (PVC), vinyl A cheap and versatile *thermoplastic* polymer, so widely used that '*plastics*' often means PVC. It is impervious to water, oil, and many chemicals, and can be a *vapour barrier*. Usual grades of PVC are 'rigid' *unplasticized* (PVC-U, formerly uPVC), rubbery *plasticized*, or chemically modified (MuPVC, cPVC, hiPVC). In fire PVC is almost *non-combustible* and does not burn easily itself, but it melts at a low temperature, and under intense heat (or incineration) gives off poisonous gases. It has many uses: as floor tiles and sheet flooring, rainwater gutters, drain pipes, cold-water service pipes, electrical conduit, coatings over wood window surrounds and foam (or *expanded PVC*). Components or systems made of PVC need to allow for its high *thermal movement*. Pipes have *push-fit* or *solvent-welded joints*, although PVC can also be joined with threaded joints, by heat shrinking, or by *heat fusing*. As cable insulation PVC is slowly giving way to *cross-linked polyethylene*.

polyvinyl fluoride (PVF) A weather-resistant *plastics* used for *organic coating* of steel or aluminium *cladding panels*.

polyvinylidene fluoride (PVF$_2$) An *organic coating* for metals, factory-applied at high temperature. It is a *fluoropolymer*, highly resistant to chemicals and exposure, but it is easily scratched and needs careful *pro-

tection during delivery and after installation as *touching-up* on site is done with lower-quality materials.

pommel A ball-shaped *finial*, e.g. on top of a gate post or stair *newel post*.

ponding Shallow pools of water on a horizontal surface with low spots or irregularities which create a *backfall* and prevent proper drainage. It can occur on a *flat roof*, outdoor *hard surfaces*, or a *screed to falls*.

Poole's tile A clay *single-lap* roof tile with two waterways. It resembles a *double Roman tile* but with a central ridge that extends only half way up.

poplar Poplar trees should not be allowed near buildings. See **trees**.

popout, knockout A disc in the side of a plastics *accessory box* which is thin enough to be easily pushed out for a *conduit* entry hole.

popping The *blowing* of plasterwork.

pop rivet An aluminium *fastener* with a hollow shank and flat head. It is inserted into a drilled hole, then a steel pin inside the shank is pulled with special pliers, expanding the shank until tight in the hole, when the pin breaks from its necked head with a 'pop' sound. All work is done from one side. Pop rivets are mainly used to join *sheetmetal* and can be removed by drilling out.

pop-up waste A captive plug for a *basin*, operated by a lever or knob.

porcelain enamel A luxurious finish used on cast-iron baths, similar to vitreous *enamel* but fired at a higher temperature.

pore Any small hole in a surface, e.g. the sap tubes seen in the end grain of *hardwoods*.

pore treatments The use of chemicals to reduce *dampness* in masonry walls by spraying or brushing on silicone resin solution on the outside. The wall absorbs less rainwater, but the escape of water vapour is also slower. See **water-repellent treatments**.

porosity The quality of being porous, i.e. permeable by fluids. Open holes in the surface of porous materials allow *absorption* of water or vapour, or create *suction* for paint and plaster.

porous pipe A pipe which lets in water, used for *subsoil drains* (*C*). It can be made of *no-fines concrete* throughout, or with watertight dense concrete in the bottom third.

portal frame A frame of two columns with one horizontal roof beam between them, or two sloping *rafters* that join in the middle. They are usually in concrete or *structural steelwork* and are used for single-storey factories and warehouses with spans of 30 m or more. The column/rafter joints may have *haunches* for extra strength.

Portland cement The most usual *cement* in building, originally so called because when hard its colour and texture were thought to resemble *Portland stone*. It was first patented in England in 1824, shortly after Louis Vicat in France made artificial hydraulic lime, in 1818. Their forerunner was *Roman cement*. The raw materials are a mixture of clay and lime which is heated above 1200°C in a rotary kiln. The clinker formed is ground to a fine powder which is used as *ordinary Portland cement*, in *blended cements*, etc.

Portland stone A limestone from Portland on the south coast of England, used for facing important buildings. It weathers well and forms strong contrasts between the parts which get wet and stay white and those which are sheltered and may blacken.

possession Legal control of and responsibility for a *site* or building, without the other rights of *title*. The builder is given possession to allow *commencement* of work and keeps it until *handover* back to the client upon *practical completion*.

post, pole, column A large vertical member in a *timber frame*, usually thicker than a *stud*.

post-and-beam construction A loadbearing building frame with vertical posts (columns) and horizontal beams, carrying floors and roof, often partly braced by non-loadbearing infill walls.

post-contract stage The time after work has been completed on site, when the *final account* is settled, a busy time for the *contractor's surveyor*.

potable water *Drinking water*.

potato peeler A machine that scrubs off the skins of potatoes. It needs a separate *waste* pipe to prevent hot waste water solidifying the starch in its discharge.

pot floor A *rib and block suspended floor*.

pot life *Working life* of glue.

poultice attack, p. effect The corrosion of *aluminium* kept in contact with a wet, porous material. It can destroy *aluminium foil*, e.g. on *insulation*.

pour A *bay* of a concrete slab or wall cast in one operation. Its edges may form a *movement joint* or a *construction joint*. See **placing of concrete**.

pour and roll One method used to apply *built-up roofing*, in which the hot *bonding compound* is poured in front of the *bitumen felt* as it is unrolled.

pouring of concrete *Placing of concrete*.

pouring-grade sealant Very liquid *sealant* for narrow *filled joints*.

pouring rope A *joint runner*.

powder-activated tool A *cartridge tool* or fixing gun.

powder coating A tough, durable factory-applied *organic coating* on metals, such as aluminium or galvanized steel, available in many colours. Powder coatings are either *thermoset* plastics such as polyester, polyurethane, acrylic, and epoxy, which are sprayed on, followed by heat *curing*, to give a film thickness of 50 to 100 *microns*; or they are *thermoplastics* such as *polyethylene* and *nylon*, which are applied by dipping components in a hot fluidized bed, creating a film 200 to 300 microns thick. Any holes should be made and any cutting done before the coating is applied. See BS 6496 and BS 6497.

powderpost beetle (Lyctus) A *borer* which usually attacks *hardwoods*, leaving *pinholes* about 1.5 mm dia. and a fine, dust-like flour.

power Low-voltage electricity for general purposes, e.g. to drive motors, electric tools, for heating, etc. It is supplied from *socket outlets* or a *motive power* circuit. In a building the power circuits are usually separate from those for *lighting*.

power and lighting installation The low-voltage *electrical services* which supply motive power, socket outlets, lighting, etc.

power float, machine trowel, rotary f., helicopter A motor-driven machine with three steel *floats* for *floating* and *trowelling* concrete floors or *screeds*. The concrete is *screeded* off flat and allowed to reach *initial set* before work starts. The many passes of the blades leave a hard smooth surface.

power panel A *distribution board* for power circuits but not lighting.

power point Usually a *socket outlet*.

power roller A *pressure-fed roller*.

power tool Hand-held portable *electric tools* are the commonest power tools used on building sites, followed by *pneumatic tools*. They include *circular saws*, *drills*, *grinders*, *routers*, *planers*, and *sanders*, and are often similar to larger bench- or floor-mounted *woodworking machines*. Power tools quickly pay off their cost, achieving fast, accurate work that requires little or no finishing. Since, like all tools, they are dangerous, safe working procedures, protective clothing, safety *goggles*, and general caution in use are essential, as well as the control of *noise* and *dust*. See diagram, p. 346.

pozzolan, pozzolana A natural volcanic silica dust, described by Vitruvius, the Roman architect, as producing astonishing results when crushed and mixed with *lime* and rubble. It sets hard, even under water, making *Roman cement* (originally from Pozzuoli in Italy).

practical completion, substantial c. The stage of work on site when a building is ready for *handover*. It does not have to be perfect, only reasonably fit for occupation, and minor *defects* may still be put right provided they cause no inconvenience to the owner. The issue of the *certificate of practical completion* starts the *defects liability period*, although it is not uncommon for the *commissioning* of services to continue for some time afterwards.

preamble An introduction to each *trade* in *bills of quantities*, with details of the rules of *measurement* and what is included in *descriptions*, particularly any departures from the *Standard Method of Measurement*.

pre-boring Drilling a *pilot hole*.

precedence of documents The order in which *contract documents* are to be used if they disagree. A usual sequence is the contract *agreement, conditions of contract*, then *drawings*, first large-scale, then small-scale, the *specification*, and finishings *schedules*. The *bills of quantities* as well as the *programme* may form part of the contract. If documents of equal precedence contradict each other, the architect is asked for a *clarification*.

pre-contract stage The time before a builder starts work on site, taken up in *tendering*, arranging *sub-contracts*, programming, etc.

prefabricated building All buildings are to some extent prefabricated, since many building components are brought to the site completed (doors, trussed rafters, etc.). The word prefabricated is therefore usually reserved for building *elements*, such as complete walls, roofs, or floors, which are not made *in-situ* but in a factory, or on-site then moved, often of *precast concrete* (*C*).

prefabricated scaffold, frame s. A *scaffold* of steel or aluminium tubes with snap-on *couplings* on their ends or at other useful points so that it can be put up quickly and safely by unskilled labour.

prefabrication primer A *pretreatment primer* applied after the *blast cleaning* of structural steelwork, formulated to reduce fumes from welding and cutting.

pre-filter, dust collector, sand filter An air-conditioning device to remove large solids or dust from fresh air before it reaches the *filters*.

pre-finished *Self-finished*.

preliminaries (USA **mobilization costs**) The project administration

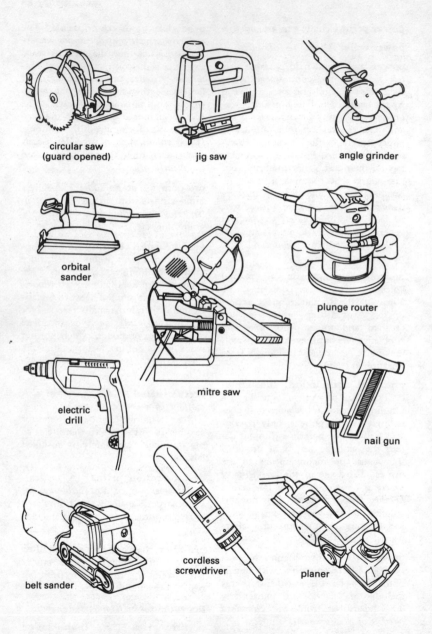

circular saw
(guard opened)

jig saw

angle grinder

orbital
sander

plunge router

mitre saw

electric
drill

nail gun

belt sander

cordless
screwdriver

planer

Power tools.

and site equipment required for work to go ahead or to be properly completed, but not part of the work. The cost of preliminaries to the *main contractor* requires close *financial control* and is often partly reimbursed by an *advance*. Some of the things included are: office *overheads*, giving notices, paying fees, taking out insurance, preparing a *programme*, temporary works (for shelter, access, protection, welfare), water, electricity, telephone, *contractor's plant* (*C*), supervision, setting-out, testing of materials, prevention of nuisance, watching, cleaning, removing rubbish, handover (but not commissioning), as-built drawings, repairs during the *defects liability period*, etc.

preliminary work Work *by others* on site before a contract commences, such as demolition, bulk excavation, pile driving, diversion of a gas main, etc.

premixed plaster Most plaster is supplied factory-mixed, either wet in drums, *ready-mixed*, or dry in bags, and usually containing *lightweight aggregate*. A dry mix may be based on *retarded hemihydrate* gypsum or Portland *cement*, so that on site only water is added. The high thermal *insulation* of these plasters makes them useful in cold climates and for the fire *encasement* of structural steelwork. *Thin-wall plaster* is one type.

prepainting *Pretreatment priming.*

preparation, surface p. Work on a surface to make paint, adhesive, plaster, or sealant stick properly. The preparation of a *substrate* for paintwork generally includes removing *loose material* and filling small holes. New timber, such as joinery delivered *in the white*, may need only *knotting, priming, and stopping*, and must be dry. Steel is usually given *blast cleaning* quickly followed by a *pretreatment primer*,

while aluminium needs *degreasing* and a pretreatment primer. Masonry and cement-based boards should be dry, and when new an *alkali-resisting primer* is used. Old surfaces for *repainting* may need *bringing forward* if damaged or sanding down if glossy. The final steps in preparation are *masking* and *dusting*. Paintwork preparation requires good *workmanship* to prevent *defects* such as blistering, exudation, and peeling. Sealants and adhesives require similar preparation to that for paintwork, but the maker's instructions must be followed. Preparation of plasterwork *backgrounds* involves keying, controlling *suction*, a bonding treatment, or a coat of *spatterdash*.

prequalification The examination of each contractor's qualifications before contract *negotiations*, particularly with *open tendering*.

preservative A substance with which timber is treated to improve its natural resistance to attack by fungal *decay* or *borers*. Preservative may also work as a decorative *stain*. Many timbers with good natural durability resist the application of preservatives. All *softwoods* for external work should be factory-impregnated with preservative before priming or delivery to site. In general, organic preservatives do not cause corrosion of metal in timber *connectors* or sheet *roofing*, but water-soluble inorganic preservative salts in damp timber can attack galvanized steel or aluminium unless the preservative contains a corrosion inhibitor. Acids or alkalis in inorganic preservatives may migrate to the surface of wood and form *hygroscopic salts*. Preservatives can lack *compatibility* with paint or adhesives, and water-repellent preservatives make *putty* slow to harden. See BRE Digest 201, BS 1282.

pressed brick A *clay brick* made by

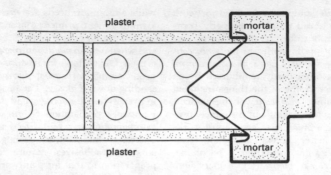

A cross-section of a pressed-steel door frame.

squeezing clay at high pressure in a steel mould. Stiff plastic clay is usual, except for *dry-press bricks*. It usually has a *frog*, sharp *arrises*, and smooth surfaces, although it can be sand-faced or post-textured (stippled, combed) after pressing. About half UK bricks are pressed, e.g. *Fletton bricks*. See diagram, p. 48.

pressed glass *Glass blocks* or *vitreous glass* shaped by pressing.

pressed metal Usually galvanized-steel sheet 1.6 to 2.5 mm thick, stamped out by a large factory machine with expensive dies, and used to make *door frames*, window *lintels*, etc. See diagram.

pressure The force needed to hold a joint tightly while *adhesive* sets, so that the *glue line* is narrow, or during nailing. Screw presses and *cramps* are cheap and simple, but do not have follow-up pressure, which can be applied with a hydraulic press. See also *C*.

pressure circulation The forced circulation of air or water with a fan or a circulating pump.

pressure cutout A *sensor* that operates a switch at a preset pressure.

pressure-equalized joint A joint which keeps *driving rain* from leaking in through *curtain walls* by having air spaces inside the framing that are ventilated to the outside air. As there is equal pressure on both sides there is no force to push rainwater through the joint.

pressure-fed roller, power r. A paint *roller* supplied with paint by a pump and hose. Less time is wasted than with a roller and tray.

pressure gun A *caulking gun*.

pressure impregnation Usually *vacuum-pressure impregnation*.

pressure-maintenance vessel A *diaphragm tank* after a *booster*.

pressurized escape route A stair or corridor on a *protected route* into which fresh air is blown to prevent smoke entering. If pressurized escape routes are used, the stair and lobbies do not have to be located on outside walls. The pressurization fans are started by a *fire detector* working through the fire alarm system, which also opens the *smoke outlets* and stops the air conditioning fans or switches them to pressurization service. A reliable source of power is needed, as pressurization is an *active* *fire protection* system. A standby generator may be desirable, as well as wiring that is run where fire cannot damage it.

pressurized structure An *air house*.

prestandard (ENV) (German, 'Vornorm') A *European Standard* used in fast-developing areas and where the safety of goods or people is not involved.

pretreatment primers, chemical conversion coatings, prepainting Chemically active liquids that react with a metal surface to create a thin film of high adhesion. They need an absolutely clean surface and for steelwork are best used for *shop priming* after *blast cleaning*. Little protection is given against corrosion until the *primer* is applied and the full *paint system* completed. The most common are *etch primers*, but *bonding treatments*, *phosphating*, *passivation*, *wash primers*, etc., are also used.

priced bill, p. BQ For large projects where *tenders* are worked out using *bills of quantities* the contract can be 'with quantities' by having contractors enter an *extended* price in the each column for each item. The *collection* of prices from items has to agree with the tender price and the priced bill becomes a *contract document*.

price index A list of prices for *items* of building work, usually published monthly. The changes from month to month are used to calculate *fluctuations*.

pricking up Applying the first coat of *plaster* on wood or metal *lathing*.

primary beam A beam or girder that carries secondary beams or other heavy loads.

primary fixing (1) A *fixing* that is built into the *structure* of a building for the later attachment of *secondary fix-ings*, usually with some allowance for adjustment, e.g. *concrete inserts* and *fixing channels*.

(2) A *fastener* used to hold profiled sheeting to the roof or wall structure.

primary flow and return pipes The pipes in which water circulates between *boiler* and a hot-water storage *cylinder*. The flow pipe is that by which the water leaves the boiler, returning to it by the return pipe usually because of *gravity circulation* alone.

primary gluing Gluing in *plywood* manufacture and other veneering work, as opposed to *assembly gluing*.

prime contractor The *main contractor*.

prime cost (PC) An item in *bills of quantities* for which a price is already stated, usually for materials to be obtained from a *nominated supplier* for *fix only*. The contractor is reimbursed the amount to be paid, plus an allowance for profit, and keeps any $2\frac{1}{2}$% discount allowed for prompt settlement of accounts. If a PC is used for a *nominated sub-contractor*, a separate agreement may be needed on *attendance*, to be provided by the *main contractor*.

primer A first *coat* put on a surface to help something else to stick. The most usual is *priming paint*, but primers are also used under *bitumen felt* for roofing, for the *bonding compound* to adhere to boarding, and before applying *sealants* and *adhesives*, particularly on porous surfaces.

primer-sealer *Priming paint* also used to control the porosity of a surface.

priming The action of putting on a *primer*, usually on a previously unpainted surface. Wood *end grain* may need double priming.

priming coat, prime c. A coat of *primer*.

priming paint A *wood primer*, *metal primer*, or *masonry primer* applied so that the *undercoats* will stick. It is the foundation of the *paint system*. For both wood and steel, priming paint alone gives very little protection. Primed joinery or metalwork should not be left out in the weather or the damage done will reduce the life of the whole paint system.

principal rafter The sloping top member of a roof *truss*, which carries the *purlins*. The purlins carry the intermediate *rafters*.

priority circuit A circuit providing power for essential services, either from an *uninterruptible power supply* or 'brownout' power from a generator. Normal circuits are unloaded by automatic *load-shedding*.

prismatic girder A sturdy roof beam, often tapering or curved, made with three steel tubes in the form of a deep triangle, joined by diagonal braces.

privacy latch Usually an *indicating bolt*.

private automatic branch exchange (PABX) The most usual type of *private branch exchange*. Outgoing calls can be dialled and connected automatically from each extension and incoming calls are routed through an operator. Internal calls between extensions are automatic. Meters may be installed for charging, and particular extensions fitted with a *trunk bar*.

private branch exchange (PBX) A telephone exchange for an office building, hotel, etc., such as a *private automatic branch exchange* or a system with *direct dialling-in*.

privatization The sale to the public of the *statutory undertakers* which provide gas, water, power, and telephone services in Britain. They receive the profits.

processed shakes Common sawn *shakes* surface-textured to look split.

product, building p. A *material* that has been manufactured or processed for use in building.

product data sheet A manufacturer's publicity sheet, usually coloured illustrations of a product or services, giving details such as sizes, composition, completed projects, and test authorities.

profile (1) (USA **batter board**) One of a pair of *hurdles* for *setting-out* placed close to but clear of the building, between which a string *line* is stretched.
(2) A square steel rod used by *bricklayers* to hold the end of the string *line*. It is clamped upright on the first *courses*, usually at a corner, and has holes 75 mm apart so that the line can be quickly raised one course at a time.
(3) A *template* for shaping a plaster moulding.

profiled metal decking Galvanized steel *profiled sheeting*, shaped for use as *composite decking* or *permanent formwork* under a concrete floor slab.

profiled sheeting, metal decking, tray d., troughed s. Steel or aluminium sheets stiffened with regular longitudinal folds, to give trapezoidal (troughed) ribs of a deeper cross-section than *corrugated sheeting*, either self-finished or with coloured *organic coatings*. Occasionally other materials such as asbestos cement, GRP, and PVC are feasible. They are used for roofing and industrial wall cladding, being *self-supporting* over short spans. The flat tops of each rib make them easy to fix and give a good surface for fixing *insulating boards*. They come in long sheets, usually laid full length eaves to ridge without *headlaps*, but they need a *sidelap* of one or two ribs, depending on where they are used and the profile shape. See BS 4868 (aluminium) and BS 5427 (steel).

programme (USA **progress chart**) A schedule of work that the builder plans to carry out in order to complete a project. The master programme is usually a critical path *arrow diagram*, with simplified *bar charts* used for coordinating *sub-contractors*. On large projects the contractor is usually required to submit a programme early in the job. It then becomes a *contract document*, which is used for such purposes as reporting progress at each *site meeting*. See diagram, pp. 352-3.

programmer A *clock timer* for several circuits or complex switching arrangements, as in a *building management system*.

progress certificate Usually a *monthly certificate*.

progress payment Usually a *monthly instalment*.

progress report A report on work that has been completed on site, usually given as a percentage of each key *activity* completed, enabling a comparison to be made between real progress and the intended *programme* position. Progress reports are given at *site meetings* and often reveal *slippage*.

projecting hinge A *butt hinge* with wide *flaps* and a *knuckle* that sticks out, allowing a gap between a fully open door and the wall.

projecting scaffold A *scaffolding* built out from an upper storey, not reaching to the ground.

projection plastering, mechanical p., spray p. The application of gypsum plaster, render, or stucco by pumping it through a nozzle supplied through a short hose from a *plastering machine*.

Gypsum plasterwork is 5 to 25 mm thick, usually in a single coat of 13 mm on walls or 8–10 mm on ceilings. After spraying, the plaster is levelled with a *featheredge rule*, then later trowelled to a smooth finish. Slower-setting render or stucco is sprayed in the morning and finished in the afternoon. See diagram, p. 333.

project management See **project manager**.

project manager, client's representative (USA **construction program m.**) A person appointed by a *client* to plan and control a large, complex project and coordinate the building team through all stages of design and construction, as a professional service aimed at saving time and cost.

project representative A *clerk of works*.

prop A post to give temporary support under a load such as floor *formwork*. Timber props can be a simple length of wood, set on *wedges* and possibly with a *headtree*. Telescopic steel props have screw adjustment, a flat steel base-plate, and a plate or *split head*, or may be of *push-pull* type.

propane A *bottled gas* used for portable heating. The high heating value of the gas provides a very hot flame, used for heating *torches* and *blowlamps*, although it is not as hot as *oxy-acetylene*.

propeller fan, axial flow f. A fairly compact and efficient but noisy *fan*, mainly used for high-discharge *mechanical ventilation*.

protected membrane roof An *inverted roof*.

protected opening A hole in a *fire wall* or floor with a *fire door* or *fire shutter*, so that the protection of the *compartment* is unbroken.

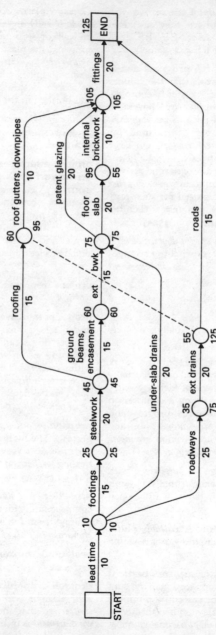

ARROW DIAGRAM

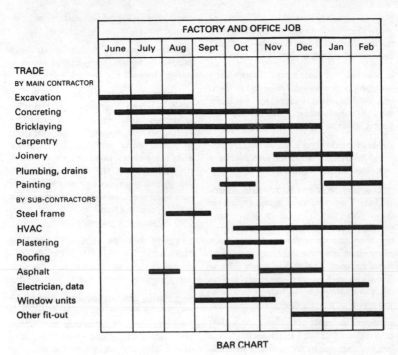

BAR CHART

Programmes.

protected route A corridor or stair on an *escape route* that is surrounded by fire-resisting construction, to protect people leaving the building from smoke and fire. It is reached through a *fire lobby* with self-closing *fire doors* and can be *pressurized*.

protected shaft A vertical services *duct* for a gas *riser* in a high-rise building, required by the *Building Regulations*. It must have ventilation to the outside at its top and bottom, and *fire stops* at each floor.

protected stair, enclosed s. (USA **fire tower**) An *escape stair* in the *protected route* of a tall building.

protection Measures against damage and injury to buildings or their occupants, as well as of building works and the *safety* of operatives. Protection can be aimed at preventing loss from *fire, lightning,* or abuse (*mechanical protection*), or to increase the *durability* of components likely to suffer decay or corrosion from the effects of *exposure, ultraviolet,* sun, wind, rain, or frost. Bitumen roofing may have dual-purpose *heavy protection.* Electrical installations involve many types of protection, against shocks or interference, and their circuits and accessories themselves need both mechanical protection and *protection gear.*

protection fan A *debris-collection fan.*

protection gear Electrical devices to

protect circuits, mostly from *overcurrent*. They include *circuit breakers* and motor *starters*. They may be grouped on a *distribution board*.

protection of finishings Temporary covering to prevent damage to finishes, removed before *handover*, e.g. peelable *film*, plywood casings, plastics sheet or board, and the *manifestation* of glass.

protective device Usually small electrical *protection gear*, such as a *fuse* or a *miniature circuit breaker*.

protective finishes to metals Many surface coatings can protect metals from *corrosion*. The most important purely protective finishes are *galvanizing, sherardizing, cadmium plating*, and *aluminium/zinc coatings*, or *sprayed metal* for large steel components. Combined protective and decorative finishes are *chromium plating*, nickel plating and many types of *organic coating*.

protective rail Usually a wooden board fixed to a wall to protect it from impact damage, as from chairs or trolleys.

proud Projecting beyond its surroundings.

provisional bills *Approximate quantities*, subject to *remeasurement*.

provisional sum (PS) An amount of money mentioned in *bills of quantities* to provide for *items* or work for which an exact price cannot be obtained or which are not clearly foreseen and will be subject to *remeasurement*. Some, such as the cost of insurance, site telephone for the *employer*, or photographs, do not affect *preliminaries*, while others, such as *daywork* and contingencies, may do so.

pry bar A *nail puller*.

PTFE *Polytetrafluoroethylene*.

P-trap A tubular pipe *trap* with an ordinary U-shaped seal and a final horizontal outlet.

puddle flange A flat collar round a pipe to make a watertight seal where it passes through concrete.

pugging (1) A thick layer of heavy infill between the joists of a timber *separating floor*, either on top of the ceiling or on *pugging boards*. It both improves *sound insulation*, by reducing airborne *transmission*, and increases the *fire resistance* of the floor. Dry sand, ashes, and high-density *slag wool* have also been used, as well as plaster *coarse stuff* containing hair.
(2) Mixing in a *pug mill*.

pugging boards, sound boarding Horizontal boards fitted closely between floor *joists* inside the floor to carry *pugging*.

pug mill, forced-action mixer A mixer with rollers, knives, or *twin-shaft paddles* used for breaking up lumps and blending (pugging) very stiff mixes of soil, clay, mortar, paint, etc.

pull A handle for opening a door with a *dead lock* or a drawer, etc., and called a door pull, drawer pull, or *sash lift* as the case may be.

pull box A *draw-in box*.

pull-cord switch, cord s., pull s. A *switch* fixed to the ceiling and operated by a hanging cord, thus safe to use in *bathrooms*. It also avoids the soiling of walls by dirty fingers.

pulley A wheel with high flanges shaped so that a rope can run easily on it.

pumice A natural *lightweight aggregate*, of 580 to 880 kg/m³ density, which can give a moderately strong concrete (2 to 14 N/mm²). It is used in *flue blocks*.

pump A machine for lifting or forcing

fluids or wet mixes. In building *services* pumps are mainly used for *circulating* heating water. They usually have *anti-vibration mountings* to reduce equipment noise. Large pump sets driven by electric motors need supply from a *motive power* circuit.

Pumps for the delivery of concrete, mortar, projection plaster, render, screed mix, etc., are usually mobile. On large projects they save time and cost once the pump, storage bins, mixers, pipes, hoses, etc., are in place.

pumpability The ease with which a wet mix of concrete, plaster, etc., can be forced through a *pump*.

pumping-grade concrete A concrete *mix* with good *pumpability*, having more sand than usual and *plasticizers*, so that it can be placed with a *concrete pump*.

pumping main The pipe on the outlet side of a pump.

punch A hand tool with a steel shaft and a pointed or flat tip, struck with a *hammer* to drive *lost-head nails*, to make holes in sheet roofing or cladding, to start drill holes, for *wasting* the surface of stone, etc.

puncheon (1) A short post part way along a *truss*, as in a *queen-post truss*.
(2) (USA) A slab of wood roughly dressed and used for flooring.

punch list (USA) A *schedule of defects*.

purchasing officer A buyer in a builder's head office in charge of buying *materials* for all sites, to be delivered on dates given by each *site manager*.

purge To clear a pipe or system of unwanted material by *bleeding*, flushing with fresh air, etc. Large oil-fired boilers may use pre-firing and post-firing firebox purges as a safety precaution.

purlin One of several horizontal roof beams parallel to the eaves and ridge, carried on the main framing members, such as *trusses* or *portal frames*. It carries *profiled sheeting* or *roof decking*, and may be a heavy timber on-edge or for longer spans steel *Z-purlins*. A horizontal timber supported by *principals* and carrying *common joists* is also a purlin.

push bar A rail at hand height on a door, as protective *door furniture*.

push button A button that closes (or opens) an electrical circuit, used to operate automatic controls, *fire alarms, lifts*, or an *emergency shutdown*.

push-button control The simplest automatic *lift control* system, which answers one call at a time, giving preference to calls from the lift car over those from the landings.

push-fit joint, gasketed j., push-on j. A pipe joint that is quickly made by simply pushing the *spigot* into a *socket*, with a sliding or rolling *joint ring* to make a watertight seal. There are many different types. Most push-fit joints can allow *thermal movement* by withdrawing the joint slightly after it is made, usually 10 mm, or to a manufacturer's mark, although drain pipes may have a soluble paper spacer ring. The joint is flexible and if the pipeline carries gas or water under pressure it needs *thrust blocks* and *anchor blocks* (this can be a danger during *demolition* work). Push-fit joints for plastics water pipes have a grip ring and they do not come apart unless the pipe fitting has a screw thread for dismantling.

push plate A metal plate on the *shutting stile* of a door at hand level, to protect the door face from dirt or minor damage by people's hands.

push-pull prop A telescopic steel *prop* that does not slide apart when pulled, used for the temporary bracing of *column forms, cladding panels,* etc. Only two are needed for a column, set at right angles on plan, instead of four ordinary props. The prop ends are bolted to the work, the bottom to an anchor previously cast into the floor slab (which has to be remembered).

putlog Short horizontal beams in a *putlog scaffold,* supported at one end by *brickwork* and at the other end by a *ledger.* Putlogs for *tubular scaffolding* have a flattened end to fit into *putlog holes.* In the Middle Ages heavy timbers were used as putlogs.

putlog adaptor A flat steel plate bolted to the end of a steel tube, making it a *putlog.*

putlog clip A lightweight scaffold clip to hold a *putlog* on its *ledger.*

putlog hole A raked out *bed joint* in brickwork or a large square hole in stone masonry to carry the end of a *putlog.*

putlog scaffold A *scaffold* carried on *putlogs,* usually a *bricklayer's scaffold.*

putty, glazier's p. A plastic substance made of *whiting* (chalk powder) mixed with *linseed oil,* which hardens on exposure to air, used for *glazing* glass into wooden window frames. First the *bedding* putty is put in the *glazing rebate,* then the glass *pane* is pushed in until the *back putty* is formed by squeezing the bedding to the right thickness. The pane is held by *glazing sprigs,* and finally a weatherproof fillet of *face putty* is made on the outside. Putty usually needs seven days before it is hard enough to take paint, but within twenty-eight days it should be painted to protect it from becoming brittle in later years. Linseed-oil putty is also used by painters for filling holes and cracks, but *stopping* shrinks much less.

putty knife A steel knife with the tip of its blade rounded on one side, straight on the other. A *stopping knife* and a *glazier's chisel* are similar.

PVC *Polyvinyl chloride.*

Pyran, Pyrostop Glasses for wire-free *fire-resisting glazing.*

Pyrotenax cable A *mineral insulated cable.*

Q

quad, quadrant moulding A *quarter round*.

quadrant A curved, metal *casement stay*.

quality assurance (QA) Measures taken to ensure that materials and services meet customer needs, by careful attention to design, production, and installation backed up by a registration, *certification*, and inspection service. Quality-assessed goods carry a *conformity mark*, in Europe the *CE mark*. Identical *standards* are BS 5750, ISO 9000 and EN 29000.

quality management (QM) The planning, procurement, process control, traceability, testing, customer service, design updating, and staff training policies and procedures for *quality assurance*. QM systems vary in size and type for manufacturing or services (architects, project managers, software).

quantities The amounts of building materials and of work put into a building, listed in *bills of quantities*. The quantity (length, area, volume, weight, number) in each *item* is multiplied by a *unit price* to give an amount of money. A contract 'with quantities' has a *priced bill*.

quantity surveying The preparation of *bills of quantities*, the settlement of payments from the *client* to the *contractor*, and the *arbitration* of disputed points after completion of the work.

quantity surveyor A person trained in accounting for building materials, construction costs, and contract procedures who costs feasibility studies, ad-

vising the client on the selection of a *contractor*, the preparation of *bills of quantities*, and *financial control* during the work and in the *post-contract stage*. The profession, mainly restricted to the UK, originated in the nineteenth century, when groups of competing contractors shared the cost of employing one person to draw up bills of quantities as a common basis for tenders. Chartered quantity surveyors belong to the *Royal Institute of Chartered Surveyors*.

quarry, floor q., q. tile A clay *floor tile* made from ordinary clays, usually 'as-dug' or treated in a *pug mill*, formed by extrusion and fired in a kiln. There are many different sizes and shapes, such as *interlocking*, rectangular, and square. They have attractive red-yellow colours, are usually hard wearing and non-slip, but are noisy and cold. Quarries should be washed with detergent and clean warm water. They are porous and should not be polished, but can be sealed with *linseed oil*.

quarry sap Moisture in freshly quarried stone. After it has dried out the stone is seasoned and stronger but harder to work.

quarter bend A *pipe fitting* that turns 90°, or for drain pipes 92$\frac{1}{2}$°, to give a 2$\frac{1}{2}$° fall to a horizontal pipe joined to a vertical *stack* pipe.

quarter point The two points a quarter way along a beam from each end. Two-point lifting at the quarter points usually produces low bending stresses.

quarter round, quad, quadrant moulding A *moulding* such as a

quarter-sawn timber

quarter circle, usually of solid timber.

quarter-sawn timber Boards sawn with the *growth rings* more or less at right angles to the *face*, so that most shrinkage occurs in the board's thickness. An exact right angle is not possible; less than 45° is usual for floorboards, but not for *fully quarter-sawn*.

quarter-space landing A small square landing between straight stair *flights* at an angle of 90° to each other.

quartzite A hard-wearing silica stone used for large floor tiles, up to 900 mm square. It has good resistance to chemicals, frost, and grease.

queen closer A half-width *special* brick, used as a *closer*. In Britain it is 46 mm wide, but still 215 mm long by 65 mm high.

queen-post truss A *truss* with no central post, unlike the *king-post truss*, but

A right-angle quoin in stone masonry.

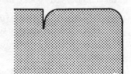

quirk

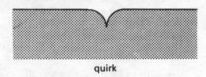

quirk and bead

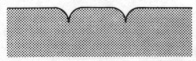

double quirk

Quirks.

with two *puncheons* on either side of centre which can be replaced by long steel bolts if the members act only in tension.

Quetta bond A type of reinforced brickwork rarely used. It is laid *brick-and-a-half* thick to form vertical voids for steel *reinforcement* rods and concrete. It is similar to *rat-trap bond*, but the bricks are laid flat.

quick bend A *bend* with a small radius, the opposite of a *slow* or swept bend.

quicklime Calcium oxide (CaO), burnt lime that has been calcined (heated) until anhydrous (waterless). It reacts strongly with water to give hydrated *lime*.

quilt A *blanket*.

quilted figure, blister f. An elaborate *figure* in timber, consisting of apparent knolls in *birch* or maple due to uneven *annual rings*.

quirk A V-groove with rounded edges, resembling the centre of an open book, used in timber *joinery*. See diagram.

quirk bead A rounded edge with a *quirk* beside it, often with a vertical inner side. A double quirk bead has quirks each side.

quoin An outer corner of a wall, made with squared stones, or bricks, or blocks laid in alternating courses as a *header* in one wall and a *stretcher* in the other. This *bonds* the corner and forms *toothings* for other walling. Quoins are usually decorative and project slightly. They may have smooth *dressings, rusticated ashlar*, or bricks of a different colour. They are also used at door and window openings. See diagram.

quoin block, corner b. An L-shaped building *block* used to make a corner.

quotation, quote A price to do work, usually given by a *sub-contractor* to a *contractor*. It is an *offer* and if accepted a *contract* is formed.

R

rabbet A *rebate*.

raceway (USA) *Trunking*.

racking back, raking b. Laying bricks in steps at the end of a wall to support the bricklayer's *line* while courses are filled in. See diagram.

radial circuit A *branch circuit* which is fed from one end only. With a 32 amp fuse it can serve unlimited socket outlets in a floor area not exceeding 50 m².

radial-sett paving Paving with *granite setts* laid in semi-circles within arm's reach of the paving layer.

radial shake Timber *shake* radiating outwards from the *heart*.

radial shrinkage The drying *shrinkage* of timber at right angles to the *growth rings*, usually about two thirds of the *tangential shrinkage*.

radiant heating Space *heating* directly from a heat source. Hot gas, electric, or wood fires give off concentrated infra-red radiation and are comfortable even if the room air is cool. Other sources are at medium temperatures, such as radiant panels, *radiators*, *panel heating*, and *room-heaters*. Low-temperature radiant heating is usually *concealed*.

radiation detector A *flame detector*.

radiation protection The use of heavy concrete in walls and floors, *lead*

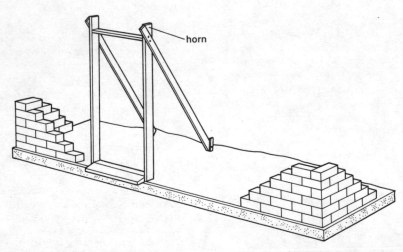

horn

Racking back. The corners of the wall are raised first, gauged for course height, and plumbed. Door and window frames are set, plumbed, and braced. A line between the corners is used to guide bricklaying.

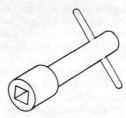

A radiator key. The small hole is about 5 mm across.

sheet linings, or *barium plaster* finishings, for protection against X-rays.

radiator A *heater* that warms by radiation. A fixed radiator for hot-water *central heating* can be a large flat panel of pressed steel, often with rows of fins, usually factory painted. Radiators were formerly of cast iron with *heat-resistant paint*. They have valves for control, *balancing*, and *bleeding*, and a thermostat can be fitted. Although they warm by radiation, being at a low temperature most of their heat goes into the air and they could therefore be called *convectors*.

radiator key A small spanner that releases air from a radiator through a special vent. The process is known as 'bleeding' the radiator. See diagram.

radio-frequency heating The use of microwaves to heat hot-setting *adhesives* inside wood assemblies. *Curing* is thus quicker, which speeds up factory production and saves clamps. It is used on synthetic *resin* adhesives, for *finger joints* in structural softwoods, and to make wood building boards.

radius rod In plastering, a stick that carries the *mould* used to form a circular arch, pivoting on a nail driven through the other end.

radon A poisonous, heavy, radioactive gas emitted from granites. It can be excluded from housing built over gran-

ite either by a gastight underlay below the *ground slab* or by the ventilation of the *underfloor space*. Because it is dense it can accumulate in a room with too few *air changes*. In Britain the main problem areas are in Devon and Cornwall. Advice before building is given on the BRE radon 'hotline', tel. (0923) 644707. See BRE report BR 211. See also **phosphogypsum** and *C*.

rafter A sloping roof beam, usually from eave to ridge. In traditional timber framing the rafters share loads in common; a full-length *intermediate rafter* runs from eave to ridge but a *jack rafter* goes only part way, meeting with *hip* or *valley* rafters. The term rafter is also applied to all types of *trussed rafter*, the sloping beam of a *portal frame*, and the *principal rafter* of a truss.

rafter filling, beam f., wind f. Brick infilling between rafters at *wall-plate* level.

ragbolt A thick roughened bolt cast into concrete as a *primary fixing*.

ragging A *textured finish* to interior paint given by passing a special *roller* with loose flapping rags while the coat is still soft.

raglet, raggle, raglin, reglet A narrow horizontal groove or *chase* in a wall to receive the turned-in edge of a *flashing*, which is fixed with mortar. Straight-sided raglets in brickwork are made by raking out the face of *bed joints*. They can be *dovetailed* if cut into masonry or formed in concrete.

rail (1) A horizontal joinery member, framed into vertical *stiles*. A panelled door *leaf frame* has top, bottom, and intermediate or lock rails.

(2) The upper continuous part of a *balustrade*, a *handrail*.

(3) Any horizontal timber, as in a fence, or to hang curtains on, etc.

railing

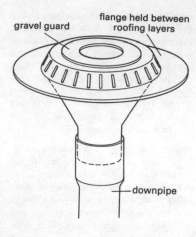

gravel guard — flange held between roofing layers

downpipe

A rainwater outlet for a flat roof.

railing An open fence with posts and horizontal or vertical rails.

rain See **driving rain**.

rainwater goods Roof *gutters*, *downpipes*, and fittings or accessories, such as *offsets* or *shoes*. They are usually sold as part of a system and may not be interchangeable with parts from a system by another maker. Usual materials are PVC, powder-coated aluminium, and galvanized steel. The layout and joints of rainwater goods should allow high *thermal movement*.

rainwater header, r. head, r. hopper (USA **conductor h., leader h.**) The enlarged funnel on top of a *downpipe* to increase the flow of rainwater by keeping the downpipe filled. Headers are outside the wall in eaves gutters and box gutter *scuppers*, but in built-up roofing they can be a *dishing* or form part of a *rainwater outlet*.

rainwater outlet A hole by which rainwater leaves a roof or gutter. It may discharge into a *rainwater header*

or a *downpipe*, or pass through a *parapet* wall. The outlet has a *spigot* that goes inside the downpipe, usually a slide fit to allow *thermal movement*. Eaves gutters have a *drop end*. Flat roofs have a funnel header, with a *ballast grating* and *luting* flange for roofing felts, often flexibly connected to prevent them cracking. For maximum flow they should be located in an open area of roof. Insulated and electrically heated models are used to prevent blockage by ice and snow. See diagram.

rainwater pipe A *downpipe*.

rainwater plumbing, external p. The *trade* of installing *flashings*, *supported sheetmetal roofing*, *plumber's metalwork*, roof gutters, and other *rainwater goods*. It is a *plumber's* work.

rainwater shoe A bend on the bottom of a *downpipe* forming an outlet spout.

rainwater spout, gargoyle A roof outlet that throws rainwater clear of walls to protect them from scouring or becoming saturated and prone to frost damage.

raised countersunk head screw A wood screw with a head slightly domed on top and countersunk underneath, used for fixing metal *threshold strips*.

raised fibres The fibres of wood rise from the surface when it dries naturally or after the application of a coat of paint or varnish. To achieve a glossy paint or varnish surface the fibres should be removed with fine *sandpaper* after every coat but the last.

raised floor An *access floor*.

raising piece A *packer* or short length of material inserted to bring a component to its required level.

rake, batter The angle of a member to the vertical.

raked-out joint A mortar *joint* which

has been cleaned out of mortar for a small part of its depth. Brickwork joints are raked out 20 mm before *pointing*.

raking cutting Cutting not at a right angle, e.g. of bricks or tiles.

raking flashing A *counter-flashing* at a sloping *abutment*, where the side of a roof meets a wall or chimney. The top edge of a raking flashing is parallel to the roof slope and let into a sloping *raglet*, usually formed by a saw cut.

raking out Removing mortar to make a *raked-out joint*.

raking riser A stair *riser* with an outward slope which partly overhangs the *tread* below and gives more toe space. The tread is deeper than its *going*. See diagram, p. 437.

raking shore, inclined s., shoring, raker A long *baulk* of timber, or several heavy timbers, erected to hold up temporarily a wall of a building. It is carried on a *sole plate* of old railway sleepers. It abuts against the building a little below each floor and is driven up hard under a *needle* at that level. See diagram, p. 406.

rammed-earth construction (USA) *Pisé*.

ramp (1) A bend in a *handrail* or coping which is concave on the upper side, the opposite of a *knee*.
(2) A short, steep length of drain pipe.

random ashlar (USA) Walling of *coursed squared rubble*.

random courses *Courses* of varying depths.

random paving Thin stones of irregular shape laid to fit together.

random rubble walls *Rubble walls* built of stones of irregular shape and size and not in courses.

random slates Roof *slates* of varying widths, not *sized*, laid in *diminishing courses* or at random. Their rustic appearance is often attractive. The best-known English randoms are Delabole and Westmorland.

range A row of similar appliances, usually sharing the same *services*, e.g. basins, commercial cookers. They can be on a wall or freestanding.

range masonry, coursed ashlar, r. work (USA) *Regular coursed rubble*.

ranging line A string *line* between profiles, used for *setting out* a straight line, such as the face of a wall.

rasp A toothed steel bar with a handle. It is a file that roughly shapes wood. See diagram, p. 337.

rate A *unit price*.

rate of growth The number of *growth rings* per cm of timber measured radially, one of the *gross features* used in *stress grading*. In general, *narrow-ringed* timbers are the strongest.

rate of pay, wage r. The amount of money given in exchange for the work done by a building *operative* for an agreed number of hours. Additional payments may be made for *overtime, bonus*, etc. The earliest recorded legal decision on rates of pay in the building industry in England was given by the London Assize of Wages of 1212, for *freemasons* and others.

rate-of-rise detector A *fire detector* which reacts to a sudden increase in temperature. It responds more quickly than a *fixed-temperature heat detector*.

rating The *nominal* maximum capacity of a piece of equipment, e.g. the stated power of an electric motor, the duty load of a *lift*, the largest bit for an electric *drill*, or the breaking capacity of a fuse.

rat-trap bond A brickwork *bond* laid with bricks on edge in *Flemish bond*, giving a wall 215 mm thick, with a cavity behind each *stretcher*. Its cost and performance are between those of a 215 mm *solid wall* and a 275 mm *cavity wall*.

raw linseed oil Refined *linseed oil* which has not been *boiled* or *blown*. Unrefined linseed oil is never used in paints.

ray, wood r., xylem r. Small strips or ribbons of attractive *grain* in timber, sometimes visible, especially in the *silver grain* of quartered oak. Rays occur in the *xylem* or growth wood at right angles to the *annual rings*, not in the medulla or pith, and the term 'medullary ray' is therefore obsolescent.

razor socket A *shaver socket*.

readymix concrete, ready-mixed c. Mixed but unset *concrete* brought to site from a batching plant, usually in an *agitating truck* (*C*). Close coordination is needed with the plant operator (batcher), usually only through the *general foreman*. For all structural work, test *cubes* need to be taken on site. See also *C*.

ready-mixed mortar A factory-mixed product delivered to site wet. There are two main groups: (a) Lime-sand-mortar *coarse stuff*, for bricklaying *composition mortar*. It is re-mixed on site with cement and more water. A *cement mixer* is needed, but there is a saving in costly cement. Before use it can be stored for a long time in its large steel delivery bin. (b) Normal mortar with a *retarder*, used for bricklaying, plastering, rendering, and screeds. It may be pumped directly to the work or into plastics tubs lifted by crane. It has to be used promptly or it will go off (setting in a tub in about three days), but no cement mixer is needed on site. See BS 4721.

rebar Abbreviation for *reinforcement bar*.

rebate, rabbet (Scotland **check**) A step-shaped rectangular recess along an edge, e.g. a *glazing rebate* in a window, or the rebate in a *door frame* into which the door leaf closes. See diagram, p. 207.

rebated joint An *ogee joint*.

rebated weatherboarding *Weatherboarding* of wedge-shaped cross-section with a *rebate* along the inner face of its lower edge so that the top (thin) edge of the lower boards fits into the rebate of the lower (thicker) edge of the upper board. *German siding* and *shiplap boards* are also both rebated but are not specially so called.

receiver A tank, pit, or vessel for temporary storage, e.g. of compressed air, refrigerant, dangerous effluent, etc.

receiver clock, secondary c. A clock worked by a *time distribution* system.

receptacle (USA) A *socket outlet*.

receptor (1) A *channel* section fixed to a wall, floor, or ceiling to hold the edge of a *demountable partition*.
(2) The material into which a *fixing* is driven. Wood is a good receptor for nails and screws.
(3) (USA) A *shower tray*.

recess A hole formed in a surface, either open, or as a *housing* for a built-in part, which often ends up *flush* with the surface.

recessed-head screw One of many different types of *crossed-slot* screw heads, e.g. Pozidriv (PZ), Supadriv, *Phillips*, etc. Each type should be driven only with its own special *cross-top screwdriver* of the right size. Compared with a *straight slot*, they are neat and easy to drive without slipping and gouging the work surface.

recessed joint A mortar *joint* set back about 6 mm from the face of the wall, either a *raked-out joint* or one made by *tooling*. It is used for brickwork, and for stonework to prevent *flushing*, and is strongly shadowed.

recessed luminaire A *luminaire* concealed in the ceiling and giving light from only part of its surface, none of which falls directly on the ceiling. It should have a good *reflector* and may need a baffle against *glare*.

recirculated air Air-conditioning *return air* which is sent back to the central plant.

recirculated hot water Domestic hot-water *secondary circulation*.

recoating time The time allowed for the *drying* of one coat of paint before *overcoating* it with another coat. Oil paints can usually be recoated after about twenty-four hours, unless put on too thick or in cold weather, but only after they are *hard dry*. Emulsion paints dry more quickly and can be recoated earlier.

reconditioning The high-temperature steam treatment of *hardwood* which has been *kiln-dried*. The process was developed for Australian eucalyptus timbers which suffer *collapse*, but is also used in France to correct over-drying of outer layers. It should bring timber to *equilibrium moisture content* throughout its thickness and reduce *warping*.

reconstituted marble Waste chips of marble from quarrying, bonded together with cement or synthetic resin and powdered marble, cured, then sawn into slabs.

reconstructed stone *Cast stone*.

record drawing Usually an *as-built drawing*, or for services 'as-fitted'.

rectangular on plan Block-shaped on *plan*, e.g. a pump house, shed, or transformer sub-station.

red lead (lead oxide) A poisonous red *pigment* consisting of mixed oxides of lead, mainly Pb_3O_4. It is an *inhibiting pigment* once used widely in *primer* for steel and wood, and still used with *putty* to seal joints in cast-iron or galvanized-steel gutters.

red oxide (iron oxide) A paint *pigment* used in *metal primer* to keep out water. It does not act as a rust inhibitor, as *red lead* did.

reducer (1) A *taper* pipe reducing in diameter in the direction of flow.
(2) Paint *thinner*.

redwood (Pinus sylvestris) A *softwood* which grows throughout northern Europe and Asia, used for *carpentry* and joinery. It is easy to work, durable, machinable, and takes *nails*, *screws*, and *adhesives* well. The word redwood is usually confined to timber imported to Britain, often from the Baltic.

reed thatching The most durable *thatching*, done with 'best reed' (Arundo phragmites), although the more pliable sedge (Cladium mariscus) is used for ridges.

reference panel A *sample panel*.

reference specification A *specification* that does not describe everything in detail but refers to *standards* or *codes of practice*.

reflectance The proportion of *luminous flux* falling on a surface which is given back again, expressed as a percentage or by a *Munsell* number. Light colours and high *gloss* produce high reflectance.

reflective glass A type of *solar-control glazing*.

reflective insulation Insulating materials with a shiny surface, e.g. *aluminium foil*, which have *low emissivity*. They reduce heat loss from radiation.

reflector (1) A light-coloured or shiny metal surface behind a *lamp* to direct light back or to disperse it and reduce *glare*.
(2) A *luminaire* with a reflector, such as an industrial *fluorescent tube* with batwing sides. It is used as a *downlighter* for *high bay lighting*.

reflux valve See **check valve** (*C*).

refractory mortar Mortar able to withstand very high temperatures. One suitable mix is three parts *grog* and one part *high-alumina cement* (*C*).

refrigerant The working fluid in a *refrigerator*, which alternates between being a liquid and being a vapour. Compressor refrigerators using the *vapour compression cycle* have different refrigerants from *absorption refrigerators*.

refrigerated storage A *cold store* or a *freezer*.

refrigeration unit, r. machine A *chiller*.

refrigerator A machine used to produce or maintain low temperatures. They work either on the *vapour compression cycle* or by *absorption*.

refurbishment, rehabilitation Bringing back an old building to a useful condition, using original materials or modern imitations, but usually without major *conversion*. The work can be limited to simple maintenance and *restoration* of the exterior with renovation and redecoration of the interior, or go as far as almost total demolition, with only *façade retention*. There are

particular dangers in removing *lead paints* and *asbestos* or in *cleaning down* masonry. Special *fire precautions* and *hot-work permits* may also be needed.

refuse chute A vertical pipe for refuse disposal in high-rise housing and multi-storey buildings. The chute should be round and at least 375 mm dia., preferably larger. Rubbish is put in through a hopper, usually in a public area, and drops into a receiving chamber under the building. See BS 1703.

register (1) An outlet into a room from an air-conditioning duct, with a *damper* and *grille* or diffuser. It is an *air terminal unit* for *supply air*.
(2) A chimney *damper*.

reglet A *raglet*.

regrating The *cleaning-down* of stone masonry.

regular coursed rubble (USA **coursed ashlar, range masonry, range work**) Coursed *rubble walling* in courses of different heights, generally from 75 to 150 mm.

regularized wood Wood that has been machined by planing or thicknessing on two or four sides to provide a true, flat surface. The aim is not strength or beauty but to make the cross-section uniform throughout its length, and there may be rough unplaned patches.

regulus metal An alloy of *lead* and 10% antimony. It is easier to work than pure lead and resists *creep* (*C*).

rehabilitation *Refurbishment*.

re-heat unit, terminal r.u. An *air terminal unit* with a heating *battery* for the final adjustment of *supply air* temperature before discharge. It is usually controlled by a room thermostat.

reinforced bitumen felt A lightweight *bitumen felt* made of fibre saturated with *bitumen*, with an embedded layer of jute *hessian*. It is used as an *underlay* on unboarded roofs.

reinforced concrete *Concrete* with steel *reinforcement*, a versatile material widely used in the loadbearing *structure* of buildings. See *C*.

reinforced masonry Brick, block, or stone work reinforced with steel, such as wire mesh in the *bed joints* or bars in vertical holes filled with concrete.

reinforced screed A cement *screed* reinforced with light steel *fabric*.

reinforcement Small amounts of a strong material (steel bars, fibres) added to a weaker material (concrete, bitumen, plastics), usually during manufacture, giving a *composite*.

reinforcement bar, reinf., rebar, steel A steel rod used as *reinforcement* for concrete. The placing of bars inside *formwork* is done by *steelfixers* working to the structural *drawings* and *bending schedules*. The bars must have earth *bonding*. See **cover**.

reinforcement schedule A *bending schedule*.

relamping The replacement of the *lamp* in a *luminaire*. Commercial and office buildings are often fully re-lamped at the one time, the relamping being based on the economic cost of *light loss* as lamps get old. This can result in less disruption to people, allow cleaning of the luminaire, and ensure that the right types of lamp and starter are used.

relative humidity The weight of water vapour in air compared with saturated air at the same temperature, usually given as a percentage. In winter, cold air from outdoors has a lower relative humidity once it is heated to room temperature, as warm air can carry more water vapour than

cold air. The feeling of dryness from the lower relative humidity can be corrected by a *humidifier*, although relative humidity has little effect on *comfort* provided it remains between 40 and 70%. Buildings need *ventilation* to remove warm moist indoor air. If it comes in contact with a cold surface it can reach 100% relative humidity, resulting in *condensation*.

relieving arch, safety a. A strong arch built into masonry above an opening to carry some of the load otherwise carried by a *lintel* or arch.

reloading Restoring power to a circuit after *load shedding*.

relocatable partition A *demountable partition*, used in large offices. It can be moved often and easily.

remeasurement The *measurement* of actual *quantities* after work has been done, usually for contracts with a *schedule of prices* or *provisional bills*.

remedial work Work to make good a *defect*, improve existing ground, etc.

remote-entry system Usually an *intercom* from an entrance door whereby a tenant can allow people to get in by operating an *electric striking plate*.

render The mortar, or other mix, used for *rendering* or *stucco* work. Cement-based render needs similar *curing* to concrete.

render and set Plaster *two-coat work*.

render, float, and set Plaster *three-coat work*.

rendering Plastering with *render*; or the *finish* produced. Cement *mortar* and *resin mortar* render are not affected by dampness and generally keep out water. Mortar is used for all *undercoats*, as well as the *final coats* for external walls. Rainwater is thrown off external rendering by having a rough final surface texture, such as *roughcast* and *Tyrolean finish*, or by using *mouldings*, as in *stucco*. Indoor rendering is used in *wet areas* to receive tiling.

rendering coat A first coat of *plaster* on a wall, or on *lathing*, usually about 10 mm thick. In *three-coat work* it has the same mix as a *floating coat*.

renovation The repair and improvement of a building, including some demolition or conversion. See **refurbishment**.

renovation plaster Pre-mixed plaster made with *lightweight aggregate* and either *retarded hemihydrate* gypsum or cement and lime, for use on old walls likely to suffer dampness or condensation. It is much stronger than a normal plaster.

repainting, maintenance painting, overpainting Painting on a previously painted surface. Surface *preparation* includes the removal of old cracked, peeled, or flaked paint, as well as repairs to the *substrate*, re-priming and *bringing forward* undercoats, rubbing down *gloss* coats, and applying finish coats.

repointing (1) Replacing *mortar* that weather has eroded from *face joints* in brickwork, stonework, etc., to reduce the rate of deterioration of *wall ties* or as part of *refurbishment*. See BS 6270.
(2) *Pointing* after *raking out*.

repressed brick An extruded *wirecut brick* which is later pressed in a mould to true its sides.

resaturant A product used for reviving the *bitumen* in old roofing felts.

re-sawing The sawing of *flitches* into *square-sawn timber* or building *scantlings* – the second stage of *conversion*, after *breaking down*.

rescheduling Programme *updating*.

resealing trap A *trap* that allows momentary *unsealing* to equalize high or low pressure in the *discharge pipes*.

reserve A *fire reserve*.

reshoring *Back propping*.

resident engineer (RE) The site representative of the structural or mechanical services *consultant*, who should deal only with the *site manager*.

residual-current circuit-breaker, current-balance c.-b. An *earth-leakage circuit-breaker* that compares the current flowing in the two wires of a circuit and switches the circuit off if they become unequal, indicating leakage to earth.

resilient flooring *Floor finishes* which are soft to walk on and do not create *impact noise*, e.g. cork, linoleum, rubber, and flexible plastics.

resilient mounting An *anti-vibration mounting*.

resin (1) Natural pitch, from the sap of pine trees.
(2) **synthetic r.** A group of *polymers* used in high-performance plastics, adhesives, waterproofers, binders, and paints, for instance *acrylic*, *alkyd*, *epoxy*, *polyester*, and *silicone*, as well as melamine-, phenol-, resorcinol-, and urea-formaldehyde resins. Some resins are poisonous – in general they should not be allowed on the skin or in the eyes, nose, and mouth. Breathing in resin vapours or fumes is also dangerous to health.

resin mortar A mixture of sand and synthetic *resin*, used as *levelling compound* for special plaster finishes, repairing masonry, etc.

resin pocket, pitch p. An opening between growth rings in timber containing natural resin. It is a defect in softwoods.

resistance Restraint to the flow of heat, electricity, sound, etc., the reciprocal of *conductance*.

resorcinol-formaldehyde (RF) An expensive synthetic *resin*, one of the few able to make *weatherproof adhesives*.

resource smoothing, load s. The adjustment of a *programme* to give a steady need for construction plant and equipment and site labour.

response A tender *submission*.

responsibility Duties and obligations. In the administration of *contracts* responsibility may become *divided* or *transferred*. See also *C*.

rest bend A *duckfoot bend*.

restoration Full *refurbishment* to bring back to the original condition, such as the cleaning of metals, repointing brickwork, and redressing stonework.

restricted tendering *Selected tendering*.

retaining wall See *C*.

retarded hemihydrate plaster, Class B p. (to BS 1191) The basic material for all *gypsum plaster*. It is *hemihydrate plaster* containing a *retarder* such as *keratin*.

retarder, retarding admixture An *admixture* which slows down the initial rate of setting, usually by inhibiting hydration. Different types are used for cement or plaster.

retempering, knocking up Adding extra water to and re-mixing concrete, plaster, or mortar that has partially dried or set and is too stiff to use. Retempering is regarded as bad *workmanship* and usually not allowed; smaller *batches* may be needed.

retention Amounts of money deducted from each *monthly instalment* made to a builder, usually 5% of the amount on the architect's *certificate*, as a way of making sure that the builder continues working in *performance* of the contract. Usually one half of the retention is released (paid) at *practical completion* and the other half at *final completion* after the *defects liability period*. See **surety**.

reticulated finish An irregular network of lines on the face of stone.

retrofit To fit extra equipment or strengthening to an existing building.

return (1) **r. end, r. wall** A short change of direction at the end of a wall, usually at right angles. Returns of brick walls are important for two reasons. First, the return adds strength because of its buttressing effect. Secondly, if the tip of the return is restrained and there is *movement* in the main length of wall, the corner may crack.

(2) The opposite of *flow* or *supply*: said of water or air going back to a boiler, chiller, or treatment plant, to complete its *circuit*.

return air Air exhausted from rooms, blown through return-air ducts to the central plant, where it is mostly recirculated, except for part of the volume which is rejected outdoors to make way for fresh air.

return fill *Backfill*.

return latch, catch bolt, spring l., tongue The part of a door *latch* which engages in the door frame, forced outwards by a spring. Its end is *bevelled* to slide back into the lock as the door closes. When in the 'shot' position in the *striking plate* it holds the door closed, until it is withdrawn by turning the door handle, or key, allowing the door to be opened. See **follower**.

return pipe

return pipe The pipe by which the water returns to a boiler from a radiator, hot-water cylinder, etc.; the primary return pipe to a *calorifier*; or a pipe from a cooling *coil* back to the *chiller*, etc. A *flow pipe* begins the circuit.

return wall A *return*.

re-use of formwork *Formwork* can be re-used to spread its cost. Steel *panel forms* or steel form linings can last for 1000 re-uses, film-faced plywood for about a dozen, and *chipboard* for one or two.

reveal The visible part of a *jamb* in the opening for a door, window, etc., not covered by the frame, although it may have a *reveal lining*.

reveal lining The finish over a *reveal*, often a *sub-frame* or *surround*.

reveal pin, r. tie A strut across a *reveal* tightened by screw adjustment, or wedges, which ties a *scaffold* to the building.

reverberation The re-echoing of sound between wall, floor, and ceiling surfaces. It reaches the hearer after the direct sound and may confuse speech. See next entry.

reverberation time The time in seconds for sound within a room to die away by 60 *decibels*, which is the difference between normal conversation and the threshold of hearing. The reverberation time is proportional to the room volume divided by the total *absorption* of room surfaces and contents (such as people). Sound absorbers can be introduced into a room to lower the reverberation period, so improving listening conditions and reducing noise levels.

reversible lock, double-handed l. A *lock* that can be *handed* both ways by removing and turning over the *latch* or mounting the lock either way up.

reversible movement *Cyclic movement*.

revision (1) A re-issue of a *drawing*, usually indicated by a revision letter (A, B, C) after the drawing number. The list of revisions in the *title block* gives the date of each re-issue and brief details of the *amendments*.

(2) A minor amendment to a *programme*, but see **updating**.

revolving door, tambour A door with four (or three) leaves on a central pivot, plus two partial curved outer screens, one on each side, used at entrance lobbies to offices, hotels, etc. They exclude draughts, thus reducing the load on *air conditioning*, and keep out noise, while allowing people to flow in and out. Although heavy they are stable, as pressure is balanced by each leaf. Automatic motor drive is usual for the larger sizes. Two ordinary *swing doors* are also needed, one each side, or a mechanism to collapse the leaves and allow a free passage for fire evacuation or bulky objects.

rewirable fuse An old type of *fuse* with a ceramic holder for fuse wire.

rewirable wiring Electrical *wiring* that is drawn in to *conduit*.

rework To make good work that has been incorrectly done or contains a *defect*.

RIBA The *Royal Institute of British Architects*.

rib and block suspended floor A lightweight concrete floor slab with stiffening ribs formed by the spaces between hollow floor blocks. See **hollow-tile floor** (*C*).

ribbed-sheet roofing *Profiled sheeting*.

ribbon board, ledger A horizontal beam fixed to a wall or housed into *studs*, to carry the ends of floor *joists*.

370

ribbon cable A flat electrical cable used for under-carpet distribution wiring between telephones or computer equipment.

ribbon course An ornamental course of slates or tiles of a different colour.

rich mix A *fat mix.*

RICS The *Royal Institute of Chartered Surveyors.*

ride (1) A door which touches the floor when it opens is said to ride. This is overcome by trimming or raising the leaf (by re-hanging or adding washers to *lift-off hinges*) or, if the floor slopes, by *cocking* the hinges.
(2) A joint which touches at one or more high spots or points is said to ride.

rider shore A short, topmost *raking shore* which rests not on the ground but on a *back shore* lying on top of the highest full-length raking shore. The rider may be omitted for a low building. See diagram, p. 406.

ridge The horizontal line on top of a double pitched roof, between the two slopes. Wind forces at the ridge can be double those on the general areas of roofing.

ridge board, r. piece A horizontal board set on edge at which *rafters* meet, usually about 230 mm high and 20 to 40 mm thick.

ridge capping A covering over a *ridge.* It may be *ridge tiles*, folded sheet metal, *bitumen felt*, etc., depending on roof construction. The ridge capping is particularly vulnerable to the weather.

ridge course The last course of tiles (or slates) next the ridge. Each tile head should be well supported. Courses should be set out for full-length tiles, but a ridge course can be of cut (or purpose-made) shorter tiles.

ridge end A *special* for the end of a run of *ridge tiles*, e.g. a *bonnet.*

ridge stop A sheet metal *flashing* where a ridge meets a wall at right angles. It is dressed over the top of the ridge, up the wall and turned into a *raglet.*

ridge terminal A neat and compact outlet from a soil *vent pipe* or a gas *flue* built into the ridge of a *trussed-rafter* roof. There is no need for a *pipe flashing.*

ridge tile A roof tile for covering a *ridge*, often suitable as a *hip tile*, usually *half round.* It is either 300 or 460 mm long, has *spigot-and-socket* joints, and is traditionally bedded in mortar, with *dentil slips* for some deep troughs, although *dry ridges* are available for many tile patterns. The *ridge end* may overlap a top hip tile. See BS 402 and BS 5534.

ridge vent A ventilator to let moist air escape from a *cold roof.*

rift-sawn timber *Quarter-sawn timber.*

rigid damp-proof course A *brick damp course* or a *slate damp-proof course.*

rilled finish Continuous shallow parallel grooves across a surface.

rim latch A door *latch* which is fixed to the surface of a door leaf (surface-mounted, planted) on the edge of the *shutting stile*, not *morticed.* Its metal case usually matches its *box staple* on the door jamb.

rim lock A lockable *rim latch.* It gives security from one side only.

ring In pipe fittings, a *joint ring.*

ring beam (1) A horizontal *tie beam* in a blockwork perimeter wall.
(2) The circular beam round the

bottom of a *dome*, to take its outward thrust, usually at the *springing* point.

ring main, r. circuit An economical method of *wiring* power branch circuits that gives two current paths, allowing smaller, cheaper cables. The cable is run from the house consumer's unit and *looped in* to each *socket outlet* or other power point, then returned and also connected at the other end, usually in *sheathed wiring*, twin 2.5 mm² with earth. It is usual to have about ten socket outlets on a house ring circuit, although an unlimited number are allowed, including those on *spurs*, provided they are all within a floor area of 100 m². Office buildings can also be wired to the same rules. The whole ring circuit is protected by a 32 amp fuse, and each socket outlet has its own 13 amp fused plug.

ring shake, cup s., wind s. *Shake* along one or more *growth rings*.

ring-shanked nail, improved n. A nail formed with upstand rings on the shank, which greatly increase its withdrawal load. Like ordinary nails they do not make fully rigid joints. They may be rust-proofed.

ring transformer A device that detects magnetic fields, used in three-phase wiring to detect current imbalance and trip the *protection gear*.

rip To saw wood parallel to the *grain*.

ripsaw A coarse *handsaw* with teeth sharpened to a flat chisel point, set alternately left and right, used to *rip* wood. See diagram, p. 402.

rise (1) A vertical dimension, such as the difference in level between two stair *treads* or between stair *landings*. The rise of stair treads should all be equal, which may involve an allowance for *screed* thickness at the landings. See diagram, p. 437.

(2) The height of a roof, from its support to the highest point, or of an arch from *springing point* to *intrados*.

rise and fall *Fluctuations*.

rise-and-fall table A circular *saw bench* which can be raised or lowered relative to the fixed saw.

riser (1) **rising main** A vertical distribution pipe or cable, usually run inside a services riser *shaft* or *duct*, to bring electrical power, cold water, or gas to the upper storeys of a building. It may also be a *fire riser*.

(2) The front face of a step. The *rise* of stair risers should not exceed 220 mm. It is usually vertical in a timber stair. See diagram, p. 437. See also **raking riser**.

riser shaft A vertical services *duct*.

rising-butt hinges *Hinges* which cause a door to rise about 10 mm as it opens. They have a helical bearing surface between the two flaps. The door therefore tends to close automatically as well as to clear a carpet when opening. The closing action is not strong enough for use on a *fire door*. During *hanging* the top of the door leaf has to be relieved (cut) along the inside edge of the head rail, near the hanging side, so as to clear the frame. Rising butts are *handed* and usually *lift-off*. Steel rising butts require regular oiling and cleaning. See diagram, p. 229.

rising damp The movement of water up a wall from the ground by *capillary action*, making it very damp at its base but less so higher up, a complaint that is commonest in houses built before 1875, when *damp-proof courses* became compulsory in the UK. In new houses with the necessary damp course it is rare unless *bridges* exist or the damp course has failed. Where it seems impracticable to insert a damp course by *sawing* or *injection*, it may be possible to hide the damp area of wall and reduce its ill effects. In an old house with no damp course that has suffered the evaporation of the dampness into the house for years, the wall at the level of the evaporation becomes impregnated with *hygroscopic salts*. See BS 6576 and, for diagnosis and treatment, BRE Digest 245.

rising main, r. pipe A services *riser*.

rising spindle A *spindle* that moves up as the tap or valve is opened. It takes up more space than a *non-rising spindle*, but shows how much the valve is open.

risks Possible future events that may lead to extra costs, delay, or *accidents*. The *contractor's* risks, e.g. for errors in *setting-out* or pumping water from excavations, should be allowed for in a *tender* price. Hazardous or unsafe methods of construction are usually forbidden by law. See also **fire**.

riven slate *Slate* that is split, not sawn. It may have a mottled surface.

robotics Construction robots can do difficult, hazardous work or precise tasks. They can climb and inspect the walls of high-rise buildings, paint with toxic solvents, tile walls, or go underwater and into confined spaces.

rocket tester, smoke r. A canister which gives off dense, lasting smoke, used for finding leaks or sometimes as a *drain test*.

rock excavation Excavation in ground too hard to dig without the use of *breakers* or *blasting* (*C*). It is measured *extra-over* normal excavations.

rock face Building stones with a rough dressed face, for *rubble walls*.

rock pocket (USA) Concrete *honeycombing*.

rock wool *Mineral wool* made from

rock, used for heat *insulation*. It is rotproof, non-combustible, and can contribute to *fire resistance*.

rod A board on which dimensions are marked for *setting-out*, such as a *storey rod* for brickwork or stairs.

rodding Inserting *drain rods* into a pipe, usually to clear a blockage, although drain rods are also used for *drawing in* cables. The term is also used for systems with a high-pressure water pump, a long hose, and a special nozzle which sprays backwards, both pulling the hose up the drain and flushing out stones, grit, grease, and water. The hose is inserted in the uphill direction of the drain. See diagram, p. 374.

rodding eye An *access cover* at the head of a branch drain, formed by bringing a steep branch back up to ground level and sealing it with a cover plate.

rodding point A *junction* on a long drain run, with a pipe up to the ground surface at 45° through which a *drain rod* can be inserted. This eliminates the need for an *inspection chamber* (*C*).

roll (1) A semi-cylindrical shape, used on a *bath* rim or a *double Roman tile*.

(2) A joint between the sides of two sheets of *supported sheetmetal roofing*, either a *hollow roll* or a *wood roll*. It can be made by hand, unlike a *standing seam*.

(3) Any long sheet of flexible material wound upon itself or a former, e.g. wallpaper, a damp-proof course, bitumen felt, etc.

roll-capped ridge tile A *ridge tile* with a roll added along its apex.

rolled lean concrete Dry lean *concrete* compacted with a roller, used for building roadbases and in reservoir walls. See *roller-compacted concrete* (*C*).

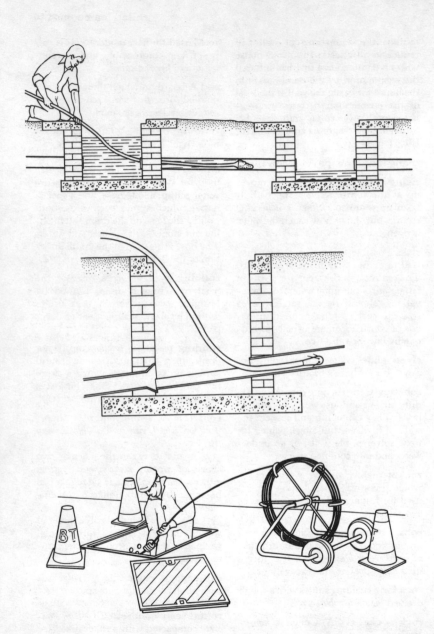

Rodding.

roller (1) An absorbent cylinder for the *application* of paint, used like a *brush*. Paint is either picked up from a tray or supplied by hose for *pressure-fed rollers*. The roller can give textured effects, such as *ragging* and *stippling*, or be shaped to fit corrugated roofing. (2) An *indenting roller* (*C*).

rolling grille A security grille of horizontal rods, bent to a pattern and joined at intervals by metal links. It works like a *rolling shutter*.

rolling shutter, roller door A door that is opened by sliding vertically between guide rails on each *jamb* and coiling round a roller across the head of the opening. Large industrial types have horizontal metal slats with interlocking edges and are operated by winch and *long arm*, endless chain or electric motor. They can be made fire resisting. Lighter types, for hand operation, can have a flexible shutter of profiled steel sheet.

roll-on filter A *fabric roll filter*.

Roman cement Properly speaking *hydraulic cement* made from *pozzolan* mixed with *lime*, but in England also a cement made in the nineteenth century by burning lumps of marl found in London clay.

Roman tile A British *single-lap tile*. The single Roman tile is not standard and has one waterway. The *double Roman tile* is standardized and has two waterways separated by a central roll.

roof New house roofs are mostly framed with *trussed rafters*, although *rafters* are still used. Office and commercial buildings usually have a reinforced concrete *slab*, while factories and warehouses use *portal frames* or girders with *northlights*. Roofs can be *flat* or *pitched* and have many different types of *roofing*. See diagram, p. 376.

roof abutment An *abutment*.

roof boarding Close-boarded *roof decking*.

roof cladding *Roofing*.

roof conductor A thick copper tape round a roof *parapet* and joined to any metal equipment on the roof, forming an *air termination network* for lightning protection. Aluminium tape is used above aluminium roofs.

roof covering *Roofing*.

roof cutting The art of setting out a traditional framed roof and sawing the *rafters* and other timbers to fit and lock together.

roof decking Sheet material on top of roof framing forming a firm base to carry flexible roofing such as *supported sheetmetal* or *bitumen felt*. It may also give some heat *insulation*. Materials used for roof decking include plywood, planks, chipboard, woodwool, strawboard, and ribbed steel.

roof drain A *rainwater outlet* from a roof, usually discharging into a *downpipe*.

roof drip An *eaves drip*.

roofed-in stage A *stage* of construction at which the roof is on and gives some shelter to site workers without the building being fully *enclosed*.

roofer A *roofing specialist*.

roof extract unit, r. extractor A large fan mounted on the roof of an industrial or commercial building, usually with a *kerb* to keep out the weather.

roof guard A *snow board*.

roofing, roof covering, waterproofing, weatherproofing The weatherproof skin on top of a *roof*. The main materials are traditional *slates*, concrete or clay *tiles*, membranes of *bitumen felt*,

roofing felt

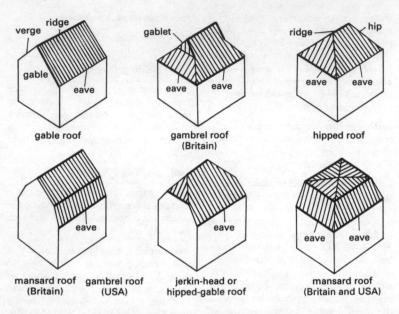

gable roof

gambrel roof
(Britain)

hipped roof

mansard roof
(Britain)

gambrel roof
(USA)

jerkin-head or
hipped-gable roof

mansard roof
(Britain and USA)

Pitched roofs, usually of spans from about 4 to 9 m.

supported sheetmetal roofing, and steel folded to self-supporting corrugations or profiles.

roofing felt *Bitumen felt.*

roofing nail Any nail for fixing *tiles* or *slates* to roof *battens* or for fixing the battens to *rafters*. Nails for tiles and slates are not heavily loaded but should be not less than 40 mm long. Those for battens need to be longer: the batten thickness plus 30 mm. Galvanized steel is usual but less durable than aluminium, copper, or *silicon bronze*. In exceptionally corrosive conditions such as chemical works or animal houses special stainless-steel nails may be needed. Corrosion-resistant oak pegs are used for *peg tiles*. *Springhead roofing nails* have been the traditional nail for corrugated roofing, although for profiled sheeting *roofing screws* or *hook*

bolts may be needed for longer spans or exposed locations. Nails are sometimes used to fix *bitumen felt*.

roofing punch A sharp *punch* for making nail holes in *corrugated sheeting*.

roofing screw A large metal screw for fixing *corrugated* or *profiled sheeting* on roofs. It is usually of steel for steel roofing and aluminium for aluminium roofing, even if it is *powder-coated*. It should always be driven through the top of the roofing corrugation, and usually has a synthetic *rubber* seal to further prevent leaks. Many roofing screws are *self-drilling* and can be driven into wood battens or steel purlins with an *electric screwdriver* and just as easily unscrewed.

roofing specialist, roofer A skilled site *operative*, or a specialist sub-con-

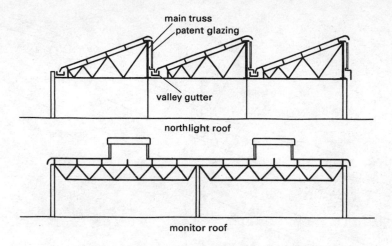

main truss
patent glazing

valley gutter

northlight roof

monitor roof

Factory rooflights

tractor, who puts on *roof tiles, built-up roofing, membrane roofing,* or *supported sheetmetal roofing.* See **rainwater plumbing**.

roofing tile A *roof tile*.

roof insulation See **cold roof, warm roof**.

rooflight, skylight An area of *glazing* in a roof, either an *opening light,* a *fixed light,* or simple *rooflight sheet*. It can be a *barrel light, domelight, lantern, monitor, northlight, patent glazing,* or *saucer dome*. A plastics rooflight may also act as a *fire vent,* but for safety should have steel mesh under it to prevent anyone falling through it. See diagram.

rooflight sheet Corrugated clear *plastics sheet glazing* made to the same profile as roof sheeting, for house and factory light panels. Safety wire mesh may be needed under fragile sheeting to prevent anyone falling through it.

roof screen A *fire barrier*.

roof slab A reinforced concrete slab that supports a *flat roof*.

roof space, r. void The cavity inside a roof, above the top-storey ceiling, either empty apart from *services* or containing an *attic* or *loft*.

roof tile, roofing t. A concrete, burnt clay, or even pressed-steel tile for covering roofs. Since clay tiles are the oldest and pleasantest to look at, the other types often copy their shapes and colours. Roof tiles are of three general types: (a) *plain tiles*, (b) *single-lap tiles*, (c) *Italian* or *Spanish tiling*. The last two are both expensive and heavy but are very decorative. See BS 5534. See diagram, p. 378.

rooftop unit A packaged *unit air conditioner* mounted on top of a roof and serving the room or rooms below it.

roof truss A *truss* or a *trussed rafter*.

roof waterproofing, r. weatherproofing *Roofing*.

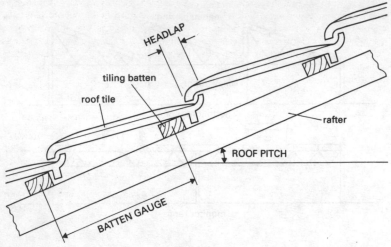

Roof tiles.

room air conditioner A small *unit air conditioner* usually mounted in an outside wall (often in or below a window) and serving one room.

room-heater A closed stove for burning anthracite or other smokeless solid fuel, usually with *heat-resisting glass* in the fire door. Room-heaters resemble slow combustion stoves and use the recirculation of gases to achieve high combustion efficiency. Many have *back boilers* that can heat six radiators and domestic hot water. See **wood-heater**.

room-sealed appliance A gas heater or boiler which draws its air from the outside, as well as discharging its burnt gas outside, to protect occupants in a room from the danger of leaking gas. A domestic heater *balanced flue* is room-sealed, as are appliances connected to a *ducted flue*.

root The part of a *tenon* which widens out at the shoulders.

ropiness, ropy finish The condition of a paint surface in which *brush* marks have not flowed out, owing to a lack of *flow*, the high porosity of the *substrate*, or poor *brushability*. Ropiness in dried *undercoats* is removed by *flatting-down*.

rose (1) A decorative plate through which a door-handle *spindle* passes.
(2) A decorative plate or boss through which an electric light flex hangs from a ceiling (ceiling rose). It has three terminals for wiring as a *lighting point*.
(3) A shower *sprayer*.

rot *Decay*, usually *dry rot*.

rotary-cup burner, rotating-c. b. An industrial boiler *burner* that sprays liquid fuel oil on to a cup spinning at high speed, aiming to break up the liquid into very small particles that burn completely. A blower supplies air to take the atomized fuel into the boiler firebox.

rotary cutting A timber *veneer* cutting method in which the hot soaked log, or *flitch*, is taken from the cooking vat, mounted on a powerful lathe, and turned at a speed of 30 to 60 rpm (depending on log diameter) against a long knife. The knife cuts off a continuous sheet of veneer at a speed of about 1 m/sec; 90% of veneer is cut in this way, including most of the veneer from which *plywood* is made. Rotary cut veneer may have 'lathe checks' on the *loose side*.

rotary float A *power float*.

rotary planer A *planer*.

rotary veneer Veneer made by *rotary cutting*.

rough ashlar A block of stone *rough-hewn* at the quarry.

roughcast, slap dash, wet dash (Scotland **harl**) A plaster *dashed finish* in which the final coat, a *rendering* mix containing pebbles, is put on with a *roughcast machine* and left rough. It is porous and allows water vapour to escape, and its rough texture tends to shed rainwater, so it is mainly used on external walls.

roughcast machine, plastering m., r. applicator, Tyrolean m. A sheet steel or plastics pot with stiff steel bristles on a shaft turned by a hand crank, used to flick wet *render* or *spatterdash* on to a wall. See diagram, p. 332.

rough-cast glass Translucent flat glass with one surface textured by rolling.

rough cutting A quick and simple cutting of bricks or similar materials, by striking them with a hammer, a *bolster*, or the edge of a trowel, leaving a ragged edge, suitable for common brickwork that is to be plastered. Rough cutting work is *measured* as an area and for *facework* is presumed to have a skin of facing bricks *measured*

separately as *fair cutting*. A 327 mm *solid wall* with facing would have 224 mm of rough cutting; in this wall a length of 4 m would be measured as 4 × 0.23 × 0.92 m². The fair cutting for the same job would be 4 m.

rough floor (USA) A timber *subfloor* to carry floor boards.

rough grounds *Common grounds*.

rough hardware (USA) Builder's hardware which is not seen, e.g. bolts, nails, fasteners, and other metal fittings, but not *finish hardware*.

rough-hewn stone Building stone which has been shaped by chipping, but not sawn or rubbed smooth.

roughing in, r. out Doing the first rough work. In plumbing, this is installing water, gas, or other pipes as far as the points where they will later join their fixtures. In carpentry it is roughly shaping a piece of wood or doing the *carcassing* and *first fixings*. In plastering, it is laying the core for a large *moulding* or *dubbing out* a wall.

rough opening A hole in the building *fabric* for a door or window frame, or a services *penetration*.

rough string A stair *carriage*.

rough-terrain fork-lift A *fork-lift truck* (C) which can travel over rough ground, often used on open building sites to move materials from a storage area to near their final position.

rounded angle An edge that has been smoothed to a small radius, e.g. a *pencil arris*.

round timber, t. in the r. Logs that have not been sawn.

rout To cut a groove in wood with a *router*.

router A hand-held *power tool* for grooving timber, with a *cutter block*

rovings

spun at high speed by an electric motor. Specialized versions of this machine up to 3 hp are used for cutting *rebates* for hinge *butts*, or *weatherstripping*, and even for lock *mortices* 110 mm deep.

rovings Thick parallel bundles of *glass-fibre* strands, used in the form of yarn or woven into fabric, usually of plain square weave.

rowlock cavity wall, all-rowlock w. A wall built in *rat-trap bond*.

rowlock course A *brick-on-edge* course.

Royal Institute of British Architects (RIBA) The main organization for professional *architects*, architectural students, and their training. It also provides services to architects and the construction industry, publishes on architecture, and has a bookshop.

Royal Institution of Chartered Surveyors (RICS) An organization concerned with land, property, and buildings, and the training and professional qualifications of land, mining and *building surveyors*.

rubbed bricks, rubbers Soft, smooth clay bricks usually without a *frog*, suitable for rubbing to shape, for traditional *gauged brickwork*. They are easily cut with an abrasive wheel.

rubbed finish A finish to hardened concrete obtained by rubbing down with a carborundum stone or similar *abrasive*.

rubber (1) A flexible material containing natural or synthetic *elastomeric polymers*. It can be moulded to shape, extruded, formed into sheet or strip from solid or *foam* material, or used as a liquid as in *latex cement* and *chlorinated rubber paints*. Common synthetic rubbers are *ethylene propylene diene*

rubber, hypalon, neoprene, and polyisobutylene, all of which are *thermosets*.
(2) A *rubbed brick*.

rubber and cork tile A warm, resilient, non-slip *floor tile* made of granulated *cork* and latex rubber formed under pressure, 4.8 or 6.4 mm thick.

rubber buffer, r. silencer A door *slam buffer*.

rubber flooring A good but expensive *thin floor finish* used for high-traffic areas, made of natural or synthetic solid rubber, obtained as smooth or *studded* sheet or tiles, which may be *sponge-backed*. It is laid by specialists using *epoxy* or *neoprene* adhesives, and can have *heat-fused* joints. Rubber is also used in *cement-rubber latex* jointless floor and in *rubber and cork tiles*.

rubber ring The usual name for a *joint ring*.

rubbing down (1) **flatting d.** Smoothing high spots off a surface with *sandpaper* as part of its *preparation* for painting.
(2) Cleaning masonry with an abrasive *rubbing stone*.

rubbing stone A small block of *abrasive* used for cleaning or smoothing hard materials such as concrete, stone, or bricks.

rubbish, builder's r., building r. Waste from building operations on site, such as demolition *rubble*, offcuts from building *trades* (trade rubbish), droppings from *plasterwork*, packaging, etc.

rubbish bucket, kibble, skip A large steel bucket for moving rubbish by crane.

rubbish container A *skip*.

rubbish pulley A *gin block*.

rubble (1) Broken bricks, old mortar, and similar material. Clean rubble from demolition, without any old timber, may make suitable *hardcore* (*C*) for *fill*.

(2) Walling stones with irregular faces, which are not smoothed to give fine joints like *ashlars* but are sometimes squared and coursed in *rubble walls*.

rubble ashlar An *ashlar*-faced wall, backed with rubble.

rubble wall Stone masonry built with *rubble*, which differs from *ashlar* in being roughly cut and so has mortar *joints* as much as 25 mm thick. The stones can be up to 250 mm high. *Random rubble* and *snecked rubble* are the main types. They are either uncoursed or occasionally coursed, but *squared rubble* is usually coursed since uncoursed squared rubble would be the same as snecked rubble. See BS 5390.

rule A *straightedge* of any length or material, used for measuring, drawing straight lines, setting them out on site, or for *screeding*.

rule off To strike off excess wet plaster or concrete, creating a flat surface, by pulling along a *screed board* guided at each end by *screed rails*.

run (1) A horizontal *dimension*, e.g. between the supports for a stair *string* or a roof *rafter*, the *going* of a stair *tread*, or the *total going* of a *flight*.

(2) A temporary traffic way, such as a *barrow run*.

(3) Pipes or cables in the main line of flow, e.g. between branches.

(4) To lay pipes or cables, or to *draw in* wiring.

(5) A defect in *paintwork* in which a narrow ridge has flowed down, advancing from a small bulb or teardrop, after the paint has begun to set. It can be a single tear or complete *curtaining*.

(6) To pass plaster or *lime putty* through a sieve.

runby pit A *lift pit*.

rung A round or D-shape horizontal bar used as a step in a *ladder*.

run moulding, horsed m. A *fibrous plaster* moulding formed in place by running a *template* along to shape the wet plaster.

runner (1) A guide rail, e.g. for a drawer or for a *lift car*.

(2) A horizontal beam, e.g. between the *hangers* of a *suspended ceiling*.

(3) A *running tie*; but for thatching see **withe**.

(4) A rotor in a centrifugal fan, pump, turbine, etc., with blades forming a cylindrical shape, spinning on a shaft.

running bond *Stretcher bond*.

running tie, runner A long timber joining several *common joists*, rafters, studs, etc. to keep them in place.

rust The *corrosion* of steel or iron, which is very difficult to halt once it has started. Protective coatings must be applied to an absolutely clean surface, with all rust removed. Many paints for use on steel contain *inhibiting pigments*.

rusticated ashlar *Ashlar* stonework on which the face is left rough, bulging out between the *joints*, the stones being cut back at the edges by bevelling or rebating to a *drafted margin*. Often *quoins* are made of rusticated ashlar.

rusticated brick, rustic b. A *facing brick* with a surface which has been *textured* before drying and firing by covering it with sand, by impressing it with a pattern, or by other means. Rusticated bricks may also have variegated colours.

rustic joint A *joint* sunk back from the surface of *rusticated ashlar*.

r-value, thermal resistivity A figure

R-value

which measures how well a material resists the flow of heat by conduction. It is the reciprocal of the *k-value*.

R-value, thermal resistance The *insulation* value of a given thickness of material, calculated from its *r-value* by its thickness. The units are square metres by degrees Celsius per watt $(m^2.°C/W)$.

S
___

sabin The unit of sound *absorption*, equivalent to the absorption of 1 m² of open window (after Dr Sabine, pioneer in acoustics).

sacrificial form (USA) *Permanent formwork*.

saddle (1) **s. piece** A piece of *plumber's metalwork* about 460 by 460 mm, dressed to shape and fixed under roof *tiles* at vulnerable points, such as where a *ridge* joins a roof slope or at a wall *abutment*. A chimney saddle, also called a cricket in the USA, may replace a *back gutter*.
(2) **pipe s., service clamp** Two half pipes which are bolted over a water main or sewer to make a connection. Sealed attachments are used to drill a hole and screw in a fitting without loss of pressure in a *live main*.
(3) A large *roll* on a flat roof, dividing it into bays.
(4) A *holster*.
(5) A *saddle clip*.

saddleback board A narrow board chamfered on both edges, fixed on a floor across a door opening to stand above the flooring each side. If a door is hung to close on this board it will clear the carpet when open.

saddleback coping A *coping* with a sharp apex and flat flanks which are weathered away at a slope from each side. See diagram.

saddle clip A U-shaped pipe *fixing* with *lugs* on each end.

saddle-jib crane (USA **hammerhead c.**) A *tower crane* with a horizontal *jib* and *counterjib*. A *trolley* or crab runs along the jib, with steel ropes down to the hook. By trolleying in or out the hook radius can be changed.

saddle piece A *saddle* of plumber's metalwork.

saddle scaffold A *scaffold* built over a roof *ridge*, sometimes supported by

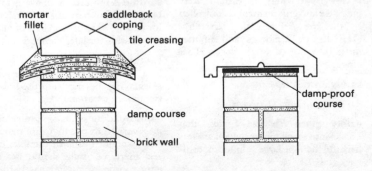

Saddleback coping.

standards at the side of the building. It is used for repairing chimneys.

saddle stone The apex stone of a *gable*.

saddle trusses A set of *trussed rafters* used where a roof turns a corner. They sit at right angles and on top of the standard trusses in one direction and become smaller towards the ridge.

safe (1) A *tray*.
(2) A security strong box.

safeguarding Preventing theft, vandalism, or damage on a building site after working hours by means of a *hoarding*, watching, and lighting.

safety Precautions are needed to prevent injury on building sites. The most common accidents are people falling and objects being dropped. They are prevented by using guard railings, *debris-collection fans*, nets and harnesses, securing *ladders*, ensuring the stability of *scaffolds*, etc. Other safety precautions include the wearing of protective clothing, e.g. *helmets*, *goggles*, ear-muffs, respirators, overalls, and steel-capped boots; guards over moving machines; and safe working procedures. Contractors with twenty-one or more employees must appoint a safety supervisor to see that site work is carried out safely. In Britain safety guidance and enforcement is controlled by the *Health and Safety Executive* (HSE). Site accidents and injuries must be reported to the local HSE office.

safety arch A *relieving arch*.

safety gear Lift *safety gear*.

safety glass *Toughened glass* that breaks into small pieces, which are less harmful than the large splinters of ordinary glass.

safety ladder, jacket l. A fixed verti-

cal metal *ladder* backed with hoops forming a cage through which a person climbs.

safety net A large net placed to catch falling tools, materials, or workers, protecting them and those below.

safety railing A *guard rail*.

safety signs Buildings, construction sites, and other places are required by law to have safety signs of standard shape and colour: red circle and cross-bar on white for prohibition (don't), black triangle round yellow for warning (danger), blue circle with white letters for mandatory (must do), and green rectangle with white letters for safe condition (e.g. exits). See BS 5378 and **fire safety signs**.

sagging Paint *curtaining*.

salt air Salt in seawater spray from breaking waves can cause severe corrosion of metals, particularly galvanized steel *wall ties*, steel *reinforcement* and copper. Most damage is done within 1 km of the sea, and little beyond 30 km.

salt-glazed ware Glazed clayware used before *vitrified clayware* became available in the UK, although it is still used elsewhere.

sample A few articles of a larger number, or a small part of a material, taken for testing, or to give an idea of quality, or put aside for later comparison.

sample panel, reference p. A small area of brick *facework* built before bricklaying begins and used as the standard for *workmanship*.

sand Building sands are grains of fine *aggregate* from 0.2 to 3 mm diameter. They should be hard, durable, clean, and inert, i.e. they should not swell after mortar or plaster has hardened. Any dirt or grease in a sand reduces the strength of concrete, mortar, or *coarse stuff*, or makes it *sweat out*. Silica sands, the commonest, are completely inert from the builder's viewpoint. Coarse, medium, and fine sands are ordered for different uses: bedding, concreting, bricklaying, render, screeds, etc. Apart from their size sands may be *sharp* or *soft*.

sand bedding A layer of sand, usually 50 mm thick, placed on an excavated *formation*, to provide a flat surface for a *ground slab*.

sand blasting *Blast cleaning* with sand, using protective dust extraction, normally under factory conditions because of the risk of *silicosis*. See **wet blasting**.

sand box A box filled with sand supporting a *formwork* prop and made so that the sand can be let out to allow *stripping*. The best types are made of steel pipe, with a close-fitting steel plunger and a screw plug.

sand-dry surface A stage in the *drying* of a paint film when it is *surface dry*. This is tested by sprinkling *silver sand*, which must not stick to and injure the surface when brushed off one minute later.

sanded felt A *bitumen felt* which is surfaced with sand.

sander, sanding machine A *power tool* for *dressing* and *surfacing* materials such as wood with a moving piece of *sandpaper*. Sanders can be *belt*, *disc*, or *orbital* types, and may be *bench* tools, hand tools, or *floor sanders*. Powerful sanders often have *dust* extraction, which is particularly needed for *hardwoods*.

sand filter (1) A water filter with sand as the filter medium. See *C*.
(2) An air *prefilter*.

sanding Rubbing a surface with *sand-*

paper, by hand or a machine, which smooths it by removing a thin layer of material and any high spots. In painting, the same process of smoothing paint surfaces is called rubbing or *flatting-down*. See **sander**.

sanding sealer A specially hard first coat of *sealer* which seals or fills but does not hide the *grain* of wood. The surface can be sanded after the sanding sealer is dry.

sand-lime brick A type of *calcium silicate brick*.

sandpaper Tough paper coated with powdered *abrasive*, such as emery, garnet, or glass. Different *grit* size or grades of coarseness are used for *sanding*, *flatting-down*, rubbing down, *wet rubbing*, polishing, etc. Sheet sand paper is usually 280 by 230 mm, but it is also produced in rolls of varying widths.

sandstone A sedimentary rock made up of naturally cemented quartz grains, used for *dimension stone*, pavings, cladding, and decoration. Its surface can be sawn, polished, or textured. It can be imitated by *cast stone*.

sandwich construction See **composite construction**.

sanitary accommodation The rooms or 'apartments' in which *sanitary fittings* are installed. Being *wet areas* they usually have tiled floors and walls. The fittings required in public buildings are listed in BS 6465.

sanitary connector A *pan connector*.

sanitary cove A *cove* between a stair tread and riser, for easy cleaning.

sanitary fitting, s. appliance Baths, basins, bidets, sinks, and other equipment for washing and cleaning, including *soil fitments* such as WCs and urinals. They usually have a fixed bowl with an outlet or *waste* that is connected to the *sanitary pipework* and are supplied with water, often hot and cold, from taps or mixers (except for soil fitments, which usually have other arrangements). They can be made of any strong, watertight, and durable material, from decorative *vitreous china* to utilitarian *stainless steel*.

sanitary pipework, s. plumbing The *internal pipework* for taking away *foul water* discharge from *sanitary fittings*, which may be separated into *soil water* and *waste water*. It connects to the buried *drainage* system. Various systems are in use, such as the *single-stack system* and *one-pipe system*. The usual materials are copper or *plastics pipes*. See BRE Digests 248 and 249, and BS 5572.

sanitary shoe (USA) A *cove*.

sanitary waste disposer A *garbage disposal sink*.

sapele (Entandrophragma cylindricum) A dark red stripy African *hardwood*. Easily worked and durable, it is used for *joinery* and plywood *veneers*.

saponification The breakdown of *oil paints* when attacked by the alkalis in new and damp *masonry*, concrete, render, or mineral-fibre building boards. The oil in the paint and the alkali combine to form a soft mass chemically similar to soap. It is avoided by using *alkali-resistant primer* under oil paints on masonry less than six months old, or by painting with *emulsion paints*, which are oil-free.

sap stain *Blue stain*.

sapwood, alburnum The outer wood of a tree, which before the tree was felled was living tissue and is softer and paler in colour than the *heartwood*. Unless the timber has been *kiln-dried*, the sapwood usually decays more easily

than heartwood, as it has more food for *borers* and *fungus*, but it also absorbs preservative more thirstily and is not always weaker than the heartwood.

sarking A flexible roof *underlay*, or in Scotland roof *boarding*.

sash (1) A timber, metal, or plastics surround for *glazing*. It fits into the window *frame* and can be hinged, sliding, or fixed. A *casement* is a hinged sash. See **secondary sash glazing** and the following entries.
(2) A *sash window*.

sash balance A spring-loaded device for a *sash window*. The two main types have either a spiral tube or a spring tape, and they are always fitted in pairs, one each side. They are used instead of *sash weights*, cords, and pulleys.

sash bar, s. astragal A *glazing bar*.

sash cord, s. line Braided rope about 8 mm dia. nailed to the sides of the sashes in traditional timber *cased frames*. They pass over a top pulley and are tied to a hidden cast-iron *sash weight*. The window is easy to slide and stays at any height. The frame is made so that sash cords can be replaced.

sash cramp A *cramp* with a slide bar 600 to 1500 mm long and adjustable jaws like a small *vice*, used to apply *pressure* during gluing.

sash fastener, s. lock (USA **s. fast**) A two-piece window fastener to hold a *sash window* closed. One piece, on the *meeting rail* of one sash, swings across and engages with the other piece, on the meeting rail of the other sash.

sash fillister A *fillister*.

sash lift A *pull* or hook on a *sash window* for opening it.

sash pulley A pulley set in the top of the *stile* on each side of a *sash window*. The *sash cord* passes over it.

sash ribbon A steel spring tape for a *sash balance*.

sash stop A moulding nailed to the *jamb* of a traditional cased frame *sash window*, on the inside of the inner sliding sash, to keep it in the frame and stop draughts. Before removing it to replace a *sash cord*, a careful knife cut along the joint can limit damage to old paintwork.

sash weight A cast-iron rod, hidden in the *cased frame* of a *sash window*, hung on *sash cords*. There is one on each side. The pair should weigh the same as the *sash* they counterbalance.

sash window (1) **double-hung s.w., hanging s., vertical sliding w.** A window with two *sashes* that open by sliding vertically. Both have *sash balances* or traditional *sash weights* so that the sashes are easy to push up and stay put at any height.
(2) A horizontally sliding sash window is a *sliding window*.
(3) A hinged sash window is a *casement window*.

satin paint Paint with *eggshell* or low-gloss finish, usually *emulsion paint*.

saturation, chroma In describing a *colour*, the intensity of a *hue* compared with a neutral grey of similar *lightness*.

saturation coefficient (1) In Britain, the ratio between the volume of water absorbed by a brick in five hours' boiling and the volume of its pore space.
(2) (USA) The ratio of water absorbed by a brick soaked in cold water for twenty-four hours to that absorbed when the brick is boiled for five hours.

saucer dome A glass or plastics one-piece *domelight*.

saw A *tool* with a toothed steel blade for cutting materials, mainly wood, metals, or plastics. The slot made by a

saw is its *kerf*. It is as wide as the blade plus the outward *set* of its teeth. Saws are *edge tools* and need sharpening and setting to maintain them in good working condition. Skill is needed to use *handsaws* accurately. Power saws are dangerous and can leave 'machine burns' on wood. See diagrams on p. 388 and on pp. 346 and 402.

saw bench A *woodworking machine* with a steel table and a *circular saw* that passes upwards from below through a slot. A *planer* and *doweller* are often attached. It is not portable and probably more dangerous to use than a saw that cuts from above.

sawdust cement A mixture of cement and wood sawdust, used to make *nailing strips*. It can have high shrinkage, and the natural acids in some timbers can attack concrete.

sawfile A file of circular cross-section about 1.5 mm diameter and of varying length, held taut by its ends in the steel frame of a *hacksaw, coping saw*, etc. It can cut hard materials to intricate shapes.

saw horse A four-legged stool with a flat work top about 920 × 100 mm, used in pairs to rest timbers on while they are being sawn. They are often made on the job by a carpenter.

sawing *Conversion* of timber, or its later cutting or trimming.

sawn damp course To eliminate *rising damp* in old brick buildings with no *damp-proof course* (dpc), or where the existing dpc has failed, the first possibility is to investigate whether a bed joint can be sawn through at a suitable level to insert a new dpc. The work is done in short lengths and the subsidence to be expected is about 1.5 mm. The method can be used for brick walls up to a maximum of 530 mm thick, beyond which *injection* is used.

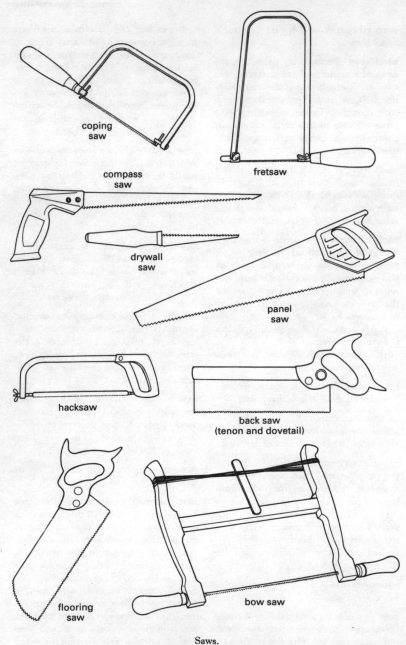

coping saw

fretsaw

compass saw

drywall saw

panel saw

hacksaw

back saw
(tenon and dovetail)

flooring saw

bow saw

Saws.

388

sawn stone *Dimension stone* that is cut to shape with an abrasive wheel.

sawn veneer In the USA *quarter-sawn* oak is still cut on large-diameter segment saws, which make a cut only 1 mm thick, with a *veneer* of about the same thickness or slightly thicker. Even with this narrow *kerf*, the loss of wood in sawing can amount to half the log volume, but the method is used because it makes sheets of strong veneer for *plywood* which polish well.

saw set, swage An adjustable tool which gives the correct *set* to the teeth of a saw.

sawtooth roof A *northlight roof*.

scabbing *Scabbling*.

scabbler A powerful machine for *scabbling* the surface of a horizontal concrete slab, which should be at least 100 mm thick. Both piston and rotating flail scabblers create noise and *dust*.

scabbling, hacking Roughening the surface of concrete or masonry and removing *laitance*, either at a *construction joint* or to provide a *key* to make plaster or a screed stick. Horizontal concrete slabs are done with a *scabbler*, while a *bush hammer* or *comb hammer* is used on vertical surfaces. Gentler shot or grit blasting is needed for thinner precast concrete work.

scaffold, scaffolding A temporary platform carried by a framework erected on site, to give access for long jobs (bricklaying, painting, services) and to carry materials. Scaffold stagings or *tower scaffolds* are usually of steel or aluminium, either of a *prefabricated* frame type that clip together or of a *tubular* type joined with fittings, but wood is also used, and bamboo in Asia. Scaffold frameworks are also used for *façade retention* or as *falsework* (*C*). In the UK scaffold firms and *scaf-*

folders must be licensed, scaffolds inspected before use and each week, and records kept (on the blue Register Form 91, HMSO). The local offices of the *Health and Safety Executive* give guidance. See BS 5973 and BS 5974.

scaffold boards, s. planks Softwood boards which form the working deck of a *scaffold*. They must be at least 230 mm wide and 40 mm thick and may be up to 3.70 m long. Their ends are bound with hoop iron to prevent them splitting. They are also used as *toeboards* or *splashboards*, and to form *barrow runs*.

scaffold clip A fitting for *tubular scaffolding* which firmly grips only one of the two tubes it joins, e.g. a *putlog clip*.

scaffold coupler A strong fitting that firmly grips two different tubes of a *tubular scaffolding*, such as the joint between a *standard* and a *ledger*.

scaffolder, scaffold hand A skilled worker who erects, changes, or takes down *scaffolds*. The *Construction Industry Training Board* runs training schemes for all levels of scaffolders.

scaffold fittings The pieces used to make *tubular scaffolding*, such as *baseplates, clips, couplings, putlog adaptors*, etc.

scaffold ties Tubes between an *independent scaffold* and a building to give sideways support, often attached to a *reveal pin*. They should not be removed, e.g. by bricklayers or plasterers, or the scaffold may collapse.

scaffold tube A *brace, ledger, putlog, standard* (pole), transom, etc.

scagliola marble A coloured wall finishing made of hardened gypsum plaster, hand-polished with a stone and coated with linseed oil, using an old Italian process.

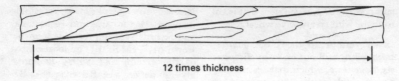

12 times thickness

A glued scarf joint at 1 in 12.

scale (1) The ratio between the size of a drawing or model and the object it represents. Floor plans are usually 1 : 100 scale; large details may be *full-scale*.

(2) A programme *time scale*.

(3) *Furring*.

scaled dimension A size measured by scaling from a *drawing*, thus less reliable than a *figured dimension*.

scalloping A decorative edge made with a row of half circles joined side to side, with or without a gap.

scallops Short *withes* placed to hold *thatch* at the *verges* of a roof and held down by *spars*.

scantling (1) According to BS 6100, either *square-sawn* softwood between 47 mm and 100 mm thick and between 50 mm and 125 mm wide, or square-sawn hardwood of non-standard size.

(2) (USA) See **dimension lumber**.

scarf joint, scarfed j. A joint between edges of boards or veneers glued together after bevelling at 1 in 6 to 1 in 24. See diagram.

schedule (1) A list of finishings, colours, doors, keys, reinforcement bars, lintels, etc., and their locations, used as a *contract document*.

(2) A *programme*, or list of maintenance work.

scheduled ancient monument See **ancient monument**.

schedule of defects, s. of outstanding works, snagging list (USA **punch l.**) A list of the work remaining to be completed before the *certificate of practical completion* is issued at *handover* or before the *final certificate*. The *architect* and the *builder* may have two separate lists.

schedule of prices, s. of rates A document used for *tendering* where exact *quantities* are unknown. It gives a description of each sort of work to be done, but no exact quantities, only a general figure to indicate the scale of the work. The tenderers write down their *unit prices* for each item, which is later multiplied by the actual quantity of work from *remeasurement*. Although the sum is not known in advance it is regarded as a *fixed-price contract*.

scissors lift A mobile work platform that lifts vertically. Some models have a motor and steering operated from the platform itself. See *C*.

scissors truss A simple roof *truss* formed of four main members, two of which are normal *rafters*. The other two extend from the wall plate to the middle of the opposite rafter. These two members intersect at the middle of the span, giving a scissors-like appearance to the truss. They give a good ceiling height at mid-span and are usually made of timber.

scope of works A brief description of the nature and extent of a *trade* or an entire project in a *specification*, plus details of the *limits of works* and any work that is excluded.

scoring, score and snap Cutting a line into the surface of sheet material, then bending it across a straight edge until it breaks. This is usually quicker than saw cutting, but may leave a sharp or ragged edge. It is used for window glass, slates, some floor or wall tiles, plasterboard, etc.

scouring Compacting the surface of plasterwork with a *float*, which is pressed flat and worked in a circular manner. Sometimes water is sprinkled on with a brush at the same time. The finishing process with a steel trowel is called 'ironing in'. Not all plasters are suitable.

scraped finish A finish to plasterwork made by scraping with a *feather-edge rule* as the finishing coat begins to harden, which must be judged carefully. For external *stucco* the aggregate is well exposed and the finish can be patterned if a serrated tool is used.

scraper, paint s. A broad steel blade with a handle, used for removing thin layers of material such as old paint from a surface.

scratch coat A first coat of plaster, with its surface keyed to bond with the second coat. The scratch coat is allowed to harden before the second coat is applied.

scratcher, comb A *drag, devil float, or V-notched trowel.*

scratching Roughening the surface of a coat of fresh *plaster*, to give a *key* for the next coat.

scratch tool A *plasterer's small tool* for completing plaster *enrichments*.

screed (1) **floor s.** An in-situ flooring of cement mortar laid to an accurate flat surface by *screeding* (or by *self-levelling*), usually as the base for a separate *floor finish* or *topping*. Types such as the *separate* (bonded), *unbonded*, and *floating* screed vary from 40 to 75 mm thick, while *monolithic screeds* are thinner. The screed mix has from $1:3$ to $1:4\frac{1}{2}$ cement:sand. Suppliers can provide ready-mixed screeds suitable for placing by barrow or by *screed pump*. Screeds are laid on the *structural floor* and usually have no *reinforcement*. When used as the *sub-floor* for *thin floor finishes* they are laid in large areas, and their surface is compacted and smoothed by a *power float*. It is expensive to lay a screed to high standards of *tolerance* for *level* and *flatness*. Drainage from a flat roof or wet area is frequently by a *screed to falls*. Laying screeds is a *wet trade* and screeds need *curing*, then time for *drying-out*. Suspended timber floors are not an ideal base for a screed, but can be used if they are rigid and have a *damp-proof membrane* on top and adequate ventilation from below. See BS 8204. See diagram, p. 392.

(2) **synthetic s.** *Levelling compound.*

screed board, rule, straightedge A straight board used for *screeding*. It is slightly longer than the distance between *screed rails*. It may also be used for an *off-tamp finish* and large types are *vibrating*.

screeding Passing a *screed board* across a surface while also working it lengthways to-and-fro, so that *plaster*, concrete or the mortar of a *screed* is levelled, as all material that is higher than the *screed rails* is struck off. See diagram, p. 393.

screed pump A machine for mixing the material for a floor *screed* with water and pumping the liquid mix through a hose to where it is laid. The material (screed mix) is a premixed powder, supplied in bags or from a

Monolithic: 12 to 25 mm thick, 30 m² maximum bay

Separate, bonded: 40 mm minimum, 15 m² maximum bay. Hardened concrete with hacked or scabbled surface

Unbonded: 50 mm minimum, 15 m² maximum bay. Damp-proof membrane (or building paper in a dry location)

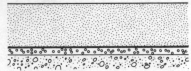

Floating: 60 to 75 mm thick. Underlay of building paper or damp-proof membrane. Resilient material (expanded polystyrene)

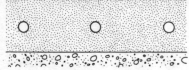

Heated: 60 mm thick minimum. Underfloor heating pipes

Types of screed and their thicknesses.

storage silo. Screed pumps have a *twin-shaft paddle mixer* and either a screw or a compressed-air delivery pump, and are usually mobile. See diagram, p. 394.

screed rail A temporary board, *edge form*, steel tube, or even strips of plaster or concrete between *dots*, which are placed along both sides of a surface and used to guide the ends of the *screed board* as it is drawn along them.

screed to falls A *screed* laid with its top surface sloping towards a drainage outlet, such as a *gulley* in a bathroom.

screed tester See *C*.

screen (1) A rectangular frame about 2 × 1 m, with mesh built into it for separating pebbles from sand. See also *C*.

(2) An *insect screen*.

screw A *fastener* with a *thread* formed on its shank and a slotted *head* so that it can be tightened or dismantled with a screwdriver. Screws are usually made of metal but may be made of *nylon*.

screw cap An *Edison screw cap*.

screw cup A conical pressed metal ring used as a washer under the head of a *countersunk head screw*.

screw-down valve A valve (or tap) with a plate or disc across its waterway that is moved straight up or down by a *spindle* to open or close it. Since it has higher resistance to flow than a *full-way valve*, it is used only where pressure is more than adequate. Screw-down taps are used for basins and sinks.

screwdriver A hand tool for turning *screws*, or an *electric screwdriver*.

screwdriver bit A replaceable tip for an *electric screwdriver*.

screwed and glued joint A strong and reliable joint between two pieces of wood made with both *adhesive* and screws.

screwed pipe The cheapest and heavi-

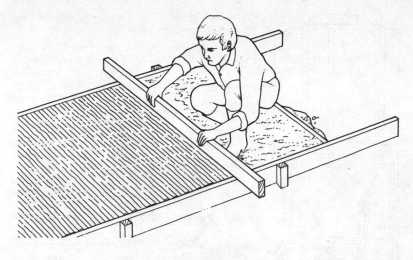

Screeding.

est gas or water pipe, made of *dead-mild steel* (*C*), sold in 6 m lengths. A *taper thread* is made on the outside, and the pipe is rotated tightly into malleable iron *union* fittings, with *jointing tape* over the threads to make a seal. They are bulky and difficult to install and, for uses like house water pipes, have been superseded by *copper pipe* and *stainless-steel plumbing* with *capillary joints*.

screw eye A metal loop at the head of a screw, used as a *primary fixing*.

screwgun An *electric screwdriver*.

screw nail A *drive screw*.

screw plug (1) **disc p.** An expanding *drain plug*, made of two dished steel plates back-to-back with a thick rubber ring between their rims, which is squeezed outwards against the inside of a drain pipe by a central screw.

(2) A threaded *access cover*.

scribe To cut a line lightly on a surface with a pointed *scriber* to the outline of a template, or to mark a *scribed joint* but without deep *scoring*.

scribed joint, cut and fit An exact joint cut for the edge of one joinery member to fit snugly against a surface, such as a wall, floor, or ceiling. The position of the cut is marked by *offering-up*, then running a *scriber* or a pencil along the edge, held at a suitable distance by any handy spacer such as a block of wood. The joint is then cut along the line, giving a *coped joint*. Prefabricated *joinery fittings* are often made 50 mm oversize to allow for a scribed joint. Timber *skirtings* can be scribed to an uneven floor.

scriber A fine pointed tool like an *awl*, or a pair of dividers, for accurately marking out *scribed joints* for cuts. A scriber should not be used to mark folds in sheet metal, which is weakened by any *scoring*.

scrim (1) Fine woven mesh usually of synthetic fibre, used to bridge the joints of *tapered-edge plasterboard* and *gypsum*

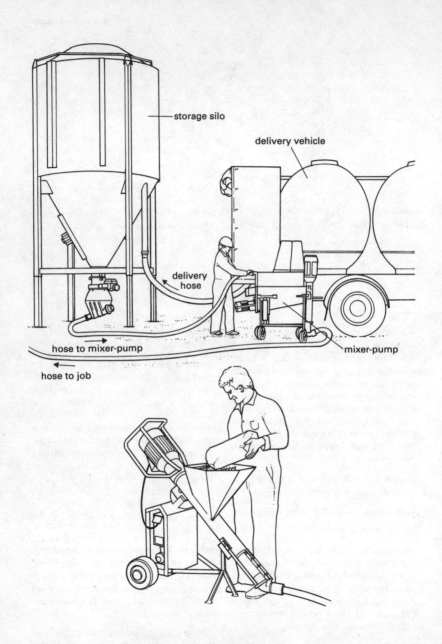

storage silo

delivery vehicle

delivery hose

hose to mixer-pump

mixer-pump

hose to job

Screed pumps.

394

baseboard and their corners. It comes as rolls of *joint tape* or *corner tape* and is finished over with *joint cement*.

(2) Coarse canvas, cotton, or metal mesh used to reinforce *fibrous plaster*.

scrimming Setting *scrim* over joints in *dry linings*, masonry, or carpentry, as a base for plastering, painting, or papering.

scroll A spiral that turns under the ending of a *handrail*.

scroll saw A power-operated *fretsaw*, either floor-standing or bench type.

scumble glaze, s. stain Transparent paints used in *scumbling* for enriching, softening, or otherwise modifying the colour of the coat below the finish coat.

scumbling The achievement of broken colour effects by removing or texturing a wet coat of paint to expose part of the *ground coat* beneath it.

scupper A hole in a *parapet* wall to let rainwater pass from a *box gutter*. It may discharge into a *rainwater header* or be a safety overflow.

scutch A *bricklayer's* tool which resembles a small hammer without a striking face. A cross-peen on both ends of the head is used for *fair cutting* or *rough cutting*.

seal (1) A *joint* that prevents leakage, such as a pipe *joint ring*, door or window draught seals or *weatherstripping*, seals for *smoke control*, etc.

(2) To block up with any type of material, including *mortar*, *sealer*, etc.

(3) **water s.** The water in a plumbing *trap* that prevents foul air passing out of the drain. Usually 50 mm deep, it is deeper in the *deep-seal trap*.

sealant, building s., sealing compound A sticky, viscous liquid that is put into a joint, where it stiffens, becom-ing rubbery and adhering to the sides. Sealants exclude water, wind, and grit. Fully *elastomeric sealants* are usually *two-part* products, which undergo complete *curing* no matter how thick they are, and can accommodate large amounts of movement, allowing smaller joints, which partly balances their high cost. *One-part* sealants and *mastics* are easier to use. Different types are used for *glazing materials*, as adhesives, for *bedding*, for *pointing*, and in *movement joints*. The stiffest sealants, described as 'strip-grade', are semi-solid, the next, 'gun-grade', are applied with a *caulking gun*, and these are followed by 'knife grade' and 'pouring grade'. They include *acrylic sealants*, *butyl sealants*, *polysulphides*, polyurethane sealants, and *silicone sealants*. Joint surface *preparation* may include cleaning and the use of a *primer*. Sealants should not be applied to movement joints in extremes of temperature; the joint should be near the middle of its size range. Manufacturers will usually give technical advice, or have skilled *applicators*. Guidance on sealant selection is given in BS 6213.

sealant remover A *solvent* for removing *sealant* from finishes, tools, clothes etc.

sealed system A pipework circuit for *low-pressure hot-water* central-heating systems, with glandless *circulating pumps* and a closed *expansion vessel*. The sealed system contains a fixed volume of water, kept separated from the air, and thus has a low rate of corrosion, which is further reduced by *corrosion treatment* of the water. Sealing of *small-bore systems* allows higher water pressure and temperature. This makes for economy.

sealed unit A *double-glazed* window with its two panes of glass (or clear plastics) sealed together in the factory.

A suitable window frame or surround is needed to protect the edge *sealant* from sunlight, rainwater, and moist indoor air, with masking, drainage, ventilation, and *vapour barriers*. The preferred type of edge sealants are permeable to water vapour, and the space round the edges is vented to the outside (or 'cold', dry side), allowing any moisture to escape from between the panes. The use of impermeable sealants is less desirable, as moisture can become trapped. During manufacture the space between the panes is filled either with clean, dry air or low-conductivity inert gas, such as argon or *sulphur hexafluoride* (SF_6). Sealed units have a *U-value* of 2.8, compared with 5.7 for ordinary single glazing. The use of a *low-emissivity coating* on the outside of the inner pane can reduce this to 2.0. See diagram, p. 131.

sealer A coat of paint or clear varnish that forms a barrier on a surface, either to satisfy *suction* of the substrate, to keep moisture in the substrate from affecting the paint, or to keep dirt from staining a decorative surface, such as a timber table or floor. Some clear sealers have anti-*ultraviolet* treatment. Sealers may also act as a *primer* or as a *vapour barrier*.

sealing coat A coat of *sealer*.

sealing compound (1) A fluid bitumen applied cold to *bitumen felt* for sealing the laps in *built-up roofing*. It is similar to *bonding compound*.

(2) A *sealant*.

sealing ring A *joint ring*.

sealing strip Preformed compressible *polyurethane foam*, impregnated with bitumen or waxes and resins, used in *movement joints* that require a very flexible seal. See **expansion strip**.

seam, welted s., welt, lock joint A joint between two strips of *supported* *sheetmetal roofing* made by turning the edge of each vertically upwards, bringing the two upturned parts together and folding them over each other, either once (single lock) or twice (double lock). Sloping seams down a roof (long joints) may be left as *standing*, but *cross-joints* are usually dressed down flat.

seamer Pliers with special jaws, or an electrically powered machine for rolling and folding the edges of metal sheet roofing, to form *seams*.

seamless tube Metal pipe made by *extrusion* (*C*) from the solid, or by making a hole through a hot billet and rolling down to size. When seamless tube is bent it does not crack as easily as tube with welded seams formed of folded or spirally wound strip.

seam roll A *hollow roll*.

seasoning (1) The drying of *green timber* to reduce the *moisture content*, which makes it strong and hard. Timber is either naturally *air-dried* or *kiln-dried*, although *water seasoning* has been used. Seasoning can lead to *defects* such as *case-hardening, checks*, and *warps*.

(2) The drying of the *quarry sap* in building stone.

seat cut (USA) A *foot cut*.

secondary circulation, recirculation A system for keeping hot the hot water in pipes of large *domestic hot-water* installations, such as those in hotels or offices, by having a small pipe that runs back from the furthest tap to the boiler. Hot water is pumped round this circuit, and the pipes are all insulated, limiting the temperature drop in the circuit to about 10°C. The only *dead legs* are the short branch pipes off the circuit to each basin, bath, shower, etc.

secondary fixing (1) A *fixing* that holds the final component to a *primary fixing*, such as the bolts that go into an *anchor* or *fixing channel*.
(2) Roofing *fasteners* which attach sheeting at *sidelaps* or *endlaps*.

secondary glazing *Double glazing* formed by adding an extra *pane* of glass inside the existing glazing. Householders can do their own secondary glazing on wooden frames by using synthetic rubber profiles, which are screw-fixed to the wood and clip over the edge of the new pane, or with *edging strips* and clips. The cavity needs *breather holes* to the outside, to prevent *condensation* on the outer pane. Secondary glazing needs to be removed for window cleaning about once a year. See diagram, p. 131.

secondary sash glazing, applied s.g. Professional *double glazing* with a lightweight hinged or sliding metal *sash*, fixed inside an existing window frame.

second fixings Fittings, trim, and accessories which are fixed after the plastering, e.g. *skirtings*, linings, cupboards, and most *joinery* except doors. For *services* second fixings can include attaching taps, wiring in *cable tails* to light switches, etc.

seconds Second-quality materials, particularly clayware such as *bricks* and drain pipes.

secret fixings Fixings which cannot be seen at the surface, used in *joinery*, recessed *luminaires*, air outlets, etc. They improve appearance and reduce opportunities for vandalism.

secret gutter (USA **concealed valley**) A nearly hidden roof *valley gutter* in which the metal gutter lining is covered by the tiles or slates. Unlike an *open valley*, it is easily blocked by leaves and so is not recommended by BS 5534.

secret nailing, blind n., edge n. Nailing which is not seen on the surface. Tongued and grooved, or *rebated*, boards can be slant-nailed through the tongue or rebate. Cladding *fixing strips* are secret-nailed in a similar way.

secret wedging Fixing a joinery *stub tenon* by inserting wooden wedges into sawcuts in its end before it is inserted into a *blind mortice*. As the tenon is driven in, the wedges are forced in by the pressure of the blind end of the mortice on them. If the mortice is *dovetail*-shaped the assembly is called foxtail wedging.

section (1) A *work section*.
(2) A *bay* of structural concrete or floor screed.
(3) *Cross-sectional area* (*C*).

sectional insulation *Moulded insulation*.

sectional tank, bolted t. A storage tank made of square panels of galvanized pressed steel or *glass-reinforced plastics*, about 1.2 m square, which are bolted together through flanged edges, usually turned outwards, and sealed with jointing compound. The panels and internal bracing can be handled into inaccessible areas without lifting equipment and built to any size, although usually limited in height to four panels. They are easily unbolted for removal or enlargement.

section manager, s. engineer A *site engineer* in charge of work for part of a large project, under a *site manager* or *construction manager*.

section mould A *template* of sheet material cut to the cross-section required for a member.

security Measures to prevent abuse, intrusion, theft, or vandalism, either in completed buildings or on building sites. The subject covers *locks, entry systems*, physical barriers such as fences and grilles, watching and floodlighting at night, *intruder alarm systems*, and *fire protection*.

security fence *Anti-intruder chain-link fencing.*

security glazing High-strength flat glass (or clear plastics such as *polycarbonate*) for *anti-bandit* or *bullet resistant* windows, used at bank counters or jeweller's windows. Glass security glazing is usually *laminated* or *toughened*, and some types have embedded metal mesh.

security lock, thief-resistant 1. A lock which is made (to BS 3621) to prevent determined illegal entry but which allows people to leave easily in a fire.

sedge (Cladium mariscus) A tall grass from marshy places, used to form the ridge of *thatch*. It is made flexible by wetting.

SE duct A *ducted flue* rising through a multi-storey building, connected to *room-sealed appliances*. Fresh air is drawn in at the bottom and burnt gas is discharged from the top. It may be fan-assisted, to overcome the effects of wind gusts. See **U-duct**.

seepage pit A *soakaway* (*C*).

segregation The separation of *fines* from the coarse *aggregate* in concrete during *placing* or as a result of *overvibration*. It may cause *honeycombing*.

selected tenderer (USA **invited bidder**) In selective rather than *open tendering*, a contractor from a restricted list who is asked to give a *tender* for work.

self-centering formwork Formwork on *floor centres*.

self-cleansing Not holding dirt. A self-cleansing drain or *trap* allows small solids to be flushed away, avoiding a blockage. A self-cleansing façade is washed by the rain, saving on cleaning costs.

self-climbing tower crane A *climbing crane*.

self-curing The *curing* of concrete by internal moisture and its own *heat of hydration*, encouraged by adding a cover or leaving *formwork* in place.

self-drilling screw A hard steel screw with a tip in the form of a drill *bit*. Its thread is *self-tapping* and it usually has a *crossed-slot head* so that it can be driven home completely in one action with an electric drill.

self-embedding screw A *self-drilling screw* with *countersink* teeth formed under the head. It can be driven home flush into hard timber with an *electric screwdriver*.

self-extinguishing A property of some combustible materials, such as silicone gaskets or plastics conduit, which can catch fire but go out within seconds and so have low *fire hazard*.

self-finished, factory-f., pre-f. Having finishes applied during manufacture, such as *organic coatings* on metalwork, *anodizing* on aluminium, and PVC over a softwood core for doors or windows. Self-finished building components need *protection* and careful handling during delivery and fixing in place. The protection is usually not removed until *handover*.

self-finished felt A *bitumen felt* with metal foil (usually aluminium) factory-bonded to its top surface, for use as a *self-protective* cap sheet.

self-illuminating exit sign, self-contained e.s., emergency e. indicator An *exit sign* that has its own internal batteries and a charger, so that it lights up even when power is not available. See BS 5499.

self-levelling screed, self-smoothing s. A thin floor *screed* or *levelling compound* which flows out to a smooth surface. It can be placed by *screed pump* and light raking, without *trowelling*. Large areas can be laid quickly. It is suitable for carpet or lino, or can be left bare.

self-protective material A material with its own *solar protection*, e.g. *self-finished felt*, or plastics which contain carbon black *fillers* to slow down attack by *ultraviolet*.

self-siphonage The sucking-out of the water seal in a *trap* by a sanitary fitting's own discharge. A test done from a completely filled fitting should not cause total *unsealing*.

self-supporting Independent or *free-standing*, said of a structure such as a chimney, wall, or scaffold; also said of roof sheeting that is *profiled* to carry its own weight and other loads between *purlins*.

self-tapping screw A screw for use in sheetmetal, with an upstanding thread, tapered at the point. It is put into a pre-drilled hole and cuts its own thread as it is driven. Usually it has a *crossed-slot* head and is made of hardened steel.

semi-detached house (USA **double h.**) One of a pair of houses, built simultaneously with a *separating wall* between them. It is therefore attached on one side only, unlike a *terrace house*.

semi-skilled worker Someone who has learnt his or her work by helping others and is rather less skilled than a *tradesman* or craft operative, usually having risen from *labourer* in the trade.

semi-solid core door A medium-duty *flush door* with a core of *flaxboard* or *chipboard*.

sensible heat Heat that can be felt as radiation or measured directly with a thermometer. Compare **latent heat**.

sensor A device which reacts to a change in the local environment, used in *automatic controls*, alarms, *fire detectors*, etc. Sensors include microwave movement detectors, pressure mats, press buttons, devices to pick up ultrasound, infra-red, light, etc., and proximity devices that work by electrostatic or induction effect.

separate screed, bonded s. A *screed* laid on hardened concrete which has been prepared by *scabbling*, washing, drying, and *grouting* to give a good *bond*. It should be at least 40 mm thick. See diagram, p. 392.

separate system of internal pipework Drainage of *foul water* and *rainwater* in different pipes, as opposed to a *combined system*.

separating floor, party f. A floor between two different dwellings, offices, etc. In addition to *fire resistance*, to meet requirements in the *Building Regulations* for resistance to *impact noise* and sound *transmission* it may need a *floating floor*, ceiling *pugging*, or an *isolated ceiling*. See BRE Digest 334.

separating layer An *underlay* put down on a concrete slab before a *screed* or *floor tiles* are laid. It avoids damage to them from initial *drying shrinkage* in the concrete or *thermal movement* from *underfloor heating*.

separating wall, party w. (USA **common w.**) A wall between two different dwellings, shops, etc. It is usually a *fire wall* and must give *sound insulation*. For lightweight dwellings it

1	preliminaries	A	preliminaries/general conditions
2	demolition and shoring	C	demolition/alteration/renovation
3	earthwork	D	groundwork
4	piling	E	in-situ concrete/large precast
5	concrete in situ		concrete
6	precast concrete and hollow floors	F	masonry
7	brickwork and partitions	G	structural/carcassing metal/timber
8	drainage and sewerage	H	cladding/covering
9	asphalt	J	waterproofing
10	pavings	K	linings/sheathings/dry partitioning
11	masonry	L	windows/doors/stairs
12	roofing	M	surface finishes
13	timber and hardware	N	furniture/equipment
14	structural steelwork	P	building fabric sundries
15	metalwork	Q	pavings/planting/fencing/site
16	plastering, wall tiling, terrazzo		furniture
17	sheet metal	R	disposal systems
18	rainwater plumbing	S	piped supply systems
19	cold-water supply and sanitary plumbing	T	mechanical heating/cooling/refrigeration systems
20	hot-water supply	U	ventilation/air-conditioning
21	gas and water mains		systems
22	heating	V	electrical supply/power/lighting
23	ventilating		systems
24	electrical	X	transport systems
25	glazing	Y	mechanical and electrical services
26	painting and decoration		measurement
27	provisional sums; items by specialists	Z	building fabric reference specification

Sequence of trades. A traditional sequence (*left*); the Common Arrangement (*right*).

can be a *double-leaf separating wall*. Details are given in the *Building Regulations*. See also BRE Digest 333.

septic tank A small sewage treatment plant, usually for one house, where there is no *sewer* (*C*). It is a tank through which *foul water* flows slowly enough for it to decompose and be purified. It is divided into two or more chambers, separated by *scumboards* (*C*), with an opening below. There are no moving parts to wear out and break down. The water capacity is from seven to twenty-four hours' discharge from the house. Treated effluent (the over-

flow) can be used for watering plant or tree roots. The chambers need periodic *desludging*. See BS 6297.

sequence arrow The arrow of an *arrow diagram*

sequence of trades The order in which the building *trades* carry out their work, therefore the chapters of a *specification* and *bills of quantities*. The traditional sequences are followed when building a new house, but that may not apply to a large project. The *CI/SfB classification* is one sequence and the *Common Arrangement* is another. See table above.

Serpula lacrymans The only fungus which causes *dry rot* in Britain.

servant key A key that opens one lock in *master keying*, although several servant keys may all be able to open a low-security common lock.

servery A storage cupboard or counter used to keep food at a suitable temperature before it is served. Different dishes are kept in a *hot cupboard*, an *ambient counter*, a *chilled counter*, a *cold cupboard*, etc.

service cable The power cable joining the electricity company's supply *main* in the street to the *intake unit* in a building, whether run underground or as a *drop wire*.

service clamp A pipe *saddle*.

service lift, goods l. A *lift*, often small, used only to carry goods, food, books, etc., but not people, simplifying its safety requirements.

service pipe A fairly horizontal gas or water pipe from the street *main* to the premises receiving the supply. It is called the *communication pipe* up to the boundary, then becomes the *supply pipe*, which for gas can be carried up to 2 m inside the building in a steel or plastics duct.

services Water, gas, and electricity, air conditioning, heating, communications, drains, etc. for buildings, supplied or collected by pipes, ducts, and cables. Building services are broadly grouped as *mechanical and electrical* installations and *plumbing* installations. For a large city project they can represent half the total cost and need much *coordination*. In the completed building they also need planned *maintenance*. In the USA public services outside buildings are called utilities.

services core, mechanical c. The arrangement of vertical *services* in a tall building, with all pipes, ducts, and cables grouped in *shafts*, usually beside the *lifts*, forming a compact area surrounded by thick concrete walls. This controls the number of *main* runs, keeps *equipment noise* out of office spaces, and improves *fire protection*.

services duct See **duct**.

servicing valve A *stopvalve* on the water pipe to a *cistern*, enabling it to be repaired without turning off the entire water service.

serving A sheath over an electrical cable, either as an outer casing to protect against corrosion or as cushioning under a soft outer casing.

set (1) A group of similar objects forming an assembly, suite, or unit.
(2) The hardening of a *binder* in concrete, adhesive, plaster, paint, sealant, or the like, which happens soon after it is mixed or exposed to the air.
(3) The slight overhang given to the point of each tooth of a *saw* to make the *kerf* just wider than the saw blade and enable it to run easily.
(4) A *nail punch*.
(5) A *doorset*.

set back (1) A withdrawal of the *building line* from the street *frontage*, either at ground level or for the upper storeys of a tall building. It is required by town planning laws, to ensure that enough daylight reaches the street.
(2) A door lock *backset*.

set coat A *final coat* of plaster.

setdown A shallow hole in a floor such as a doormat *sinking*.

set-off A reduction of the price for work that has not been done according to the specification or drawings.

set screw A *screw* with a square head, used to tighten a pulley wheel on to a drive shaft. A headless set screw is a *grub screw*.

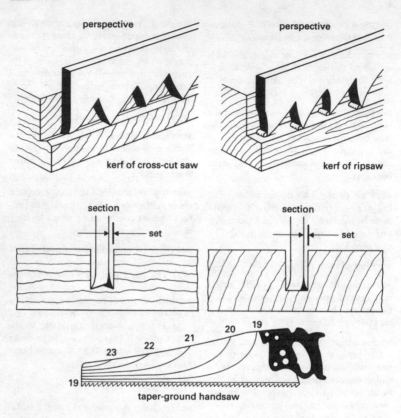

perspective | perspective

kerf of cross-cut saw | kerf of ripsaw

section → set | section → set

19 20 21 22 23 19

taper-ground handsaw

Setting handsaws. The saw is taper-ground, and the contours on it show the tapering thickness of the blade in standard wire gauge.

sett A *granite sett*.

setter out, marker out An experienced tradesman who does *setting-out*.

setting (1) The first hardening of concrete, mortar, plaster, adhesive, paint, etc., after which they should not be disturbed.

(2) The laying of bricks, blocks, stones, lintels, etc., in a wall.

(3) The brickwork which supports and encloses a furnace or a boiler.

(4) The third preliminary operation in hand *sharpening* a saw, after *topping* and *shaping*, and preceding the actual sharpening. The teeth are bent alternately to one side then the other. Setting can be done by hand with a *saw set*, but saws are best sharpened by machine. Handsaws are set as *ripsaws* or *cross-cut saws*, as are many power saws, but others have *chisel-set teeth*. See diagram.

setting block A *glazing block*.

setting coat *Finishing coat*.

setting-out, marking-o. The placing of marks to show where work is located. The cost of setting-out is a *preliminaries* item. Setting-out for bulk excavations requires only pegs, which are placed by the *site engineer*, often using surveying instruments or *lasers* (*C*). Concrete foundations and ground floor walls or columns are set out from *hurdles* and marks for upper floors plumbed up from them. Some of the risks in setting-out are damage to hurdles, knocking of instrument legs, errors of arithmetic, not allowing for floor slab or *screed* thickness, using an *offset* as the real mark, etc. All early setting-out should be checked by a quick and reliable method, then double checked by somebody else.

setting shrinkage The *drying shrinkage* of concrete between the time it is placed in forms and its *initial set*.

severe exposure Site conditions of high winds and heavy rainfall. It can be estimated using the *driving rain index*.

sewer bricks (USA) Low-*absorption*, abrasion-resistant bricks, used for the same purposes as British *engineering bricks*.

sewer chimney (USA) A *backdrop*.

sewer connection The connection of buried *drainage* pipes to a *sewer* (*C*) in the street. The contractor puts in an application to the *local authority*, who make the connection and charge a fee. Work such as excavation and demolition is often done by the contractor. Extra insurance may be needed if the work is *off-site*.

SfB (Samarbetskomitten för Byggnadsfrågor) A Swedish body which started a system for classifying building documentation. See **CI/SfB, International SfB**.

SF₆ *Sulphur hexafluoride.*

shaft A vertical services *duct, lift shaft, fire-fighting shaft,* etc.

shake (1) The separation of wood fibres parallel to the *grain* during felling or seasoning due to stresses in the standing timber. Shake is a *gross feature* which is usually not an allowable *defect*. It can be *heartshake, radial shake, ring shake,* or *star shake*.
(2) A *shingle* split on at least one face, or a *processed shake*.

shanked drill A large *drill bit* with a small diameter shank to fit a small *chuck*.

shaping The second operation in hand sharpening a *saw*, in which the teeth are filed to uniform shape and size.

sharp arris An *arris* before it is *eased* (softened).

sharpening Restoring the cutting edge of an *edge tool* by removing some metal with a *grinder*, often done automatically by machine, which is fast and accurate compared with hand-sharpening, even using tool guides.

sharp sand *Sand* with grains which are coarse and angular, not rounded. It contains no clay, feels gritty, and is clean and suitable for making either *concrete* or plaster *undercoats* and for *bedding* paving slabs. Compare **soft sand**.

shave hook A *scraper* with a T-blade (triangular, pear-shaped, etc.) on a shaft with a handle. It is pulled towards the user. See diagram, p. 314.

shaver socket, razor s. A *socket outlet* that is an exception to the rule forbidding socket outlets in bathrooms. Electric shock is prevented by an isolating transformer, to BS 3052.

sheariness An oily opalescent surface of a *gloss*-painted *film*, a defect caused by uneven porosity in the *undercoats*, by lack of *compatibility* between paint and materials, or from not keeping a *live edge*.

shear plate A compact timber *connector*, usually 67 mm dia., placed between timbers in circular grooves cut with a *hole saw* and held with a 19 mm dia. bolt.

sheathed wiring Electrical cables for *wiring* to *socket outlets* or similar light-duty loads. They have copper conductors inside PVC or other plastics *insulation*, enclosed in an outer sheath of similar material, and can be single-*core*, twin-core, or *twin and earth*.

sheathing boards *Boarding*.

sheathing felts *Bitumen felts* made for use as *underlays* for metal roofing and as *isolating membranes* under mastic *asphalt* roofing or flooring. They can be black or brown (inodorous) and are not used for *built-up roofing*.

sheathing paper *Building paper* on roof *boarding*.

she bolt A cheap and versatile *form tie* made from *reinforcement bar* with bumps that form a rough thread, over which a special nut is tightened.

shedding *Load shedding*.

shed dormer A *dormer* with its roof sloping in the same direction as the main roof, away from the *ridge*. It may have *cheeks* or be an *eyebrow dormer*.

sheen A *gloss* seen at glancing angles on an otherwise flat paint finish.

sheet Thin flat materials produced by rolling, pressing, or extrusion, including *bitumen felts*, plastics sheet, and sheetmetal. See the following entries.

sheet glass, drawn g. Glass made by vertical drawing, 2 to 6 mm thick, with slightly uneven surfaces that distort vision. It is little used today except in agricultural buildings. See **flat glass**.

sheeting Sheet *cladding*.

sheetmetal Copper, aluminium, or zinc which is thicker than 0.15 mm, thinner than 9.5 mm, and 450 mm wide or more. *Foil* is thinner, *strip* is narrower. Flat lead sheet is rolled, not cast, and is usually described as *milled sheet* or strip. Steel sheet is material thinner than 3 mm.

sheetmetal work General *metalwork*, *plumber's metalwork, supported sheet-metal roofing*, or other purpose-made work in sheetmetal, such as air-conditioning *ducts* or smoke *flues* and accessories. Sheetmetal to be shaped by folding should not be marked with *scribers*. It is joined with welted *seams, pop rivets*, and *self-tapping screws*.

shelf angle A galvanized-steel *angle* section bolted on outside a building to support cladding *brickwork* above a horizontal *movement joint*. A gap is left above the last course of bricks beneath the shelf angle, weatherproofed with a thin strip of flexible *sealant* to let the joint move.

shelf life The time for which an adhesive, paint, sealant, etc., can be stored (usually unopened) before it becomes unusable. Compare **working life**.

shellac A natural *resin* formed by insects on jungle trees, dissolved in alcohol to make *French polish* or *knotting*, which dry by *solvent release*.

shell bedding The usual way of laying *hollow blocks*, with narrow strips of mortar along the outer and inner edges of the *bed joints* and *perpends*.

shell boiler A commercial-size *boiler* with water enclosed inside a mild-steel shell, often with *fire tubes*. It works at water pressures up to 10 *bars*.

shell shake A *ring shake* which shows on the surface of sawn timber (shelly timber).

sherardizing A protective coating of *zinc* on small steel items such as nuts, bolts, and nails, which are rolled for ten hours in a drum containing zinc dust and sand, heated to 380°C. The coating is quite thin, but the zinc also diffuses into the steel to form an alloy / of up to 95% zinc. It is a dull grey, can be more durable than *galvanizing*, does not peel off, and gives less increase in size and distortion.

shield bolt A masonry *anchor* consisting of a steel cylinder with a bolt inside. The bolt is tightened to force the inner end of the shield outwards against the sides of a drilled hole and so it acts as an *expansion bolt* (*C*).

shim An *insert* in *veneer*. See also *C*.

shingle (1) Rounded *aggregate* of variable size and shape, without sand.

(2) A thin rectangular piece of timber about 400 × 130 × 6 mm, used like a *tile* for covering roofs or walls. It usually reduces in thickness from *tail* (butt) to *head*. Shingles were used in ancient Rome and split shingles (or *shakes*) were common in medieval Britain. In the USA shingle roof fires are fought from inside the house.

shingle dash A *dry-dash finish* showing *shingle* aggregate.

shingling (USA) Roofing felts laid with *laps*.

shiplap boards (USA s. siding) *Weatherboards* of rectangular cross-section with a *rebate* cut on each edge, fitting into corresponding rebates on the neighbouring boards. White, opaque, extruded PVC 'planks' are widely preferred to the less durable wooden shiplap boards. PVC has more *thermal movement* than wood, especially if backed with insulation material, but it should remain sound for twenty years without painting. Timber needs yearly attention.

ship spike, boat s. A steel spike for fixing large timbers, forged from square bar and having a wedge point.

shoe (1) A short length of pipe at the foot of a *downpipe* to direct the flow away from the wall.

(2) An iron or steel socket enclosing the end of a *rafter* or other loadbearing timber.

(3) A fitting for *patent glazing* which both holds the lower end of a *glazing bar* to the building framework and prevents the glass sliding.

shoe mold (USA) A *base shoe*.

shoot (1) To *plane* the edge, or a *mitre* cut, of a board or panel, to remove roughness from sawing. Edge shooting may be unnecessary after cutting with a fine-toothed *mitre saw* with *carbide tips* or for joinery timber which has been *regularized*.

(2) A *drain chute*.

shop A workshop or factory used for manufacturing components.

shop drawing A manufacturer's drawing for purpose-made work (e.g. joinery fittings or metalwork) to be made in a factory or workshop, not on site. Shop drawings are often drawn on to *background drawings*, and before any work commences they usually must be submitted to the *architect* for approval.

shop priming Factory-applied *priming*. It has better adhesion than site priming because of thorough surface *preparation* and protection from the weather. On both steel and timber the final paint coats are applied on site, which should be done promptly.

shop work Work done in the manufacturer's shop, not *sitework*.

shore A *raking shore*, *dead shore*, or *flying shore*.

shoring Giving temporary support

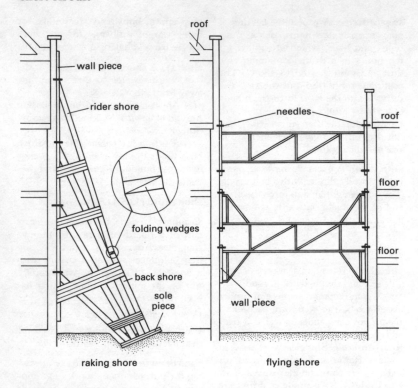

raking shore

flying shore

A raking shore and a flying shore. Both types bear against vertical wall pieces held in place by needles of timber or steel into the wall.

with *shores* to a building being repaired, altered, or *underpinned*. See BS 8004. See also diagram.

short circuit An electrical fault caused by two wires (live, neutral) being electrically joined, which produces a very high *overcurrent* in the circuit. It should cause the *fuse* to melt or other circuit *protection gear* to trip.

short-grain timber *Brash timber*.

short oil A low *oil length* in paint or varnish, from having a low ratio of oil

to resin. The *film* formed tends to be hard but brittle.

shotfired fixings, explosive f. Hard steel pins with a sharp point which are driven with a *cartridge tool*. There are many different sizes and shapes, and they are mostly driven into materials like steel or concrete.

shothole A worm hole 2 to 3 mm dia. left in wood by a large *borer*.

shoulder The surface at the root of a *tenon* which abuts on the wood outside the *mortice*, thus stiffening the joint.

shoulder fitting, s. nipple (USA) A *barrel nipple*.

shoved joints (USA **pick and dip**) The usual method of *bricklaying* for walls. First the bed course is spread and excess mortar gathered on the trowel. Then a brick is picked up in one hand and its end *buttered* with mortar still on the trowel. The brick is laid on the bed course and shoved against the last brick laid, forming the vertical *perpend*. The perpend should be properly filled, but often is not. The final positioning of the brick can be aided by tapping it with the trowel, but with skill and the right mortar this should not be necessary. Mortar squeezed out of the joints is again gathered with the trowel and the cycle repeated.

shower divertor A *bath/shower mixer*.

shower enclosure A prefabricated waterproof cubicle to provide privacy and prevent water from splashing around, usually made of *glassfibre-reinforced polyester* or *acrylic*.

shower head The sprayer outlet for a shower mounted on the wall or on the end of a hose as a handshower.

shower room A room with one or more showers, a *wet area*, subjected to hot and cold water, but which also dries out. It usually has high-quality impervious finishings, such as wall and floor tiles, over a water-resistant backing, such as cement *rendering*. Gypsum *plaster* should not be used in shower rooms, and glazed screens should be in plastics or *safety glass*.

shower rose A large low-pressure *shower head*.

shower tray, s. base, s. receiver A large robust waterproof tray set in or on the floor of a *shower room*, with a *waste* for connection to the drainage system. It is usually made of plastics, ceramic ware, or pressed steel, 760 to 915 mm square.

shrinkage A permanent decrease in size. Shrinkage in building materials may cause *cracking* or *warping*, particularly if uneven. Timber shrinkage during drying is about 0.4% along its length, about 4.5% parallel to the *growth rings*, and about 3% at right angles to them. Concrete and mortar have *drying shrinkage* and *carbonation* shrinkage. See **movement**.

shuffle glazing A window without a frame, glazed into a wall, etc., by sliding the *pane* first one way for its edge to enter a deep groove (or pocket), then the opposite way into a normal groove on the other side, and finally into a third direction into another normal groove. Space to manoeuvre the pane is needed on the fourth side, which has to have a *glazing fillet*, although it may be possible to hide it above a *suspended ceiling*.

shunt duct One of two parallel lengths of air-conditioning *duct* going to adjacent rooms. They reduce *crosstalk* and prevent the passage of fire.

shutter (1) A wooden or steel cover which can be closed over a window, mainly to give the building occupants extra security and warmth at night.
   (2) **shuttering** A panel used as concrete *formwork*.
   (3) A *fire shutter*.

shutter bar A pivoted bar for holding window shutters closed.

shutter bolt An *espagnolette bolt*.

shuttered socket A *socket outlet* with three holes, two for the insertion of power prongs and one for an earth prong. Small shutters block the power holes and are opened only when the earth prong is inserted. The earth prong is slightly longer than the others and must be inserted first. The plug can then be pushed in. There is less danger of electric shock to children. Standard shuttered sockets are rated at 13 amp (3 kW at 240 volts) and are used for U K domestic *ring mains*.

shutter hinge A *parliament hinge*.

shutting jamb, closing j. A *jamb* to receive a *shutting stile*.

shutting stile, lock s., slamming s., closing s. That *stile* of a door leaf which carries the lock, on the opposite side to the *hanging stile*.

shy Undersized compared with *nominal dimensions*, the opposite of 'full'.

sick building A building which causes illness in its occupants, with symptoms like irritation in the nose, throat, and eyes, or on the skin, as well as shortness of breath, nausea, dizziness, and fatigue, the so-called sick building syndrome (SBS). It is usually minor and disappears after people leave the building, but it may also cause sleeplessness. People tend to be similarly affected in numbers exceeding normal levels of sickness. SBS is commonest in air-conditioned office buildings and has been found to affect particularly women, people with uninteresting jobs, those working in government departments, or those who have no control over the temperature and lighting. Low rates of *ventilation* are often mentioned as a contributory factor (or poor air distribution and the recycling of exhaust air). Local air conditioning by *fan coil unit* has the worst record, followed by central heating with *induction units* or fan coils, then *all-air* buildings with *high-velocity* ducts. Mechanical venti-

lation by an *extract system* usually gives the best result, often better than natural ventilation.

Allergy to indoor air contaminants is the main factor in SBS. Allergies cannot be cured, only avoided. These contaminants can be in the form of fine dust, fumes, fibres, odours, or gases which are not removed by air-conditioning *filters* or by cleaning. Individual contaminants may act in a combined way. There are many sources: traffic outside, soft furnishings, people, office equipment, cigarette smoke, and 'off-gassing' from building products, formerly of *formaldehyde*. House dust mites and spores from fungi, both of which produce allergies, can be reduced by the spray/extract cleaning of carpets and by lowering indoor relative humidity to below 45%.

Indoor lighting which is dull and even, or lack of natural daylight, is also a factor, allowing the body's circadian rhythms to shift compared with normal day/night cycles. This can be prevented by a varied lighting environment with plenty of daylight, or by strong illumination in the mornings (2500 lux), followed by usual indoor lighting for the rest of the day. Computer *visual display units* have been closely studied, but are generally thought not to be linked to SBS, unless used for longer than six hours a day (which is frequent).

Buildings can also become temporarily unhealthy places from a fairly obvious cause, such as micro-organisms, often carried in aerosol-size water droplets, including everything from the common cold to *legionnaire's disease*, as well as *humidifier fever*, airborne *asbestos* and *glass-fibres*, fumes from car parks, or by wind blowing in smoke from boilers. This latter group of building-related illnesses is usually taken as being separate from the sick building syndrome, although they may occur at the same time.

side A broad surface of *square-sawn timber*, usually called a *face*.

sidefill Gravel laid each side of a pipe *drain*, which becomes part of the surround. It also holds the pipes steady during the *drain test*.

side form Concrete *formwork* for the sides of a beam.

side gutter A small gutter running down a roof slope at the intersection of a *chimney* or *dormer* or other vertical surface with the main roof.

side-hung door, s.-h. window A door or window with hinges at the side.

sidelap A *lap* between roof tiles, sheet cladding, etc., made by placing the edge of one on top of a tile or sheet already laid, to make a weathertight joint. Sidelaps are parallel to the ribs of *profiled sheeting* or *corrugated sheeting*. For *single-lap tiles* each sidelap can be as little as 37 mm. Corrugated sheets used for roofing usually have a sidelap of two full corrugations.

sidelight, winglight A *fixed light* such as a *flanking window*.

siding (USA) *Cladding*.

sienna Paint *pigment* obtained from hydrated mineral iron oxide and described as raw sienna when it is the yellow-brown, clean crushed mineral and as burnt sienna when it is calcined to a rich orange-brown. It is not very opaque, but its transparency can be made use of in oil *stainers*. Sienna is an *earth colour*.

sight size The area of glass seen inside a window frame, the full *glazing size* being concealed by the rebates. It is the size that admits *daylight*.

silencing pipe A vertical tube from a *ballvalve* to the bottom of a *cistern*, with its outlet always below water level. It has a small anti-siphon hole near the top to stop *back siphonage*, which prevents completely silent filling.

silicates Very common and often chemically complex rock-forming minerals containing the radical SiO_3, e.g. sodium silicate surface *hardeners* for concrete, and *calcium silicate bricks* and building boards.

silicon bronze Probably the most durable but also one of the most costly metal alloys, used for making nails, screws, anchors, etc., for boats or roofs where steel rusts quickly. It is mainly copper, with only 3–5% silicon, plus lesser amounts of beryllium or manganese for high hardness. Usually it is quite strong and easy to weld.

silicone paint Expensive but very durable paint made with *silicones*. Silicone paints are heat-resistant and some are blended with other polymers for *stoving*.

silicones Synthetic polymer *resins* that are chemically inert, temperature-stable, and electrically insulating. They are used in sealants, paints, *water-repellents*, lubricants, rubbers, etc.

silicone sealants Versatile *one-part* building *sealants* which cure to a tough silicone rubber that will withstand extreme temperatures, from −60°C to +200°C, usually supplied as 'clear' or white. When squeezed out from their sealed cartridge, a dried *curing* agent becomes activated by moisture from the atmosphere. Hard, medium, and soft versions all stick well to materials such as glass, metals, or ceramics, but a *primer* should be used on porous surfaces. They can be applied at temperatures between frosty and hot and cure in five to seven days. Resistance to sunlight is good, but some types are not suitable for use on *sanitary fittings*, as they suffer *biodegradation*. In general they are not paintable. See BS 5889.

silicosis A severe lung disease from inhaling rock *dust*, e.g. from dry *sand blasting* or using masonry drills.

silk Trade term for low to medium *sheen* or eggshell *gloss*.

sill (1) A *window sill*.
(2) **s. member, sole plate** The lowest horizontal member of a *frame*, *framed partition*, *framed construction*, or window frame. A door sill is usually called a *threshold*.

sill anchor A bolt cast into the concrete foundations of a timber house. It passes through a hole drilled in the wall *sill* to hold the frame down against the wind.

sillboard (Scotland) A *window board*.

sill cock (USA) A *hose union tap*.

sill height The vertical distance from *finished-floor level* to the top of a *window sill*.

silver brazing, s. soldering Joining metals with silver solder, an expensive alloy of copper, silver, and phosphorus, which is slightly stronger than most *brazing* alloys and far stronger than soft *solder*. It is usually done with an *oxy-acetylene flame*, as silver solder softens at over 600°C, but it can be done with an ordinary propane lamp for the *capillary jointing* of copper tube, for which no flux is needed. Silver solder is also used to repair cast iron, which must be ground completely clean and fluxed. See BS 1845.

silver fir (Abies alba) Many firs go by this name, most of them from central and southern Europe. They are broadly grouped with *whitewoods*.

silver grain The fine pale-grey, shin-

ing flecks of wood *ray* seen in the edge grain of *quarter-sawn* oak or beech.

silver sand Fine rounded white sand used to make *mortar* to match concrete colours in *patching*.

simplex control A system that operates a single *lift*.

single bridging One row of *bridging*, usually at midspan of floor *joists*.

single-coat plaster *One-coat plaster*.

single glazing A window with only one sheet of glass, giving higher heat losses than *double glazing*. Its average *U-value* is about 5.7 W/m².°C.

single-hole basin A *single-taphole basin*.

single-hung window A *sash window* of which only one sash, usually the bottom one, is movable and which has *sash balances* or weights.

single-lap tile, interlocking t. A *roof tile* laid so that its lower edge makes a lap with tiles in only one course below.

single-layer roof A *single-ply roof*.

single-lever mixer A *lever mixer*.

single-line diagram, one-l. d. An electrical circuit diagram simplified to show only one line for each cable rather than a line for every wire.

single-lock welt A *seam* joint for sheet metal made by folding the edges so that they hook on to each other. It is suitable for the cross-joints in *supported sheetmetal roofing* and can be used as an *expansion joint*, as it holds less securely than a *double-lock welt*. See diagram.

single-outlet combination tap assembly A *mixer* tap.

single-phase supply Electricity supplied at 240 volts, with only one *live* phase, plus the *neutral*. It is used for

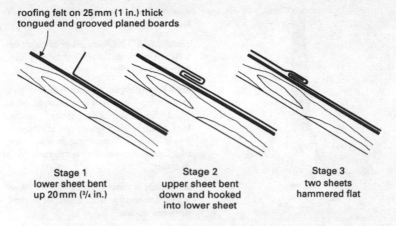

roofing felt on 25 mm (1 in.) thick
tongued and grooved planed boards

Stage 1
lower sheet bent
up 20 mm (³/₄ in.)

Stage 2
upper sheet bent
down and hooked
into lower sheet

Stage 3
two sheets
hammered flat

A single-lock cross-welt in supported sheetmetal roofing.

houses, offices, and light-duty commercial installations. See **three-phase
supply**.

single-pipe system A *one-pipe
system*.

single-pitch roof A *monopitch roof*.

single-ply roof *Membrane roofing*
using high-performance materials such
as *bitumen-polymer* reinforced with
polyester fibres, some synthetic *rubbers*
such as *ethylene propylene diene rubber*,
hypalon, polyisobutylene, all *thermosets*, but also *thermoplastics*, such as
polyvinyl chloride. They are mostly
used for *flat roofs*, laid on decking,
over insulation, or as an *inverted roof*.
Properly laid by skilled roofers a
single-ply roof can be expected to last
about thirty years.

single-point heater An *instantaneous
water heater* for a sink.

single-pole switch A simple *switch*
that can make or break the connection on only one *pole*, to control an
electrical circuit. Relays can also be
single-pole.

single-stack system *Sanitary pipework* with a large-diameter vertical
stack to carry *soil water* and *waste water*
drainage and provide *ventilation*. There
is no *vent pipe*, nor any *branch vents* as
in the *one-pipe system*. It is mainly used
in low-rise buildings and has limitations, including minimum slopes for

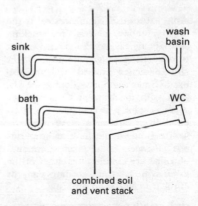

sink

wash
basin

bath

WC

combined soil
and vent stack

A single-stack system. The WC pipe must
join the stack at least 200 mm below the
bath (or any other) branch pipe.

rubber

A sink plunger.

branches, the need for *deep-seal traps*, and the fact that a bath waste branch must not be connected in the 200 mm height above a W C branch connection. See BRE Digests 248 and 249. See also diagram, p. 411.

single-taphole basin, s.-t. sink A basin or sink with only one *taphole*, for cold water only, or a *monobloc mixer*.

sink, kitchen s. Stainless steel is the most popular material for making sinks, being easy to clean and resistant to damage, although ceramics, GRP, and enamelled pressed steel are also used. Sinks can have double bowls, each with its *waste* and one or two *drainers*. Hot and cold water is supplied from taps or a *mixer*, and the wall behind should have an impervious *splashback*. The greatest enemies of a sink are the blocking either of the *drain* with fat from cooking pans or of the *trap* with solid rubbish.

sink grinder A *garbage disposal sink*.

sinking A shallow hole, such as that in a concrete floor for a *doormat*, or a

mortice for hinges in a door frame or leaf.

sinking-in The absorption of paint by a substrate or an *undercoat*, which may cause loss of *gloss*.

sink plunger, force cup A tool with a large rubber cup and a wooden handle, used for unblocking sinks, basins, baths, etc. With properly installed pipework and self-cleansing *traps*, such blockages should be very rare. See diagram.

sink unit A *joinery fitting* to take a *sink* on top, usually with cupboards and drawers under.

sintered pulverized-fuel ash (SFA) A *lightweight aggregate*, such as Lytag, made from power-station *fly-ash* (*C*), with a density from 640 to 960 kg/m^3 and used in structural lightweight *blocks*, concrete, and screeds.

siphonage (1) The sucking-out of the water *seal* in a *trap* by the water falling down the *stack* or running down a drain. It can be *self-siphonage* or be caused by other fittings. With proper pipe sizes, layout, and *ventilation* siphonage should not occur. A check for any trap *unsealing* is a usual part of the *commissioning* of internal pipework.

(2) *Back siphonage*.

siphonic closet A *water closet* with a normal *pan* and a double *trap*, usually made as a *close-coupled* suite for a 9 litre *cistern*. Siphonic closets have a long, quiet flushing action and are not suitable for *dual-flush cisterns*. Compared with *washdown closets* they are slightly more expensive, and the second bend can be difficult to clear if blocked by bulky objects.

site, building s. (USA **jobsite, worksite**) A plot of land for building or on which work for a building project is in hand, often an untidy process. The builder is responsible for the safe-keeping of the site and the works on it from *possession* until *handover*. Laws on *noise control* and *safety* apply to building sites in many countries.

site accommodation Space for offices, storage, changing clothes, taking meals, meetings, toilets, washing, etc., for the work or welfare of people on site, the contractor's staff, the *clerk of works*, etc. Portable *jack-leg cabins*, sheds or huts, adjoining buildings, etc., may be used. Temporary buildings within the building being worked on, or closer than 10 m, should be constructed of non-combustible materials and have at least thirty minutes' *fire resistance*. The cost of site accommodation is a *preliminaries* item.

site agent A *site manager*.

site boundary The edge of a building *site*, usually with a fence or *hoarding* for general security and to keep out children. It also marks the point between what is *on-site* and what is *off-site*.

site clerk, checker, timekeeper A builder's employee who does the routine office work on site, such as checking and recording deliveries, keeping timecards, noting the weather, dealing with telephone calls, etc.

site constraints Site limitations imposed by the presence of existing property or services, such as underground rail tunnels, the need to provide support for adjoining buildings, restrictions on trespass by crane jibs, a particular need for quiet working, etc.

site diary A record kept by the *site manager* of what happens each day on site: the work in hand, its progress and what labour and plant were used, weather and lost time, *daywork*, and details in support of *claims*, etc.

site engineer (USA **field superintendent**) A qualified but junior person employed on site by the contractor for a large project to do the *setting-out* of the main building structure and the *external works*, to direct machine excavations, or to run a small section of work.

site instruction (USA **field order**) A written instruction issued by the *architect* or another consultant, or their representative, addressed to the contractor, giving details about work to be done or precautions to be taken. A site instruction is often a minor *variation*.

site investigations, ground i. A study of the land for a proposed building to determine the depth of suitable ground for *foundations*. For low buildings this should include checking Ordnance Survey and geological maps, visiting the *local authority*, and observing other sites nearby. See BRE Digest 347 and *C*.

site manager, s. agent The contractor's organizer on a building site, with overall responsibility for the work, including the administration of the contract and the coordination of *sub-contractors*. Contact with site *operatives* is usually made through the *general foreman* and major orders for materials through the company *purchasing officer*.

site meeting A *meeting* held on site between representatives of the *contractor* and the *employer* to review progress and information needed, and to decide on action and who is to take it. It is usually held monthly to coincide with inspections by the *architect*, who chairs the meeting, with the *site manager* taking minutes.

site practice *Good practice*.

site roads Roads between the street

and buildings on site, thus part of the *external works*. They are often partly built early in the *programme* to give *access* for deliveries, the final wearing course being laid at the end of the job.

site security Measures to keep intruders out of building sites and prevent theft. They include *hoardings*, watching and lighting, identification badges and *permits to work*, signing for portable tools, lock-up tool boxes, painting and numbering plant and equipment, checking all vehicles leaving the site, and notices warning of dismissal for *theft* (*C*).

site services Temporary connections of power, water, drains, telephone, etc., for use during the time work is done on site. They are priced as *preliminaries*.

sitework Work in the building *trades* on site, particularly the fixing in place of materials, which may be delivered and kept in storage until they are to be installed and labour is available. Since site (or field) conditions place practical limits on accuracy, *tolerances* for fit may be needed. Finishes that could be damaged require *protection*.

Sitka spruce (Picea sitchensis) A strong *softwood* from the island of Sitka on the north-west coast of Canada, also grown in plantations in the UK. It is pinkish white to very pale brown and has a silky sheen. Sitka is the finest *spruce* and obtainable in large sizes for structural timbers, although also used for joinery.

SI units The current version of the metric system. Its base units are the metre, kilogram, second, ampere, kelvin, candela and mole. The derived units – radian, steradian, hertz, newton, joule, watt, pascal, volt, ohm, coulomb, lumen, lux, etc. – are defined to work together without the need for correction factors. Metric multipliers are also used with the SI system (see p. x).

size A liquid *sealer*, usually transparent, with which wood or plaster is coated so that *adhesive*, *paint*, or *varnish* applied over it will not be too much absorbed. Glue size is made from adhesive diluted with water. Varnish size and *gold size* contain mainly varnish with extra *thinner*.

sized slates Roof *slates* of uniform dimensions, not *random slates*.

skeleton construction A loadbearing *frame* of columns and beams to carry each floor and non-loadbearing *panel walls*. It is usually of reinforced concrete, steel, or timber.

skeleton core door An early type of *hollow-core door*, framed with a timber grid.

skeleton stair A *stair* with *treads* but no *risers*, giving an *open rise*.

skew (1) Oblique.
(2) A *kneeler*.

skewback The upper surface of a *springer*, or the springer itself, which slopes to carry the first *arch* stone.

skewback saw A *handsaw* with its back slightly concave curved.

skew corbel A *gable springer*.

skew flashing A sloping roof *flashing* inside a low *gable coping*.

skew nailing, toe n., tusk n. Driving *nails* in at an angle near the edge of a timber so that they come out of another edge and go into another timber. The joint made is not very strong. Where possible alternate nails are driven in opposite directions.

skew table A *kneeler*.

skid-mounted equipment Heavy

mechanical equipment that stands on steel runners. It is pushed, pulled, or lifted into place.

skids Short lengths of wood used for packing walling stones to the correct height during placing.

skilled operative, craft o., s. worker Either a *tradesman* or a leading hand in a modern *trade* which has expanded too quickly for the apprenticeship system to become general (usually riggers, steelfixers, roofers). Many such workers, who are not called tradesmen, may nevertheless have the skill and adaptability of a tradesman. Most of them work from drawings. Many skilled men have risen from *labourer*, and may have *National Vocational Qualifications*.

skim coat, skimming c. A plaster or render *final coat* about 3 mm thick put on with a steel trowel to fill holes which would make decoration difficult, such as *blowholes* in smooth concrete or the voids in *woodwool slabs*.

skinning The formation of a skin on the surface of a *paint* or *varnish* from the oxidation and *drying* of a thin top layer, which then protects the contents below. The skin may form in the closed tin and is removed and discarded before the paint is used.

skip (1) **holiday** An area left unpainted, thus a defect of *application*. An *undercoat* can be given a slight *tint* to reveal skips when applied to another, differently tinted undercoat.

(2) A container for lifting fresh *concrete* by crane, with a funnel bottom and gate, opened to let it out slowly, controlled by pulling on a bar. A fabric apron can be clipped under the outlet to catch falling stones during lifting.

(3) A flat steel container of several cubic metres capacity for the storage and removal of building rubbish. It can be parked in the street, for which a licence from the local authority must be obtained.

(4) A crane *rubbish bucket*.

skirt A *downstand* round the edge of a *foundation*, usually to prevent the ground below from being washed away.

skirting (1) **s. board** (Scotland **base plate**; USA **baseboard, mopboard, washboard**) A board set vertically round the foot of a wall as *trim* to protect wall plasterwork from kicks. It can be of wood, PVC, or other materials, and is fixed with nails or *gun-grade adhesive*. In *wet areas* skirtings can be *coved tiles* or formed in *terrazzo*.

(2) A roofing *upstand* on a parapet wall, particularly if formed in *mortar* or *asphalt*.

(3) Part of a wash *basin* that stands up along its back edge.

skirting block An *architrave block*.

skirting trunking (USA **plugmold, wireway**) A hollow *skirting* containing cables, usually of PVC or *powder-coated* pressed steel. It has a backing plate fixed to the wall and may have separate compartments for power, telephone, data, and control cables, with outlets for each service and a clip-on frontplate. It should be kept clear of *radiators* or heating pipes, in the same way as similar trunking in *architraves* and *dados*.

sky component, s. factor In the *daylight factor method*, light that comes directly from the sky through a window to a point inside a room, calculated from a *standard overcast sky*.

skylight A *rooflight*.

slab (1) A flat piece of precast concrete. *cast stone*, granite, limestone, slate, narble, *terrazzo*, etc., used for *pavings* and floor finishings.

(2) A concrete *floor slab*.

(3) A piece of timber with one sawn face and one *waney*, curved face.

slab-on-grade (USA) A ground slab.

slab urinal A flat wall slab of stainless-steel sheet with a channel below and sloping floor treads, to BS 4880.

slack A simplification to the definition of *float* in the *critical path method*. It is the difference between the latest and earliest date when an *event* occurs. On the *programme* it can be before a *beginning event* (beginning slack) or after the *end event* (end slack).

slack side The *loose side* of veneer.

slag, blastfurnace s. A glassy by-product from steelmill (or other) blast-furnaces, granulated by cooling with water and ground to the fineness of cement. It has *pozzolanic* properties, making it a useful *binder* in *blended cement* and other building materials. See the following entries.

slag brick Bricks made from *slag* crushed and mixed with *lime*.

slag strip (USA) Edge trim to stop gravel *ballast* being blown off a roof.

slag wool Filaments of blastfurnace *slag* used to make a fine *mineral wool* with high fire resistance, thermal insulation, and sound *absorption*. It is heavy but brittle.

slaked lime, calcium hydroxide (Ca(OH)$_2$) *Hydrated lime*.

slam buffer (USA **mute, rubber silencer**) A small rubber 'nail' with a dome top that is pushed into a hole drilled in the *shutting jamb* of a door to reduce noise if the door slams.

slamming stile A *shutting stile*.

slamming strip A *lipping* along the edge of the *shutting stile* of a door leaf.

slap dash *Roughcast*.

slat A thin strip of wood, plastic, or metal in a *louvre*, blind, etc.

slate (1) A dark grey natural stone made up of many thin layers which can be split (riven) into thin sheets, then axed to size and holed by punching, or sawn in any direction. Slate is quarried in North Wales and other areas and was once widely used for roofing, copings, and sills. A natural slate roof can be expected to last 100 years or much more and requires little maintenance. Slate's main uses today are for paving and flooring slabs.

(2) Any rectangular sheet of roofing material, whether of slate, stone, *cast stone*, fibre cement sheet, bitumen felt, metal, etc., including *strip slates*.

slate hanging, weather slating Vertical slating on a wall.

slating The *trade* of roofing with slates, or hanging slates as a wall covering. Roof slates are laid on *battens* to a double-lap pattern and usually nailed down. See BS 5534.

sleeper (1) A *valley board* on the *rafters* of the main roof. It carries the feet of the *jack rafters* and replaces a *valley rafter*.

(2) A *sleeper plate*.

(3) See C.

sleeper clip (USA) A *floor clip*.

sleeper plate A *wall plate* on a *sleeper wall*; or a similar wooden plate on a concrete floor carrying floor boards or *joists*.

sleeper wall, basement w. A brick or block wall to carry the *joists* of a timber ground floor, usually laid in *honeycomb bond* to allow ventilation.

sleepiness A reduction in *gloss* as a paint film dries, a defect due to insufficient drying of an *undercoat* before *recoating*, or to wet weather.

sleeve, expansion s. A short length of pipe built into a wall or floor to form a neat hole through which another pipe passes. This allows *movement* of the inner pipe from expansion and contraction, needed to avoid damage or creaking noises in the pipes or wall. A *fire stop* may be required in the gap between the pipe and sleeve, and for plastics pipes a special *fire stop sleeve*. See diagram.

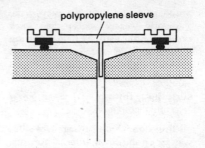

polypropylene sleeve

A sleeve coupling with rubber sealing rings for joining vitrified clay pipes.

sleeve coupling, s. connector, sleeved joint, socket coupling A *fitting* used for jointing plain-ended (un-socketed) drain pipes, which are pushed in from each direction. The sleeve coupling is two *sockets* back-to-back, made to suit the outside diameter of the pipes to become *spigot* ends. This saves the expense of using pipes with socket ends, and for buried drains allows a narrower trench. Poly-propylene sleeve couplings with *push-fit joints* are used for *vitrified clay* pipes. Stainless steel or cast-iron bolted sleeve couplings with rubber gaskets are used for joining cast iron pipes. See diagram.

sleeve piece, ferrule, thimble A short tubular pipe fitting, often a length of thin-walled brass tube, soldered to a lead pipe. Copper and lead should not be joined.

slewing Rotation of a crane jib.

sliced veneer Timber *veneer* made by a machine with a knife up to 5 m long which cuts straight slices in the same way as a hand *plane*. The *flitch* is first heated in a cooking vat so that when cut it is wet and steaming. The slices can be as thin as 0.2 mm but are usually about 1 mm. This method

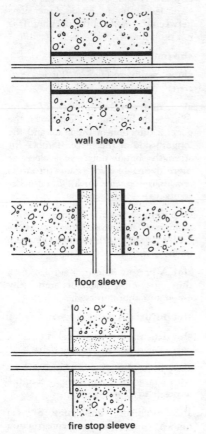

wall sleeve

floor sleeve

fire stop sleeve

An expansion sleeve.

produces no sawdust but may make a slightly weaker veneer than *sawn veneer*. It is the usual method of cutting figured face veneer, for which it is preferable to rotary cutting. Sliced veneer has a *loose side* and a *tight side*.

slide bolt, thumb s. A small *barrel bolt*, often let in *flush*.

sliding door A *door* that moves sideways or up and down. Internal horizontal sliding doors usually have an overhead track and floor guide, while external *patio doors* often have a floor track and overhead guide. Small sliding doors should have end buffers or *cushion action*, and are well suited to automatic operation. Vertical sliding doors are usually *rolling shutters*.

sliding-door lock A *lock* for a *sliding door*, such as a *hook-bolt lock*.

sliding-folding door A *folding door*.

sliding window, s. sash, slider A window which opens by sliding. The term is usually applied to a window sliding horizontally, whereas a window sliding vertically is traditionally called a *sash window*.

slip (1) A thin piece of material, such as a wood *fixing fillet*, a *dentil slip*, a piece of plain roof tile, etc.
(2) Fluid grout made with cement and water or plaster and water.

slip brick A *brick slip*.

slip feather A *loose tongue*.

slip joint A joint between pipe ends made merely by passing a pipe into a socket, without screwing, soldering, or solvent welding.

slip mortice An *open mortice* or a *chase*.

slippage of dates Time lost in a building *programme* when a *critical activity* is behind schedule and completion is delayed. To catch up, more labour may be needed, or extra resources such as materials or equipment, sometimes with *updating*. Slippage is common, but the disruption it causes *sub-contractors* will not usually give them the right to a *claim*.

slipper bend A *channel bend*.

slip-resistant See **anti-slip paint, non-slip floors**.

slip sill A door or window *sill* of the same length as the inside dimensions between the *jambs* of the opening. It is not built into the walls, as is a *lug sill*.

slip tongue A *loose tongue*.

slope concrete (USA) *Blinding concrete*.

slope of grain One of the *gross features* of sawn timber, the angle between the general direction of the *grain* and the longest side of the piece. Timber for strength should not have a slope of more than 1 in 8 for beams (or struts less than 100 mm thick). For thicker struts it should be less than 1 in 11.

slop hopper, s. sink A robust hospital *soil fitment* large enough to empty buckets into, with a 100 mm dia. *trap* and a flushing *cistern*.

slot A narrow hole in a surface, or a thin gap. Screw-head slots are usually either straight or crossed.

slot diffuser A *linear diffuser*.

slot drain A gutter concealed under a narrow gap between paving slabs.

slow bend, long sweep bend A *bend* with a large radius or long sweep, the opposite to a *quick bend*.

slow-burning construction (1) (USA) Heavy-timber construction which is more fire-resisting than bare steel frames.
(2) Construction with materials

treated with fire-resistant surfacing or impregnation.

slumping The downwards flow of a *sealant* after application. Gun-grade sealants should be non-slumping. For the concrete *slump test* see *C*.

slurry A fluid cement–water mix, sometimes containing sand. See also *C*.

slurrying The *protection* of the finished surface of a stone facing against staining by covering it with a weak mix of white lime and stone dust, which is washed off before *handover*.

slushed-joint method (USA) A method of *bricklaying*. The bricks are laid in courses with gaps between them and these are then filled with mortar or liquid *grout*.

small-bore system A *central-heating* system with radiators (or other heat exchangers) fed by pairs of copper or stainless-steel pipes of about 15 mm outside diameter, through which hot water is driven by a *circulating pump*. Small-bore and *minibore* are used for heating houses, usually in a *sealed system*. They have superseded older systems with large-diameter pipes and *gravity circulation*.

small-bore unit A packaged *boiler* for a *small-bore system*, complete with *circulating pumps* and controls.

small tools (1) Hand or *power tools* either belonging to each tradesman or provided by the builder. They are usually locked in a tool box or shed at night, or taken home by their owner, to help prevent *thefts from site* (*C*).

(2) A *plasterer's small tool*.

smalti Cubes of cast glass cut to give a sparkling surface, available in a wide range of colours and regarded as the best *tesserae* for *mosaic* work.

smart building See **building management system**.

smoke Smoke has no fixed chemical composition, but contains *combustion gases* given off from all organic *combustible* materials. Inhaling this part of smoke is the main cause of deaths in fire. Early smoke from building contents (often not covered by regulations) is usually the most deadly. Burning plastics may give off dense or poisonous smoke. Hot smoke rises in air because of its buoyancy and spreads across ceilings. People have survived fires by crawling along the floor to an exit or by keeping doors shut.

smoke alarm A domestic fire alarm with a *smoke detector* and an alarm sounder in one compact unit, usually with batteries (which need regular replacement), required by law in every new home for sale in England and Wales. Smoke alarms are inexpensive and have greatly reduced home fire deaths, e.g. from fires started by appliances left on at night to use cheaper power. Mains-powered types should be connected by an electrician, to the *Institution of Electrical Engineers Regulations for Electrical Installations*. If installed on the ground floor they must be loud enough to wake people sleeping upstairs. No danger comes from their radiation, which is a tenth that of a colour TV.

smoke chamber The space inside a fireplace *chimney* formed by the *gathering*, the *throat*, and the *smoke shelf*.

smoke control *Fire precautions* aimed at restricting the spread of smoke (containment) and allowing or forcing it to leave the building (dispersal). Smoke control may be required by the *Building Regulations* for corridors and stairs that are *protected routes*. Containment is usually by *smoke doors* or *pressurized escape routes*, and dispersal through *smoke outlets.* See BRE Digest 260 and BS 5588.

smoke detector A *fire detector* that reacts to smoke and raises a fire alarm. Smoke detectors work by the scattering or obscuring of light, e.g. the *optical smoke* and visible smoke types. The *ionization chamber* type detects both smoke and *combustion gases*. All of these are compact *point detectors*, but optical-beam smoke detectors work across large distances, as in an *atrium*. See **smoke alarm**.

smoke door, s. control d., s. stop d. A *doorset* which prevents cold smoke from passing into areas such as fire *escape routes*. It need not be tested for *fire resistance*, and has *brush seals* or *blade seals* round its edges. A *fire door* that can also act as a smoke door has the letter 'S' added to its designation, e.g. FD 20S. Smoke doors are tested to BS 476 part 31.

smoke explosion See **backdraught**.

smoke extract fans *Fans* which blow hot or cold smoke out of a building, either directly or through a fire-resisting shaft or duct.

smoke logging The filling of rooms, corridors, or other areas with smoke, which may prevent or hinder access to *escape routes* and *fire-fighting shafts*.

smoke outlet An opening through which smoke or fire disperses to the outdoor air. It can rely on the natural buoyancy of hot smoke (*fire venting*) or on mechanical *smoke extract*.

smoke pipe A *flue*.

smoke rocket A drain *rocket tester*.

smoke shelf A ledge across the back of a fireplace *smoke chamber* which is curved upwards to reflect downdraught from wind gusts back up the *flue*.

smoke venting *Fire venting*.

smoothing (1) Finishing a concrete surface with a *steel trowel*.

(2) *Resource smoothing*.

smoothing compound *Levelling compound*.

snagging (1) Doing the last minor jobs and adjustments before the *handover* of a completed building.

(2) The catching of clothes on door *lever handles* or similar ironmongery.

snagging list A *schedule of defects*.

snake A long thin steel coil wound very tightly on polypropylene rod. It can be pushed up drains to unblock them or into *conduits* to pull back a *draw cable*.

snap header, half bat A brick *bat* about 100 mm long, with its uncut end used as a *header* face.

snapping line, snap l. A *chalk line*.

snap-ring joint An O-ring *push-fit joint*.

snap tie A *form tie* made of flat steel bar with notches enabling it to be broken off inside the concrete after formwork is stripped.

sneck In *rubble walling* a small squared stone not less than 75 mm high, but smaller than a normal *stone*.

snecked rubble (Scotland) A *rubble wall* built of squared stones of irregular size. Its peculiarity lies in the use of *snecks*.

snib, check lock A button that holds the latch or bolt of a *lock* in one position, to remain either shot (to lock the door) or withdrawn.

snib latch Usually an *indicating bolt*.

snow board, s. slats (Scotland **s. cradling**) (1) Horizontal wooden slats fixed across the top of a *box gutter*, with gaps between them to allow melting snow to drain away without blocking the outlet. They are carried by *gutter bearers* on each side, and may also be used as a walkway.

(2) **s. guard, roof g.** A wire fence or board on edge about 200 mm high fixed at least 100 mm up the roof slope above the eaves gutter to prevent heavy masses of snow slipping off and breaking the eaves gutter or its supports and falling on people. The clearance between the roof surface and the bottom of the snow board should be at least 50 mm. See BS 5534.

snow cradling *Snow boards.*

soaker A small piece of *flashing*, usually lead sheet roughly the size of a roof tile, used to make a watertight joint at a valley, hip, or abutment and usually made weathertight by overlapping. Although it is *plumber's metalwork* it may be laid and *dressed* to shape by the roofer, who needs it in good time.

soap A *special* brick of normal length but narrower, with its width equal to its height. In the UK today it is 215 × 46 × 46 mm.

socket (Scotland **faucet**) (1) (USA **bell**) An enlarged end to a pipe, made as one piece with the pipe *barrel*, inside which the plain end of another pipe can fit to make a *spigot-and-socket joint*.

(2) An inside-threaded tubular *pipe fitting*, used as a connector between lengths of *screwed pipe*. See diagram, p. 177.

(3) A *mortice* in ironmongery to receive the cone of a *pivot*, the ball of a *ball catch*, etc.

socket coupling A *sleeve coupling*.

socketed iron pipe A *spigot-and-socket* cast-iron pipe.

socket former A tool used to enlarge the end of a copper pipe to the bore needed for the socket of a *capillary joint*. It is inserted and expanded by pumping fluid into its hydraulic ram. The tool is expensive, but a hammer-driven socket former can be improvised from a short piece of stainless-steel tube of the same diameter as the copper tube to be expanded. Joints made this way are less conspicuous than any other type and the method enables short ends of tube to be used up, but it takes more time than other methods. Before starting, the copper is softened by *annealing*. If the metal work hardens during widening it has to be annealed again or the socket will split and be ruined.

socket inlet An electrical *plug* that goes into a *socket outlet*.

socket joint A *spigot-and-socket joint*.

socketless pipe An *unsocketed pipe*.

socket outlet (1) (USA **receptacle**) A power point for *low voltage* fixed to the wall and containing two or three holes into which the metal prongs of a *plug* are inserted. In the home, *shuttered sockets* should be used. Socket outlets are suited for small power up to 32 amps but not heavy *motive power*, and they are usually run off a *ring main*. Ordinary socket outlets are not allowed in bathrooms in the UK, but *shaver sockets* can be used.

(2) The part of a *plug-in connector* for gas which is fixed to the wall.

sodium-vapour lamp (SON) A high-efficiency lamp with a warm yellow light, used mainly in streets or for floodlighting large buildings.

soffit, soffite The under-surface of any spanning or overhanging part of a building, a floor slab, lintel, door frame head, cornice, stair, beam or arch, or its *lining*.

soffit board, eaves lining (USA

plancier piece) A horizontal sheet fixed under the *eaves*, concealing the *rafters* and the underside of the *roofing*. It runs between the back of the *fascia* and the face of the outer wall. It may have *ventilation* holes for a *cold roof*. See diagram, p. 147.

soffit form, s. shutter The *formwork* for a soffit, of a concrete beam or similar *encasement* to steelwork. A slab soffit form is usually a *deck*.

softboard *Insulating board*.

soft-burnt ceramics *Ceramics* such as *clay bricks*, tiles, etc., which have been fired at a low temperature and therefore have high *absorption* and low strength.

softened angle An *eased arris*.

soft landscaping, organic l. The live plants in *landscaping*, trees, shrubs, grass and flowers, with their *irrigation*, as compared with *hard surfaces*.

soft sand *Sand* with small rounded grains of uniform size, smooth to the hand when squeezed, unlike *sharp sand*. It is used for plastering or making mortar for *rendering, pointing,* bricklaying, etc., but not for concrete.

soft solder A tin/lead, tin/silver, or tin/copper mixture which melts below 200°C, used to join copper, brass, or galvanized steel. It is made in different grades, either as wire with a *flux* core or as triangular bars. Where soft solder is not strong enough, *hard solder* is used. See **lead-free solder**.

softwood The common description of *timbers* which belong to the botanical group *gymnosperms*, most of which in commerce are conifers. Softwoods are one of two main groups of timbers, the other being *hardwoods* Some softwoods, such as yew, are harder than most hardwoods. Softwoods often contain *resin*, which makes them water-proof. Softwoods generally have less *moisture movement* than hardwoods, are easier to work, and are less decorative, so that if plentiful they are used for utility, not appearance. Two main groups of softwoods are *whitewoods* and *redwoods*.

soil *Soil water*.

soil branch A branch pipe leading to the *soil stack*.

soil cement (USA) *Cob* construction, or *soil stabilization* (*C*).

soil drain A *drain* for carrying *soil water* to the *sewer* (*C*) (when it becomes sewage). It is of at least 100 mm dia. and buried in a trench. After laying, it is given a *hydraulic test* (*C*) before the trench is *backfilled*.

soil fitment, s. appliance A *sanitary fitting* such as a *water closet, urinal,* bed-pan sink or washer, or a slop sink, often with a pan and a flushing *cistern*. Soil fitments must be connected to a *soil pipe*.

soil pipe The *sanitary pipework* that carries *soil water*, either alone or with *waste* water or *rainwater*.

soil stack A vertical pipe which takes discharge down from the parts of a building above ground into the *soil drain*. Its upper end passes above the roof and is left open to ventilate the drain. Formerly, cast-iron pipes were usual, but *plastics* are commoner in new work.

soil vent A *vent pipe* for a *soil stack*.

soil water Discharge from *soil fitments*, carried away in *sanitary pipework*. *Soil water* and *waste water* are both *foul water*.

solar collector, s. panel A flat absorber plate facing the mid-day sun. Absorber plates have pipes filled with a frost-resistant water/glycol mix which becomes heated. For *solar hot-water services* they are *glazed* and of either the boxed or the in-roof type, although unglazed types are used for swimming pools and *heat pumps*. Metal absorber plates have *low-emissivity coatings* of electroplated black chromium or black nickel that reduce re-radiation heat losses and can get as hot as 200°C. Most systems are of the 'passive' type, with only simple controls.

solar-control film Thin plastics *films* put on a window *pane* to create *solar-control glazing*.

solar-control glazing, anti-sun glass Window glass that reduces the sun's heat or glare, and in some types also *daylight*. Reflective types can reject 50% of the heat, but they stop more light than *low-emissivity coatings*. Reflecting (surface-coated float or surface-modified tinted float) glasses and stick-on *films* perform well because they warm up less than heat-absorbing *tinted glasses*, which stop the most light, become hot, and give off inwards a third of the absorbed heat; if used for the outer pane of *double glazing* this falls to a sixth. Heat-absorbing types get very hot and expand and contract more than clear glass. They therefore need to be fixed with synthetic rubber *gasket glazing* or *sealants* and made of *toughened glass* so that shadows from the window frame, flag poles, etc., that fall on them do not cause *thermal stress breakage*.

solar dial The twenty-four-hour dial of a *clock timer* or *programmer*.

solar gain, heat g. Heat from the sun that enters a building and becomes a *heat load* on air-conditioning cooling plant.

solar heating The use of heat from the sun, which, falling on a surface at right angles to its rays, on a cloudless day and with high sun angles, averages nearly 1 kW/m². The simplest solar heating is provided by the good *orientation* of buildings. More complex methods involve *double glazing*, *Trombe walls*, and *solar hot-water services*.

solar hot-water service A *solar collector* and a storage *cylinder* used to produce *domestic hot water* in warm sunny climates, such as France south of the Loire, Israel, and Australia, or in other climates to pre-heat boiler *feed* water. See BRE Digests 205 and 254.

solar load The cooling load from *solar gain*.

solar protection Material placed on top of *bitumen felt* to block out damaging *ultraviolet* from the sun.

solar reflective surfaced roofing A *self-finished felt*.

solder Any easily melted alloy used for joining metals. *Soft solders* are mainly lead/tin alloys or *lead-free solders*; *hard solders* are copper/zinc alloys. See also **silver brazing, soldering**.

soldered dot, lead d. An intermediate fixing for lead *supported sheetmetal roofing* or cladding, made by forming a shallow dishing in the timber *substrate*, into which the lead sheet is *dressed* and fixed with a brass screw and washer, then filled with solder to leave a *flush* surface. On vertical *dormer* cheeks they are placed at 500 mm spacings both ways. See diagram, p. 424.

soldering Making a joint with *solder*, which softens gradually when heated, making it easy to work. The surfaces to be joined must be prepared by cleaning and a suitable *flux* used.

soldering iron A copper *bit* fixed to a shaft with an insulated handle. The bit

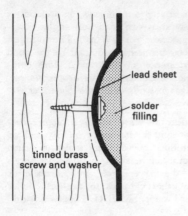

lead sheet

solder
filling

tinned brass
screw and washer

A soldered dot.

is heated electrically or by gas and used
for *soldering*.

soldier (1) A short upright member,
e.g. as *grounds* for fixing a *skirting*. See
also *C*.
(2) An upright brick showing its *face*
in brickwork.

soldier arch A *flat arch* of uncut
bricks on end.

soldier course A *course* of bricks on
end, e.g. for a wall *capping*.

sole The·smooth under-surface of a
plane or *planer*, out of which the *cutting
iron* or *cutter block* projects.

solenoid valve An electrically operated *stopvalve*, which can be opened or
closed automatically.

sole plate (1) **footplate, ground sill,
mudsill, s. piece** (USA **abutment
piece**) The *sleepers* laid on the ground
under the feet of *raking shores*.
(2) A *sill*.

solid bedding Adhesive or mortar *bedding* that completely fills the space
underneath a brick, floor tile, or similar
component.

solid block A building *block* that contains no formed holes, even although it
may be porous or made from a *cellular*
material.

solid brick A *brick* without deliberately made holes, such as *frogs* or *perforations*. Solid *wirecuts* are often obtainable only by special order.

solid bridging, s. strutting (USA
block b.) The *strutting* of timber
floor *joists* by regular short lengths of
joist cut to fit between, and at right
angles to, the joists, usually at mid-
span. Solid bridging needs more precise work than *herring-bone strutting*,
but is stiffer and forms a valuable *fire
barrier*, for which reason it may also
be used in hollow wooden *partitions*.

solid-core door, s.d. A *flush door* with
a *core* of 100% solid material, such as
chipboard, flaxboard, plastics foam, or
laminated softwood, thus suited to
heavy duty. Laminated softwood solid
cores include three layers of tongued
and grooved boards, vertical on the
two faces and horizontal in the middle.
They usually have plywood *facings*,
but sheet steel is used on *fire doors*.

solid drawn tube Metal tube made
by *extrusion* (*C*), and thus *seamless*.

solid floor (1) **s. slab** A concrete floor
slab built without *hollow blocks* or *void
formers* (*C*) but with solid concrete
throughout its thickness.
(2) A *plank-on-edge floor*.

solid frame An expensive timber door
frame with the *stiles* and *head* made
from a single piece, which is *double
rebated* to form the *door stop*.

solid glass door An *all-glass door*.

solid masonry unit Any *brick* or
block, whether *perforated* or cellular
provided that its cross-sectional area in
every plane parallel to the bed is 75%
or more of the bed area. Otherwise it is
a *hollow brick* or *hollow block*.

solid moulding, stuck m. A *moulding* cut in the timber, not planted.

solid-moulding cutter A single piece of steel used in a *spindle moulder*, with perfectly balanced cutters that run at high speeds and do not need frequent sharpening. They are accurate in cutting but costly to buy. See **carbide tips**.

solid newel stairs *Spiral stairs* of stone in which the inner end of each step is shaped to form a nearly continuous cylinder with the inner ends of the other steps. Partly because they take up very little space they were the commonest means of access to upper floors until about 1550, when the framed wooden stair was introduced. They were also popular because they were easily defended by a single swordsman. If the owner of the house was right-handed, the stair was built with the newel on the left looking down.

solid panel A *flush panel*.

solid partition A *partition* which has no cavity. It may be of *bricks, blocks,* a *panel product,* etc.

solid plastering Plaster formed in its final location, put on a wall or ceiling as *coats,* not prefabricated in the shop as is *fibrous plaster.* See **one-coat plaster, two-coat work, three-coat work**.

solid roll A roofing *wood roll*.

solid slab A concrete *solid floor*.

solid stop A *door stop* in a *solid frame*.

solid strutting *Solid bridging*.

solid timber Timber made from one piece, without *planted* mouldings.

solid wall A *brickwork* or *blockwork* wall laid without a *cavity* or in a *bond* that does not leave holes between the bricks or blocks. It may suffer from dampness by *driving rain* entering through pores and small holes. Using standard-size *bricks,* solid walls are either 102.5, 215, or 327 mm thick, corresponding to the traditional descriptions of a *half-brick, one-brick,* and *brick-and-a-half* walls. Blockwork solid walls can be of any thickness, but are often metric *modular.* See **faced wall**.

solid wood floor A *plank-on-edge floor*.

solum, footprint The ground below the lowest floor of a building. If for any reason this is not covered with *oversite concrete,* it must be damp-proofed in accordance with the *Building Regulations,* in Scotland with hot pitch or asphalt, or south of the border with thick plastics sheet covered with 50 mm of fine concrete (BS 2832).

solvent, organic s. Liquids such as acetone, alcohol, petrol, trichlorethane, *white spirit,* etc., often mixed together, which dissolve solids. They are found in paints, paint removers, degreasers, and adhesives. To add more solvent to a solution is to 'cut' it. Most solvents and *thinners* are *flammable liquids,* and many of them are also poisonous. They can enter the body through the lungs or the skin. Breathing in their vapours should be avoided, for example by adequate ventilation while they are evaporating. Lids should be kept on tins and waste containers. Contact with the skin should also be avoided and protective clothing or breathing apparatus worn if appropriate. Solvents should not be used or stored near *hot work,* and after using them anyone should wash before eating or smoking. The legal duties of employers and employees for handling solvents are given in the Control of Substances Hazardous to Health Regulations (COSHH) and in information sheets available from any local

solvent release

office of the *Health and Safety Executive.*

solvent release *Paints* or sealants which dry or set by the evaporation of a *solvent* and are non-*curing* use solvent release. Because the solvents evaporate, paints made with them dry quickly, except in the coldest weather, and the solvent is absent from the *film.* Since the same solvent can re-dissolve the set material, solvent-release paints can be softened by later *coats.*

solvent-welded joint A permanent joint between *thermoplastic* plastics tubes, made after smearing both spigot and socket with an appropriate jointing fluid containing a solvent, merely by inserting the spigot into the socket and turning it slightly to ensure even distribution. The joint is strong enough to be handled within minutes and can take pressure after several hours if kept dry so that the solvent evaporates. Different jointing fluids are used for PVC and ABS.

soot door (USA **ashpit d., cleanout d.**) A door at the foot of a *chimney* through which the chimney is swept and the soot is removed.

soot pocket An extension of a *chimney* below the smoke inlet, usually fitted with a *soot door.*

sound boarding *Pugging boards.*

sound distribution Loudspeakers throughout a building, usually in the ceiling, with wires to central electronic sound equipment, used as a *communications* system for public messages, paging, background music, etc.

sounder A bell, buzzer, gong, horn, or siren, particularly one used as a *fire alarm.* It may be activated zone-by-zone, to allow phased evacuation of a large building.

sound insulation, noise i. The reduc-tion of the sound transmission from one space to another. In buildings, the sound that passes through a wall decreases as the wall's mass increases (the '*mass law*'). Heavy *fire-resisting* construction usually gives good sound insulation and overcomes the *coincidence effect.* Flexibility can help, particularly in thin or lightweight materials, as does *discontinuous construction,* e.g. in a *double-leaf separating wall* or *double glazing.* Small gaps greatly reduce the sound insulation of floors and walls, as may windows, doors, etc. The *Building Regulations* require reasonable levels of sound insulation for separating walls and floors. Guidance on suitable types of construction can be found in Approved Document E of the Building Regulations and in BS 8233. Improving the sound insulation of floors can be by sand *pugging* in the ceiling, by lightweight pugging and a *floating floor,* or – the most effective and expensive – with an *isolated ceiling.* See BRE Digests 143, 337, 338 etc., video AP 39 and diy leaflet XL 4.

sound knot A tight, solid, undecayed *knot,* at least as hard as the wood round it, unlike an *unsound knot.*

sound-level meter An indicator driven by a microphone and amplifier to measure loudness. The sound frequencies are analysed by filter circuits, and the meter is usually able to give readings in *dBA* or *Noise Ratings.*

soundproofing A term which has very little meaning and is therefore falling out of use in the same way as '*fireproofing*'.

sound-reduction factor A factor which gives a measure of the reduction in intensity of the sound passing through a wall, floor, door, or window at any given frequency, thus its effect on *sound insulation.* It can be expressed in *decibels.*

426

southlight roof A *northlight roof* of the southern hemisphere, with the vertical glazing facing south.

soya oil A *drying oil* used in making paints.

space Of a *saw*, the length from one point of a tooth to the next.

space heating Building *services* for *heating* rooms or other spaces, but not for *domestic hot water*.

spacer A block of material which keeps two things apart, such as *glazing bars, reinforcement bars*, and those in wall *formwork*.

spacer-lug tile A wall or floor tile with *lugs* on the edge, to form joints of the right width.

spade A hand tool for *digging*.

spall (1) To break off the rough edges of stone with slanting blows of a *chisel* or *spalling hammer*.
(2) **gallet** A flake of stone, e.g. used for filling the joints between stones in *rubble walls*.

spalling The crushing of an edge or corner of masonry or precast concrete, or the general flaking of a surface.

spalling hammer, spall h. A very heavy mason's hammer with a small flat face, used to work hard stones to size by breaking off *spalls*.

spandrel (1) The wall under a window, or a *spandrel panel*.
(2) The wall between the *arches* of an *arcade*. It is roughly triangular with two curved sides.
(3) The triangular infilling under the *outer string* of a stair.

spandrel panel An opaque panel in a *curtain wall*, under a window sill on one floor and extending down to the window head of the storey below.

spandrel step A triangular stone for a step, with its edges carved to lock together, giving a flush stair *soffit*.

Spanish tiles (1) (USA) **mission t.** Clay *roof tiles* shaped like a half cylinder slightly wider at one end than the other. The *over-tile* and *under-tile* are of roughly the same shape, the under-tile being larger. These tiles form a beautiful roof which is flexible in both *sidelap* and *headlap*. However, the amount of timber needed is considerable, as 80 × 50 mm timbers must be fixed on the slope of the roof (over felt on *boarding*) to secure the under-tiles. Compare **Italian tiling**.
(2) (USA) A *single-lap tile* which slightly resembles the tile called in Britain a single *Roman tile*.

span roof An ordinary *pitched roof*.

spar (1) Angular pebbles of limestone or other white rock, used as *ballast* for membrane roofing, *spar dash*, etc. See **fluorspar**.
(2) A *rafter*.
(3) **brotch, buckle** In *thatch*, a split piece of hazel or willow about 600 mm long. Eight spars are obtained from a 50 mm dia. branch. They are pointed at both ends, doubled, and driven into the thatch over *runners* or *scallops*.

spar dash A *dry-dash finish* showing white *spar* pebbles.

sparge pipe A perforated pipe for flushing a *urinal* stall, etc.

sparrow peck (1) A texture given to plaster by pitting it with a stiff broom.
(2) A texture given to stone by *picking*.

spar varnish *Phenolic varnish*.

spatterdash A rich mix of *cement* and sand (1 : 1½ to 1 : 2) which is mixed very wet, with or without a *bonding agent*, and flicked on in small blobs by a *roughcast machine*. It is put on smooth surfaces with little *suction*, such as brick-work or dense concrete, to provide a *key* for the first coat of *plaster* or *rendering*. If spatterdash does not give enough adhesion, a *bonding plaster* may have to be used. In another method, the same mortar is mixed with some *polyvinyl acetate* emulsion, which is then vigorously brushed into the surface and stippled immediately to form a key (BS 5262).

special (1) **s. brick** A *brick* which is not the usual rectangular shape. Specials are made to relate to standard working sizes for *brickwork*, with allowance for the mortar joints (format size). If a special is usually held in stock it is called a standard special, e.g. a bullnose, bull header, or chamfer stop brick; otherwise it is a purpose-made or 'special' special. Bricklaying with specials usually requires extra *labours*. Many specials are described and named in BS 4729.

(2) Any purpose-made or non-standard pipe fitting, tile, etc.

special attendance Use of the *main contractor*'s plant or equipment for a short time by a *sub-contractor*, usually charged per hour, unlike *general attendance*.

special brick A *special*.

specialists Firms which do a small part of a building job requiring special skill, usually as *sub-contractors* selected by the *main contractor* who is responsible for their work. They may cover a whole *trade*, such as *roofing* or *glazing*, or individual work involving *sealants*, blinds, lighting, signs, etc.

special risks *Force majeure*.

specification A written description of the workmanship and materials for a particular job. It is usually one of several *contract documents* that have to be read in relation to each other, such as the drawings and bills of quantities. A good or 'tight' specification gives detailed information on the accepted steps in each *trade* and has back-up clauses to ensure that work is done properly, although this may not happen without proper *supervision*. Specifications are written by a specifier, who may be an *architect* or consulting engineer, and are directed at the contractor. Job specifications are usually written mainly in terms of the *methods* to be used or the *performance* levels required, but they can also be based on *reference* to standards or the direct use of trade names. Use of the *National Building Specification* can help in both clarity and brevity.

specifier, specification writer A person who writes *specifications*.

specular reflection A sharp mirror image from a shiny surface.

spigot An end of a pipe of the same outside diameter as the rest of the pipe (or nearly), inserted into a *socket* in the next pipe to make a *spigot-and-socket joint*.

spigot-and-socket joint (USA **bell-and-spigot j.**) A joint between lengths of pipe, flue, gutter, etc., which are made with enlarged *sockets* on one end, the other end having a *spigot*. Drain pipes are usually in plastics, clay, or cast iron, with sockets and spigots shaped for *joint rings*. Older types of clay pipe, jointed with rigid cement *mortar* joints, have largely given way to *sleeve couplings*. Copper or stainless-steel socket and spigot pipes usually have *capillary joints*, and thermoplastics can be *solvent-welded*.

spike A pointed metal shaft, or a large *nail*.

spindle (1) An axle, e.g. of a *tap* or valve. Its inside end is threaded to raise or lower the *stem* as the spindle is turned. Spindles can be *non-rising* or *rising* and have a *gland* to prevent leakage.

(2) A square steel rod used to operate a latch. The door handles are fixed to each end of the spindle, which passes through the door and turns the latch *follower*.

(3) A turned piece of wood, such as a circular *baluster*.

spindle moulder A fixed *woodworking machine* which has a vertical shaft rotating at about 7500 rpm carrying a *cutter block* with two or more specially shaped knives or a *solid moulding cutter*. Since it is a very versatile machine its operator should be skilled as well as careful.

spine wall A wall through the centre of a building, parallel to the longest sides.

spiral duct, spiro d. A cylindrical air-conditioning *duct* formed from a strip of galvanized-steel sheet wound round and joined by a seam.

spiral grain The grain produced by fibres growing spirally round the trunk of a tree, thus making the wood difficult to work.

spiral stair The usual name for a circular *stair* with *winders* that radiate from one *springing point* on plan (the shape is helical, not spiral). These stairs can be made of stone, with a *solid newel* (as was customary in the past), or of decorative cast iron, or of galvanized and brightly coloured *powder-coated* steel, or timber or precast concrete. Main uses are for compact interior staircases or economical outdoor fire *escape stairs*.

spiriting-off The last passes in *French polishing*, in which all traces of oil are removed by a rag damped with methylated spirit drawn quickly and often over the surface.

spirit level A *level*.

spirit stain A dye dissolved in alcohol (methylated spirit), usually with shellac or other dissolved resin as a *binder*. It is used for decorating a wood surface, but emphasizes the *grain* less than a *water stain*.

splashback, backsplash A durable, easily cleaned area at the back of a sink, basin, cooktop, or cooker, either an *upstand* or several rows of wall tiles.

splashboard (1) A plank set on edge along the inside of a *scaffold* to protect brick *facework* from being stained by splashed mortar.

(2) A *weather moulding* planted outside across the foot of an external door.

splash lap That part of the *overcloak* of a drip or roll in *supported sheetmetal roofing* which extends (usually 40 mm) on to the flat surface of the next sheet.

splat A cover strip over the joints of *wallboards*.

splay A sloping cut across the full width of a surface, often at 45°, a large *chamfer*, or even a *cut corner*.

splay brick, cant b. One of several *standard special bricks* with bevelled edges, corners, or faces, or shaped to make a *return* or plinth stop.

splayed grounds Timber *grounds* with a bevelled or rebated edge to provide a *key* for plaster, where the grounds also act as a *screed*.

splayed heading joint A joint between the ends of floorboards which are not cut vertically (square) but at 45°, so that one overlaps the other.

splayed skirting A *skirting* with its top edge bevelled, not moulded.

splay knot A *knot* in timber cut nearly parallel to its length.

splice A *butt joint* which is strength-

splice bar

ened to take tension as well as compression, e.g. by bolted *fishplates* (*C*), a *crimped coupler* (*C*), *splice bars*, etc.

splice bar A bar two *bond lengths* long, overlapping two *butt-jointed* bars.

spline A *loose tongue*.

split A crack in wood or *veneer* which passes through it, unlike a *check*.

split brick A brick cut lengthways to reduce its height to less than normal.

split course A low *course* of *split bricks*.

split-face block A concrete *block* or *cast stone* made to twice normal width and later split in half, giving one rough-textured face to each block.

split gasket A pair of extruded PVC or synthetic rubber strips used one on each side of glass for *gasket glazing*, usually made to form neat *trim*. The window frame should compress the gasket and have slots to allow drainage.

split head A U-shaped top end for a *prop*, to carry and hold upright a timber beam or *joist* placed on edge.

split pipe A pipe cut lengthwise; a channel.

split-ring connector A timber *connector* with a steel ring held between two members in circular grooves made with a *hole saw*. A bolt is passed through a central hole and screwed up tightly on to large washers at each end to make a rigid joint.

split shake A timber *shingle* which has been split, not sawn.

split system (1) A system of heating with both radiators and air-conditioning (or *warm-air* space-heating). Compared with systems that heat by warm air alone, the air is cooler and more comfortable and less *air handling* is needed.

(2) A *refrigerator* with its cooling evaporator some distance from its compressor and condenser.

splittable activity An *activity* in a *programme* that does not have to be completed before the next activity can start.

splitter damper An air-conditioning *damper* to divide the air going to two different ducts.

spoil, muck Earth from *excavations*. It can be used as *fill* if suitable, stockpiled for *backfill*, or treated as *excess spoil*.

spokeshave A *plane* with handles straight out from each side, used in both hands for shaping curved surfaces such as wooden spokes. See diagram, p. 67.

sponge-backed rubber flooring, foam-backed r.f. Solid rubber sheet about 3 mm thick backed with sponge (foam) rubber 3 to 9 mm thick, which improves its resilience and reduces *impact noise*. It is obtainable in rolls 1.80 m wide and 23 m long and is laid in the same way as solid *rubber flooring*.

sponging A *textured finish* given to fresh paint using a sponge roller.

spot board, mortar b. A platform on which mixed *mortar* is loaded so that it can be used by a bricklayer or plasterer. It is usually plywood about 1 m square and often placed on a stand 600 mm high or taller. It should be wetted before use so that water is not absorbed from the mortar.

spot gluing Fixing with *adhesive* put on in small *dabs*, used for sheet finishings such as *gypsum plasterboard*.

spot item Work for building alterations or repairs that does not fit into a single *trade* and is thus priced on the labour and materials required, fixings, and any *making good*. The *description* of a typical example could read: 'Remove door, brick up opening, and extend wall finishings.'

spotlight, spot A *luminaire* with a narrow-angle beam of light, particularly used as a *downlighter*.

spotting A defect in which small areas of a painted surface have a different colour or gloss from the rest.

spotting in, spot finishing *Rubbing down* and refinishing small defective patches in a coating (BS 2015).

spout An outlet pipe from a *tap*, roof gutter, *downpipe*, etc.

sprag Informal term for a *nail*.

sprayed asbestos A coating formerly applied by *flock spraying*, to improve the fire resistance of steelwork, superseded by *sprayed mineral insulation*. See **asbestos**.

sprayed metal A coating put on by *metal spraying*.

sprayed mineral insulation, firespray Insulating materials applied by *flock spraying* and similar methods, usually over metal pipes or *structural steelwork*, either for fire protection *encasement* or to reduce heat transfer to or from the surroundings. Expanded *vermiculite* or *mineral wool* fibres may be used, or any *asbestos-free material*.

sprayer, spray nozzle A *shower head*, or a nozzle for a *spray tap*.

spray gun (1) A hand-held nozzle which emits atomized paint for *spray painting*. Except in *airless spraying*, it works on compressed air supplied through a hose and has a paint container under the nozzle.

(2) Other types are used for *metal spraying* or for *projection plastering*.

spray painting Painting with a *spray*

gun, which is very much faster and easier than brush or roller work, provided that the areas to be painted are large and continuous, such as the walls and ceilings of a room. A very smooth finish can be achieved with most types of coating, and it is the only practical way to apply *solvent-release* paints. Adjoining surfaces need *masking* against *overspray*. The paint is usually mixed with extra *thinners* to make it more fluid. Breathing apparatus and suitable *ventilation* may be needed to protect the operator from 'spray mist', or toxic fumes and poisonous vapours, such as from solvent-release paints and those containing *epoxy*, *polyester*, or *polyurethane*. *Lead paints* are not allowed to be sprayed in the UK. Priming by spraying is not advisable − brushing gives better penetration.

spray plastering *Projection plastering*.

spray tap A nozzle that gives off a fine spray, to reduce waste, especially of hot water in domestic sinks and basins or industrial *ablution fittings*. See *legionnaire's disease*.

spread and level (USA **wasting**) To dispose of *excess spoil*, usually as *fill* in an area of low ground, in order to save carting the spoil off site.

spreader bar A lightweight strut, e.g. across the bottom of a pressed-metal *door frame* to hold the *jambs* at the correct spacing. It may sit on the structural slab and be covered by the floor *screed*.

spreading rate The surface area covered by a unit volume of paint or adhesive.

spread of flame *Flame spread*.

sprig A small *nail* with no head, e.g. a *glazing sprig*.

spring (1) **edge bend** (USA **crook**) A variety of *warp* in timber from stresses released by sawing. It curves boards sideways.

(2) A *bending spring*.

springer A *springing stone*.

spring hanger An *anti-vibration* pipe support.

springhead roofing nail A galvanized steel *nail* with a thin steel cup under its head. It is driven through the top of a corrugation in sheet roofing to hold the sheet down firmly and cover the hole in it.

spring hinge A *hinge* with a spring that closes a door. It can be single-acting, or, for swing doors, double-acting. It should have a *check*, particularly if the door closes against a stop. See **floor spring**.

springing The intersection at each side of an arch between its lower surface (*intrados*) and the face of the wall or pier which carries it.

springing line The horizontal line that joins the two *springings* of an arch.

springing point The start of a line for *setting-out*. For a *geometric stair* the springing point is the start of radiating lines to the front faces of the stair treads. *Winders* have a common springing point.

springing stone, springer The stone on which an arch is built. It takes vertical loads and outward thrust on its sloping *skewback* top surface.

springwood, earlywood That part of the *annual ring* which is formed in the spring. It is usually paler, less dense, and weaker than *summerwood*.

sprinkler system An auto-suppression *fire-extinguishing installation* that sprays water from sprinkler heads, operated by flame and hot smoke rising and spreading under the ceiling. The first sprinklers to gain discounts in fire insurance premiums, made by Parmalee in 1874, had a *fusible link* with slow response times. Today's sprinklers owe much to the work of Grinnell in the USA and the Wormalds in England. Current types have a bulb containing an easily boiled liquid, or are one of several fast-response types. They can be made to close and re-set themselves once the fire is out. Early systems were mostly used to protect factories or stores from fire damage. They are the most reliable *active fire protection* measure known, with a near perfect success rate in fully sprinklered premises, according to records kept in Australia since 1888. The city of Fresno, California, has justified their use by a large reduction in the cost of fire brigade manpower. More recently they have been seen as protecting the environment from damage caused by major industrial blazes. Sprinklers may contribute to life safety if they put out the fire or give people longer to escape, but a more powerful *fire-venting* system would be needed to move cooler smoke. The first legislation to require sprinklers was introduced under the 1922 Fire Escapes Act in Victoria, Australia, a law later overturned. Later legislation on 'high buildings' in 1957 not only remained in force, but was in due course tightened. Today sprinklers are written into the regulations of over 500 cities worldwide.

Except for outdoor *drencher* types, building sprinkler heads are mounted on the ceiling (or where hot smoke will collect) and spray downwards over a wide area, operating at 68°C (or other temperature), thus in the early development stage of a fire. The volume of water they deliver is small compared with normal fire-fighting using hoses, and this reduces water damage to the building fabric and contents. Sprinkler systems regularly put out fires, with only minor damage and without creating news. Accidental discharge outside a fire area is very rare and damage or inconvenience can be limited by planning (e.g. by using water-resistant floor finishings, by drainage, by using basement sump pumps with wiring where fire cannot reach, etc.). When a sprinkler begins to release water, the flow in the fire main drives a *gong* and usually sets off a signal to the fire brigade, thus further acting as a *fire detector* and a *fire alarm*. Only the fire brigade, after checking the whole building, should turn off the water supply. Sprinkler pipes are usually hung under the floor slab, for easy modification or maintenance, but systems can be almost fully concealed, so as not to disfigure historic buildings. In an *atrium* fire their spray of water can cool *smoke* and send it downwards, but in general the arrival of water from overhead has been recorded as preventing panic. Sprinkler systems are usually 'wet', with pipes full of water, although for unheated buildings in a cold climate they can be 'dry', with compressed air in the pipes and with water released only in a fire. The best sprinkler systems have two independent water supplies, one from the water mains, the other from a *fire reserve*.

Although sprinklers in a tall building may range from 1% to 1.25% of total project cost, this is partly offset by *trade-offs*, such as larger *compartments* or a reduction in passive fire resistance. They have controlled *fires* started by the worst of unforeseen events, including arson. During construction the system should be *commissioned* early, before *fit-out*, when the danger from burnable materials and *hot work* is high. Sprinkler installations in the UK conform to the Loss Prevention Council publication: 'Rules for Automatic Sprinkler Installations'. See BS 5306.

sprocket A *cocking piece*.

sprocketed eaves *Eaves* given an upwards tilt by *cocking pieces*.

spruce Many species of Picea, *softwoods* mainly exported from North America, which are very light, typically weighing 432 kg/m³ at 12% moisture content. The finest spruce is *Sitka*.

spud A *dowel* at the foot of a door post to fix it to the floor.

spun pipe A *centrifugally cast pipe*.

spur (1) A *socket outlet* from a single cable branching off a *ring main*, often used to install extra power points. The connection must be made in a joint box and there should not be more than one twin socket outlet per spur, nor should more than half of them be on spurs.

(2) A short concrete post set in the ground to carry a wooden post above ground. The post lasts many years longer than it would if it was sunk in the earth.

square (1) **try s.** An L-shaped metal, or metal and wood tool for setting out right angles. See also **bevel**, **steel square**.

(2) A square small *pane* of glazing, a square *tile*, etc.

(3) (Australia, USA) An area of 100 ft² (9.29 m²).

squared log A *baulk* of timber.

squared rubble A *rubble wall* containing stones of varying size which are squared and not *snecked*. It is usually *coursed* at every third or fourth course and may be built almost as carefully and with as thin joints as *ashlar*.

square-edged timber Sawn timber with a plain edge cut at right angles, without *wane* and not *tongued and grooved*.

square hook A fixing device with a screw thread and an L-shaped hook.

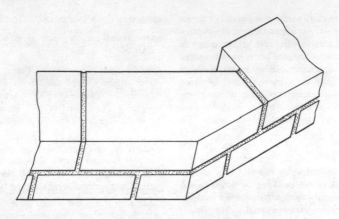

A squint corner made with 30° squint bricks.

square joint A *butt joint* made at right angles.

square-sawn timber Timber sawn to a rectangular cross-section, with or without *wane*, such as *flitches* or *resawn* battens, joists, or boards.

square-turned baluster, s.-t. newel A *baluster* or *newel* post with *mouldings* cut on four faces by any method except *turning*.

squash-court plaster Special *hard plaster* made to take squash-ball impacts and other knocks, and to resist *condensation*. Such plasters are usually made from *retarded hemihydrate* gypsum with *polymer* modifiers, or from *glass-reinforced concrete*.

squatting closet, asiatic c., squat pan A vitreous china *water closet* pan let into the floor as a *soil fitment*, usually with a high-level flushing *cistern*. Although used in public buildings in Continental Europe, they are rare in the UK.

squint, s. brick A *special* brick with cut corners on one end, forming a right-angled tip, used to give a flush finish at a bonded *squint corner*. *Standard special* squints are made at 30°, 45°, and 60°.

squint corner, s. quoin A projecting corner of a building which is not at a right angle. It is strongest if made with bonded brick *squints*, *external angles*, or from whole blocks (with projecting corners), rather than a weaker, cut, *butt joint*. See diagram.

stability (1) The *dimensional stability* of a material, or its resistance to thermal or moisture *movement*, under normal conditions or in a fire.
(2) The ability of a building element in *fire-resistance grading* not to collapse, which may depend on time or on the *critical temperature* of its materials.

stabilizer A steel 'foot' lowered to the ground from a rubber-tyred truck or tractor which carries a crane, concrete pump, or digger boom, to steady it while reaching sideways. It can be an *outrigger* (*C*).

stable door (USA **Dutch d.**) A door cut through horizontally at about half its height, with each half separately hung.

stack (1) A vertical pipe to carry the main discharge from *sanitary pipework* or for its ventilation. Branch pipes enter a stack at *junctions*. Stacks can be for *soil water* and *waste water*, or a *single stack*.
(2) A chimney stack, containing one or more *flues*.

stack bonding Not a *bond* at all really, but bricks laid one on top of the other, forming straight rows.

stack effect The *draught* that is created by warm buoyant air as it rises in a tall confined space. A fire fanned by the stack effect gives off hot gases, which create more draught. The stack effect makes a *chimney* work properly and can be used for *fire venting*. However, it can spread fire in a stairway, lift shaft, atrium, floor void, light well, etc. In services *ducts* and *cavity walls* the effect is prevented by *fire stops*.

stack pipe A *stack* which carries discharge, usually of *foul water*.

stack vent A *stack* carried above a roof, used for *ventilation*.

staff bead A metal *angle bead*, or a traditional external corner run in *hard plaster*.

Staffordshire blue brick An extremely hard, dense, deep-blue *clay brick* which can be obtained *wirecut* or *pressed*. With its remarkably high compressive strength it is one of the best *engineering bricks* in Britain.

stage A major point in the progress of a building project. For the contractor the first stage is the *pre-contract* work of *tendering* or other negotiations, followed by the *contract* period and *post-contract* work. Typical stages in site-work are: *commencement* or ground breaking; out of the ground; *topping out*; *roofed-in*; *lock-up*; *dry trades*; and *completion* or handover.

stagger To arrange nails, bolts, or other fixings alternately so that, for example, nails do not come opposite other nails in the next row.

staggered joints, break j. Joints arranged in alternating *courses* so that each one is over the middle of the brick, block, tile, or sheet below or next to it.

staggered-stud partition A *framed partition* formed of two separate 'half-walls' back-to-back, each with rows of vertical timber *studs* and sheet or board *linings*. It is a type of *discontinuous construction* which gives *sound insulation* nearly as good as a solid brick wall. See BRE Digest 347.

staging A step-up *scaffold*.

stain (1) A solution or suspension of dye or other colouring matter in a *medium*, used to decorate a surface by penetrating it. It has less *hiding power* than paint, allows more water vapour movement, and generally gives less protection. Exterior timber joinery can be finished with *high-build* or *low-build* stains, which are usually applied over a *preservative*. True stains are *water stains*, *oil stains*, or *spirit stains*, according to the medium. Varnish stains are not true stains, but *stainers*, since they do not penetrate the surface and merely leave a coloured coating on it.
(2) *Blue stain*.

stained-glass windows Windows of glass which is coloured by firing.

stainer A coloured pigment ground in an oil paste, such as a paint *medium*, which can be added in small amounts to ready-mixed *paints* to modify their colour. Stainers have intense staining power but are not always very opaque. Oil-bound stainers are unsuitable for *emulsion paints*.

stainless steel An *alloy steel* (C) with

varying amounts of alloying metals, but often with chromium and nickel, commonly 18% Cr and 8% Ni. It has good resistance to corrosion, abrasion, heat, and damage; and when polished its smooth surface does not easily retain dirt. It is not always stain-free, but resists *bimetallic corrosion* with copper. As a building material it is used for shopfronts, door furniture, roofing, and *stainless-steel plumbing*. See BRE Digest 349.

stainless-steel plumbing Pipe made of austenitic *stainless steel* up to 65 mm dia., used for water supply and *small-bore systems*. Both *capillary joints* and standard *compression fittings* can be used. It is worked like *copper pipe* but is stiffer and more difficult to bend. Air pockets should be avoided in stainless-steel plumbing, as chlorides can accumulate at the water surface line and cause *stress-corrosion cracking* (C). See BRE Digest 83.

stair, staircase A series of steps with or without landings, including necessary handrails and balustrades, for access between different floors or levels, or as a fire *escape stair*. Stairs with straight *flights* may have a *dogleg*, less often a *double return*, although flights can be *geometric* or *spiral*. Minimum recommended dimensions of house stairs are 900 mm overall width, a *rise* of each tread of 240 mm, and a *going* of 215 mm. In a straight *flight* twice the rise plus the going should not add up to more than 700 mm, nor less than 550 mm. See the stairs code, BS 5395. See also diagram.

stair clearance The distance, measured at right angles, from the *nosing line* of a stair to the ceiling or to any beams. It is required in addition to the vertical *headroom*. Stair clearance should be at least 1.50 m, the minimum

for a tall person hurrying downstairs.

stairhead The top of a *stair*.

stairlift, inclinator An armchair which carries the occupant upstairs. See BS 5776.

stair shaft The walls which enclose a *stair*.

stairway A *stair* used as a circulating way or as a fire escape. Open stairways can allow fire to spread. In non-domestic buildings they are usually isolated with *fire doors*.

stake A timber with one end pointed for driving into the ground.

staking-out *Setting-out*.

stallboard A strong *sill* and its framing beneath a shop window, over the *stall riser*.

stallboard light A *pavement light* near a *stallboard*.

stall riser In a shopfront, the vertical surface of polished granite, glass, concrete, tiles, wood, marble, etc., from the pavement to the *stallboard*.

stanchion A steel *column*. See *C*.

stanchion base A *baseplate*.

stanchion casing Concrete *encasement* to a steel *column*.

standard (1) Usual or accepted. Standard methods may be described in a document such as a *British Standard*, *International Standard*, *European Standard*, etc. These documents describe products of agreed dimensions which have passed appropriate tests, are of assured quality, etc. Standard sizes make building cheaper and planning for building easier since makers can plan for and make fewer sizes.
 (2) A main upright member in a framework or a *scaffold*.

standard form of contract A booklet

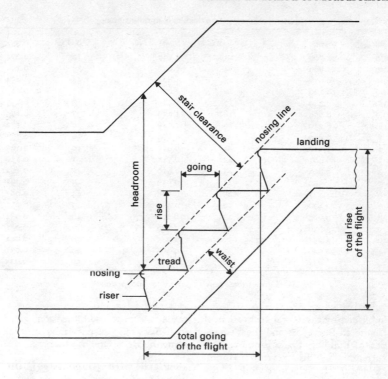

A flight of stairs.

containing a contract *agreement* and a set of clauses for the *conditions of contract* which are filled in and adapted as required for a particular job. In Britain alone there are many different standard forms of contract for building work. For private work they are issued by the *Joint Contracts Tribunal,* but other forms exist for use with public authority clients.

standardization Agreement between producer and consumer under the authority (in Britain) of the *British Standards Institution* on certain tests, dimensions, tolerances, and qualities of a certain product for certain purposes. An agreement when reached is published

as a *British Standard.* Building, after coinage, was probably the first subject to be standardized, and building standards have been used ever since systematic building began. Standards for sun-dried bricks are mentioned in the Code of Hammurabi (1700 BC).

standard knot (USA) A knot in timber of 38 mm dia. or less.

Standard Method of Measurement (SMM) The 'bible' followed by *quantity surveyors* when preparing *bills of quantities,* to enable their user, the *estimator,* to work quickly and accurately. The Standard Method of Measurement of Building Works,

437

standard overcast sky

Standard wire gauge diameters in millimetres

SWG	(mm)	SWG	(mm)	SWG	(mm)
7/0	12.700	13	2.337	32	0.274
6/0	11.786	14	2.032	33	0.254
5/0	10.973	15	1.829	34	0.234
4/0	10.160	16	1.626	35	0.213
3/0	9.449	17	1.422	36	0.193
2/0	8.839	18	1.219	37	0.173
1/0	8.230	19	1.016	38	0.152
1	7.620	20	0.914	39	0.132
2	7.010	21	0.813	40	0.122
3	6.401	22	0.711	41	0.112
4	5.893	23	0.610	42	0.102
5	5.385	24	0.559	43	0.091
6	4.877	25	0.508	44	0.081
7	4.470	26	0.457	45	0.071
8	4.064	27	0.417	46	0.061
9	3.658	28	0.376	47	0.051
10	3.251	29	0.345	48	0.041
11	2.946	30	0.315	49	0.031
12	2.642	31	0.295	50	0.025

(With grateful acknowledgement to *Specification*)

published by the Royal Institution of Chartered Surveyors and the Federation of Building Trade Employers, is currently in its seventh edition (SMM7). Its chapters follow the *Common Arrangement*, each section giving rules for *measurement* and the phrasing of *descriptions*. Many bills of quantities have some parts that do not exactly follow the SMM, and it is usual to warn estimators about these departures in the *preambles*. Other standard methods of measurement exist.

standard overcast sky A model of the *daylight* from a cloudy sky. The *illuminance* varies: 128% of average from overhead, 55% from the horizon.

standard sand A constant and reliable sand, used to make mortar for tests on *cement*. Leighton Buzzard sand is used in Britain, Leucate sand in France, and Ottowa sand in the USA.

standard special brick A *special brick*

made to a shape and dimensions given in BS 4729, which can be ordered using the BS reference numbers.

standard wire gauge (swg), imperial swg An old-established way of specifying the thickness of steel saws, sheet, wire, tube, cut nails, and some non-ferrous metals such as copper. The Birmingham gauge (bg), used for sheet steel, generally differs from the same number of the swg by less than 20%. There are many other British and American 'gauges' for sheet metal, wire, and plastics sheet or tube, including the Birmingham wire gauge, which is not the same as the Birmingham gauge. See table above.

standby equipment Equipment that comes into action in an emergency or following a breakdown, such as diesel-powered generator sets, fire *boosters* and fans in a *pressurized escape route*, and self-contained battery lighting units.

standing ladder A self-supporting ladder, usually with a swing-back trestle and level treads, unlike a *builder's ladder*.

standing leaf A leaf of a *folding door* which is bolted in a closed position, as opposed to the *opening leaf*.

standing seam A *seam* in *supported sheetmetal roofing*, usually running from ridge to eave. It stands up vertically – no disadvantage, since water will flow parallel to it. In traditional hand-formed seams the two ends of the sheets to be joined are bent up respectively 40 mm and 30 mm. A 40 mm high *tingle* (or clip) is bent up with them and nailed to the roof boarding with copper nails under one of the sheets. The tingle and the two ends are together folded through 360°, forming a final upstand about 18 mm high. The tingles measure about 80 mm by 20 mm and are placed every 300 mm to hold the sheets in place. Alternatively, standing seams can be made by machine, which is the usual method for *longstrip roofing*. Standing seams, and *rolls*, are not used for surfaces where people can walk.

stand oil *Linseed oil* polymerized originally by standing in the sun so that it hardens more quickly when used in paint, a process used by the ancient Romans. The same effect is produced in *boiled oil*.

standpipe (USA) A *fire riser*.

staple A metal U-shape, either a nail with two points, used as a *fastener*, or a loop for padlocking a door or gate, as a *hasp and staple*.

staple gun, stapling machine A power tool for driving *staples*. *Hardboard* or other *fibreboard* can be fixed firmly to wooden studs with staples, with less labour and less damage to the board than by hammering, and for difficult fixing (overhead), much more easily. Staples with diverging points should be used. Machines fix hardboard four times as fast as hand hammering does. The staple gun was the fore-runner of the *nail gun*.

star drill (USA) A *plugging chisel*.

star shake Several *heartshakes* together.

starter (1) **s. bar** A steel *reinforcement bar* that projects from a *construction joint* to lap bars in the next section. Since starters form dangerous spikes sticking out of the concrete, the ends of wall and column starters may need to be covered to prevent facial injuries or serious injury in falling accidents.

(2) **lamp s.** An automatic switch that helps to create the high current needed to strike the arc in a *discharge lamp*. Starters for *fluorescent lamps* have different ratings from 5 to 80 W.

(3) **motor s.** An electrical device in the supply circuit to an *electric motor* to protect against *overcurrent* at low speeds or from a stalled motor. It is needed on a large single-phase motor and most three-phase motors.

starting course (USA) An *eaves course*.

start time In the *critical path method*, the term used for *commencement*.

starved substrate A *hungry* substrate.

stat Abbreviation for a *thermostat*.

static electricity, electrostatic Force fields from electric charges that can come from synthetic fabrics in carpets and computer equipment, particularly a visual display unit (vdu). It is generally aggravated by a dry climate or the low *relative humidity* in heated buildings in winter, which make surfaces less conductive and may contribute to the *sick building* syndrome. It can be prevented by having shielded and earthed electronic equipment and by *antistatic flooring*.

station roof, umbrella r. A roof carried on a single row of stanchions. It is therefore *cantilevered* (*C*) on one or both sides.

statutory undertaker (UK) An órganization with a duty laid down by law to provide a *service* to the public. Many of the statutory undertakers that concern building have disappeared since the *privatization* of gas, telecom, water, and electricity services, becoming service companies.

staved lumber core (USA) *Block-board*.

stay (1) A horizontal bar which strengthens a *mullion* or a leaded light.
(2) A bar which holds together two opposite outside walls and prevents them falling apart.
(3) A *casement stay*.

steam stripper, wallpaper s. A portable boiler connected to a flat metal nozzle which blows steam on to a wall, making it easy to strip *wallpaper*.

steel An alloy of iron and carbon. It is heavy, strong, tough, and stiff, but difficult to work and, except for *stainless steel*, easily attacked by *corrosion*. Steel is used in *structural steelwork* and concrete *reinforcement*, or to make tools and in *ironmongery*. Steel structures need to be protected from fire, as steel has quite a low *critical temperature* of 550°C. See also *C*.

steel casement A *steel window*.

steel-clad flag A concrete slab with a bonded steel facing and *anchors* underneath, usually 305 × 305 mm and 25 or 50 mm thick, made for heavy industrial flooring and laid on thick mortar *bedding*. See **anchor plate**.

steel conduit *Metal conduit*.

steel-cored glazing bar In *patent glazing*, a specially shaped *glazing bar* with a core of rolled steel, for strength, enclosed within a *lead sheath*, for corrosion resistance.

steel fabric See **fabric**.

steel-fibre-reinforced concrete A *fibre-reinforced concrete* containing crimped strands of steel about 50 mm long and 0.5 mm dia.

steelfixer (USA **ironfighter**) A skilled worker who does *steelfixing*, often working as part of a gang of *subcontractors*. Steelfixers work from *drawings* and *bending schedules*.

steelfixing (USA **bar setting**) The work of fixing steel *reinforcement* in place, inside *formwork* or · in the ground. It may be made up as *cages* held together with *binding wire*, and is usually kept in place with *bar chairs* and spacers. Inspection of the completed work may be required before concreting, which at least one *steelfixer* should attend. Steelfixing is a modern *trade* involving heavy work done outdoors in all weather conditions, under pressure of time and requiring accuracy. Care is needed to avoid dirtying bars with mud or formwork *release agents* (*C*).

steel lathing Expanded-metal or steel-wire mesh used as *metal lathing*.

steel square, framing s., roofing s. A *square* graduated with data for calculating the length of rafters and angles in traditional *roof cutting*.

steel-tool pointing Tooling brickwork joints with a *jointing tool*.

steel tray *Profiled metal decking*.

steel trowel A flat rectangular steel blade about 260 × 110 mm with a handle in the middle of its back, used for applying or smoothing coats of

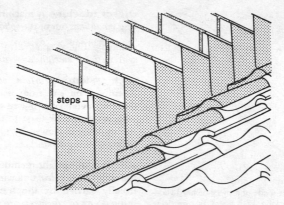

A stepped flashing prepared by a plumber and ready for building into the brickwork above roof level. The height between steps is equal to the height of a brick course. The length of the steps varies with the roof slope.

plaster, finishing concrete, spreading materials, etc. See diagram, p. 332.

steel window, s. casement Rolled-steel sections joined at the corners by welding, then *galvanized* or finished with a *powder coating*. On-site glazing is done with semi-elastomeric glazing material or *metal-casement putty*.

stem A threaded rod in a *gate valve*, raised as the *spindle* is turned, to open the valve which is attached to it.

step One unit of a *stair*, consisting of a *tread* and a *riser*. It may be a *flier*, *tapered*, or a *winder*. See diagram, p. 437.

step joint A joint between a roof *rafter* and its *tie beam*, in which the tie beam has a long shallow triangular notch to receive the end of the rafter. For extra strength a double step may be used.

step ladder A single section of shelf ladder. See **steps**.

step-on, step-off The short distance of level travel at the top or bottom of

an *escalator*, usually two treads long, for people to step on to or step off from.

stepped flashing A *counter-flashing* built into *raked-out joints* of brickwork at a sloping *abutment*. For a chimney stack it is usually made in sheet lead, shaped with a *step turner*, and wedged in after brickwork is completed. For a *cavity wall* it may be attached to a *cavity tray*, built in as brickwork proceeds. See diagram.

stepped footing A *strip footing* built in sections, each with a level bottom, joined by steps, used on a sloping site.

stepped scarf joint, splayed s.j. A *scarf joint* with a step in the middle, used for heavy timbers that act in compression.

step roller track A steel *channel-*section guide rail for the moving stair treads of an *escalator*.

steps A ladder with shelf *treads*, not round rungs. It is used at a fixed angle,

step turner

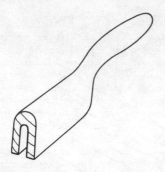

A step turner made of a piece of hardwood about 30 cm long, with a slot 25 mm deep and 6 mm wide.

so that the treads are horizontal. It can be standing, swing-back (pair of steps), or have a top work platform. Full definitions are given in BS 6100.

step turner A hardwood tool for shaping *stepped flashings*. It has a sawcut 6 mm wide in one edge, which is slipped over the edge of the flashing and used to turn it through 90° so that the flashing is shaped to *tuck in* to the bed joint of brickwork. See diagram.

sterilization, disinfection The treatment of new water supply pipes to destroy bacteria that could be dangerous to health, done before *commissioning*, e.g. by a water service company. The pipes are flushed out, filled with water and chlorine chemicals to give at least 50 parts per million of chlorine, and allowed to stand for three hours. A sample is taken and if it passes a test for the smell of residual chlorine, all pipes are flushed with fresh water and put into service. See BS 6700.

stick-and-rag work *Fibrous plaster*.

sticker A small separator, usually a strip of wood, laid with others of uniform thickness between planks of freshly cut wood to encourage *air drying*.

sticker machine A machine for cutting *mouldings* out of the *solid*.

sticking The shaping of a moulding with a *sticker machine* or *router*.

stick system A *curtain wall* mostly assembled on site from *mullions* and *transoms*, which are bolted to a lightweight framework fixed to the face of the building, followed by glazing and fitting of opaque panels.

stile (1) An upright member forming the edge of a *doorleaf* or a window *sash*, usually carried by the frame. Side-hung doors or *casements* are supported from a *jamb* by their *hanging stile*, the other side being either a *shutting stile* or a *meeting stile*.
(2) One of the two side members of a *ladder*.

stillage area An industrial floor for short-term storage, usually a concrete *ground slab* which has to be accurately laid to level.

Stillson A plumber's *pipe wrench*.

stipple To dab paint with a bristle or rubber *stippler* soon after a coat is put on, giving a textured finish similar to *orange peeling*, or to break up the colour of a coat with *flecks* of a different colour.

stippler, stippling brush A brush or other tool for breaking up the texture of a wet coat, or to remove brush marks and give the surface a uniform, slightly granulated finish. Sometimes it is a large brush whose stock, apart from the handle, resembles a short plank of about 170 × 140 mm into which is set an area of soft bristle some 60 mm long. Other stipplers are of rubber.

stippling A *textured finish* given to paint with a *stippler* or some types of *roller*.

stirrups (1) **ligatures** Reinforcement *binder bars* round the main bars in a beam, forming a *cage*. Sufficient *cover* is needed over stirrups to prevent corrosion.

(2) **s. strap, hanger** A steel strap built into brickwork or fixed to a post or beam, to hold a horizontal member up to it. One example is the steel strap which holds the tie beam up to a king post in a truss.

stock (1) Available, easily bought.

(2) *Converted* timber, also called stuff.

(3) The body or handle of a tool, usually of wood or plastics.

(4) A tool which holds a *die* for cutting an external thread.

stock brick A *moulded brick* made from plastic 'soft mud' clay found in the south of England (Essex, Hampshire, Kent, Surrey, Sussex), which is moulded wet (by machine or hand), then dried before firing. Handmade stock bricks are moulded on a bench using a rectangular frame that fits over a 'stock board'. The stock board has a 'kick' to form the *frog* and is nailed to the bench top. The mould is *coated* with sand or sawdust to prevent the clay sticking. Clay, shaped into a 'clot' by rolling in more sand or sawdust, is thrown into the mould. Excess clay is cut off with a two-handed wire bow, the stock board preventing the mould from sliding off the bench. *Clamp*-burnt stocks were classed as 'common' or 'facing' and 'hard' or 'mild'. See **London stock brick.**

stock brush A brush used for wetting a wall before plastering, to control the *suction* and prevent excessive absorption of water from the plaster.

stock lumber (USA) Wood sawn to market sizes.

stockpile Usually a heap of loose material, such as *spoil* for backfilling or disposal, or *topsoil* for re-use.

stomper A *consolidating rammer.*

stone (1) Cut rock such as sandstone, limestone, or granite, or *cast stone*. The traditional material for the best *masonry*, stone is also used for facings, carving, window mullions, and pavings. A block of stone may be rough *rubble* or accurate *dimension stone*, or cut to a particular shape, such as a tapered *arch* stone or *keystone*.

(2) A carborundum or other natural or artificial *hone* for putting a cutting edge on to a *chisel* or plane *cutting iron*, rubbing down concrete, etc.

(3) Coarse *aggregate*.

stone facing A thin *veneer* of stone used as external *cladding* to imitate *stonework*.

stone flour, stonedust Fine granular material made of crushed stone, used for *aggregate*, to give a matt finish on asphalt, etc.

stone paint *Masonry paint.*

stoneware Highly *vitrified clayware* used for *sanitary fittings* and brightly coloured frostproof claddings.

stonework Stone *masonry* built throughout its thickness of solid stone (not just *facings*), either *dimension stone* or a *rubble wall*.

stool, stooling (1) The raised flat ends of a *weathered* concrete or stone *lug sill*, which act as a horizontal bed for the brickwork above.

(2) A *window board*.

stop (1) A decorative conclusion to a stuck moulding.

(2) **s. end** A *special* brick to end a moulding, e.g. a *plinth stretcher* or *splay brick*.

(3) A *door stop*, or a buffer.

(4) A *fire stop* or draught stop.

stop bead, plaster s. *Metal trim* at the edge of an area of plasterwork. It forms a neat break against a different wall finishing.

stopcock A *stopvalve* that turns off a water or gas supply, operated by an official *turn cock*. Stopcocks for water allow flow in one direction only and prevent *backflow* from the house and the possible pollution of the mains thereby. If the stopcock is installed the wrong way round no water will flow through it. Stopcocks for gas do not have the non-return function.

stop end (1) Concrete *formwork* at the end of a wall or at the edge of a slab. If it forms a *construction joint*, it can consist of drilled, or split and notched, timbers that allow *reinforcement* to pass while enabling easy *stripping*, fine steel mesh which is left in place, etc.
(2) **gutter end** A closed end to a roof gutter, with full-height sides. For *supported sheetmetal* gutters, its corners should be *dog-eared*.
(3) The closed end of a *cavity tray* at the end of a run, a vertical *damp-proof course*, etc.

stop end and key A concrete *joggle joint*.

stopped chamfer, stop c. A *chamfer* which often ends with a triangular shape that joins it to an *arris*.

stopper (1) A device for blocking a pipe during a *drain test*, usually a *screw plug* or a *bag plug*.
(2) *Stopping*.

stopping (1) Filling holes of any sort.
(2) **stopper** Stiff paste for filling deep holes, wide cracks, etc., as *preparation* for paintwork. It may be of the same composition as *high-build filler* or traditional *hard stopping*.

stopping knife, putty k. A *glazier*'s knife for smoothing putty, also used by painters for putting *hard stopping* into holes or cracks. It is like a *chisel knife*

with one rounded edge and one splayed edge.

stopvalve A water or gas *fitting* used to cut off or allow flow in a pipe. It is used for isolation, not as a *discharge valve*, and is more simply made. A *cock* is one example. One-way stopvalves have an arrow or the word 'inlet' to show the direction of flow.

storage tank A *tank* or a *cistern*.

storage water heater A *cylinder* in which *domestic hot water* is heated and stored until needed. It is either gas-fired or electrically heated at night on low-cost off-peak power and controlled by a *thermostat*.

storey The part of a building between one floor and the next above it, usually measured from *finished-floor level*. Thus the seventh storey is the part of the building from the seventh floor level to eighth floor level.

storey-height unit A component that is as tall as the height between two storeys, e.g. a *cladding panel*, *lift shaft module*, etc.

storey rod, gauge r. A *batten* cut to the exact height of a *storey*, with *setting-out* dimensions marked on it, as for each *course* of brickwork, the window *sills*, window *heads*, and *stair* treads.

storm cellar, cyclone c. A cellar in which the occupants of a house take shelter against the violent cyclones and tornadoes of central and southern USA.

storm clip A saddle-shaped metal clip, fixed outside a *glazing bar*, to hold the glass down in *patent glazing*.

storm door (USA) An additional, outer entrance door, used in winter to keep out cold and wind.

storm lobby, s. porch An enclosed porch in front of an entrance door, usually of lightweight materials such as aluminium-framed window units.

stormproof window A *casement window* with additional protection against driving rain, for example *hood mouldings*, or gaskets with *drained and ventilated joints*.

storm window A *coupled window* or an *internal dormer*.

stoving, infra-red drying (USA **baking**) The drying of durable *enamel paints* by heat, generally above 65°C, the limit for *forced drying* but below the temperatures used for vitreous *enamel*. Heating can be either in a convection oven or by radiant heat lamps.

straddle pole A sloping *scaffold* pole along a roof in a *saddle scaffold* from a *standard* to meet the other straddle pole at the ridge.

straddle scaffold A *saddle scaffold*.

straight arch A *flat arch*.

straightedge A straight, parallel-sided piece of metal or seasoned wood, used in most building *trades* for ruling straight lines in *setting-out*, or for *screeding*, checking *flatness tolerance*, etc.

straight flight A *stair* flight with rectangular *fliers* only.

straight grain *Grain* parallel with the length of the timber.

straight joint (1) A *butt joint*.
(2) A *perpend* immediately above another one, a mistake in brick bonding unless used for *breaking the bond* or in *stack bonding*.

straight-joint tiles *Single-lap* roofing tiles made so that their edges in successive courses run in one line from *eaves* to *ridge*.

straight-peen hammer Any *hammer* which has, opposite the striking face of

the hammerhead, a blunt wedge parallel to the shaft of the hammer.

straight-slot head A *screw* head with a single slot cut right across it, as is traditional for *wood screws*.

straight tee, bullhead connector A pipe or duct *fitting* with a single branch joining sharply at right angles, not in a curve, as with a *sweep tee*.

straight tongue An ordinary *tongue* projecting from the edge of a board, held by a *rebate* cut in the solid timber, to make a *tongued and grooved joint*.

straight tread A *flier*.

strainer A coarse mesh screen, such as for a pump intake filter.

straining beam Any horizontal *strut* (C), particularly that between the heads of the queen posts in a *queen-post truss*.

straining piece, strutting p. A horizontal timber fastened to the middle of a horizontal *flying shore* as an abutment from which the shorter sloping *struts* (C) at each end obtain their thrust.

straining sill A timber lying on and fastened to the *tie beam* of a *queen-post truss*, between the queen posts or between the queen and princess posts, to keep them in place.

strand A bundle of *fibres*, or of wires in a steel wire rope. Strands of *glassfibre* are made of very fine filament, usually in multiples of 204, and are used to make *rovings* or mat.

S-trap A *tubular trap* on the waste of a sink, basin, water closet, etc., from which the outlet pipe leaves vertically downwards. See diagram, p. 474.

strap anchor A steel plate joining two floor *joists* which butt at a support. It is fixed on those joists which are anchored to the walls with *wall anchors*. In this way the walls are given good sideways support by the floor joists.

strap bolt A fastening for timber consisting of a metal strip with holes drilled through one end and a *threaded rod* at the other end.

strapping Common *grounds* of timber or metal attached to a beam or column to provide fixing points for the lining boards of a *casing*.

strawboard, compressed straw slab A *panel product* made of straw hot-pressed on edge, encased in strong paper, sometimes containing voids for 20 mm conduit. The straw can have boron insecticide treatment. Facings are plain or showerproof, or for a gypsum *plaster* skim coat. Cut edges should be protected from moisture, mould, and insects with a PVA liquid sealer. Strawboard is best used in a dry location. It gives good heat *insulation* and some *fire resistance* as a wall or ceiling *lining* or as an unframed *partition*. See BS 4046.

straw marking Blemishes on the surface of sheet *glass* that has been stored in constant contact with a small amount of moisture, into which alkalis escape from the glass. When the alkalis become strong enough they attack the surface. This used to happen when glass was packed in straw. It can occur from *condensation* inside *double glazing*.

straw thatching The most widespread type of *thatching*, using yellow-coloured 'red wheat' straw.

street furniture Equipment for the *external works* such as planters, park benches, traffic bollards, litter bins.

stress-graded timber Timber that has been graded for strength by either *visual grading* or *machine grading*. In the UK visual grades are marked GS

(general structural) or SS (special structural) and machine grades MGS and MSS. Stress-graded timber is required for structural uses by the *Building Regulations*. The grading no longer applies once the piece of timber is cut.

stress-grading machine A machine that loads timbers slightly as they are run through rollers, measures the amount they bend, and automatically puts on a grading mark.

stretch A run of single-tier *patent glazing*. Its area in square metres is given by the distance between the end *glazing bars*, times the bar length.

stretched coverings Fabric or vinyl sheet stretched taut between battens along the edges of indoor walls or ceilings, without being glued down. They have a flat, even surface.

stretcher A brick, block, or stone laid longways in a wall to form part of a *bond*. The same wall may also have *headers* and *closers*.

stretcher bond, common b., half b., running b., stretching b. A *bond* for brickwork or blockwork made up entirely of *stretchers* (except for *headers* at corners, *snap headers* at ends, etc.), usually laid with *staggered joints*. It is the simplest and today the most usual way of building walls.

stretcher course A *course* of stretchers.

stretcher face The longer *face* of a brick, visible in a wall.

striking off Removing excess mortar, concrete, or plaster while it is still wet. For flat surfaces this is usually done by *screeding*.

striking of forms *Stripping*.

striking of support The removal of *props* from under *formwork*.

striking plate, keeper, strike, strike plate A piece of *door furniture* consisting of a plate with a rectangular hole, fixed to a door *jamb*. When the door closes, the spring-loaded *latch* first slides against the striking plate and is forced back into the lock, but when the door is fully closed the latch enters the hole and keeps the door closed.

striking time The minimum time allowed before *striking of support*.

striking wedges *Folding wedges*.

string, stringer A sloping board up each side of a *stair* to carry the ends of the treads and risers. It may be a *close string* or a *cut string*, and either a *wall string* or an *outer string*.

string course, band c. (USA also **belt c.**) A decorative, thin, horizontal course of brick or stone along the wall of a building. It can be projecting, to throw off rainwater and cast sharp shadows. It often continues the line of the window *sills* or *dripstones*. In concrete construction its place is sometimes taken by the visible edge of a floor slab.

stringer A *string*.

stringline A *line*.

string wall A wall that supports the treads of a *stair*.

strip (1) Softwood under 50 × 100 mm, or hardwood 50 × 50 mm to 50 × 140 mm, or in USA *lumber* less than 2 × 8 in. (50 × 200 mm).
(2) Copper, zinc, or aluminium which is thicker than foil (0.15 mm), thinner than 9.5 mm, and narrower than 450 mm.
(3) Lead sheet.

strip board A term sometimes used for *blockboard* in which the core blocks are not glued to each other.

strip diffuser A *linear diffuser*.

strip flooring *Parquet strip*.

strip footing A shallow foundation some 600 mm wide or more, to carry a brickwork wall, usually of concrete reinforced with *footing fabric*. It must have a level bottom, but can be *stepped*. See **trench-fill foundation**.

strip lamp A long, thin, incandescent lamp.

strippable *Peelable*.

stripper (1) *Paint remover*.
(2) A site operative who dismantles *formwork* for concrete, then cleans and prepares it for re-use.

stripping (1) **striking** The removal of *formwork* after concrete has set and gained enough strength to support itself. First wedges or screws are *eased* and *wrecking strips* removed. Stripping is made easy by tapered shapes or collapsible forms.
(2) Removing old paint with a *heat gun* or chemical *paint remover*, or by other suitable means. Stripping *lead paint* requires special precautions.
(3) Excavating *topsoil*, usually 150 mm deep, and placing it in a *stockpile* for later use.
(4) *Weatherstripping* between a door or window and its frame.

stripping time The number of days or hours that *formwork* must be left in place before it is removed from set concrete. For non-loadbearing vertical surfaces it may be 24 hours, but it is much longer for beams and slabs.

strip sealant Preformed flexible tape or a section of solid or cellular material used for *glazing* or to exclude draughts, seal joints, etc.

strip slate, s. tile Fibreglass reinforced bitumen made to the shape of several *slates* joined side-by-side. Sizes vary slightly between manufacturers.

strip tie A *vertical twist tie.*

strongback An additional deep beam for extra stiffness, used on large panels of *formwork*, or to reduce stresses when lifting long beams or concrete piles.

strop A single lifting rope or strap passed round a load and hooked to itself.

struck joint (1) A *weather-struck joint.*
(2) A brickwork bed joint pressed in at the lower edge, sloping in the reverse direction to a *weather-struck joint*, used mainly for interior work, in which rain does not need to be thrown off.

structural Concerned with strength, as, e.g., the parts of a building which carry loads in addition to their own weight. See **structure.**

structural clay tile (USA) A *hollow clay block.*

structural concrete Strong concrete in foundations, columns, beams, and slabs, usually *reinforced* with steel bars, to BS 5328.

structural drawings The *drawings* showing a building *structure.*

structural fire precautions Safety requirements to protect people, such as the division of buildings into *compartments* and the provision of fire-resisting walls and floors, as required by the *Building Regulations.*

structural floor level The rough surface of a concrete floor *slab*, kept down below *finished-floor level* (usually 50 to 70 mm), to receive a *screed* or thick floor tiles on mortar *bedding.*

structural gasket A *gasket* that grips a *mullion* and presses against the glass pane in a *curtain wall.* The mullion may allow the internal drainage of rainwater.

structural glass (USA) Glass made into panels or tiles, often coloured, used as a *finishing* on external walls.

structural glazing Glass held with strong *sealants* and carrying some wind load. Sealant may be used to bond it to concealed metal rails, to give an all-glass facade that has good weather resistance and is easy to maintain.

structural lumber (USA) Sawn timber 2 × 4 in. (50 × 100 mm) or larger, usually *stress-graded* and under calculated loading.

structural steelwork Steel beams and columns cut and welded by specialist steelwork fabricators and erected on site to form a building frame. It is used on large projects, such as tall office blocks or wide-span warehouses, and can save construction time compared with a *reinforced-concrete* frame. The *main contractor* usually does work such as the foundation *bases* and *encasement*, arranges for earth *bonding*, etc. See **universal section.** See also *C.*

structural timber Timber, sometimes made into *trusses, trussed rafters*, or *glued-laminated* beams and columns, for small spans and moderate loads.

structural trades The group of trades that put up the main *structure*; for a reinforced concrete building, mainly the *formwork, steelfixing*, and *concreting.* The skills needed at this stage of work on site are different from those of the later *finishing trades.*

structure (1) The loadbearing frame or *fabric* of a building, its walls, floors, and roof, but not *finishings* or *joinery.* Materials used include *reinforced concrete, structural steelwork, brickwork*, and timber. It is separated at ground level into the *substructure* below and the *superstructure* above.
(2) A construction, such as a framework or a roof *plantroom.*

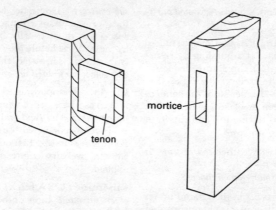

mortice

tenon

A stub tenon. The mortice is slightly deeper than the tenon.

strut A structural member that works in compression, usually horizontal or sloping, unlike a *prop* or *shore*.

strutting (1) Using *struts* as supports, e.g. *dead shores* or *earthwork support*.
(2) *Solid bridging* or *herring-bone strutting*.

stub (1) A *nib* of a tile.
(2) A short length of pipe, usually a *tubular* or a *tail*.

stub tenon A *tenon* which is inserted into a *blind mortice*. If it is wedged this can only be done with *secret wedging*. See diagram.

stucco Smooth *rendering* with decorative *mouldings* shaped to imitate the carved columns and entablature of classical stone masonry, which help to throw off rainwater. Stucco, usually painted and seen on the outside of old buildings, was originally of lime and sand, but from the nineteenth century made with cement or cement-lime mortar.

stuck Usually, fixed with glue, but see **stuck moulding**.

stuck moulding A *moulding* cut out

of the solid with a *router* or *spindle moulder*, the opposite of a *planted* moulding.

stud (1) An intermediate vertical timber (or folded steel sheet) in a *framed partition*, usually 100 × 50 mm. Studs are placed at about 450 mm spacings. The end members (posts) are heavier. Studs run between top and bottom *wall plates*, and have *noggings* fixed between them. Bracing is with steel straps, close boarding, or a *dry lining* material, which is usually nailed on each side.
(2) A threaded steel rod with no head, used as a screwed fixing.
(3) A heavy steel pin with an enlarged head, fixed to steel floor beams by *stud welding* to give the joint with a concrete slab some strength in *shear* (*C*).
(4) A disc or cylinder of material, or a timber *pellet*.

studded-rubber flooring, stud-surfaced r.f. Non-slip *rubber flooring* with its surface moulded into closely spaced rows of low discs.

stud gun A *cartridge tool*.

stud partition, s. wall A *framed partition*.

stud welding Fixing a metal stud to the steel frame of a building by *resistance welding* (*C*). Small threaded studs can be rapidly fixed on site and used to carry joinery. Larger studs for connections between steel and concrete are usually fixed during the fabrication of the *structural steelwork*.

stuff (1) Plaster, either *coarse stuff* or *finish plaster*.

(2) **stock** *Square-sawn timber*.

stumps Posts set in the ground under the ground floor of a traditional timber house in Australia, originally of jarrah or red gum. They lasted about fifty years, after which re-stumping was needed. If stumps are used today, they are normally of precast concrete. The usual stump size is 100 × 100 mm and they stand on small pad *footings* at 1 to 2 m spacings each way.

styrene/butadiene copolymer The material of latex synthetic rubber, used in *bitumen-polymer sheet* roofing and *emulsion paint*.

sub-base, base (USA) A *skirting board*. See also *C*.

sub-basement A second *storey* below the ground, the second *sub-level*.

sub-board A *distribution board*.

sub-circuit A *branch circuit*.

sub-contract A contract between the *main contractor* and a *sub-contractor*, usually for specialist work such as asphalting or air conditioning. The main contractor is responsible for the work and is paid for it under the 'head' contract with the *client*. Payment under the sub-contract should then follow promptly.

sub-contractor A specialist employed by the *main contractor* to perform a

sub-contract. 'Domestic' sub-contractors, or the main contractor's 'own' specialists, must usually be approved by the *architect* before they are engaged (sub-letting). However, the *client* can also select *nominated sub-contractors*.

sub-floor A smooth, level *floor* which is used as a base for sticking down thin *floor finishes*. Any rigid and stable material such as concrete or timber underflooring is suitable. It should be clean, smooth, and dry before *adhesives* are applied.

sub-frame (USA **buck**) A timber or steel *surround* built into a wall as brickwork is raised to form an exact opening for a prefabricated door or window. Any wall plaster is finished against the sides of the sub-frame.

sub-letting The employment of a *sub-contractor*.

sub-level Floors below street or ground level, usually numbered − 1, − 2, − 3, etc. A *basement* would be on sub-level − 1.

sub-master key A *master key* that can open locks only in a particular limited area, as on one *storey* of a building.

submission, response The total *tender* submitted by a contractor, with a completed *form of tender*, *priced bills*, a tender *bond*, product data sheets, etc.

submittal (USA) A document provided in addition to usual *bid* documents, such as calculations, references of completed work, or test certificates.

subsidence, settlement The downward movement of the ground surface. See *C*.

sub-sill A cover on top of the wall under a *window sill*, with a projection to throw water away from the building, usually fixed to the *sub-frame*.

substance, surface density Mass per unit area. Sheet lead 2.50 mm thick has a substance of 30 kg/m² and 6 mm sheet glass 15 kg/m².

substantial completion *Practical completion.*

sub-station (1) A room or building containing electrical equipment such as high-voltage *switchgear*, with transformers to reduce incoming power to a *low voltage* at which the consumer can conveniently use it. It may be provided by the electricity company or the consumer.

(2) Part of a *district heating* system which is supplied with high-pressure hot water or steam and distributes low-pressure hot water to buildings.

substrate, ground The surface which is to receive a coating, finishing (e.g. paint), adhesive, etc. (but not plaster). It may require treatment to control *suction* as well as surface *preparation* and dusting, and generally must be clean and dry before *application*. Plasterers usually refer to a substrate as a 'background'.

substructure The part of the building *structure* below ground level, the *foundations* and basements or *sub-levels*. It is usually of reinforced concrete and often protected by *tanking*. Substructure work is always a *critical activity*; once it is completed the building is *out of the ground* and the *superstructure* can be started.

sub-suite A group of locks operated by a *sub-master key* or their own key.

subway A passage below ground for people to walk through, sometimes used as a *main services duct* but in the USA also an underground railway.

successful tenderer (USA **s. bidder**) The contractor whose offer is accepted by the *client*, thus forming

a *contract*. The client is under no obligation to accept any tender, but usually accepts the lowest one.

suction The ability of a porous surface to take up water and very fine solid particles, which allows adhesion of wet mortar to bricks or tiles or plaster to its *background*. Low-*absorption* materials (including hard smooth concrete, glazed tiles, and engineering bricks) have low adhesion and may need a *bonding treatment* before plastering. Very porous brickwork, blockwork, or dry concrete may need wetting to reduce suction or *priming* with *polyvinyl acetate*. Excessive absorption of a *substrate* for paint can be reduced by a *sealer*. 'Control' of suction means its reduction.

sugar soap An alkaline *paint remover*.

suite (1) A set of interconnected rooms.

(2) A set of fittings made to match each other, such as a bath, basin, and WC.

(3) A set of several locks which can each be opened by a *master key* in addition to their own key.

sullage Domestic *waste water* from a basin, bath, or sink, but not *soil water*.

sulphate attack The weakening of concrete or mortar due to a chemical reaction between the tricalcium aluminate in *cement* and soluble sulphates in clay, in clay bricks, or in *flue condensation*, which are carried into the concrete or mortar by continuous saturation with water. The reaction produces ettringite (hydrated calcium sulphoaluminate), which occupies a larger volume and is weaker than the cement product it came from. In the UK sulphates are present in the strata of London clay, Lower Lias, Oxford clay, Kimmeridge clay and Keuper marl, and in bricks made from these clays. They include calcium sulphate (gypsum), magnesium sulphate, and sodium sulphate. About 50% of UK clay common bricks contain sufficient sulphates to be troublesome. Sulphate attack at first hardens brickwork *mortar*, then slowly turns it into a white powder, although the surface may be hardened by atmospheric *carbonation* and appear to be normal. Brickwork exposed to frost is doubly affected by a combined action. Attack can be reduced by keeping brickwork dry, with wide *copings* and correct *damp-proof courses*, which are usual even if sulphates are not present. Concrete in contact with sulphate-bearing ground is usually made with *sulphate-resisting cement* (*C*), or can be protected by *tanking* or a coating of bituminous paint. See BRE Digests 362 and 363.

sulphate expansion The swelling of *brickwork* due to *sulphate attack* on the mortar.

sulphur hexafluoride (SF$_6$) An inert gas used to fill high-voltage electrical *switchgear* or *double glazing*.

summary A list of *trade collections* in *bills of quantities* which are added together to give the *estimate* for the work or the *tender* price.

summerwood, latewood The denser, darker wood in an *annual ring*, formed in summer. In *even-textured* timber it looks similar to *springwood*.

sump pump A small pump, usually with an electric motor on a drive shaft to keep it above water level, operated by a *float switch* (*C*) for occasionally emptying a *sump* (*C*) in a basement below the level of the drains.

sun Heat from the sun can cause considerable *thermal movement* in cladding and roofing, but is useful for *solar heating*. Sunlight gives high levels of

illumination. Exposure to the sun's *ultraviolet* degrades materials such as *bitumen* and many plastics.

sunken fence A *ha-ha*.

sunken gutter A *box gutter* sunk below the roof surface.

sun shade A *brise soleil*.

super Abbreviation for superficial, and thus 'area'.

superficial void A *blowhole*.

super hardboard The most dense and water-resistant *hardboard*. It is used as a lining to *formwork*, as a floor finish, for boat building, and so on.

superheated steam Dry steam from a *superheater*, which can be piped great distances, as it is too hot to condense and give up its great *latent heat*. It can lose its heat in a *desuperheater*.

superheater A set of extra tubes in a steam boiler to heat steam after it is no longer in contact with the water, increasing its temperature but not its pressure.

superplasticizers Concrete *admixtures*, such as naphthalene sulphonates and melamine formaldehyde, which extend the range of normal *plasticizers* giving very high *workability*. They allow large reductions in water content for high early strength, even with *flowing concrete* (*C*). See BS 5075.

superstructure The parts of the *structure* above ground-floor level, which carry the building *enclosure*. Greater accuracy is usually required than for the *substructure*.

supervision The direction of site workers by an experienced contractor's *foreman* or *charge hand*, who are responsible for producing good *workmanship*, often helped by *inspection*.

supply air The air blown into a room

by *air-conditioning* equipment, leaving the supply air ducts through *air terminal units*. It comes from the central plant and contains mainly recycled *return air* plus a proportion of fresh air. Supply air temperature is usually no more than 6–8°C below room temperature in summer, nor more than 10–15°C above room temperature in winter.

supply and fix (USA **furnish and install**) A description of a system whereby the same firm, usually a manufacturer, installs as well as supplies materials. The system can save arguments about who is responsible for satisfactory performance.

supply only (1) Bought from a supplier, not *supply and fix*.
(2) Brought to site by one *trade* and handed over to another trade for fixing, e.g. a plumber's flashing fixed by the bricklayer.

supply pipe The consumer's part of the *service pipe*, usually the length between a building and its site boundary or the stopvalve, whichever is nearer the *main*.

supported sheetmetal roofing, flexible metal r. Metal sheet roof coverings which are fully supported by a *deck* of timber boarding. They usually require an *underlay*. Lead or copper sheet can be laid only this way, and for stainless steel or aluminium there can be a saving in metal thickness compared with *self-supporting* roofing. Traditional work is done in fairly small panels joined by *wood rolls* down the slope and *welted* joints or *drips* across it. Larger panels are used in *longstrip roofing* with *standing seams*.

surcharged drain A *drain* filled to its crown, consequently under a pressure that is higher than atmospheric. If the surcharging increases until the sewage

reaches the tops of the manholes, their covers may be lifted off and the sewage flow into the streets.

surety Usually a *performance bond*, but see also **retention**.

surface box (USA **valve b.**) A small cast-iron lid and frame set in a footway or other surface, with a shaft leading down to a buried stopvalve, which can be operated with a *loose key* or *turn cock*. See BS 5834.

surface coefficient The heat loss per degree of temperature difference between a surface and the air or other fluid surrounding it, per unit of area and unit time. It is the reciprocal of the surface resistance. The surface coefficient is added to the *C-value* to give the *U-value*.

surface damp-proof membrane or **course** A *damp-proof membrane* laid on top of a *ground slab*, usually of *mastic asphalt* and trowelled smooth as a base for *thin floor finishings*.

surface dressing A *levelling compound*.

surface dry A stage in *drying* at which a paint film has a *sand-dry surface*, although it may not yet be *hard dry* beneath.

surfaced timber, dressed t., wrot t. Timber planed on one or more surfaces.

surface filler A *filler*.

surface-fixed hinge A *non-mortice hinge*.

surface hardener A *hardener* for concrete.

surface planer, surfacer A stationary *planing machine* with a steel bed plate in two halves at different adjustable levels, but parallel and flat. The *cutter block* between the two halves is

surface preparation

set so that the tops of its blades are level with the upper plate. Pieces of wood are pushed forward by hand, and the bottom is cut away and smoothed. With care, bevelling and chamfering can be done. The machine is particularly dangerous and should not be used by untrained people.

surface preparation *Preparation*.

surfacer (1) A *surface planer*.

(2) A thin, pigmented, high-build *filler* or *sealer*, or both, put on for the final smoothing of slightly uneven *substrates* in preparation for painting. It is usually sanded smooth after drying.

surface resistance The reciprocal of the *surface coefficient*.

surface retarder A *retarder* put on *formwork* to make the surface of the concrete set more slowly. If the formwork is *stripped* early, the surface is soft and can be brushed away easily, either to make an *exposed aggregate finish* or to leave a rough surface on which plaster will stick.

surface spread of flame *Flame spread*.

surface voids *Blowholes*.

surface-water drain Any pipe for water from the ground or *hard surfaces*.

surface waterproofer A *water-repellent treatment*.

surfacing The treatment of a surface by shaping, texturing, or smoothing, or by applying a *coating*, facing, or finishing material.

surfacing materials *Finishings*, including linings and coverings, or treatments for *solar protection*.

surround (1) A frame or decorative trim round an area of *finishing*.

(2) A material that encloses and protects another, e.g. concrete *encasement* to

steelwork, *peagravel* round a buried drain, etc.

surveyor A *quantity surveyor*, planning and development surveyor, land surveyor, etc.

suspended ceiling A *ceiling* usually of fully *dry construction* carried by the floor above it on *hangers* fixed to the slab *soffit*. The hangers support a lightweight grid of galvanized steel or aluminium main runners and crossrunners. These stiffen the ceiling, which is normally of laid-in tiles, removable for access. A suspended ceiling may be simply a *false ceiling*, but more often hides building *services* such as air-conditioning ducts and plumbing pipes, and is integrated with the lighting fittings and air outlets and inlets. Sometimes a suspended ceiling improves the *fire resistance* of the floor above or its *sound insulation*, if solid and free of gaps. Ceiling surfaces exist in an extraordinary variety, perhaps the commonest type being *mineral-fibre tiles*, but *plaster tiles* or *plasterboard* slabs are also used. Perforated metal strips with an absorbent quilt backing can absorb as much sound as an *acoustic-tile* ceiling and, like open-grid *cell ceilings*, are very light and need few supports. See B S 8290. See also diagram.

suspended floor A floor which is supported clear of the ground, usually on walls or columns.

suspended scaffold A *flying scaffold* or a *projecting scaffold*.

suspended slab A reinforced concrete *suspended floor*, not a *ground slab*.

suspension (1) An *emulsion* (*C*). One liquid carrying another liquid suspended within it as small separate droplets, as in an *emulsion paint*.

(2) Small solid particles distributed through a liquid. Most paints are a suspension of this sort.

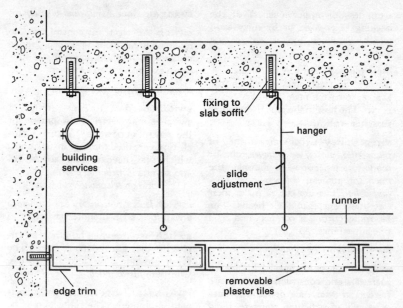

building services

fixing to
slab soffit

hanger

slide
adjustment

runner

edge trim

removable
plaster tiles

A suspended ceiling.

suspension of works The stopping of work on site, but without *determination* of the contract. The contractor is usually entitled to do this if not paid in accordance with the *conditions of contract*.

swale A shallow drainage channel across an area of lawn. It has gently rounded edges that allow mowing without scalping the grass.

swan neck (1) A combination of a *ramp* and a *knee* in a stair handrail.
(2) An *offset*.

swatch A pile of samples of *veneers*, each sheet being taken from the centre of its *flitch*. The term is also used for a pile of samples bound at one edge to form a book.

sway A hazel or willow sapling about 20 mm dia. and 2.70 m long, used for

holding *thatch* down. It is laid horizontally across the *rafters* under the thatch and fixed to them by iron hooks or tarred cord.

sweat To unite metal parts by holding them together while molten solder flows between them, as in a *capillary joint*.

sweating (1) See **sweat**.
(2) The defect of separation of the liquids in a paint; one of them appears at the surface of the film.
(3) A *gloss* developing in a dry paint or varnish film after *sanding*.
(4) Surface *condensation*.

sweat joint (USA) A *capillary joint*.

sweat out Gypsum plaster which, after it has set, appears damp and mushy is said to sweat out. The defect may be caused by cold weather, by the

455

dampness or imperviousness of the backing brickwork, or by dirty sand which demands too much mixing water.

sweep tee, swept t. A *tee* for plastics, copper, or steel pipe, in which the branch joins the main run in a gentle curve. The flow in the pipe is smoother than in a *straight tee*.

swept valley A roof *valley* formed of *plain tiles*, *slates*, or *shingles*, cut or made to a taper so as to eliminate the need for a metal *valley gutter*. *Tile-and-a-half tiles* are cut to make their *tails* narrower than their heads. The laborious cutting is compensated by a fine appearance.

swg *Standard wire gauge.*

swimming pool Special choice of materials and construction details are needed in swimming pools. The bath is usually of structural concrete, lined with *ceramic tiles* or *chlorinated rubber paint*, and with *non-slip* surfacing to the poolside deck surround. Metalwork is restricted to polished *stainless steel*, and *sealants* are only those types suitable for contact with treated pool water. Indoor pool buildings filled with warm saturated air need water-resistant wall and ceiling finishes, and a *warm roof* is usually recommended (see BRE Digest 336). Outdoor pools need a skimmer to remove leaves and precautions against frost. A hotel rooftop pool may seem to be a luxury, but it can make an excellent *elevated gravity tank* for fire fighting.

swing door A door which can open in both directions and therefore has no *door stop* in the frame. It is usually opened by pushing and may have *double-acting hinges*, or be carried on *floor springs*, or be automatically rotated.

swinging scaffold A *flying scaffold*.

swing-up door An *up-and-over door*.

switch An electrical control used to make or break a circuit. It can operate under load (current flowing) and so is more expensive than a *coupler*.

switch and fuse A *switchfuse*.

switchboard (1) Any group of switches and electrical equipment for the control of power or lighting. Cubicle types are used for large installations, which may have a *main switchboard* and several *distribution boards*.
(2) A branch *telephone* exchange.

switch box An *accessory box* built into a wall for a flush-mounted switch.

switchfuse A switch built into an enclosure that also contains fuses. The fuses are not built into the switch. Compare *fused switch*.

switchgear Large *switches* and associated equipment. The contacts may be surrounded by air (air break) or by arc-quenching *sulphur hexafluoride*.

switchroom A *plantroom* to house an electrical *switchboard*.

synthetic Man-made. Most synthetic building products are made from chemicals. They include *fibres*, *resins*, and *rubbers*.

synthetic fibre, man-made f. The first synthetic *fibre*, artificial silk, was made in 1885 from nitrocellulose; viscose was made in 1910, cellulose acetate in 1911, nylon in 1936, and other *polymers* more recently.

system building Either complete prefabricated *volumetric buildings*, or only the outer walls and their cladding, which are factory-finished, saving time on site, reducing *wet trades* and the need for skilled labour. They are *industrialized building methods* and have been used for factories, schools, and housing.

T
---

tab A *tingle*.

table form Complete *formwork* for a concrete floor slab, with a deck and props, made into a single unit. It is lifted into place between walls by crane and after use rolled out sideways on wheels to be lifted again.

tack (1) A sharp, short *nail*.
(2) A *tingle*.
(3) Stickiness of a paint film while it is *drying* or from *after-tack*.

tack-free A stage in the *drying* of a paint film when it is no longer sticky.

tack rag Cheese cloth or other cotton fabric damped with slow-drying varnish for *dusting* off a surface after *rubbing down* and before putting on the next coat. Tack rags should be kept in an airtight jar so that they do not harden.

tacky Sticky, an early stage in the *drying* of paint or adhesive.

tag A copper strip folded several times and used as a wedge for holding copper sheet into a masonry joint.

tail (1) A short length of pipe or cable, to make a connection to a fitting.
(2) The lower edge of a slate or tile.
(3) The built-in end of a cantilevered stair tread.

tailing in, t. down Counterweighting the inside end of a cantilevered member by building the wall above it of bricks, stones, or any heavy material.

tailpiece (USA) A *trimmed joist*.

take-off The result of *taking-off*.

taker-off An experienced *quantity surveyor* who specializes in *taking-off*.

taking-off The recording of *dimensions* from drawings or schedules, with a description of the work. It is the first main stage in preparing *bills of quantities* and is followed by *working-up*. Exact rules should always be followed when taking off so that the work can be continued without misunderstanding, even by another person.

takspan Swedish pine roof *shingles* 430 to 510 mm long by 100 to 250 mm wide, and only 3 mm thick, made like *sliced veneer*.

tall boy A hood about 1.50 m high, usually made of galvanized sheet steel, added to the top of a chimney to reduce downdraughts.

tally slates *Slates* sold by number, not by weight.

tambour (French, drum) Any drum shape, e.g. a circular wall carrying a dome, a cylindrical lobby round *revolving doors* to prevent draughts passing, or a stone in a circular column.

tamped finish An *off-tamp finish*.

tamper A long *screed board* used for finishing concrete.

Tanalized timber The trade name for a treatment process using an all-round *preservative*, put on by *vacuum-pressure impregnation*. It protects building timbers against *fungus* and *borers*.

tang (1) A *fishtail fixing lug*.
(2) The pointed part of a steel tool such as a file, rasp, knife, or *chisel* which fixes the blade to its handle.

tangential shrinkage The *shrinkage* of timber parallel to the *growth rings*,

457

usually about 50% greater than *radial shrinkage*.

tank A storage vessel for liquid, which may also allow *expansion*. Unlike *cisterns*, tanks may be made to take pressure, in which case they are cylindrical. Large tanks in buildings are usually bolted steel *sectional tanks*.

tanking A waterproof skin, usually three coats of *asphalt work*, that completely covers the outside of a basement. Concrete or brick *mechanical protection* is needed on the outside of asphalt. The asphalt may be in several coats, but in general horizontal work is 30 mm thick and vertical work 20 mm thick. Tanking for horizontal surfaces under basements is laid on the *blinding concrete*, before the *ground slab*. Vertical surfaces can be tanked as a separate operation after concrete work is completed, although many contractors build basements from the outside in, by laying the blinding concrete, then blockwork outside where the walls are to go. Ground water under pressure often delays work and *dewatering* (*C*) or temporary *loading coats* may be needed. Finally the basement concrete slab and walls are cast, again working from the inside. For a complex *substructure* which is difficult to waterproof completely, a *drained cavity* may be used instead of tanking. See BS 8000 part 4.

tap (1) (USA **faucet**) A plumbing *fitting* on the end of a pipe from which water is drawn. Hot and cold water may come from separate taps or be blended in a *mixer*. Many different styles of tap are available, from decorative and *easy-clean* types in bathrooms and kitchens to simple *hose union taps* for use by cleaners. Their mechanism can be of traditional *screw-down* pattern, use *ceramic disc valves*, or be operated by automatic controls.

(2) An accurately threaded plug of hard steel which cuts a thread inside a bored hole. It has a square end which is turned with a *tap wrench*.

(3) To draw fluid from a *services main*, usually by connecting a *branch pipe* to it, or to do the same from an electrical main, sometimes by a *tap-off unit*.

tape (1) *Joint tape* or *jointing tape*.

(2) **t. measure, measuring tape** A graduated strip of springy steel or plastics for use in *setting out* distances, usually 2 or 5 m long and spring-loaded to wind itself into a compact case when not in use. A *builder's tape* is longer.

taper (1) A pipe or duct *fitting* with a gradual change of size along its length, either a *reducer* or an *increaser*, e.g. before or after a *centrifugal pump* (*C*), at branches off a main, etc.

(2) A narrow area of finishing that can be seen to change in width, such as floor tiles with the first row a few centimetres from a wall. It can be avoided by careful choice of layout, e.g. with edge rows no less than half a tile wide, *scribed* to the wall.

taper bend A *taper* pipe made as a *bend*.

tapered-edge plasterboard *Gypsum plasterboard* with its long edges reduced in thickness at a straight slope of about 1 in 5 on the *face* side only. After the sheets are fixed to the wall or ceiling, by nailing through the tapered edge, the joints are reinforced with *scrim* or *joint tape* and plastered over with *veneer finish plaster* or the maker's *joint cement*. The very small amount of joint cement is the only 'wet' work in what is otherwise a *dry lining*. With a suitable *trowel* and sanding down the joint is usually invisible after painting except to the most critical examination under low-angle lighting.

tapered tread A *tread* of a *geometric stair* that is wider at one end than the other. It is usually a *winder*, but a *balanced step* has an eccentric *springing point*.

taper thread A standard screw *thread* used outside screwed pipe ends to ensure a pressure-tight joint for gas or water when screwed up tight over *jointing tape*. Taper threads are used on all pipes and *fittings* except *connectors*, on which the thread is too long to be tapered. The amount of taper is 1 in 16, that is, if continued to a point, the taper would form a cone of 1 unit dia. and 16 units long. See **parallel thread**.

tap holder A bracket for fixing a draw-off *tap* to a wall.

taphole A hole for a tap (or spout) in the rim of a *basin*. There may be one, two, or three of them.

tap-off unit A device used to make an electrical connection to *bus bar* trunking in order to draw power for *branch circuits*. The tap-off unit may contain *fuses* and does not interrupt the bus bars, which are *line-tapped*.

tap wrench A tool with a square hole and two straight bar handles, used to turn a thread-forming *tap*.

tar Black, sticky, waterproof liquid distilled from coal.

tar epoxy paint A tough, durable *two-part* paint, used to protect concrete or steel, particularly under water or in contact with the ground. On steel it is usually put on over *zinc-rich* primer after *blast cleaning* the surface.

target-price contract A *contract* in which the cost of the works is estimated and a target price agreed before work starts. Any savings made for the *client* entitle the *contractor* to a bonus. It is intermediate in type between *cost-reimbursement* and *fixed-price contracts*.

tarpaulin Waterproof cloth used to give temporary protection from rain and weather to materials stored outdoors, or to roofs during repair work, now often of polyethylene sheet.

task An *activity*.

tatami The traditional Japanese straw floor mat, standardized at 6 × 3 shaku (Japanese foot, of about 303 mm). The floor sizes of temples and rooms are built to take an exact number of mats, which are not cut. See **module**.

teak (Tectona grandis) A tropical *hardwood*. The best comes from Burma, India, and Thailand. It is oily, stable, and very durable, making it useful for laboratory benches and quality *joinery*. In workability it is about equal to *oak*, but *adhesives* do not stick well, even after degreasing.

technical assistant A *quantity surveyor's* assistant, who does *taking-off*, etc.

tee A short pipe or duct *fitting* with a straight main run and a *branch*. A *straight tee* has a branch at right angles to the main run. Other types are *angle tee* and *sweep tee*. Terms sometimes used are to make a 'teed' connection, or to 'tee off' or 'tee up' a branch pipe.

tee-hinge A *T-hinge*.

Teflon *Polytetrafluorethylene*.

tegula The under-tile of *Italian tiling*.

telephone Large telephone installations may have a *frame room*, a *private automatic branch exchange*, and multi-pair cables ending at a wall or floor connector for each extension. Cables in *trunking* are separated from other wiring, in their own compartment. Any equipment connected to exchanges outside has to be approved by the telephone company, in the UK either British Telecom or Mercury Communications.

telescopic centering A formwork *floor centre*.

telescopic prop A type of *prop*.

telltale A strip of brittle material (e.g. slate) fixed across a crack, usually marked with the date of setting. It will indicate by its breakage if the crack is getting worse. Other types of telltale exist.

temperature gradient The rate of change of temperature over a distance, used in discussing *thermal stress* and *interstitial condensation*.

temperature movement *Thermal movement*.

temperature rise The heating of electrical cables by the current they carry. Cable *cross-sectional areas* (*C*) are chosen so that the maximum rated current will not cause an unsafe temperature rise in still air. Thermal *insulation* placed over a cable may prevent it from cooling, and start a fire.

tempered glass *Toughened glass*.

tempered hardboard *Hardboard* that has been made specially strong and water-resistant, used for external cladding, floor overlays, signs, etc.

tempering Reducing the brittleness of steel by heating and *annealing* after rapid cooling from red heat (when steel is hard but brittle). Tempering temperatures are below red heat and can be judged from the distinctive blue or yellow surface oxides. The temperature needs to be finely controlled to give the right toughness, and the whole piece must be treated (not just a tip) or brittle zones will occur. The term is also applied to heat-treated *toughened glass* and other materials.

template, jig, profile, templet (1) A full-size pattern of wood or metal used for testing or forming the shapes of plaster, for holding bars in place, setting angles, etc.
(2) A *pad* to spread a concentrated load.

temporary protection *Protection*.

temporary works Site installations provided by the *contractor* which are not part of the project but are needed during construction. Their cost is a *preliminaries* item and they often determine programme *lead times*. Examples are heavy timber *shoring*, steel framing for *façade retention, earthwork support, hoarding*, concrete crane bases, *scaffolding*, and protection against traffic damage.

tender (USA **bid**) An *offer* made by a *contractor* to do certain work for a price, stated in the *form of tender* and given in detail in a *priced bill*. The tender price is based on an *estimate*.

tendering (USA **bidding**) The procedure of sending out drawings and *bills of quantities* to *contractors* for them to prepare a *tender* price in competition with each other. Tenders are all opened on a set date. Following *assessment*, the *lowest tender* is usually accepted, thus forming a *contract*. Tendering may be *open* or *selective* and is usually single-stage, but can be done in *two stages*. Contractors' *surveyors* often refer to tendering as *pre-contract* work.

tenon An end of a *rail* or similar joinery member, reduced in area so that it fits a *mortice* in another member to make a *mortice-and-tenon joint*. The width of a tenon should be about four times its thickness. The word, like mortice, is from French, but tenons were also used by the ancient Egyptians on their wooden beds.

tenon saw A fine-toothed *handsaw* about 350 mm long, stiffened with a strip of metal folded over the back of its blade, used to cut wood accurately. See diagram, p. 388.

terminal (1) A *finial* or similar decoration.

(2) The top end of a lightning conductor, gas flue, etc.

(3) The end of a power line, or a *lug* on it, or a device for making a cable connection to an electrical fitting, usually tightened with small screws.

(4) An *air terminal unit*.

terminal re-heat unit A *re-heat unit*.

termination (1) An *air termination network* or an *earth electrode*.

(2) The *determination* of a contract.

termite, white ant An insect which eats most sorts of wood. It only looks like an ant and is unrelated to *borers*. It is found in tropical and warm-temperate regions (not the British Isles), living in colonies in mud mounds, with ventilating tunnels that are opened or closed to maintain a cool, damp, dark environment. Houses in these areas need *termite shields* and their timbers may also be treated with *preservatives*.

termite shield (Australia **ant cap**) A metal sheet with its edges turned down at a 45° angle, placed on top of house foundations, to prevent *termites* building mud tunnels upwards and reaching any timbers.

terne-coated sheet Steel sheet coated with lead alloy containing up to 20% tin. The coating is a good *substrate* for paint.

terrace (1) A raised, level, earth platform, with at least one upright or battered side.

(2) A trafficable *flat roof*.

terrace blind An *awning*.

terrace house One of a row of houses side-by-side with shared *separating walls*.

terracotta Italian for burnt clay, a finer-grained *ceramic* than brick or tiles, used for making *hollow clay blocks*, *ceramic veneer* wall facings, statuettes, etc. It is moulded to shape and fired at 1200°C. Terracotta can be unglazed but is more usually covered with a clear glaze or an opaque colour and should then be called *faience*. It has fairly low porosity and even if unglazed is very frost-resistant and durable.

terrazzo A smooth, hard, non-dusting *topping* of marble chippings (often green) mixed with coloured cement, laid in-situ 20 mm thick on fresh concrete. Large areas are divided into small panels by strips of plastics or metal to form *movement joints*. After laying it is *screeded* flat, trowelled smooth, and covered to allow *curing*. After three days it has its first grinding, followed by a second grinding five days later. Both grindings are followed by careful washing. Terrazzo work is a skilled *wet trade*. Terrazzo floors are easily cleaned with water and a little mild detergent, but may be slippery when wet. Terrazzo is used for floors, precast stair treads, toilet partitions, and counter tops.

terrazzo tiles Concrete floor tiles with a *terrazzo* facing, often 300 mm or more square, to BS 4131.

tessera (plural **tesserae**) The individual cube used in *mosaic* work, made of marble, *vitreous glass*, *smalti*, hard stones, highly vitrified clay, etc.

texture The structural character of wood as revealed by touch or reaction to cutting tools, largely determined by the distribution and size of various cells. It may be coarse, even, uneven, or fine.

textured brick A *facing brick* with its face roughened during manufacture to give a pleasant texture. Many textures exist: dragged, fissured, rusticated, sand creased, split face, stippled, etc.

textured finish A *high-build* wall coating which is worked with tools to a rough or patterned surface, e.g. *masonry paint* or *plastic paint*. On outside walls this can reduce rain penetration and improve durability. Decorative textures on inside walls include: *ragging*, rolling, *sponging*, *stippling*, etc.

textured rendering An external wall *finishing* made of a weak mix of lime or cement *mortar* thrown on and left untrowelled, but later sometimes scraped when hard enough. It has a rough surface that sheds rain and is porous, allowing water vapour to escape from the wall. One example is *Tyrolean finish*.

thatch A roof covering of *reed* or *straw* (or heather), laid wet by *thatching*; or synthetic 'reed'. A natural reed roof can last sixty or seventy years, but the life of a good straw roof is not more than twenty years and it needs cleaning and patching every seven years. Thatch has a high insulating value but its main disadvantage (apart from its short life, the unpleasantness of thatching work, and the difficulty of finding thatchers) is the fire risk. This can be lessened by a fire-resisting sheet *underlay*, by installing *drencher* pipes to spray water, or by soaking the reed or straw in fire-retardant solution. The last method is losing favour as it needs annual re-treatment and the chemicals tend to degrade the reed and encourage mould growth.

thatching The *trade* of laying a *thatch* roof. Wetted bundles of reed or straw (yealms) are placed in courses and held by *runners* to *sways* tied down to the rafters, building up to a total thickness of about 300 mm. *Verges* are made with *scallops* and *spars*. The ridge is made with *sedge* fixed with hazel or willow sticks (liggers). After laying, the thatch is tightened by tapping upwards with a steel-faced mallet, or 'leggett', and the surface combed and trimmed to final shape.

thermal break Heat *insulation* to minimize *cold bridges*, e.g. plastics inserts between the parts of aluminium window frames on the outside of the building and those on the inside, to reduce *condensation* and heat loss.

thermal bridge A *cold bridge*.

thermal capacity, t. inertia, heat c. The amount of heat that can be stored in something. It increases with mass and specific heat. A building with heavy walls and floors of high thermal capacity (solid walls, blockwork) takes a long time to heat up or cool down. Houses of solid masonry tend to stay at an even temperature from day to night, an advantage that is retained with *external insulation*. A room of low thermal capacity (with sheet cladding, dry linings, curtain walls) heats up or cools down more quickly. Low thermal capacity can reduce *condensation* from warm moist air striking a wall, or inside a *flue*, by allowing it to heat up more quickly.

thermal conductance The *C-value*.

thermal conductivity The *k-value*.

thermal fusion joint A joint in plastics pipe made by *heat fusing*.

thermally broken Built with *thermal breaks*.

thermal movement, temperature m. The expansion or contraction of materials due to temperature changes, which are reversible from day to night or between seasons, thus *cyclic movement*. Thermal movement is low for stone, brick, concrete, timber, and glass, medium for most metals (steel, copper, aluminium), and high for *plastics*. It occurs in all parts of a building, but mostly in pipework and exterior cladding (roofing, brickwork, windows, curtain walls), particularly if exposed to the sun. To prevent cracking or weakening from fatigue, free thermal movement should be allowed by means of construction details such as *movement joints, gasket glazing, panelling, pipe flashings*, bends in *pipework, push-fit joints*, etc. See BRE Digests 227, 228, and 229.

thermal resistance The *R-value*.

thermal resistivity The *r-value*.

thermal stress breakage The shattering of window glass from a high *temperature gradient* across its surface (not between faces) that creates internal stresses, e.g. from sun and shadow. Ordinary *annealed glass* may break at a temperature difference of 40°C and *wired glass* at one of 30°C, but *toughened glass* performs far better (250°C).

thermal transmittance *U-value*.

thermal wheel A *heat-recovery wheel*.

thermofusion welding *Heat fusing* of plastics.

thermoplastics *Plastics* which soften on heating and harden again on cooling, unlike the other main group, *thermosets*. They can be shaped or bent using gentle heat. Joints are made by the usual *fasteners*, by push-fit clips, by *heat fusing* or *solvent welding*, but usually not with *adhesives*. In a fire they may melt quickly at quite low temperatures. Thermoplastic rooflights soften at 75 to 150°C, which may help the early venting of *smoke* from a fire, though this may not always be desirable, especially if the fire is outside. Thermoplastic adhesives are used in *hot-melt glue*.

thermoplastic tile A *floor tile*, originally those made with *asphalt*. The term now usually means *vinyl asbestos tiles*.

thermosets *Plastics* which set permanently on heating, or undergo *curing*, and do not soften when reheated. Once set they cannot be reshaped by heat or solvents, unlike *thermoplastics*, and some remain strong up to quite high temperatures. Those thermosets that are rigid produce chips rather than slivers when cut, but this does not apply to synthetic *rubbers*, which are also thermosets. Thermosets are used in *adhesives*. Building components made of thermoset plastics are usually joined with threaded connections or adhesives.

thermosiphon *Gravity circulation*.

thermostat An automatic control for maintaining temperature between limits. It usually has a *bi-metal strip* (*C*) which bends on heating (or cooling) to break (or make) a circuit, or turn off (or on) a flow of gas, hot or cold water, etc., doing the work of both a *sensor* and an *actuator*. Thermostats are used to control air conditioning, central heating, and hot water services.

thermostatic mixer A convenient hot and cold water *mixer* that keeps a constant discharge temperature for showers or sinks, etc., despite wide variations in temperature or pressure in the supply system.

thickening (1) A local increase in the thickness of a cast or moulded component, often to strengthen foundation beams or column heads in concrete slabs, at fixing points in moulded *glassfibre-reinforced polyester*, etc.
(2) Paint *fattening*.

thicknessing machine, panel planer, thicknesser A *planing* *machine* which reduces, to the desired thickness, wood which may have already had the face made true on a *surface planer*. The two faces can be made parallel and planed to the correct thickness by the setting of the two feed rollers at a certain level above the table.

thief-resistant lock A *security lock*.

thimble A *sleeve piece*.

thin floor finish A *floor finish* such as *vinyl asbestos tiles*, *needle-punch carpet*, or sheet *rubber* which is only a few millimetres thick and usually fully stuck down with *adhesive*. Since any roughness in the surface on which the finishes are laid shows on their own surface, they need a smooth hard *subfloor*, e.g. a power-floated concrete *screed*, concrete overlain with *levelling compound*, rough timber covered with a hardboard *underlay*, or flooring-grade chipboard. Adhesive put on by hand with a *notched trowel* may show through as radial marks on the surface of highly polished thin floor finishings.

T-hinge, tee-hinge A *hinge* with one short flap and one long tapered flap.

thinner, diluent A clear *solvent* added to paints before *application* to make them flow easily, to form a smooth film, and to fill tiny cracks, after which it mostly evaporates. The usual thinner for *oil paints* is *white spirit*. The special thinners for *cellulose paints* must be compatible with their *medium*. Thinners are used for cleaning brushes or other equipment and are usually *flammable liquids*.

thinning ratio The proportion of *thinner* recommended for a particular *paint* in a particular use.

thin-wall plaster, thin-coat p. *Premixed plaster* that may be used in *plastering machines* or applied by hand. One type, sold wet in steel drums, hardens by the drying of its organic binder and keeps indefinitely in the unopened drum.

Thiokol A trade name for one type of *polysulphide* used in sealants.

third fixings, second f. The final *joinery* work, such as door hanging and fixing *door furniture*, which follows the first painting.

thread A spiral or helical ridge formed on the surface of a metal or plastics cylinder – the *screw*. Threads for *wood screws* are spiral (conical, or *tapered*) with a *gimlet point*, those for metal are helical (cylindrical, or *parallel*) and blunt ended.

threaded rod A long round steel *bar* with a continuous *thread* along which a nut can be wound, used as a *hanger*, a *fastener*, etc.

three-coat work (1) **render, float, and set** The best-quality traditional *solid plastering* on an uneven *background*, in which the first coat (the *rendering coat* on walls or pricking-up coat on lathing) fills the rough places and evens out *suction* and the second (the *floating coat*) forms a smooth surface for the third (the *finishing coat*). The total thickness is about 20 mm, of which the finishing coat is 3 mm or less.

(2) Paintwork or mastic *asphalt* applied in three layers.

three-hole basin A wash *basin* with *tapholes* for hot and cold water *valves* and a *spout*.

three-phase supply A low-voltage electrical installation for the supply of heavy-duty *motive power* (or other) at 415 volts. It has four conductors, the three *phases* (red, yellow, blue) plus one *neutral*.

three-pipe system A water circulation system with two *flow pipes*, one for *central heating* and a second for *domestic* *hot-water* supply, with a common return.

three-ply (1) The commonest sort of *plywood*, built of a core *veneer* and cross-bands each side to give *balanced construction*.

(2) *Built-up roofing* or a *damp-proof membrane* formed of three layers of *bitumen felt* lapped and bonded to each other.

three-prong plug A *plug* for electrical power from a *socket outlet*. The prongs can be inserted only in such a way that they connect correctly to the live, neutral, and earth.

three-quarter bat A *brick* cut straight across to reduce its length by a quarter.

three-quarter-round channel A *channel bend* with its outer edge raised so that the flow does not spill as it turns the corner.

three-way strap A steel tee-plate with its three arms shaped so as to anchor together the joint between two or three members of a wooden *truss*. It is fixed to them by *coach screws* or through bolts.

three way valve, t. port v. A *valve* for diverting a flow to either of two outlets. The one inlet plus two outlets makes three waterways or openings.

threshold, door sill A horizontal member across the foot of a doorway. On external doors it may be shaped to keep out the weather. Indoor thresholds are usually of timber and may have bevelled edges.

threshold strip A slightly rounded metal strip (brass, stainless steel, aluminium) over the joint between *floor finishes* or coverings which meet under the door leaf, so that the joint is hidden when the door is closed. It prevents carpet or tile edges from fraying and interrupts combustible flooring under

fire doors. Fixing is with *raised countersunk head screws*.

throat (1) **throating** A groove near the edge of a *soffit* to form a *drip*.

(2) A narrowing of a fireplace *chimney* to reduce pressure in the *smoke chamber*, sometimes fitted with a control *damper* or 'throat restrictor'.

throated sill A *window sill* with a *throat* under.

through lintel A *lintel* the full thickness of a wall.

through stone A *bonder* seen on both faces of the wall.

through tenon A *tenon* which passes beyond the morticed member.

throw (1) The total length over which a lock *bolt* can move.

(2) To move a *dead bolt* to the locked position from the *withdrawn* position.

(3) The distance a jet of air goes out from an *air terminal unit*.

thrust block A concrete mass cast at any bend (or end) of a buried pipe with *push-fit joints*, so that thrust from pipe pressure is taken by the ground, without forcing the joints apart. See **anchor block**.

thumb latch A simple door, gate, or cupboard *latch* operated by pressing.

thumb slide A *slide bolt*.

tie A device to hold things together, such as an *anchor*, *tingle*, or *wall tie*.

tie beam The bottom horizontal member of a roof *truss* or *trussed rafter*, equal in length to the *span* (C) of the roof. It ties together the feet of the *rafters* and in timber work is held to them with a *connector*.

tier, withe (USA) One *leaf* of a cavity wall.

tie wire *Binding wire*.

tight knot A *knot* in timber held firmly by the surrounding wood.

tight sheathing Diagonal *matchboards* nailed to *studs* or *rafters*.

tight side The side of timber *veneer* which was not in contact with the knife when it was cut off, unlike the *loose side*. Both roller-cut and sliced veneer have a tight side.

tight size, full s., rebate s. The size of the rebated opening for glass, about 3 mm more than the *glazing size* and about 6 mm more for double glazing *sealed units*.

tight tolerances *Tolerances* smaller than usual.

tile A thin, medium-sized square, rectangle, etc., of *finishing* or covering material of standard dimensions, made to fit together with others in a regular *format*. It can be of ceramic, metal, plastics, concrete, carpet, or other material. Dimensional accuracy is more important for tiles that are *butt-jointed* than for those that are lapped. There is more variety in the materials for *floor tiles* and *wall tiles* than for *roof tiles*.

tile-and-a-half tile A *roof tile* of width one and a half times that of the tiles with which it courses. It is used at *verges* and *valley gutters*, or in *plain tiles* also at *swept valleys* and *laced valleys*. For interlocking *single-lap tiles* it is left or right-*handed*, but for plain tiles simply 250 mm wide. It avoids the use of half tiles, which are difficult to fix and hazardous.

tile batten A *tiling batten*.

tile clip A fixing for *roof tiles* to prevent wind uplift, made of corrosion-resistant spring wire or strip metal. Tile clips are used for *single-lap tiles* made for clip fixing (rather than nail fixing), and at *eaves*, at *verges*, and in the middle of the roof. The clip attaches to the back of the tile and the other end is either hooked over and under the *batten* or nailed to it. Tile makers stock clips suited to their tiles

and can recommend maximum suitable roof *pitches*. See **dry ridge tile**.

tile creasing A *creasing* made with pieces of *plain tile*.

tiled valley A *valley* covered with purpose-made *valley tiles*.

tile fillet, t. listing A piece of roof tile cut and set in mortar at the *abutment* with a parapet, instead of a *flashing*.

tile hanging, vertical tiling, weather tiling Fixing *roof tiles* to an external wall as a *cladding* material.

tile listing A *tile fillet*.

tile peg, t. pin A *peg* for roof tiling.

tiler, tile layer A skilled worker who lays *tiles*.

tiling (1) *Wall tiles* or *floor tiles*, mostly used in *wet areas* such as bathrooms and including special *angle tiles* where the wall and floor tiles join; or the *trade* of laying tiles. See B S 5385 and B S 8203.

(2) *Roof tiles*.

tiling batten Timber *battens* that carry *roof tiles*. They are spaced to suit the *gauge* and nailed to the *rafters*.

tilt-and-turn window A *casement window* with horizontal and vertical *Cremona bolts* operated in opposite directions by one handle; turned fully either way one bolt releases, the other acts as a pivot; in between it locks.

tilting fillet, doubling piece, skew f. (U S A also **cant strip**) A thick horizontal board, sometimes triangular in cross-section, nailed across the rafters or roof boarding under the *double eaves course* to tilt it slightly less steeply than the rest of the roof and to ensure that the *tails* of the lowest tiles or slates bed tightly on each other. It may also be needed at a *verge* or the edges of a *valley gutter* of *supported sheet-metal roofing*. A *fascia* at an *eave*, an *over-fascia vent*, or a *cocking piece* may do the same job. See diagram, p. 147.

timber (USA **lumber**) (1) Wood for building, used as sawn *battens, baulks, boards*, etc. Timber is a natural material and varies widely in its properties. Before use in building it need *seasoning* and often also *preservative* treatment. It is shaped with woodworking *tools* and joined with *fasteners, connectors*, or *adhesives*. It has high *moisture movement* but little *thermal movement*. In buildings it usually needs protection from moisture and adequate *ventilation* to prevent *decay*. Heavy *structural timber* can provide some *fire resistance*.

(2) (USA) A piece of *lumber* larger than 4 × 6 in. (100 × 150 mm).

(3) **timbers** Of the hundreds of *hardwoods* and *softwoods* now in commercial use the following are the best known in Europe:
Temperate hardwoods: beech, boxwood, elm, oak, walnut.
Tropical hardwoods: keruing, mahogany, meranti, sapele, teak, utile.
Softwoods: Douglas fir, redwood, spruce, western red cedar, whitewood.

timber flooring *Floor boarding.*

timber-frame construction House building with wooden walls, *studs* carrying *dry linings* and often the *cladding*. In Britain, the *suspended floors* and roofs of houses have always been of timber, but for walls its use declined after the early medieval *cruck house* in favour of brick and block, which have better *sound insulation* and *fire resistance*. Long used in America and Australia, as *balloon framing, braced frames, platform frames*, or *brick veneer*, it is being revived in Britain. If the house is insulated and heated, it needs *breather membrane* outside the timber and *insulation*. Inside the insulation a continuous *vapour barrier* is needed; service pipes should not pass through walls, but be

run behind a separate *lining*. Timber must be free to shrink and have *movement joints* so that it does not take unintended loads, e.g. 'soft joints' against brickwork.

timber framing A loadbearing frame of timber, used in the early medieval *cruck house* and in the *timber-frame construction* of the present day. Nowadays in Britain timber walls are rare compared with bricks and blocks, but house roofs are still mostly framed with timber.

timbering *Earthwork support.*

timber in the round *Round timber.*

Timber Research and Development Association (TRADA) A research association founded in 1934, half of whose income is earned from commercial testing and consultancy, 40% by subscription from timber importers, and 10% in payment for work done for government departments. A recognized fire-test centre, TRADA also tests timber units for mechanical strength. Other important parts of its work are *appraisal*, window testing, and consultancy on woodworking, sawmilling, and the use of timber in domestic, industrial, and farm buildings.

time and a half The 50% additional payment usual in Britain for *overtime* worked less than three hours after normal finishing time and up to 4 pm on Saturdays. After this, *double time* is paid.

time distribution, t. system (USA **master clock s.**) A *communications* system with one accurate master clock and wiring to send signals to receiver clocks throughout a building.

time for completion, construction t., contract t. The number of weeks allowed for work on a building *contract*. It starts on the date of site *possession*, ends at *handover*, and is used to calculate the *completion date*.

timekeeper A *site clerk*.

timer A device that switches a circuit after a set delay, e.g. to switch off corridor lighting, often started by a remote *push button*. See **clock timer**.

time scale Evenly spaced parallel lines on a *programme* chart, marked with dates (days, months). On a *bar chart* the length of each bar shows the planned work duration. On an *arrow diagram* it is arrows. Progress is checked by comparing the percentage of work completed against the dates.

tingle (1) **cleat, ear, latchet, tab, tack, etc.** A strip of sheet metal about 50 mm wide, used as a fixing to hold down the edge of *supported sheetmetal roofing* or to secure a *hollow roll* or *seam*. One end is nailed to the roof timbers and the other folded into the edge of the sheet to be secured. Tingles are also used as *clips* for pipes or electric cables and to fix panes of *patent glazing* or replacement *slates*.

(2) A support at the middle of a long string *line* used by bricklayers.

tin snips Strong scissors for cutting thin metal sheet. See diagram, p. 337.

tint (1) A slight difference in colour, e.g. that given to *tinted glass* or to a paint when *overcoating*, to reveal *skips*.

(2) A paint *colour* made by mixing much white *pigment* with a little coloured pigment.

tinted glass, body-t. float Window glass with a green, grey, or bronze *tint*, used as *solar-control glazing*. The colour is usually not noticeable from inside unless next to clear glass. See **sick building**.

tinters *Stainers*.

tinting The final adjustment of the *colour* of a *paint*.

tip A sharpened end, as of a lightning *air termination network*.

titanium dioxide, t. white The most widely used white pigment for paint, outstandingly opaque and non-poisonous. It gives a pure and permanent white.

title The right of ownership to a property, shown on legal documents called title deeds. A builder is given site *possession* usually by the title holder.

title block A box marked on the edge or corner of a *drawing* which states the firm's name, the job or project name, the drawing name and number, any *revisions*, and their issue dates.

toe The lower part of the *shutting stile* of a door. See also *C*.

toeboard, guard b. A *scaffold board* set on edge at the side of a scaffolding. It is a *safety* precaution to prevent tools dropping off the edge, and reduces the staining of *facework* by splashed bricklaying mortar.

toe nailing *Skew nailing*.

toe recess A set-back skirting, about 50 mm deep by 70 mm high, along the base of a cupboard or panel under a worktop, sink, cooker, bath, etc. It enables people to stand comfortably close to their task.

toggle bolt A *fixing* device with a hinged T-bar joined by a nut. When the bolt is undone it can be inserted through a hole in *plasterboard* or *hardboard*. The T-bar is then turned and the bolt tightened. See diagram.

toggle switch A *tumbler switch*.

toilet A *water closet*.

tolerance, permissible deviation The allowable departures in weight, size, moisture content, etc., from the *nominal dimension*, *flatness*, etc., of a

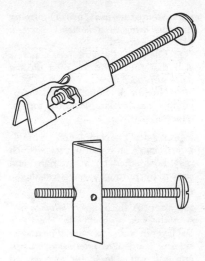

A toggle bolt.

tongue (1) A narrow strip projecting from the edge of a board, panel, or block, either a *straight tongue* or a *loose tongue*. It fits into a groove to make a *tongued and grooved joint* or a loose-tongue joint.

(2) A *return latch*.

tongued and grooved joint, tongue and groove j. A joint between the edges of boards to form a smooth wall, floor, or roof surface. It is relatively airtight even if the boards shrink. Each board is machined with a tongue on one edge and a matching slightly tight groove on the other edge, so that they fit closely when laid together. They are mostly used for *matchboarding*.

ton of refrigeration The *latent heat* needed to melt one US ton (907.2 kg) of ice. See **chiller**.

tool, hand t. Hand tools are used for working materials to size and shape, for smoothing or texturing surfaces, digging, painting, forming holes, or for *setting-out* and guiding work. Skill is needed and many are dangerous in awkward situations. Since lathes and *woodworking machines* have come to be called *machine tools* (C), traditional tools used by each *trade* are called hand tools. Many have given their names to *power tools*.

tooling (1) Jointing a brickwork *face joint* with a steel tool, usually to make a *bucket-handle joint*.

(2) *Batting*.

tool pad, pad, t. holder A combination tool consisting of a handle with a screw clamp for holding one of several small tools, such as gimlets, saws, awls, and screwdrivers of various sorts.

tooth The surface roughness of a paint *film* which has a coarse or abrasive *pigment*. Such a surface is very suitable for rubbing down and gives good adhesion to paints put on it, but is not a good top coat.

part. For example, length nominally 100 mm with a tolerance of ± 2.5 mm can be from 97.5 mm to 102.5 mm actual size (usually written 100 ± 2.5 mm). Building methods allow for tolerances in manufacturing and site work. The differences in the sizes of *bricks* are taken up in the mortar joints of brickwork. The fixings for a *curtain wall* usually have slotted holes for cast-in anchor bolts. The slots allow for tolerance in *setting-out* or in component size. Structural concrete walls and floors are usually built to within 10 mm of the dimensions shown on drawings, although *screeds* may be laid to 'tight' tolerances such as ± 3 mm.

toner A pure organic dye, without *extender*, usually a strong *colour*.

tong tester An instrument with openable jaws put round one wire in an electrical circuit, without disconnecting it, to measure the current carried by detecting the magnetic field created. Space is needed between wires to pass the tong jaws.

toother A *stretcher* brick, block, or stone which projects in *toothing*.

toothing, indenting In brickwork, blockwork, or stonework, leaving *stretchers* so that alternating courses project from the end of a wall to bond with future work. The *toothers* resemble teeth.

toothplate connector, bulldog plate c. A timber *connector* with a ring of large teeth, clamped between two members. The joint is first closed with a pull jack and later held by a steel bolt with nut and washers. See diagram, p. 99.

top-course tiles, under-ridge t. The *tiles* of the *course* next to the ridge, which may be shorter than those in lower courses to maintain the same *gauge*.

top cut (USA) In *roof cutting* a *plumb cut* at the top of a *rafter*.

top-down construction, downwards c. One of several methods for building underground between existing city buildings, by first sinking *piles* (C) or *diaphragm walls* (C), then casting the ground-floor slab, followed by excavation below it and construction of the *sub-levels* in stages. The existing buildings are well supported at all times by the new structure, without *shoring*. Combined with normal *bottom-up construction*, the method becomes *build and burrow*.

top form A *back shutter*.

top-hung window A window with a *sash* hinged at its top edge, usually opening outwards and held by a *casement stay*.

top lighting Illumination from overhead, by a *rooflight*, *borrowed light*, or artificial light.

topping (1) A wearing surface, such as *granolithic* or *terrazzo*, usually forming the top layer of a *monolithic screed*. Compare **screed**.

(2) The part of a reinforced concrete slab that forms the top of a *rib and block suspended floor*.

(3) The first operation in hand sharpening a *saw*, making level the tops of all teeth.

topping-out The completion of a building *structure*, an important *stage* which may be marked by a ceremony and shown by putting a yew tree on the roof.

top sheet A thick, heavy *bitumen felt*, smooth-faced on both sides, used as a last layer in *built-up roofing*. The underside is bonded to *intermediate sheets* and the top side is for site-applied *protection*, such as chippings, ballast, or insulation, which is also bonded to it.

topsoil The mixture of sand, clay, and humus, in Europe usually 150 mm thick, containing bacteria which release plant nutrients. It is unsuitable as a building *foundation* and is stripped at the start of *groundworks*. It may be stockpiled in low heaps that allow air and water movement, to prevent it turning sour, so that it can be used later for *landscaping*; otherwise it has to be imported to the site.

torbar A *twisted deformed bar* (C).

torch A gas burner used for heating materials, usually working on *bottled gas* or *oxy-acetylene*.

torching, rendering, tiering Mortar pointing on the underside of *slates*, over the *heads*, to prevent the *tails* lifting (shouldering or half torching). Full torching includes, in addition, the plastering of the whole underside of the slate seen between battens. It was common practice on roofs without boarding or an *underlay*, in spite of the fact that moisture is sucked into the battens by *capillary* movement through the mortar. The battens therefore decayed quickly and the practice is now condemned.

torching-on, torch and roll, flame bonding A method of laying rolls of *torch-on felts* for *built-up roofing* by heating the roll with the flame from a propane torch as it is unrolled over the previous layer of felt. The fresh, hot adhesive then bonds the two felts.

torch-on felt A *bitumen felt* coated with a bitumen adhesive which is hard until softened and made sticky by heating during *torching-on*.

torn grain *Chipped grain*.

total going The horizontal length of a *flight* of stairs, the sum of the goings of each *tread*. For curved stairs it is measured along the *walking line*.

total float The time by which a programme *activity* can be delayed without affecting the *completion date*. For a *critical activity* total float is zero.

total rise The vertical *rise* of a stair *flight* (or the complete stair), equal to the difference in level between *landings* (or floors).

touch dry A stage in *drying* of paintwork at which the surface is not sticky and very slight pressure with the fingers leaves no mark.

touchless controls Electronic *automatic controls* worked by *sensors* that react to the presence of people. They are used to operate basin taps or hand driers, to flush urinals, open doors, etc., particularly those used by the general public or disabled people, and in hospitals. They have largely superseded foot- or elbow-operated controls and floor pressure mats.

touch up To re-paint small areas of paintwork in order to make good any damage from handling during delivery and fixing in place. Touching-up is difficult to conceal, which is why *structural steelwork* is usually only primed by the fabricator and painted after erection. Small scratches in finishings like *organic coatings* are often best left alone, rather than creating a blemish with a paint that fades to a different colour. Window-unit manufacturers often supply a suitable touch-up kit, at delivery. Paint from spray cans is best collected in the plastic lid and put on with a fine brush or even a sharpened match.

toughened glass Flat glass with far greater impact strength than *annealed glass*, from which it is made by sudden cooling of the surface layer. All holes are made and shaping done before toughening. It can withstand both very cold and very hot conditions ($300°C$) and sudden changes in temperature (of $250°C$ difference). If shattered it breaks into small pieces. Thus it can be suitable for *safety*, *security*, and *fire-resisting glass*.

towel rail Bathroom towel rails are usually heated from the boiler *primary flow and return* to the hot-water storage *cylinder* and thus stay warm even when the central-heating supply to *radiators* is turned off.

tower bolt A massive steel *barrel bolt*, often used as a *foot bolt*.

tower crane A *crane* on top of a steel frame, usually self-*climbing*. It can be on a concrete base, on legs, or travel on rails. In Europe large tower cranes usually have a *saddle jib*; the *luffing jib* (*C*) is little used, although common in the USA and Australia. Small tower cranes for house building are widely used on the Continent. They usually have the counterweight at the base and a cantilever jib. See diagram, p. 472.

tower pincers

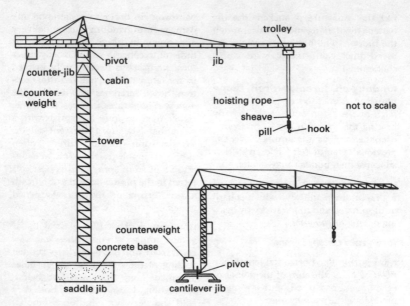

Tower cranes.

tower pincers Steelfixer's *nippers*.

tower scaffold A *prefabricated scaffold* of aluminium or steel tubes. It should be erected on a firm, level surface, and be no taller than three times its base width (including *outriggers*). Often it has lockable castors so that it can be moved by pushing on the base, after checking for overhead power cables. It can be overturned by the wind, by climbing up the outside, or by a ladder put on top to gain extra height.

town gas *Manufactured gas*, now superseded in most of Europe by *natural gas*.

town planning The coordination by town planners, who are usually architects or municipal engineers, of the interests of the community, represented by the views of economists, doctors, sociologists, and so on. In the USA town planning refers to a short-term process – city planning is the long-term process of continuously developing the master plan.

trace heating, cable h. The heating of pipes by electrical cables wound round them in order to keep heavy-fuel oil pipes warm enough for easy flow from an outdoor storage tank to a boiler room.

tracer, warning tape, marker tape Brightly coloured plastics strips (or in France, netting) laid in the ground in the *backfill* over buried cables, pipes, *ducts*, etc. to warn anybody digging that care is needed to avoid damage. See also **breaker slab**.

TRADA The *Timber Research and Development Association*.

trade One of many occupations in which *craft* skills are used and paid for, such as the traditional *building trades*.

The term also means the work done by a *tradesman* or other specialist as well as the entire building industry or a section of it, e.g. wet trades, dry trades, finishings. See **sequence of trades, work section**.

trade collection In *priced bills* a list of the page collections given at the end of each *trade*. Each trade collection is carried to the *summary*.

trade foreman The person who leads the work on site in each *trade*, often with *charge hands* to lead individual *gangs*. Trade foremen may be employees of a *sub-contractor* and attend *coordination meetings* as the sub-contractor's representatives. A large project can have a *formwork foreman*, a foreman *steelfixer*, a *foreman bricklayer*, etc.

tradeless activity A minor job that could be done equally well by one of several different *trades*, such as fixing insulation or a support bracket.

trade-off A relaxation of one part of a regulation given if the requirements are achieved by other means.

trade rubbish Waste materials, droppings, and offcuts which should be cleaned up by each *trade*. If not it is done by the *main contractor*, who demands payment for *attendance*.

tradesman, craftsman A skilled building worker who can do the work in a particular *trade*, usually with little or no supervision. On building sites tradesmen may work in a *gang* that includes *labourers*. Examples are: *bricklayer, carpenter, electrician, glazier, joiner, painter, plasterer, plumber, tiler*, etc. Before 1815, when Thomas Cubitt in London first employed tradesmen permanently, they often travelled long distances in search of work except where they were lucky enough to find employment on a semi-permanent building site such as a cathedral, castle, or abbey.

trade union An organization of people who work in the same trade, such as *carpenters, bricklayers*, or *plumbers* (craft union), or of people who work in the same industry (industrial union). Building trade unions in the UK belong to the Federation of Building Trades Operatives.

trafficable roof A *flat roof* built as a terrace or deck, to resist wear and indenting, usually with bitumen felt and *heavy protection*.

transferred responsibility A duty or obligation shifted from one person to another, which may result from *approval* or *acceptance*. See **murder clause**.

transformation piece, transition An *adaptor* between *ducts* of different shapes, e.g. from rectangular to circular.

translucent glass *Obscured glass*.

transmission Sending, transferring. So far as energy is concerned, transmission implies movement and includes heat (thermal transmission), light, or sound and noise (acoustic transmission), usually with some reduction (*attenuation*). See **insulation, sound insulation**.

transmittance (1) Thermal transmittance, the *U-value*.
(2) A correction factor for light lost in passing through *glazing*, given as a decimal fraction or a percentage.

transom A small horizontal beam, particularly the stone or timber bar separating the *sashes* of a window or separating a door from a *fanlight* over it. Compare **mullion**. See diagram, p. 474.

transom window (USA) A window over a door. It may be a *fanlight*.

trap (1) **air trap** A *pipe fitting* with a

trapped waste

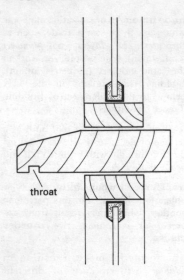

A transom (horizontal).

Traps are usually *tubular*, with a U-bend, made to suit WCs, basins, sinks, or baths. Small traps for plastics drains are often made to be unscrewed by hand for clearing blockages. See **intercepting trap, unsealing of traps**. See also diagram.

(2) A hole in a ceiling to give access to a *loft*, usually with a trapdoor or cover panel. It should be large enough for a *cistern* to pass through.

(3) A *scaffold board* which overhangs its support too far and is therefore dangerous to tread on. A safe overhang is about 100 mm.

trapped waste, w. trap A *waste* from a sanitary fitting with a built-in trap.

trapping bend A *dip pipe*.

trap vent A *branch vent*.

travel (1) The total distance by which something, such as a sliding door or lift, can move along a track.

(2) The length of the *escape route* to an *escape stair*.

dip which stays filled with water and prevents smelly, unhealthy foul air in the *drains* or sewers from escaping.

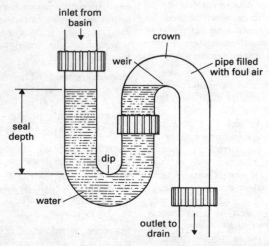

A PVC S-trap. This two-piece type is easily dismantled and re-assembled for cleaning by hand and so needs no cleaning eye. It is not prone to blockages.

traveller Equipment that can be moved along a track. See also *C*.

travertine, t. marble A hard, yellowish decorative stone, often porous.

tray, pan (USA **receptor**) A waterproof area of floor with raised edges to catch water, drained and trapped like a basin. Trays can be of plastics, sheet metal, stoneware, etc., and under showers, cisterns, cold coils, windows, etc. See **lead safe**.

tray decking *Profiled sheeting*.

tread The level part of a *stair*, or its *going*. It can be a straight *flier*, a *winder*, *tapered*, or a *balanced step*. See diagram, p. 437.

tread plate (1) **chequerplate** Steel or aluminium sheeting with rows of short raised strips, used for industrial stairs, *duct covers*, etc.

(2) A sloping area of impervious flooring in front of a *slab urinal*.

treenail, trenail A large hardwood peg driven into a hole bored through a *joint* to secure members in heavy *carpentry*. Treenails have been used for centuries and can be seen in early *laminated* beams. In a *mortice-and-tenon* joint the members may have been previously tightened with a steel *drawbore* pin before the wooden treenail is inserted. A *dowel* is similar but smaller.

trees Buildings with *shallow foundations* (*C*) on clay or silt should not be near live trees, because of the risk of *cracking in brick walls*. Some recommended minimum distances are: beech 9 m, cypress 3.5 m, elm 12 m, poplar 15 m, sycamore 9 m, and willow 11 m. A *local authority* can order a Tree Preservation Order (TPO). Tree roots are easily damaged by excavators or suffocated by *fill*. See BS 5837.

trench (1) A groove across a member, a *housing*.

(2) A long, narrow *excavation* for buried *services* or for *foundations*.

trench duct A *ground duct*.

trench-fill foundation A *strip footing* cast in the ground by filling a trench to near ground level with concrete immediately after it is dug by machine, e.g. by a *backhoe*. Since widths are usually narrow, down to 500 mm, accurate setting-out is required. The trench is shallow, well within the reach of the machine. Compared with a traditional footing and brick foundation walls, savings come from substituting concrete for brickwork, the speed of work, the absence of any need for *earthwork support* and *working space*, and from less *backfill* being needed.

trim (1) Small pieces of *finishing* materials put on to a general area, often at the edge. They include *architraves, barge boards, cappings, ceiling roses, escutcheons, fascias, joint covers*, etc. See **casing, metal trim**.

(2) To cut to size, or to fix trim, or to build *trimming* into a framework.

trimmed door A *door leaf* that is cut off at the bottom (trimming), usually after it has been fitted (hung), to clear a carpet or allow air transfer.

trimmed joist A floor *joist* which has been cut short (trimmed) at a hole (e.g. for a stair) and is carried by a *trimmer joist*.

trimmer (1) A short timber *joist* beside an opening. There is a difference between American and British senses. See **trimmer joist, trimming joist**.

(2) **t. bar** Steel *reinforcement* parallel to and round an opening.

trimmer beam A reinforced concrete beam along the edges of a large hole in a floor slab, usually a *downstand*.

trimmer joist (Scotland **bridle**) A short floor *joist* which encloses one side of a rectangular hole in a wooden floor, carrying the full-length joists which are cut off for the hole. Another trimmer joist may form the side opposite and parallel to it. These trimmer joists carry *trimming joists*. In the USA a trimmer joist is also called a 'header'.

trimming (1) Framing round or otherwise strengthening an opening through a floor, roof, or wall, whether of timber or other material. See **trimmer**. See also diagram, p. 65.

(2) **undercutting** To cut down a door leaf for a *trimmed door*. It can be done with an electric trimming saw, without unhanging the door.

(3) *Cutting off* in-situ piles, etc.

trimming joist A floor *joist* parallel to the other joists, usually the same depth, but possibly thicker. It is fixed with another, in pairs, one each side of a hole in a timber floor, and carries *trimmer joists*, which themselves carry the *trimmed joists*.

trimming machine, mitring m. A lever-operated or pedal-operated bench *guillotine* with a heavy sharp blade for cutting the ends of *mouldings* or timbers at any desired angle. For a *mitre* this is usually 45°.

Trinidad bitumen *Lake asphalt* from Trinidad.

trolley, crab A wheeled frame that runs along the jib of a *saddle-jib crane*. It has two *pulleys*, between which the hoisting ropes drop down to the hook. Other ropes are used for *trolleying*. See diagram, p. 472.

trolleying The inward or outward movement of a tower crane *trolley*, changing the hook radius. Compare **luffing**.

Trombe wall A thick external wall on the sunny side of a building, built of brick or masonry, with *glazing* on the outside. It acts as a passive *solar collector* and is controlled by opening or closing air holes in the top and bottom.

troughed sheeting *Profiled sheeting*.

trough gutter A *box gutter*.

trough valley tile A large concrete *plain tile*, purpose-made as a *valley tile*. It is not nailed down, but held by the nailed tiles each side.

trowel (1) A steel-bladed hand tool for *gauging*, mixing, laying on, or finishing plaster or concrete floors, for spreading adhesive (plain or notched) or mortar (in bricklaying). See **float, power float**.

(2) A *scraper*.

trowelling The final smoothing of plasterwork or a concrete *screed* by working with the edge of a steel trowel, done after *floating*. A *power float* is used for large horizontal areas. Trowelling a second time after waiting improves the hardness and impermeability of a trowelled concrete floor. But cement mortar *rendering* or *stucco* must be trowelled with care, as over-trowelling too early brings cement and *fines* to the surface and often leads to crazing.

truncated truss A *trussed rafter* with one end or its top cut off, so that it is shorter or lower than a standard truss.

trunk bar An electronic device which allows local calls from a *telephone* but can prevent outgoing trunk or international calls from being connected.

trunking (USA **raceway**) A cable *duct* with one removable side, used for distribution *wiring* or *bus bars* (it is sometimes called a bus duct or busway). Multi-compartment trunking for power, lighting, telephone, and data cables allows easy *wire management*. Power *riser* trunking is often on *cable tray*, going to *skirting trunking*. Cables are 'laid in' to trunking from the side, not *drawn in* as for *ducts* or *conduit*.

trunk lift A *freight elevator*.

truss (1) A roof frame, generally nowadays of steel (but sometimes of timber, concrete, or aluminium). Trusses are made up of members joined together to form triangles. Some members work in tension, some in compression. Steel trusses generally weigh 10 to 15 kg/m² of floor area for spans of 12 m or less. For every 3 m increase in span, another 2 to 5 kg/m² should be added. Bracing against wind adds a further 2 to 5 kg/m². Trusses are usually placed about 3 m apart and carry *purlins* to support the roof covering. See **girder, portal frame**.

(2) A *trussed rafter*.

truss clip A *fastener* to hold a *trussed rafter* to the *wall plate*.

trussed rafter, domestic roof truss A prefabricated truss made of light timbers joined with *nailplates* and used in most new domestic roofs in industrialized countries. It has mostly superseded the cutting of *rafters* and *ceiling joists* on site and is generally supported only on outer walls. Being close-spaced and joined by roofing *battens*, trussed rafters share loads. Spans vary from 5 m to 11 m, slopes from 15° to 35°. Hip ends are made with a girder truss that carries the hip and jack trusses, and with smaller but similar truncated trusses. Roof corners may have saddle trusses on top of the standard trusses. The complete roof structure is delivered in bundles at *wall-plate* level, lifted by a small crane on the delivery truck. After the trusses are stood up in place, *truss clips* are nailed on, as well as roofing battens and bracing. Trussed

rafters must be fully braced to give a rigid and stable roof. Omitted, wrongly positioned, or badly fixed bracing can allow distortion or collapse. Diagonal braces are fixed first, at about 45° from eaves to ridge and between wall plates. Horizontal *binders* (running ties) then go lengthways, near the ridge, under the rafters, and over the ceiling, fixed near the nailplate joints. Large trusses (of span exceeding 8 m) need chevron bracing on the truss diagonal members. See BRE Good Building Guide No. 7 and BS 5268. See **ridge terminal**. See also diagram, p. 478.

truss plate A *nailplate*.

try plane A long *bench plane*, about 560 mm, used for trueing timber.

try square A *square*.

tubular A *pipe fitting* made of tube, for example a *connector*, a *ferrule*, or a *bend* for steel *screwed pipe*.

tubular mortice lock A *bore lock*.

tubular scaffolding A *scaffold* built of 50 mm outside dia. steel tubes in plain lengths, assembled on site to suit each job, using scaffold fittings such as *couplings, clips, putlog adaptors*, etc. Ladder-shaped steel-tube beams are used for spans. Tubular scaffolding has greater flexibility than *prefabricated scaffolding* but more skill and care is needed in erection or in making changes as work proceeds. For the licensing of scaffolds, inspections, and records, see **scaffold**.

tubular trap A *trap* with a full-bore waterway, usually in the shape of a U-bend, commonly a *P-trap* or an *S-trap*. It does not get blocked easily.

tuck A recess in a *face joint* made in *tuck pointing*.

tuck-in, turn-in The narrow edge of a cover *flashing* or *bitumen-felt* roofing

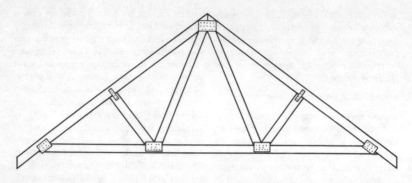

A trussed rafter.

which is bent over horizontally and wedged into a *chase* or *raglet*.

tuck pointing Projecting decorative *pointing* of brickwork, made by first cutting a *tuck* in the *face joints*, then filling it with *lime putty* (using a notched trowel) to stand out about 3 mm. It makes a crisp white line against the original mortar, which is usually coloured to match the bricks.

tumbler The part of a *cylinder lock* which prevents the rotating plug from being turned until the right key is inserted.

tumbler switch, toggle s. A simple hand-operated switch with rapid spring action to make or break a circuit without arcing and to lock in either position.

tumbling in, t. courses Sloping *courses* of brickwork on top of a *buttress* (*C*) or as a *capping* to a gable wall. The ends of the horizontal courses are cut to fit.

tundish A tray with a central drain hole, sometimes located (among other places) in a *flue* below a *soot door* to collect *flue condensation*.

tung oil A water-resistant *drying oil*

from the seeds of tropical trees (Aleurites) which grow also in China and Japan, used in hard, quick-drying floor paints and varnishes, often combined with *phenolic resins*.

tungsten-filament lamp An *incandescent lamp*.

tungsten-halogen lamp A *halogen lamp*.

tungsten-tipped drill, hard-metal t.d., masonry d. A *drill bit* with a small block of tungsten carbide brazed to its tip, used for drilling masonry and concrete, often in a *hammer drill*. See also **carbide tips**.

tunnel form Concrete *formwork* for both the walls and the floor slab of rows of flats, motel rooms, etc. Each form has two wall shutters, one each side, joined by a deck. It is *collapsible* by folding the deck and is shifted by rolling out on wheels and lifting by crane.

tupper A *bricklayer's labourer* in the north of England.

turn cock The water service's special *loose key*, used to operate a *stopcock* under a *surface box*; or the water service employee who turns the cock.

turndown A *downstand*, or a *hook intake*.

turn-in *Tuck-in*.

turning Making an object on a *wood-turning* lathe.

turnkey project A *design-build contract*.

turnup An *upstand*; also a *skirting* or a *flashing*.

turpentine A solvent distilled from pine-tree resin, formerly used to thin paints. It has been superseded by *white spirit*.

turret A *bulkhead*.

turret step A triangular stone step from which a *spiral stair* is built up. The central *solid newel* consists of the rounded ends of the turret steps laid on top of each other.

tusk nailing *Skew nailing*.

T V distribution (USA **master antenna system**) The TV antenna, amplifier, and wiring to TV outlets in a block of flats, hotel, etc.

twin and earth A light-duty power or lighting cable with two insulated conductors plus an *earth* wire inside a common sheath, used for *sheathed wiring*.

twin cable An electrical cable with two insulated conductors.

twin-shaft paddle mixer (USA **twin pug**) A *pug mill* with two horizontal shafts turning in opposite directions.

twin tenon, divided t. A *tenon* from which the central part has been cut away.

twin-thread screw A *chipboard screw*.

twin-tube fluorescent A *luminaire* with two fluorescent tube lamps.

twist The helical *warp* of timber, caused by drying shrinkage of spiral grain.

twist drill A hardened-steel *drill bit* with a pair of steep spiral grooves giving helical cutting edges, used for drilling metal or wood.

twisted fibres *Interlocked grain*.

twitcher An *angle trowel*.

two-bolt lock A common combination of a door lock turned by a key, with a *latch* operated by a knob or lever handle.

two-brick wall A *solid wall*, which if built today would be 440 mm thick (two bricks of 215 mm long plus a 10 mm joint).

two-coat work, render and set *Plastering* with an *undercoat* and *finishing coat*, used on *backgrounds* that are reasonably level and have even *suction*, e.g. *blockwork* or concrete walls.

two-part products, two-pack p. Substances that need mixing before use, usually high-performance adhesives, sealants, or paints, supplied in premeasured kits. They are fast-*curing* and can be used in a great thickness or out of contact with the air, but have a short *working life*. Unlike *one-part products* they can be used for *high-build fillers* or *high-build coatings*. In general it is unwise to breathe in their fumes or get them on the skin. Examples are *epoxy adhesive*, *polysulphide sealants*, *polyurethane*, and metalwork *etch primers*.

two-piece cleat, pipe ring A pipe bracket with a removable outer part. See diagram, p. 480.

two-pipe system (1) Drainage *internal pipework* in which *soil water* and *waste water* are carried by two separate stacks, with or without *vent pipes*. The

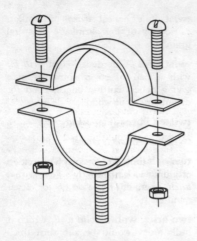

A two-piece cleat.

waste pipe discharges through a trapped *gully*, then the two flows combine.

(2) A *central-heating* circuit with a *flow* pipe as well as a *return* connected to each radiator. The radiators are thus connected 'in parallel' and the furthest radiator is therefore only slightly cooler than the one nearest the boiler, unlike in the *one-pipe system*. A two-pipe system is usually one part of a *four-pipe system*. See diagram, p. 75.

two-stage alarm system A *fire alarm* which gives an alarm in one zone and an alert in the rest of the building.

two-stage tendering Contract negotiations in which a first *tender* is called using *approximate quantities*, so that the tenderer can advise on design before the price is finalized.

two-way switch One of two *light switches* at each end of a corridor or stair, wired for the lighting to be switched on or off from either end.

tying wire *Binding wire.*

typical floor A floor in a tall building with the same *floor plan* as the other floors.

Tyrolean finish, alpine f. A textured *external rendering* for walls, usually flicked on by a hand-operated *roughcast machine*. The *render* is uncompacted and porous, allowing water *vapour* to escape outwards, its rough texture tends to throw off rain and it can absorb differences in *movement* with the wall.

Tyrolean machine A *roughcast machine*.

U

U-bend A *tubular trap*.

U-bolt A *fixing* made of a U-shaped bar with each end threaded for a nut.

UCATT Union of Construction, Allied Trades and Technicians. See **ABT**.

U-duct A *ducted flue* for gas appliances in a tall block of flats, with its air intake and burnt-gas outlet both at roof level. A down flue and an up flue run to the bottom of the building and join. *Room-sealed appliances* are connected to the up flue. The effects of wind gusts on the flues are cancelled, in the same way as by a *balanced flue*. See diagram.

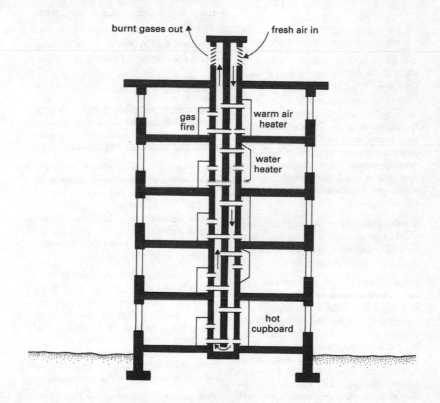

burnt gases out ↟   fresh air in

gas fire

warm air heater

water heater

hot cupboard

A U-duct.

U-gauge

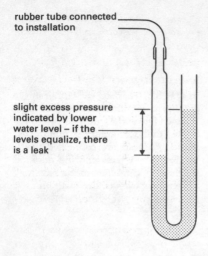

rubber tube connected to installation

slight excess pressure indicated by lower water level – if the levels equalize, there is a leak

A U-gauge used in testing for leakage in a gas or drain installation, for indicating the gas pressure, etc.

U-gauge A glass U-tube half filled with water, with one end open and the other connected by rubber tube to a drain pipe being given an *air test*. It is sensitive to small pressure differences, measured in millimetres of *water gauge*. See diagram.

ultra-lightweight aggregate *Lightweight aggregate* such as *expanded perlite* and *exfoliated vermiculite*, mainly used for thermal *insulation*.

ultraviolet (UV) Radiation from the sun and other sources of shorter wavelength than light, 290–400 nanometres (nm). It can kill living cells and damage plastics or *bitumen*. The shortest wavelengths, below 360 nm, cause yellowing and embrittlement, longer wavelengths cause fading. *Polyvinyl chloride* has poor UV resistance, *polystyrene* is particularly sensitive to 318 nm, and *polypropylene* to 370 nm.

UV is stopped by clear glass, carbon black fillers in plastics, *solar protection* coatings, and anti-UV treatments over clear plastics and in *sealers*. See **durability**.

umber The *pigment* raw umber is a natural yellow-brown hydrated iron oxide with some oxide of manganese. When calcined (strongly heated) it turns to a deep reddish-brown, burnt umber. Both are *earth colours*.

unbonded screed A *screed* that is separated from the *structural* slab, either on a *damp-proof membrane* or *floating*. See diagram, p. 392.

unburnt brick Sun-dried *adobe* bricks.

uncased fan A *fan* without a casing round the spinning blades.

uncased steelwork Structural steelwork without *encasement*.

uncoursed rubble (USA **broken-range ashlar**) *Random rubble* stone walling. *Snecked rubble* and *squared rubble* can be coursed or uncoursed.

uncovering The removal of completed work to see if work underneath was correctly done before *covering-up*. If this reveals concealed defects, the cost of uncovering, plus *remedial work* on the defects, is usually paid by the contractor. An 'order to uncover' should be given in writing.

undercloak (1) A course of *fibre cement board* or plain tiles laid along the top of a wall at a roof *verge*, supporting the sloping edge of the roof and giving the tiles a slight inward tilt so that rainwater does not overflow.
(2) In *supported sheetmetal roofing* the lower sheet of metal at a *roll*, *drip*, or *seam*. It is covered by the *overcloak*.

undercoat (1) Paint applied to a surface after *priming* and before the finish coat. It is matt, with a high content of *pigment* and *extender* and a *colour* approaching or helping that of the *final coat*. Undercoat increases the thickness and opacity of the *paint system* and therefore its protectiveness. It is smoothed by *flatting down* before later coats. If two undercoats are used, one is usually *tinted* to reveal *skips*. Varnish undercoats may have a lower *oil length* than finish varnish.

(2) A coat of *bonding plaster*, *lathing plaster*, or *browning*, in *two-coat work* or *three-coat work*, smoothed by *floating* and followed by the *final coat*.

undercuring The insufficient hardening of an *adhesive* due to low temperature or too little time for *curing*.

undercutting *Trimming* a door leaf.

under-eaves course, The bottom row of tiles in a *double eaves course*. They are special *eaves tiles* and covered by a course of normal tiles.

underfelt Carpet *underlay*.

underfired brick A *clay brick* that has not been *hard-burnt*. It has low strength and poor resistance to wear or frost.

underfloor heating, floor h., heated f. *Concealed heating* which has been used since Roman times; their 'hypocaust' had a stone floor on small pillars and hollow walls, warmed by smoke from fires below. Today underfloor heating has *copper pipes* containing hot water or *mineral-insulated cables* using off-peak power. As in any *coil heating*, the pipes or cables can be laid in a *screed* during construction at low cost, even if not intended for immediate heating. Heat *insulation* is usual beside or under heated *ground slabs*. Care is needed in selection of the *floor finishes*: slate and ceramic tiles resist heat well, others like carpet or lino are not suitable for high temperatures, and floor

surface temperatures need to be kept low to avoid discomfort. See diagram, p. 392.

underflooring A timber *sub-floor*.

underfloor space The void between a ground level *suspended floor* and the ground underneath it, often high enough to be a *crawl space*. The traditional timber floor is carried on *honeycomb-bond* walls and has *air bricks* in the external wall for ventilation. Air from an underfloor space should be kept out of living areas.

underground services *Buried services*.

underground stopvalve A water service valve under a *surface box*.

underlay (1) (USA **underlayment**) Any layer of sheet material under another material. Roofing underlays for *supported sheetmetal* are usually of *inodorous felt*; tiles and slates are underlain with plastic sheet, except on low roof pitches where *bitumen felt* (sheathing felt, reinforced felt, sarking) is used; and *thatch* can be over fire-resisting boards. For tiles or slates, underlays should be well lapped over the *ridge* and *hips*, draped over the *fascia*, not stopped short at *verges*, and doubled 600 mm wide under *valleys*, or 1000 mm wide for *trough valley tiles*, to BS 5534. See **isolating membrane, separating layer**.

(2) Hardboard or plywood placed over a rough floor to make a smooth surface suitable for laying lino, cork tiles, parquet flooring, etc.

(3) **underfelt** A felt of jute and hair, or foam plastic, put down before *fitted carpet*. It should weigh at least 187 g/m² and not act as a *vapour barrier* or cause dampness. Many carpets have an integral underlay, or *backing*.

(4) A *damp-proof membrane*, or a *radon* barrier under a *ground slab*.

underpinning Extra concrete cast underneath old foundations, to transfer the loads to a deeper level. For city buildings it can be done from a neighbouring site, along with *groundworks* for a new, deeper building, or increasingly to existing houses with severe *cracking*. See **alternate-lengths work** and BRE Digest 352.

under-ridge tile A *top-course tile*.

under-tile (1) A short plain tile laid in the first course at the *eaves* or the last course at the *ridge*, either an *eaves tile* or a *top-course tile*.
(2) **tegula** The lower tile of *Spanish* or *Italian tiling*.

undertone (1) The *colour* obtained when a coloured *pigment* is reduced with a large proportion of white pigment.
(2) The colour seen when a coloured pigment is spread on glass and viewed with light passing through it.

undervibration Insufficient *vibration* of fresh concrete, giving poor compaction, lower strength, and surface *blowholes*.

undressed timber, unwrought t. Sawn timber not planed or sanded.

uneven gain, u. texture The texture of timber with a considerable contrast between *springwood* and *summerwood*.

unfixed materials Building materials on site but not fixed represent a large outlay for a *contractor*, especially at an early stage of the job. The cost of this outlay is at least partly paid in the valuation for each *interim certificate*.

unframed door A *matchboarded door*.

ungauged lime plaster *Plaster* made with *lime*, sand, and water only, or sometimes with neat lime. Only those limes which give a compressive strength exceeding 689 kN/m² at twenty-eight days are recommended. To give early strength, Portland *cement* or *gypsum plaster* (but not both) should be *gauged* with it.

unhanging Removing a door from its frame by unscrewing the hinges or, if it has *lift-off hinges*, by raising it.

Uniform Building Code (UBC) (USA, Australia) See **Building Regulations**.

uninterruptible power supply (UPS) A rechargeable battery and *invertor* with a change-over switch, to supply *priority* power for a short time.

union (1) A pipe joint originally used for threaded steam pipe but adapted to many other types of tube, such as *screwed pipe*. The pipe ends have a *taper thread* on the outside. They are screwed into each end of a *fitting* (e.g. socket, bend, etc.) which has a matching tapered thread on the inside.
(2) **plumber's u.** A *fitting* that is used to join pipes which cannot be turned or as an *adaptor* between different types of pipe. It has at least one flanged joint or one *compression joint*, and it may also have *taper threads*.
(3) A *trade union*.

union bend A *bend* with a plumber's *union* on one end.

unit air conditioner, packaged u. Usually a steel case (the unit) containing a miniature *air-conditioning* set, with its own electrically driven fan, a compressor, an evaporator, and a condenser. The unit supplies cool filtered air to a room. Small *room air conditioners* or *window units* usually fit into an open window. Larger units, often made for *rooftop* use, may be 'split', with the evaporator (the cooler), the air fan, and filter in one box, and the hot and noisy parts at a distance, with pipes between them. Often used in hot countries, these units are quickly installed in any

universal beam

universal column

Universal sections.

room that has an electrical supply and an outside wall or roof. A small pipe for condensate should be run to a place where water can drip away without inconvenience.

unit heater, air h. A large cabinet for industrial *warm-air heating*. It is used in workshops, warehouses, and factories.

unit of bond The smallest length of a *course* of brickwork which repeats itself. In *Flemish bond* it is one and a half bricks long; in *English bond*, one brick long.

unit price, rate The price per metre, square metre, cubic metre, kilogram, etc., for building *materials* or work such as *excavation*, concreting, bricklaying, plastering, etc., or the hourly cost of labour. It is used in *estimating* or shown in a *priced bill*. An estimator may use unit prices from site *costing* or from a book such as a *price index*.

unit system A *curtain wall* made of large glazed units which are fully factory-assembled, lifted into place by crane, and bolted to the building structure.

universal beam (UB) section A *universal section* used to carry load as a beam; also as a column with off-centre loads, e.g. in a *portal frame*.

universal column (UC) section A *universal section* of 'square' proportions (similar width and depth) used as an axially loaded column, short beam, etc.

universal section, stock s. Steel beams, columns, angles, channels, etc., made in a series of standard sizes and weights. See entries above and diagram.

unplasticized polyvinyl chloride (PVC-U, formerly **uPVC), rigid PVC** A relatively stiff type of *polyvinyl chloride*, sometimes brittle in cold weather, used for drainage and water service pipes. It is often a stark white colour.

unsealing of traps, breaking of seal Loss of the water *seal* in a *trap*, allowing unwanted escape of foul air from drain pipes, or entry of air into them. It causes gurgling sounds and results from the momentum of water running down the pipes, causing low or high air pressure, either from *self-siphonage* or from *induced siphonage*. It should not happen in a correctly installed system of *sanitary pipework* with adequate *ventilation*. Unsealing from evaporation of the water can happen in a heated house while the occupants are on holiday, or in rarely used floor *gullies*. Leaving the plug in a sink or bath may help, or a couple of drops of cooking oil. Leakage of water from the trap is another cause.

unsocketed pipe, socketless p. (USA **hubless joint p.**) A *drain pipe* with plain ends joined by *sleeve couplings*. The pipes can be of cast iron, vitrified clay, etc.

unsound Defective. In *cement, gypsum plaster*, and *lime* defects are usually due to particles which expand (blowing).

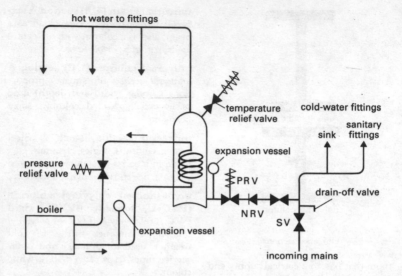

hot water to fittings

temperature
relief valve

cold-water fittings

sanitary
sink   fittings

expansion vessel

pressure
relief valve

PRV

drain-off valve

boiler

NRV

SV

expansion vessel

incoming mains

An unvented system.

unsound knot, rotten k. A *knot* which is softer in any part than the wood round it, one of the *gross features* used in grading timber.

untrimmed floor A floor carried on full-length *intermediate joists* only.

unvented system, mains pressure s. A *domestic hot-water* cylinder which takes its *feed* direct from the water mains, without the *cistern* of a traditional open-vented system. It has a non-return valve and pressure-reducing valve on the cold water feed, and a temperature-relief valve on the cylinder, sometimes with an *expansion vessel*. See BRE Digest 308 and the Model Water Byelaws. See diagram.

unwrought timber Unplaned sawn timber.

up-and-over door, overhead d.,

swing-up d. A large door that is moved upwards to open, counterbalanced by springs or weights, mostly used for lock-up *garages*. Different types swing, slide, or fold, and retract fully inside the doorway or project to form a canopy when raised.

updating, rescheduling The recalculation of a *programme*, e.g. after major *slippage* has changed the *critical path*. Strictly speaking, it is more extensive than *revision*.

UPEC classification A French system for testing and classifying floor finishings according to their resistance to wear (Usure), indenting (Poinçonnement), water (Eau), and chemicals (Chimique). They are matched to floor service conditions by a listing in the classification document.

uplift restraint strap A steel strap to

hold down roof framing against the wind.

uplighter A *luminaire* pointed at the ceiling to give *indirect lighting*, usually a *spotlight* that is easily shifted or turned, unlike *cove lighting*.

upset A tear across the fibres of wood caused by a shock, often during felling. See also **upsetting** (*C*).

upside-down roof An *inverted roof*.

upstand, upturn A vertical strip or *skirting*, such as the weatherproofing where roofing meets an *abutment* wall. It can be the *roofing* itself, particularly for mastic *asphalt*, the top part of a one-piece *apron flashing*, a mortar *skirting*, or an *upstand flashing*, and either stepped, raking, or level.

upstand beam A beam on top of a floor, instead of the more usual *downstand beam*, often a concrete kerb or wall on the edge of a concrete slab.

upstand flashing An *upstand* that forms one half of the weatherproofing at an *abutment*, the underneath layer of sheet lead, aluminium, or other *plumber's metalwork*, which is turned up the wall but not built into it. It is covered by the *counter-flashing*, which is folded down and dressed to follow the roofing.

uptime A period during which an earth-moving machine is available. The opposite is *downtime*.

uPVC Former designation of *unplasticized polyvinyl chloride*.

upward and downward collective control *Lift controls* with two landing buttons (up, down). Calls from the lift car and those from landings are answered in order on the way up or down.

upward-spray bidet A convenient *bidet*, which in the UK must be sup-plied from a *cistern*, in France through a non-return valve.

urea-formaldehyde (uf) A synthetic *resin*, one of the most common assembly *adhesives*, also used for making wood-based *building boards*. It is colourless, needs a separate hardener, and has limited *shelf life* and poorish water resistance.

urea-formaldehyde foam An *expanded plastics* used for *cavity insulation*. It can give off small quantities of toxic *formaldehyde* gas and so is allowed in brick or block *cavity walls* only with a carefully sealed inner *leaf*.

urethane A loose term for *polyurethane foam*.

urinal A non-domestic *soil fitment*, usually a *bowl urinal* or a *slab urinal*. Sparge pipes, sprayers, and outlets are mostly made of polished stainless steel.

usable life *Working life*.

use factor The *diversity factor*.

U-tie (USA) A heavy wire *wall tie* with a central kink that forms a *drip*.

utile, sipo A durable tropical hardwood with medium *moisture movement*.

utilities (USA) Public *services*.

utilization factor (U) In *artificial lighting* the share of the light given off by a *luminaire* that reaches the place where it is wanted, called the 'reference plane'. It is improved by the use of *reflectors* and prismatic diffusers. (French 'utilance'.)

U-value, air-to-air heat-transmission coefficient, thermal transmittance The amount of heat per unit area that passes through the roof, external walls, and ground floor of a building, which depends on the materials used and the site situation. A low

U-value

U-value, in watts per square metre per degree centigrade (W/m².°C), implies low heat loss. A wall or roof with good *insulation* in a sheltered site has a lower U-value than thin cladding or roofing on an exposed site, and dry materials have a lower U-value than wet ones. U-values are measured by experiment, and are equal to the *C-value* plus the *surface coefficient*. Maximum U-values required by law under *Approved Document* L of the UK *Building Regulations* for houses are 0.45 for outer walls, 0.25 for roofs, 0.45 for floors, and 0.6 for semi-exposed walls, roofs, and floors (e.g. inside a garage). For other types of building similar figures apply, except 0.45 (instead of 0.25) is allowed for roofs. A *trade-off* is allowed between a high U-value window and an insulated low U-value wall. These figures are intended to save energy and reduce heating cost. The U-value of an existing house can usually be improved by *cavity insulation, ceiling insulation,* or *double glazing.* Standard U-values are given in BRE Digest 108.

V

---

vacuum, medical v. A *medical gas* service, piped into a hospital.

vacuum cleaning plant, central v.c. A permanent installation for the vacuum cleaning of a building. Wall outlets for flexible hoses connect to smooth-bore PVC or steel pipes, an exhaust *fan*, and a *filter*, in a basement or garage. Fine dust is discharged outdoors, reducing indoor air pollution. The pipe diameter is fixed by the air velocity and the volume of air used by each operator. For bare floors or a large building, a two-stage fan may be needed. For carpet cleaning the air consumption is lower. Hose outlets are so spaced that hoses from 6 to 12 m long can reach all the floor.

vacuum-pressure impregnation Timber *impregnation* by sucking out air then flooding with *preservative* under pressure, but without *double vacuum*. See **Tanalized timber**.

valley (1) An intersection between two sloping surfaces of a *pitched roof*, towards which water flows, the opposite of a *hip*. With *thatch*, *plain tiles*, or *slates* the valley can be continuous with the slopes, that is it can be made as a *swept* or *laced* valley, or with *valley tiles*. Valleys are vulnerable to *driving rain* because they slope at an angle appreciably less than the main roof, yet carry water from the two roof slopes. They usually have a double *underlay*.

(2) The nearly horizontal line across a *northlight* roof, where the bottom of each roof slope meets the bottom of the glazing. It has a *valley gutter*.

valley board A board about 280 × 25 mm fixed on the lower ends of the *jack rafters* parallel to the *valley rafter* of roof framing. It supports a *valley gutter* or slates or tiles in a *swept valley* or *laced valley*.

valley gutter (1) *Plumber's metalwork* such as sheet lead fashioned on site, or a factory-made lining, for roof rainwater drainage down a *valley*. It can be an *open valley* or a *secret gutter*.

(2) A *box gutter* for a *northlight* valley, either *feltwork* or of a standard shape in galvanized-steel sheet, lead, zinc, aluminium, cast iron, GRP, etc.

valley jack (USA) A *jack rafter* that runs from *valley rafter* to *ridge*.

valley rafter The *rafter* that carries the heels of upper *jack rafters* in a traditional framed roof. A *trussed rafter* roof has no valley rafter.

valley tile A large purpose-made special *roof tile*, concave upwards, shaped to form a tiled *valley*, without *plumber's metalwork* and without lacing or sweeping the tiles to form a *laced valley* or a *swept valley*. It is not holed for nailing, like the tiles each side, which usually have to be cut with an abrasive wheel to make it fit firmly and not roll. Burnt clay valley tiles may be less well burnt than the *plain tiles* which course with them owing to the maker's fear of the tile warping when it is hard burnt. They may therefore not weather well. See **trough valley tile**.

value The *lightness* of a colour.

value added tax (VAT) Under the UK Finance Act 1972, new buildings are zero-rated and demolition work does not pay any VAT. Maintenance, repairs, or improvements, however, pay VAT at the full rate. Details are obtainable from HM Customs and Excise.

value cost contract A *cost-reimbursement contract* in which the *contractor* receives a larger fee if the final costs are low than if they are high.

valve (1) A water fitting which either stops the flow through a pipe (*stopvalve*) or varies it (*discharge valve*, or regulating valve). Compare **cock**, **tap**.
(2) The closing device in a valve or tap (ceramic disc, gate, globe, jumper, plug).

valve boss A raised flat area on the surface of a storage *cylinder*.

valve box (USA) A *surface box*.

valving (USA) *Water fittings*.

vanity basin A shallow oval-shaped *basin* for a *vanity top*.

vanity cabinet A bathroom wall cupboard with a mirror-faced door.

vanity top A bathroom work top with a recessed *vanity basin*.

vanity unit, vanitory u. A *vanity top* with a cupboard under.

vaporizing liquids Liquids stored under pressure which become a vapour when released. They are used as *refrigerants* and *extinguishing media*.

vapour A gas that can become a liquid when its pressure is raised or when it is cooled. Vapour chilled to below its *dewpoint* (*C*) liquefies to become *condensation*.

vapour barrier A high-quality *vapour control layer*, mostly used on the 'warm' side of insulation, to prevent *condensation*, when associated with suitable *ventilation* from the 'cold' side. Usual vapour barriers are *polyethylene sheet*,

lapped and joined with bonding tape, aluminium foil on the back of *gypsum wallboard*, or *bitumen felt*. Vapour barriers are also used round buildings, to stabilize the moisture content of clay foundations, and under freezers. See diagram.

vapour blasting *Wet blasting*.

vapour check A partial *vapour control layer*, such as some *emulsion paints*, or a *vapour barrier* material without sealed joints.

vapour compression cycle The most common refrigerating cycle, using the flow of *latent heat* into and out of a *refrigerant* during cycles in which it changes state from liquid to gas (vapour) and back again. Vapour is compressed (which heats it) and sent to the condenser, where it gives off latent heat as it liquefies, falling into a liquid receiver. A small pipe takes high-pressure liquid to the expansion valve. Expansion allows evaporation, producing low-pressure cold vapour. The cold expanded vapour passes through the large-bore tubes of the evaporator taking in heat. Low-pressure vapour arrives back at the compressor to begin the cycle again. Used in *chillers*, *heat pumps*, and *refrigerators*. See diagram, p. 492.

vapour control layer A layer of material with low *permeability* (i.e., high resistance) to the passage of *vapour*, measured by the time and pressure needed to force a gram of vapour through it, in meganewton seconds per gram (MNs/g). A full *vapour barrier* has a higher figure than a *vapour check*.

variable air volume box, VAV box An *air terminal unit* supplied from a single *high-velocity* air duct, which allows more or less air into a room past a *damper* controlled by a *thermo-*

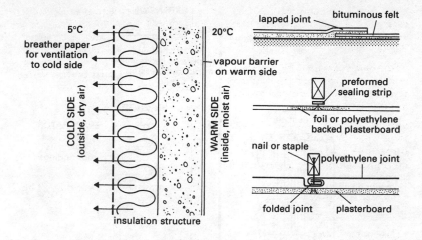

5°C

breather paper
for ventilation
to cold side

COLD SIDE
(outside, dry air)

insulation structure

20°C

vapour barrier
on warm side

WARM SIDE
(inside, moist air)

lapped joint — bituminous felt

preformed
sealing strip

foil or polyethylene
backed plasterboard

nail or staple

polyethylene joint

folded joint   plasterboard

A vapour barrier.

stat in the room, or a microprocessor. The outgoing air passes through an *attenuator* and is taken by *low-velocity* ductwork to the discharge grille. See diagram, p. 493.

variable air volume (VAV) system An *all air* system of air conditioning, which discharges through a *variable air volume box*.

variation (USA **change, modification**) A change in the quantity or quality of the work agreed to in a *contract*, as shown on the *drawings* or described in the *specification*, listed on a *schedule*, etc., or in the time allowed for *completion*. A variation usually affects the *contract sum* (extra, claim, deletion, saving), but is often not priced until the *final account* and may entitle the contractor to an *extension of time*. Variations are also used to handle matters such as drawing *revisions*, *contractual foundations*, and *fluctuations*. The rules on variations

are given in the *conditions of contract*. See **claim**.

variation notice Written notification given by the *contractor* to the *employer* that a *variation order* should be issued.

variation order (USA **change o.**) A written order from the *employer* (represented by the *architect* or other *consultant*) authorizing a *variation*.

variations Lock *differs*.

varnish A transparent coating containing no *pigment*, used mainly on decorative interior joinery. For exterior use on wood, even the best *phenolic*, *polyurethane*, or *alkyd* resin varnishes usually need replacement within a few years.

vault (1) A masonry floor or roof, in the form of an *arch* but usually wider. The Romans used *barrel vaults* (*C*), a method overtaken in the Middle Ages by *pointed vaults*. A hemispherical vault is a *dome*.

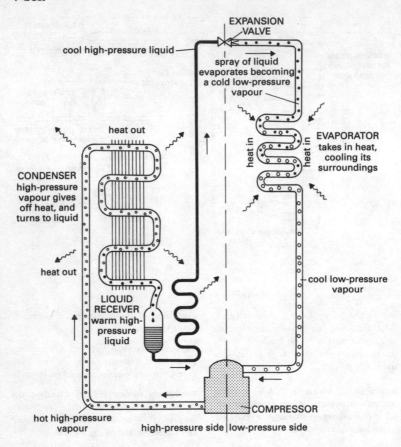

The vapour compression cycle.

(2) An underground strongroom.

V-belt, vee-belt A flexible rubber belt with sloping sides. It runs round matching V-shaped pulleys, which grip it firmly, e.g. between a motor and a fan, a pump, etc. It is both quiet and anti-vibration, thus reducing *equipment noise*.

vee joint A joint resembling the letter 'V'. It is often created by a *chamfer* on the two edges that meet, and it masks shrinkage.

vehicle Paint *medium*.

veneer (1) **face v.** A thin layer of decorative or high-quality wood used as a finishing over lower-quality wood. Face veneer is usually *sliced* or *sawn*.

(2) **ply** A layer of wood inside *plywood* or a similar *panel product*, usually produced by *rotary cutting* from a *flitch*.

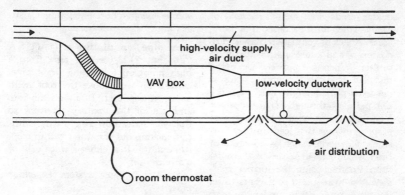

A variable air volume box.

(3) A thin finishing material, usually on the outside of an external wall. See the following entries and **brick veneer**, **ceramic veneer**.

veneered wall A wall with a *facing* and a separate backing which, not being bonded to it, cannot resist load as well as the facing can. A *brick veneer* wall has an outer *leaf* separated by a *cavity* from an inner leaf and attached to it with *wall ties* across the cavity. Compare **faced wall**.

veneering Fixing decorative *veneer* to a structural backing, usually a *panel product*. See **laminate**.

veneer finish plaster A thin coat of *gypsum plaster* for the joints of *tapered-edge plasterboard*, only 2 mm thick and applied over paper tape.

veneer tie (USA) A *wall tie* for holding *brick veneer* to wood framing.

vent (1) To release a gas or vapour into the outside air.
(2) An outlet for air and water vapour through a *vent pipe* (or stack), air flue, roof vent, *expansion pipe*, *ventlight*, etc. See **ventilation**.
(3) A *fire vent*.
(4) Glazier's term for a crack in a pane; an air bubble in glass.

ventilated lobby A *fire lobby* with natural ventilation to the outside air.

ventilating brick, ventilation b. An *air brick*.

ventilating pipe Formal term for a *vent pipe*.

ventilation Movement of air into or out of a space, to provide fresh air, to carry away moisture, odours, or dangerous gases, or to allow free flow in drains.
The ventilation of indoor areas, with the extraction of moisture from kitchens, bathrooms, and shower rooms, is required by the *Building Regulations*. Houses must have background ventilation (a slow, draught-free flow from small, permanently open ventilators), as well as a means of rapid ventilation, such as an opening window or an *extract fan*. Large buildings have mechanical *services* for heating, ventilation, and air conditioning. Indoor air quality suffers badly if ventilation rates are reduced to below one *air change* per hour, when contamination levels show a dramatic rise and *condensation* may occur. See **sick building**.
The building *fabric* needs separate ventilation to carry away water *vapour*

or dangerous gases such as *methane* or *radon*, which must be kept out of living areas. A lack of ventilation may lead to *dampness* and cause *fungus* decay in timber. When *fire barriers* are being installed in timber framing, care is needed to provide an alternative ventilation path, e.g. through a *breather paper*. Ventilation is needed when a *damp-proof membrane* is placed on the 'outside' of materials such as concrete or brick which contain or can absorb moisture. Roofing using the former *cold-deck* system required roof vents to let out water vapour from under the *roofing*. Similarly *ceiling insulation* should not block the through ventilation of a *cold roof*. The need for ventilation is also reduced by keeping warm moist air out of the building fabric or insulation by having a *vapour barrier* on the 'warm' side.

Free flow in drain pipes is ensured by having *vent pipes*.

ventilation branch A *branch pipe*.

ventilation duct An air *duct*.

ventilation pipe A *vent pipe*.

ventilation rate The number of room *air changes* per hour.

ventilator (1) Any means of ventilating a room, such as a *ventlight*.

(2) An air inlet or outlet for the *ventilation* of a building fabric or roof space, e.g. an *eaves vent*, *ridge vent*, etc.

venting layer A base layer for *warm deck* construction which allows vapour to diffuse out of roof insulation. It is a *bitumen felt* weighing 26 kg/m², reinforced with glass fibres, and perforated with 25 mm dia. holes at 75–85 mm spacings,which is laid partially bonded to the roof deck, types 3G and 3H, to BS 747.

ventlight, night vent, ventilator A small opening window which is

bottom-hung on horizontal hinges, usually above a *casement window*.

vent pipe, ventilating p. (1) (USA **vent stack**) A vertical pipe from a high point on a building drain pipe, carried up well above the roof, with an open top end. It allows air into and out of the *sanitary pipework* so that water can flow away easily, without sucking out the water *seal* in *traps* (unsealing). It is either a *stack vent* or a *branch vent*.

(2) A hot-water system *expansion pipe*.

vent soaker A *pipe flashing*.

vent stack A vertical *stack* pipe for ventilating a *one-pipe system*.

veranda An open roof to provide shade along the sunny side of a building.

verdigris Green basic acetate of copper formed as a protective *patina* over copper exposed to the air. It may change colour to brown or black from air pollution.

verge The sloping edge of a *pitched roof* above a *gable*, sometimes including the bricks which cope the gable wall. A verge can be *flush* or have an *overhang*. A verge overhang of slate or tile should not exceed 50 mm and needs an *undercloak*. Wider verges are made by cantilevering roofing *battens* or *purlins*, the ends of which carry a *barge board*. Compare **eaves**.

verge abutment An *abutment* of a roof against a gable parapet wall. It has a *sloping flashing*.

verge board, v. rafter A *barge board*.

verge fillet A *batten* fixed on a *gable* wall to the ends of the roofing battens, as a neat finish beyond which *roof tiles* overhang.

verge flashing, v. trim Roof *trim* to

waterproof the edge of *built-up roofing* at a verge and give a crisp outline.

verge tile A *tile-and-a-half tile* used at the *verge* of a roof in alternate *courses*. It is left- or right-*handed* and, unless a *dry verge tile* is used, bedded in mortar on to the *undercloak*.

vermiculated finish Stone with winding worm-like ridges on its face.

vermiculite Non-combustible, insulating *exfoliated vermiculite*, used as aggregate in *lightweight plaster* and concrete.

vermiculite-gypsum plaster To achieve a four-hour *fire-resistance grading*, which ordinarily necessitates a 150 mm concrete slab with simple plaster ceiling, a 100 mm slab can be substituted, plus a vermiculite-plaster ceiling 20 mm thick. This also greatly reduces the slab-plus-plaster weight. The *premixed plaster* is usually a $1\frac{1}{2}$ to 1 mix, which weighs only 670 kg/m³.

vertical grain The *edge grain* of *quarter-sawn* timber.

vertical shingling, hanging s., weather s. *Shingles* hung on a wall, like *tile hanging*.

vertical sliding window, v. slider A *sash window*.

vertical transport *Lifts* and *escalators*, or similar *mechanical services*, to carry people between different floors.

vertical twist tie, strip tie A stiff *wall tie* made of steel strip, used for *cavities* up to 150 mm wide. *Courses* must be accurate, as it is difficult to bend. Its ends should not have sharp edges, which may cause facial injuries.

vertical work Asphalt work in which the hot mix is trowelled up on to walls. It is slower and much harder than *horizontal work*.

vestibule A small entrance room. It is larger than a *lobby*.

vetting of tenders *Assessment of tender responses*.

vibrating screed board A long *screed board* with a motor-driven *vibrator* on top in the middle, to flatten and compact a concrete *ground slab*.

vibration (1) The unwanted result of machines, e.g. *fans* and *pumps*, vibration can be prevented from making disturbing *equipment noise* by the use of heavy *machine bases*, *anti-vibration mountings, flexible duct connectors*, etc.

(2) **concrete v.** The usual method for the *compaction of concrete* after it has been placed in *formwork* or a mould, also used by *blockmakers*. Care is needed to avoid *undervibration* or *overvibration*. Late vibration, or even *revibration* (*C*), is used to reduce *chapping* from drying shrinkage, particularly for *direct finishes*. At low percentages, every 1% of *blowholes* takes away about 6% of the strength of concrete.

vibrator Contractor's plant for *vibration* of fresh concrete. The most versatile is the 'poker' or *internal vibrator* (*C*), but other types are also used.

vice (USA **vise**) A strong metal clamp fixed to a bench and tightened with a screw, used to hold materials while being cut, shaped, or otherwise worked on.

vinyl Plastics such as *polyvinyl chloride* or *polyvinyl acetate* which are used in emulsion paints, extruded handrails, adhesives, flooring, etc.

vinyl asbestos tile, thermoplastic t. A floor tile made from *polyvinyl chloride*, plus fillers, *plasticizers*, and *asbestos* fibre reinforcement, usually 300 mm square and 3 mm thick. Like other *thin floor finishes*, vinyl asbestos tiles need a smooth, rigid *sub-floor*, which should be both clean and dry so that the *adhesive* will stick properly. They are brittle when cold, and therefore, however smooth the sub-floor is, they need heating to about 70°C before laying, so as to soften them and allow them to mould down evenly on the floor. See BS 3260.

vinyl flooring Floor coverings made of vinyl, such as *cushioned flooring, vinyl asbestos tiles*, and vinyl tiles and sheeting to BS 3261.

viscous impingement filter An *impingement filter*.

visible defect In the *visual grading* of timber, a *gross feature*, such as a *dead knot*, that makes the piece unsuitable for a particular use, such as *carcassing*.

vision-proof glass *Obscured glass*.

visual display unit (VDU) A computer screen. Those with a cathode ray tube (like a TV set) are a source of glare and eyestrain, and can be unreadable in strong light. For recommendations on lighting, see the Chartered Institution of Building Services Engineers (CISBE) Technical Memorandum 6. Tests have generally shown that they do not give off dangerous amounts of radiation, although some earlier types tended to cause *static electricity* in dry conditions and may contribute to the *sick building* syndrome. See EC Directive 90/270.

visual grading The *stress grading* of timber by looking at its *gross features*.

vitiated air *Foul air*.

vitreous china A smooth *ceramic* with good appearance, easy to clean but brittle. It is used for *sanitary fittings* such as washbasins, WCs, and bowl urinals.

vitreous enamel True *enamel*, not *enamel paint*.

vitreous glass, v. silica Glass dust pressed to shape at high temperature to form hard and accurately shaped *mosaic* tesserae.

vitrified clayware A *ceramic* material which is hard-burnt to about 1100°C, consequently vitrified right through. It has a very low water absorption (4% maximum) and needs no surface glaze when used in floor tiles, drain pipes, etc. It can be fair-cut with an *angle grinder*. Drain pipes for *spigot-and-socket joints* have precision polyester mouldings on the rough surface, to make an *O-ring* seal. *Sleeve couplings* usually have lip seals pushed straight on to plain or cut pipe ends. Tests are given in the 'dual-numbered' *standard* BS EN 295.

V-joint A *vee joint*.

V-notched trowel A tool for putting an even amount of *adhesive* on to a surface. It may produce rows of curved lines that show through *thin floor finishings*.

void (1) A large hole through a floor, a *cavity*, or a *roof space*.

(2) A hollow space inside a *voided slab* (*C*).

(3) **superficial v.** A *blowhole* in concrete.

volatile Literally that which flies away. The term therefore describes organic *solvents* and *thinners* well.

volt The unit of electrical pressure, related to the units of flow (amperes) and power (watts) in the equation: watts = volts × amperes.

voltage Electrical pressure expressed in volts. The legal descriptions are: high voltage, more than 650 volts; medium voltage, from 250 to 650 volts; low voltage, more than 50 volts direct current (DC) or 30 volts alternating current (AC) but less than 250 volts; extra-low voltage, below 50 volts DC or 30 volts AC. These figures relate to safety, but other figures for high, medium, and low voltage are common in industry.

voltage drop Distribution *line drop* (*C*).

voltage overload *Overvoltage.*

volume yield (1) The volume of *concrete* or mortar obtained from a given total weight of cement, aggregates, and water. It can be measured or calculated.

(2) The volume of *lime putty* of a stated consistency obtained from a stated weight of *quicklime*. *Fat lime* has a high yield.

volumetric building, v. system A preassembled building requiring only connection to services on delivery to site, e.g. *jack-leg cabins, bathroom pods.*

voussoir A tapered arch stone in a stone arch, or an arch brick in a brick arch.

W

waferboard A *building board* made up of random flakes of wood which are pressed flat and bonded with synthetic resin *adhesive*. It has two smooth *faces* and is similar in performance to *plywood*.

waffle form A square *pan form*.

waffle slab, Diagrid floor The most usual *coffered slab*, one in which the recesses are square on plan.

wainscot Wood panelling on a wall up to *dado* height.

waist The narrow part of a long object, in particular the least thickness of a reinforced concrete *stair* slab. See diagram, p. 437.

walk-in A *cold store*.

walking line The most usual path up a *stair* for one person at a time, and the place where the *going* is measured, usually 450 mm from the inside handrail.

walk-in wardrobe, walk-in robe A small room used for hanging clothes.

walkway A path for walking, such as a catwalk or bridge for access to mechanical *services*, boards across a roof or gutter, or a *covered walkway*.

wall Walls are built of *bricks, blocks*, and concrete, or framed and covered with *cladding* or *lining*. They may have openings, e.g. for doors and windows. See **loadbearing wall**, **partition**.

wall anchor, joist a. A steel strap screwed to the end of every second or third *common joist* and built into the brickwork to ensure that the joists give lateral support to the wall. It is fixed to the same joists as the *strap anchors*.

wallboard *Building boards* made for surfacing rather than for insulating ceilings and walls. Wallboards include *plywood, dry linings* such as *tapered-edge* and other *plasterboard*, and glossy *laminated plastics* glued to a backing of *hardboard* or plywood. Other *panel products* are used on walls. Wallboard can usually be put up much more quickly than *matchboard*, but most types are unsuitable for *wet areas*.

wall column A concrete or concrete-encased steel column partly *engaged* within the thickness of a wall.

wallcovering A *covering* on a wall, such as vinyl or paper *wallpaper*, woven *hessian*, grasscloth, felt, melamine laminate, and *stretched coverings*.

wall form One of two *shutters* held together with *form ties*, as *formwork* for a concrete wall.

wall hanger A *joist hanger* partly built into a wall. See diagram, p. 252.

wall hook A flat spike with a turned-up head, driven into a wall *bed joint* as a heavy *fastener* and used for *shores* or to carry a pipe.

wall-hung basin A wash *basin* fixed to a wall, usually on brackets. The wall surface behind the basin is normally protected by a *splashback*.

wall-hung boiler A gas boiler mounted on a wall, of very light weight because of its low water content. It often has a *balanced flue* and can be indoor or outdoor.

wall-hung lamp, bracket l. A *lumi-*

naire supported clear of a wall on a *bracket*, which is often a curved tube with the *wiring* inside.

wall-hung WC pan, corbel WC pan A *water closet* on cantilever brackets or chair carriers. It allows free access to the floor for cleaning. The *cistern* may be concealed in the wall or be part of a matching *suite*.

walling mason, waller A *mason* who sets stones in walls, bridges, etc., cutting them to shape with a *scutch*.

wall joint A hidden mortar *joint* parallel to the face of the wall.

wall outlet A *socket outlet* on a wall.

wall panel See **panel wall**.

wallpaper Printed paper or vinyl sheet with a plain or decorated surface, textured or embossed, used as a *wallcovering*. A *sealer* or *lining paper* may be needed on new plaster to prevent the discolouring of paper wallpapers. A roll (or piece) of wallpaper in Britain is 530 mm wide and 10 m long, in France 460 mm by 8.20 m, and in the USA 18 in. by 24 ft. The rolls are *butt-jointed* side by side so that the pattern matches. An old wallpaper can be overpainted if still firmly stuck to the wall. Some wallpapers are made to be *peelable*, others are removed using a *steam stripper*.

wall plate (1) **eaves p., top p.** A horizontal timber along the top of a wall at *eaves* level. It carries the *rafters* or *joists*.

(2) **w. piece** A vertical timber on a shored wall, held to it by steel spikes (wall hooks) and by short *needles* which pass through it into the wall. It spreads the thrust from *raking shores* which abut under the needles. See diagram, p. 406.

wall plug A *plug*.

wall string The *string* on the side of a *stair* next the wall, as opposed to the *outer string*.

wall tie (1) **cavity t.** A fastener across a *cavity wall* to hold the two *leaves* together. It is made of stainless or galvanized steel wire or strip, or of plastics, and has a twist or bend near the middle to form a *drip* so that water cannot pass. Wall ties are built into the brickwork *bed joints* as work proceeds and may have a large plastics washer to hold *cavity insulation* in place. Rigid ties should not have sharp *fishtail* ends, which might injure bricklayers. Usual spacing of wall ties is 900 mm horizontally and 450 mm vertically. A *double-leaf separating wall* for *sound insulation* has either flexible wall ties or none at all. See BS 1243, and for wall tie replacement BRE Digest 329. See also diagram, p. 500.

(2) **brick tie** (USA **veneer tie**), **frame tie, etc.** A similar restraint fixing between a brick wall and another component, such as a door or window frame, cladding panel, concrete column, etc.

(3) (USA) Reinforcement steel placed in the bed joint of a brick or glass block wall, parallel to its length.

wall tile A glazed ceramic tile, for use as an easily cleaned *finishing* on walls in *wet areas*, such as showers and bathrooms, or *splashbacks* behind sinks or basins. Wall tiles may have *cushion edges* and spacer *lugs*. In general wall tiles are slippery when wet and not strong enough for use as *floor tiles*. There are several different methods of tile laying.

wall unit A cupboard hung on a wall. Its door should either open above the tallest person's head, or come down to the shortest person's shoulder height, to be seen when partly open.

walnut (Juglans regia), **European w.**

wane

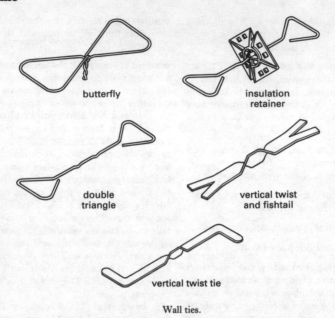

butterfly

insulation retainer

double triangle

vertical twist and fishtail

vertical twist tie

Wall ties.

A very decorative hardwood, mainly from southern Europe. The English variety has a finer *figure* and colour, but is rare. Walnut is a wood with a grey background and dark, sometimes ruddy streaks, used for carving, wood turning, and *veneers*, in which its *burrs* and crotches are valued.

wane, waney edge Bark, or the rounded surface under the bark on sawn timber. Wane is a *gross feature* in *visual grading*.

ward cut Metal cut away from a flat *key*, so that it can pass the *wards*.

wardrobe A cupboard with hanging space for clothes, or a *walk-in wardrobe*.

wards Small metal bars in a *lever lock* which ensure that only the right key can turn the lock.

ware pipes Pipes of *vitrified clayware*.

warehouse set, air s. The hardening of bagged *cement* from being stored in damp conditions.

warm-air heating A rapid space-*heating* system that blows warm air. Air can be ducted from a *central heating* boiler or blown out of a local *unit heater*, a *fan-coil unit*, or an *air curtain*. Even large spaces can be warmed very quickly, but the effect is neither as comfortable nor as efficient as *radiant heating*.

warm deck The recommended type of *flat roof*, with *external insulation* on top of the supporting deck. The deck can be a reinforced concrete slab or any other material. A warm deck needs a *vapour barrier* on the underside to prevent *condensation* in the structure. An *inverted roof* is a warm deck – its vapour barrier is the roofing membrane. See diagram.

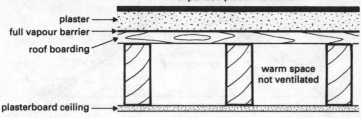

glassfibre roof board or
expanded plastic insulation

plaster

full vapour barrier

roof boarding

warm space
not ventilated

plasterboard ceiling

A warm deck.

warm façade A wall with *external insulation*.

warm roof A pitched roof with heat *insulation* above the roof space, which is not ventilated, and with a *vapour barrier* under the insulation. See **cold roof**. See also diagram, p. 502.

warning pipe An overflow from a feed or expansion *cistern* that discharges safely outdoors, deliberately in a noisy and obvious way, to bring attention to a plumbing fault. If there is a separate main overflow, the warning pipe inlet is 50 mm below it. A warning pipe should be twice the dia. of the supply pipe, laid to a fall of 1 in 10 and insulated throughout its length. The inlet end should have a turn-down that dips 50 mm below the normal water line, and the outlet has a fine mesh screen and another turn-down, to prevent freezing air entering. Hinged flaps over the outlet are unreliable and should not be used.

warning tape *Tracer*.

warning tile A *breaker slab*.

warp, warping Any distortion of timber during *seasoning* as the moisture content falls, such as *cup*, *bowing*, *spring*, and *twist*, from uneven *shrinkage* round or across the growth rings (radial, tangential shrinkage). See **gross feature**.

Warrington hammer A *joiner's hammer*.

wash (USA) A weathered slope or *weathering*.

washability The ability of a surface of *paint* to withstand washing without damage, including the ease of removing dirt.

washboard (USA) A *skirting* board.

washdown closet The most usual type of *water closet*, with a *pan*, flushing rim, and a single *trap*, normally with a horizontal outlet to take different *pan connectors* for a horizontal or vertical *soil pipe*.

washer (1) A flat ring made of rubber, plastics, leather or fibre which is held by a nut to the underside of the *jumper* of a *screwdown valve* (or tap) to stop the flow of water. A leaking tap may show that the washer needs replacing or cleaning, or that the seat it fits against has corroded and needs to be cut smooth again with a special seat-cutting tool.
(2) A steel ring to spread bolt pressure. See **diamond washer, limpet washer**, and *C*.

washleather glazing Glass bedded in simulated washleather instead of *putty*, e.g. on *swing doors*. The fixing is by *glazing fillet*.

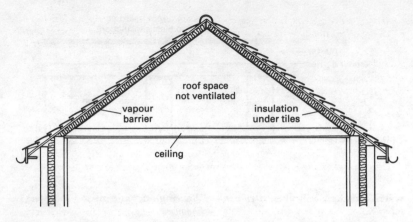

roof space
not ventilated

vapour
barrier

insulation
under tiles

ceiling

A warm roof.

wash primer, mordant w. A *pre-treatment primer* for cleaning new *galvanizing* before painting. The method was once out of favour.

waste (1) Wasted material from cutting standard lengths or full-size sheets to fit on site, the cost of which is usually described as *cutting waste*.
(2) A short threaded outlet pipe from the plug hole of a basin sink or bath. It has a flange that goes on the inside, over a sealing washer, and a threaded tail to take a flanged *back nut* on the outside, which is tightened to secure it. A *trap* is usually screwed on to the threaded tail.
(3) *Waste water*.

waste appliances Fittings which discharge only *waste* water, including baths, basins, sinks, cleaner's sinks, shower trays, drinking fountains and wash troughs.

waste branch, w. pipe A gently sloping *branch pipe* to carry water away from a basin, bath, or sink. It has a *trap* after the exit from the basin and discharges into a *waste stack*. A long waste branch may need a *branch vent*.

waste disposal unit, w. disposer A unit for flushing away domestic waste (but not tins and bottles), e.g. a *garbage disposal sink* or a *Garchey sink*.

waster (1) A *mason's* chisel, either one with a claw cutting edge for *wasting* stone or a chisel 19 mm wide. See diagram, p. 281.
(2) A *facing brick* with minor defects or blemishes which is therefore downgraded to a backing or *common* brick.

waste stack A vertical *stack* pipe for waste water. It carries discharge from the *waste branches* to the buried drains and is part of the *sanitary pipework* system.

waste tail A pipe *tail* on the outlet from a basin, bath, sink, etc.

waste water Used water from sinks or washing, not containing *soil water* or industrial effluent.

wasting (1) Removing excess stone roughly with a *waster* before dressing a block as *dimension stone*. See **picking, boasting**.
(2) *Spread and level*.

watching Having a watchman present for *site security* outside working hours.

water bar A galvanized steel bar 25 × 6 mm, bedded on edge in a groove along the top of a precast concrete *subsill* in a window opening. It fits into a corresponding groove on the underside of the *window sill* of a fixed wooden window frame, to prevent draughts and rainwater entering through the narrow gap in a storm. This type of weatherproofing is increasingly superseded by *sealants*. See also *C*.

water-based paint, aqueous p. *Emulsion paint*, or in the past *distemper*. Other products such as *sealants* and *adhesives* may also be water-based.

water blasting, w. jetting Forcing a jet of water (with or without sand) out of a fine nozzle under very high pressure. Water blasting is used for *blast cleaning*, cutting masonry, producing an *exposed aggregate finish*, stripping *lead paint*, etc.

water channel A groove in a *patent glazing* bar to drain *condensation*.

water check, fillet A kerb standing above a *flat roof*.

water-checked casement A *casement* with grooves round the outside of all *stiles* and *mullions* to break any *capillary* path for rainwater.

water chiller set A *chiller*.

water closet (WC) A toilet fitment usually made of vitreous china. The two main types of WC pan are the *washdown* and the *siphonic closet*, which have a flushing rim, a seat and cover, and can stand on a pedestal or be *wallhung*. The asiatic or *squatting closet* is rare in the UK. All WCs need a flushing *cistern* and have a *trap* with an outgo to take a *pan connector*. They are made to BS 5503. The numbers of WCs needed in a building and other

details on *sanitary accommodation* are given in the *Building Regulations*. Special WCs for the disabled are available.

water content *Moisture content*.

water curtain Fire protection by a row of *drencher sprinklers*.

water fittings (USA **valving**) Plumbing fittings to stop or regulate the flow of water in a pipe, grouped as either *stopvalves* or *discharge valves*.

water gauge (WG) A *U-gauge* containing water, giving a simple and reliable instrument for measuring very low pressures, for *drain tests*, flue *draught*, *air conditioning*, etc. (1 mm WG is about 10 Pascals; 100 mm WG is about 1% of atmospheric pressure.)

water grain A curvy attractive *grain* often seen in *birch*, *mahogany*, and sycamore.

water hammer A tapping sound in a water pipe which may occur when a tap is suddenly closed, causing the water column to vibrate. A small water hammer damper can be made from a coil of copper pipe, which flexes and absorbs the water energy. See **non-concussive action, ballvalve**, and *C*.

water heater A *storage water heater* or *instantaneous water heater*, or a *calorifier*.

water jetting *Water blasting*.

water meter A meter owned by the water supply company usually on private property. It may have a protective chamber and have a *meter bypass*.

water paint An *emulsion paint*: formerly *distemper*.

water pipe Pipes for cold-water supply have been traditionally of *screwed* dead-mild steel or cast iron.

They are increasingly being superseded by plastics. Copper, stainless steel, and plastics are used inside buildings.

waterproof adhesive See **boil-resistant adhesive, weatherproof adhesive**.

waterproofing Asphalt *tanking* or waterproof skins of bituminous or other material are the only certain methods of waterproofing walls and floors, particularly below ground level, although a *drained cavity* may sometimes work. *Integral waterproofers* in concrete are occasionally effective, but most concretes can be made waterproof without any admixture by careful grading, mixing, and placing. Compare **roofing, water-repellent treatments**.

waterproof paper *Building paper*.

water reducer A *plasticizer* for concrete.

water-repellent treatments, surface waterproofers Colourless liquids sprayed on to masonry walls to reduce the penetration of rainwater, but rarely appropriate for modern *brickwork*. They are either *silicone* resins or polyoxoaluminate stearate in a *solvent*, against which the user must take precautions. Other causes of *dampness*, such as faulty *flashings, damp-proof courses*, or gutters, should be checked first. The surface must be cleaned (without using detergents) and dried before they are applied. They are claimed to work by lining the pores in the stones and mortar, rather than by blocking the surface, as does *pore treatment*. The wall still absorbs some rainwater, and the escape of water vapour is restricted, which may lead to frost damage or discoloration. They are not suitable for use as *tanking* or for flaking indoor stonework. Different grades are made for: sandstone, clay bricks, and stucco; limestone or cast stone;

cement-based materials; and calcium silicate brickwork. See B S 6477.

water retentivity A property of brick-laying *mortar* which prevents it losing water rapidly to bricks with a high *absorption rate*. It also prevents water coming to the surface when the fresh mortar touches bricks with low absorption. It thus allows the mortar to develop a good bond and high strength with every sort of brick.

water seal The *seal* in a *trap*.

water seasoning Soaking timber to remove sapwood sugars, then air-drying it. In the Middle Ages chestnut, water-seasoned for two years, was used for roof framing, without *preservatives*.

watershot walling *Dry walling* with stones laid sloping so that water falling on them pours to the outside of the wall. Walling is often built in this way in the English Lake District, where the rainfall is about 3.7 m per year.

water softener A chemical unit that treats water to reduce the 'hardness', caused by calcium and magnesium salts. Hard water may have an unpleasant taste, make soap difficult to lather, and produce *furring* in pipes, taps, showers, kettles, and boilers. The best-known water softeners are of the *base-exchange* type. For domestic use, some chemical manufacturers produce a low-cost water softener consisting merely of a metal basket that hangs in the cold-water tank and contains crystals of very low solubility. The crystals have to be replaced every six months. They discourage furring but cannot remove it if it is already deposited.

water spotting Pale spots on a paint film, caused by water falling on the fresh paintwork. They may or may not be permanent.

water stain (1) A discoloration (which may improve the *figure* of the wood) caused by *converted timber* getting wet.

(2) A decorative *stain* for woodwork, made of colouring material dissolved in water. It accentuates the grain more than a *spirit stain*.

water tap A draw-off *tap*.

water test A *hydraulic test* (*C*) for drains.

wayleave A permission to pass over land, as for *access* to a building site, which may sometimes include leave to lay cables, pipes, etc.

WBP A *weatherproof adhesive*.

weather (1) To make a *weathering*.

(2) The amount, measured down the slope, by which a *shingle* overlaps the next shingle but one below it. It is the same as a *headlap*.

weather bar A shaped strip of metal or plastics fixed along a *sill* to exclude rain and draughts from the bottom of a door or window.

weatherboards Horizontal boards for external wall *cladding*, usually on timber-framed buildings. Traditional *feather-edge* boards are fixed with their thin edge upwards, then overlapped by the next board, thick edge down, with any *rebate* helping to keep out rain and wind; *shiplap boards* are similar. Profiled *expanded PVC* weatherboards need no painting and are virtually maintenance-free. They may be fixed over a *breather paper*. Any *vapour barrier* should be on the inside face of the wall.

weather check A *throat*, *drip*, or *rebate* to keep out weather.

weathered See **weathering**.

weathered pointing A *weather-struck joint*.

weather fillet A *cement fillet*.

weathering (1) A change in colour,

the corrosion, or the degradation of building materials after exposure to rain, frost, pollution, *salt air*, *ultraviolet*, etc. Wood suffers from the mechanical and chemical break-up of its surface. Compare **decay**.

(2) Resistant to the weather, as are *coatings* for cladding and *weathering steel* (*C*). Concrete needs to be very dense and strong to be weathering.

(3) A building component which keeps out the weather, particularly rainwater, or drains it away, such as cappings, baffles, channels, drips, or flashings, found on roofing, door thresholds, in patent glazing, etc.

(4) (USA) **wash** A slight slope to throw off rainwater, as from a stone window sill, a coping, or brickwork mortar joints.

(5) **sill w.** A sheetmetal covering over a window sill.

weather joint A *weather-struck joint*.

weather moulding A *moulding*, usually outside the bottom of a door, to throw water off the *threshold*. It includes a *drip*.

weatherproof adhesive, weather and boil-proof (WBP) a. *Adhesive* in the highest category of *durability*, able to resist prolonged exposure to the weather, boiling water, cold, and insects, fungus, or borers. Only some *phenolic* and *resorcinol* adhesives can pass the 'boil test' in BS 1204.

weather shingling *Vertical shingling*.

weather slating *Slate hanging*.

weatherstripping, draught strip A strip of sprung metal, sponge rubber, wiping seal, etc., fixed in the gap between the frame and a door or window to prevent draughts, often so effective that it reduces *ventilation* enough to cause *condensation*. Also it may make a door or window difficult to open or close.

weather-struck joint, s. joint A brickwork *bed joint* that is *jointed* with the bricklayer's trowel immediately after the bricks are laid, with an outward slope to throw rainwater off the wall.

weather tiling *Tile hanging.*

wedge A tapered piece of wood, used in timbering or *centering*, e.g. a *folding wedge.*

wedge coping A *coping* with one edge thicker than the other, thus with its upper surface sloping one way only.

wedging-in Fixing the *tuck-in* of a flashing into a *chase.*

weephole A small drain hole for water. In brickwork it is used above a *dampproof course* (dpc) or *cavity tray*, and is made by leaving *perpends* without mortar. In this case it also equalizes air pressure and reduces rain penetration. The top of the dpc or cavity tray should be carried well above the top of the weephole, for protection against *driving rain.* A weephole is drilled through the sill of a single-glazed window to allow *condensation* to escape. See **breather hole**.

weighting out Multiplying out to obtain quantities in kg or tonnes for *extended prices.*

well (1) stair well The space contained by the walls round a stair. (2) A *lift shaft.*

well-burnt brick A *hard-burnt* clay brick.

Welsh arch (USA **jack a.**) An arch over a small opening of less than about 300 mm span, bridged by a *stretcher* cut to a wedge shape and resting on two *corbelled* bricks or stones of matching shape.

welt, welted seam A *seam* in *supported sheetmetal roofing.*

welted drip *Built-up roofing* felt turned down over the edge of a flat roof, hot-folded under itself, bonded into the base sheet, and nailed.

welted nosing A junction in *supported sheetmetal roofing* in which a vertical sheet comes up under a horizontal sheet. They are folded together and dressed down the vertical surface.

western framing *Platform framing.*

western red cedar, British Columbia cedar, giant cedar, Pacific red cedar (Thuja plicata) A straight-grained, coarse, soft, weak *softwood*, very useful for making *shingles* and cheap *joinery.* The reddish-brown shingles weather to grey.

wet and dry paper Waterproof emery-faced *sandpaper* for wet rubbing.

wet area A room, such as a bathroom or shower, with water-resistant finishings, such as *tiling.* In particular *floor tiles* should be non-slip when wet and suitable for walking on barefoot. The floor may have a *screed to falls* and a *gully. Gypsum* plaster should not be used in wet areas.

wet blasting, vapour b. *Blast cleaning* with sand and water, to reduce the danger of *silicosis* from dry *sand blasting.* It is used for cleaning masonry.

wet-dash finish *Roughcast.*

wet-edge time The maximum pause before *picking up* a paintwork *live edge.*

wet riser (USA **w. standpipe**) A *fire riser* kept filled with water under pressure. It must be protected from frost, unlike a *dry riser.*

wet rot The *decay* of timber with a high or very high *moisture content*, of 40 to 50%, or alternating wet and dry conditions remaining above 22%, caused by one of two groups of *fungus,* brown rots and white rots. It occurs in window frames, external doors, cladding, or any wood in contact with the ground. Unlike *dry rot* it does not spread into neighbouring timber, so that treatment to remove it is simpler – mainly removing the source of dampness and allowing the timber to dry out rapidly. Any infected timber that is removed should be dried and burnt, taking care not to spread the fungal spores. See BRE Digest 345.

wet rubbing The fine *flatting-down* of paintwork using *wet and dry sandpaper* while rinsing with water to remove the dust. This makes the sandpaper last longer.

wet sprinkler system The most usual *sprinkler system*, with water-filled pipes.

wet time The smaller-than-usual payment made to building workers when they cannot work because of bad weather.

wet trades, w. construction *Trades* that use concrete, mortar, or plaster, including *brickwork*, and finishings such as *screeds*, *terrazzo*, or tiling bedded in mortar. Completed work must be protected against *frost* and given time for *curing*. Before *dry trades* can start, *construction moisture* has to be removed by *drying out*, helped by suitable *ventilation*. There is therefore a trend in the building industry towards prefabrication and *system building*, giving an increase in *dry construction* and reducing the time and inconvenience of wet trades. See **winter working**.

wheelbarrow, barrow A builder's barrow usually has a single rubber tyre, two handles, and a steel or heavy plastics container. It is used to move loose materials like bricks, mortar, or rubbish in loads up to about 100 kg, or more for *motorized barrows*.

wheelhead A *handwheel*.

wheeling step, wheel s. (Scotland) A stair *winder*.

whinstone Hard quartz-dolerite or quartz-basalt, used in *granolithic finish*.

white ant A *termite*.

white blob *Manifestation* of new glazing.

white cement Portland *cement* which has been selected and ground without contamination by iron (the grey-green colour of ordinary cement) or to which white *pigment* has been added. It is mainly used in *terrazzo*.

white coat A plaster *finishing coat*.

white lead A poisonous, opaque but not very brilliant white *pigment*, formerly widely used in *lead paints*. It consists of lead carbonate and sometimes basic lead sulphate.

white lime, white chalk lime *High-calcium lime*.

whitening in the grain A streaky white unpleasant appearance which is sometimes seen in varnished or polished woods with *coarse texture*, whether filled or not.

white spirit, mineral s. (Australia m. turpentine) A clear, colourless, liquid organic *solvent*, distilled from petroleum at about 150° to 200°C, mainly used as *thinner* for *oil paint*. It is also useful for cleaning paint brushes and removing grease, tar, adhesives, and black marks made on flooring by shoes. Breathing in its vapours should be avoided, and it is best kept off the skin or promptly washed off with warm water and soap.

whitewood (Picea abies and Abies alba) European *softwoods* from a wide group of trees, including *silver fir*, yellow pine, and *spruces*, with white to

whiting

yellow heartwood and sapwood. They are easily worked and mostly used for interior *joinery*.

whiting, Paris white Crushed chalk, the cheapest white pigment, used as an *extender* in paints, for making *putty* and as a filler in *sealants*.

Whitworth thread (BSW) A British Standard screw *thread* superseded by metric threads.

whole-brick wall A 215 mm *solid wall*, the thickness of one brick length.

whole timber A square *baulk*, usually of about 300 × 300 mm section.

wicket door, man d., pass d., personnel d. A small door for one person in the *leaf* of a large entrance door to a church, factory, or warehouse.

wide-ringed timber, coarse t., open-grained t. Timber with *annual rings* which are far apart owing to a high *rate of growth*. In *softwoods*, *narrow-ringed timber* is stronger, but not always in *hardwoods*.

Wilton carpet A hard-wearing woven *carpet* with a dense cut-loop pile.

wind, winding The twist of converted timber. It is a type of *warp*.

wind beam A *collar beam*. See also *C*.

wind-driven rain *Driving rain*.

winder, wheeling step A *tapered tread* of a *circular stair* (or turning part of a flight), radiating from a common springing point at the *newel post*. A *balanced step* is safer.

wind filling *Beam filling*.

winding (1) Timber *wind*.
(2) A coil of varnished copper wire, usually round a steel core. When electric current passes a magnetic field is created which can turn a motor, move a solenoid, or transfer current to other windings of a transformer.

winding stair (1) A *spiral stair*.
(2) A circular or elliptical *geometric stair*.

winding strips Two short straight edges, tapering in thickness from 8 to 3 mm, used with *boning pegs* to mark out a plane surface for stonework.

window, w. assembly A window has an outer frame, a *sash*, and its *glazing*. There are many types: hinged *casements*, vertical sliding *sash windows*, horizontal *sliding windows*, pivoted sashes, *tilt-and-turn windows*, *ventlights*, hopper windows, etc. They are mainly of timber, aluminium, steel, or PVC. Factory-finished windows are usually made to go in a *sub-frame*. In cold weather, perimeter *heating* under windows combats the downdraught off their cool inside face.

window back The vertical panelling or *lining* material between the floor and the window *sill*.

window bar A *glazing bar*.

window board (Scotland **sillboard**; USA also **stool**) A horizontal board (or other material) fixed under a *window sill* inside a window.

window casing The traditional timber cased frame of a *sash window*.

window insulation The usual way to insulate windows is by *double glazing*, although *low-emissivity coatings* can reduce radiant heat loss.

window ledge A *window sill*.

window sill (1) A horizontal ledge below a window. The outside is *weathered* outwards and has a *throat* or *drip* underneath to prevent the top of the *breast* wall against rainwater and to prevent waterstreaks on the wall. In brickwork walls, a window sill can be of standard or *special* bricks laid *brick-on-edge*, of ceramic tiles, or a stone *lug sill*

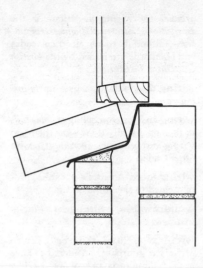

A window sill.

or *slip sill*. Timber and metal sheet are also used. See diagram.

(2) A *sub-sill*.

window stool (USA) A *window board*.

window unit (1) Several *windows* (and often a door) joined to make a large panel, usually connected together with clip-on *mullions* or *transoms*. Window units are exposure-tested to BS 5368.

(2) **w. conditioner** A *unit air conditioner* in a window.

wind shake *Ring shake*.

wind stop *Weatherstripping*.

wing (1) A long part of a building projecting from its main body.

(2) In *patent glazing*, the projecting sides of a *glazing bar* to carry the edges of the glass, which is held down by the capping.

wing light A *flanking window*.

wing nut, thumb screw A nut that can be tightened or removed without tools by turning it with the fingers.

winter working, cold-weather w.
As *cement* does not set properly at low temperatures, for safe concreting at air temperatures below + 2°C, aggregate may be heated up to 50°C and water up to 80°C. Cement should not be heated. Placed concrete needs to be given *frost protection* to keep it warm (between + 2°C and + 27°C). Bricklaying should be stopped at below + 3° and in frosty weather. Frozen bricks should not be laid, or mortar heated. Cold weather may influence the choice of *mortar designation*, and flexible *damp-proof courses* should be warmed before they are unrolled. Completed brickwork needs both frost and rain protection for up to seven days. Plastering on frozen backgrounds is forbidden, but work during frost is possible subject to precautions like those for concreting. In winter a building site becomes less muddy outside once the ground is frozen hard. See diagram, p. 197.

wiped joint A bulging joint round the outside of a lead pipe, made by wiping on heated solder with a cotton (moleskin) wiping cloth.

wiping seal A strip rubber *blade seal* or a fibre *brush seal*.

wire brush A tool used for cleaning metals before painting.

wire comb A *drag*.

wirecut brick, extruded b. *Clay bricks* made by extruding a strip of clay through a die and cutting it with wires held in a frame, followed by drying and firing. Usually they are *perforated bricks*, but 'solids' are also made. Shape is mostly very accurate, and they can be smooth or textured for *facings*. See diagram, p. 48.

wired glass *Flat glass* reinforced by embedded wire mesh which holds the pieces together if it breaks. It is mainly used as *fire-resisting glass*. See **Georgian wired glass**.

wire gauge A method of defining wire diameter by a number which stated originally the number of passes made through different, increasingly smaller, dies, to make the wire. The number for large wire is therefore smaller than for thin wire. The commonest gauge still used in Britain is the *standard wire gauge* or swg, but many other gauges exist in Britain and the USA. The Paris wire gauge increased with the wire diameter, but has been replaced by measurements in millimetres.

wire management The pre-wiring of movable equipment, or ensuring that *trunking* for wiring is big enough over its full length for all present needs, plus adequate spare space. Separate compartments should be available for data, lighting, power, telecoms, etc.

wire nail A *nail* made by cutting and shaping a piece of round or elliptical steel wire. (Brass is used for *escutcheon* pins, other metals for *roofing nails*.) Steel nails in exposed places are galvanized or *sherardized*. The commonest are round wire nails stocked in sizes from 25 × 1.6 mm to 150 × 5.9 mm, oval wire brads from 19 to 100 mm long, and clout nails usually from 19 × 2.6 mm to 50 × 3.7 mm.

wire scratcher A *devil float*.

wireway (USA) *Skirting trunking*.

wiring The cables which carry electricity in a building, installed by an *electrician*. House wiring or light-duty *power and lighting* installations are in *sheathed wiring*. For large projects, wiring is mainly laid in on *trunking* or drawn in to *conduit*, power and lighting cables being separate from telephone, alarm, and control cables, according to a *wire management* plan. The main *code of* practice for wiring in Britain is the *Institution of Electrical Engineers* regulations, although many electrical codes and standards are issued by the *British Standards Institution*.

wiring diagram Usually a *single-line diagram*.

wiring-in The connecting of *cable tails* to fittings, usually done after the main wiring and requiring *coordination* with other trades.

wiring loom A bundle of cables, prefabricated to save time on site.

withdrawable switchgear *Plug-in switchgear*.

withdrawal load *Wire nails* driven across the grain are considered to have a safe pulling resistance of about 4.3 kg per cm length per 2.5 mm dia.

withdrawn The position of a door lock *bolt* or *latch* when it is retracted into the case, by turning a handle or key. The opposite is 'shot' or 'thrown'.

withe (1) **mid feather** A half-brick wall such as a *partition* between chimney flues.

(2) **wythe, tier** (USA) One *leaf* of a *cavity wall* or *hollow wall*, or a half-brick-thick wall.

(3) **withes, osiers** Flexible sticks like those used for basket making, cut every two years from willow trees. Several are twisted together to make a runner which is used for tying *reed* on to rafters for *thatching*.

wobble saw, drunken s. A *circular saw* on a pair of tapered washers, making it slightly off perpendicular to the drive shaft to cut a wide *kerf*. It is used in *joinery* for cutting *open mortices*.

wood *Timber*, *trim*, and *wood-based panel products* such as *building boards*.

wood-based panel products *Panel*

products made of wood particles, veneers, or blocks, bonded together with
adhesive, natural resins, or cement.
They include *building boards*, such as
flakeboard, *waferboard*, and *oriented
strand board*, as well as thicker materials such as *woodwool*. See BRE Digest
323.

wood-block flooring A *floor finish*
made of wooden flooring blocks which
usually have tongued and grooved
edges. Blocks are 19 to 38 mm thick, up
to 90 mm wide, and from 150 to 380
mm long. Since the work is a *dry
trade*, the blocks should not be brought
into the building until after glazing,
plastering, and *drying-out* are completed, when they can acquire the moisture content they will have in the floor
before they are laid. Allowance for *moisture movement* is by a 12 mm edge gap,
covered by the *skirting*. Gluing down
to a concrete slab is done with cold
bitumen emulsion adhesives. *Decay* is
prevented by keeping the *moisture
content* below 20%, and by timber *preservatives*. The flooring is often finished with *floor sealer*, but impervious
floor coverings, which prevent ventilation, are best avoided. Flooding is
likely to cause permanent damage.
Many types exist, from small *wood
mosaic* to decorative *parquet*. See BS
8201.

wood borer A slow *electric drill* used
to make holes with an *auger* bit.

wood-boring weevil (Pentarthrum
huttoni, etc.) A *borer* which forms shallow channels in damp timbers, particularly if decay is already present.

wood brick A *fixing brick*.

**wood-cement particle board, w.-c.
chipboard** A *building board* made of
wood chips bonded with Portland
cement instead of the resin-based adhesives in normal wood *chipboard*. This

helps it achieve a *Class O* fire rating.
Both *faces* are smooth and suitable for
most finishings, although it is mainly
used for *floating floors*. It resists *decay*
outdoors if painted or sealed against
rain, unlike chipboard bonded with
urea-formaldehyde adhesives.

wood chipboard *Chipboard*.

wood element A wood cell, botanically called *xylem*.

wood flour Fine sawdust, sometimes
used as an *extender* for adhesives. It is
also used in explosives, as filler for
plastic wood, etc.

wood-heater A *room-heater* made to
burn wood. The wood should be as
dry as possible, when it gives off most
heat and does not blacken the glass in
the fire door.

wood mosaic, m. parquet panels A
warm, resilient, decorative *wood-block
flooring* of strips of hardwood 7.5 mm
thick and 115 × 25 mm, made up into
squares, then put together in a *basketweave pattern* as panels 475 mm square.
Each panel is held together by a sheet
of paper on its top surface. It is laid in
bituminous adhesive and the paper is
removed after laying.

wood primer The first coat of paint
on new wood, which should be at the
right *moisture content*, with a clean, dry
surface. The primer used must be *compatible* with any *preservatives* in the
wood. Primer should be brushed on,
and all *end grain* double primed. Wood
delivered primed needs to be stored
out of the weather or given its full
paint system straight away, as primer
offers little protection on its own.

**wood roll, batten r., conical r., solid
r.** A joint between bays of *supported
sheetmetal roofing* hand-made by folding the edges upwards on to a sloping-
sided or round-topped wood core. The

joint is weatherproofed either by lapping the sheets to form an *undercloak* and *overcloak* (with a *splash lap*) or by turning up the edges and nailing down a *capping*. See **hollow roll**.

wood screw A steel screw with a tapered *gimlet point*, a straight shank, and a countersunk, raised or dome head, traditionally with a straight slot.

wood slip A *fixing fillet*.

wood turning The skilled work of cutting wood to shape as it is spun round in a lathe, using *chisels* or *gouges*, e.g. to make *balusters*.

woodwool slab A *panel product* made of long thin strands of wood that are mixed with cement and compressed in a mould to bind them together. Woodwool slabs are mainly used for roof *decks* or wall *cladding*, giving *fire resistance* as well as heat and *sound insulation*. Another use is as *permanent formwork*. The surface can be *skim-coated* or plastered, which lowers sound *absorption* but increases the resistance to sound *transmission*. Since woodwool slabs are alkaline, particularly when new and damp, *alkali-resistant paint* may be needed, and since *dust* from the slabs is also alkaline, wire brushes should not be used and a face mask may be necessary. See BS 3809.

woodworking machines Bench-mounted or stand-alone *power tools* for shaping wood, such as a *doweller*, *mitre saw*, *morticing machine*, *planing machine*, *radial saw*, *sander*, *saw bench*, or *spindle moulder*. Some may need measures against *noise* and *dust*. See BS 3997. See also diagram, p. 346.

woodworm A *borer*, usually the common *furniture beetle*.

workability How hard or easy a material is to cut, shape, drill, etc. Timbers vary widely in workability.

The labour of working hard, dense, well-seasoned timber with twisted grain may be five times as much as for soft, light timber with straight grain, whether using hand tools or power tools. The following list gives the rough order of some well-known timbers, the least workable first:

Lignum vitae
Ebony, greenheart
English oak
Ash, beech, makore, Japanese oak
Pitchpine, larch, redwood
Spruce, poplar

Concrete mixes with good workability are easy to compact into *formwork* and round *reinforcement*, and give a good standard of surface finish. See also *C*.

Plaster mixes and bricklaying mortar with good *plasticity* and *water retentivity* have good workability, even on backgrounds with high *suction*. Concrete, mortar, and plaster workability can be improved with *plasticizers*.

Metals such as aluminium and copper are more workable than steel. Other materials such as building stones, plastics, asphalt, grout, etc., also vary widely in workability.

work edge, working e. (USA) The *face edge* of timber.

worker-up A *quantity surveyor* who specializes in *working-up*.

work hardening The tendency of metals like steel, aluminium, and copper to become harder and stiffer if they are bent, twisted, or hammered. Further *working* will usually break them. Work hardening is removed by *annealing*.

working (1) The swelling or shrinkage of timber causing movement as its *moisture content* becomes higher in winter and lower in summer, from changes in the relative humidity of air.

(2) Shaping materials, by cutting, grinding, bending, or folding if they

are hard, or by moulding, screeding, floating, or trowelling if they are wet or plastic.

working life, pot l. The time during which a product can be used after it has been mixed, before it sets or its *curing* system has been activated. Examples of products with working lives are *adhesives*, *two-part* paints, sealants, etc. Compare **shelf life**.

working space, extra s. Clear space round equipment to allow access, as for the removal and replacement of *filters*; or the extra ground dug out round the outside of a basement for asphalt *tanking* to be put on.

working-up The processing of data from *taking-off*. These are the two main stages in preparing *bills of quantities*. Working-up includes *abstracting* and *billing*, and is either hand-written by a *worker-up* or done by computer.

workmanship, good site practice Doing things in the correct way is good workmanship. Site *supervision* and *inspection* are usual ways of making sure that *defects* due to poor workmanship are avoided. Recommendations on workmanship in the building industry given in *British Standards* have been brought together in simple helpful language, with useful tips, in BS 8000.

work section A chapter of a *specification* that may describe a complete *trade* or a job done by a *specialist* sub-contractor.

worksite (USA) A building *site*.

workstation, work position A desk and other office furniture, usually for one person. Cables to it may run through an *access floor*.

worm hole The hole left in wood by a *borer*, either a small *pinhole* or a larger *shothole*.

woven-wire fencing *Anti-intruder chain-link fencing*.

WPB A *weatherproof adhesive*.

wrapping fabric Lightweight steel *fabric* bent round steelwork before the concrete *encasement* is put on, to hold it in a fire.

wrapping tape, pipewrap Tape of plastics or fabric with sticky waxes, wound round buried or severely exposed steel pipes to protect them against *corrosion*.

wreath That part of a stair *handrail* which is curved both on plan and in elevation. It occurs in every *geometric stair*.

wreath piece, wreathed string The curved part of the *outer string* below a *wreath*.

wrecking (USA) *Demolition*.

wrecking bar A *pinch bar*.

wrecking strip A piece of *formwork* used where setting *shrinkage* would jam the forms. It is destroyed in *stripping*, to minimize other damage.

wrench A spanner, usually adjustable.

writing short Placing *items* out of their proper category in *bills of quantities*; they are usually associated items, for example downpipe *shoes* (priced individually) given after *downpipes* (priced in metres length).

wrot timber, wrought t. *Surfaced timber*.

wye (USA **Y-branch, yoke**) A drain pipe *fitting* with a straight run and two inlet branches in the shape of the letter Y. See **junction**.

wythe A *withe*.

X

X-bracing *Cross-bracing.*

Xestobium rufovillosum The *death-watch beetle.*

XLPE *Cross-linked polyethylene.*

X-ray plaster *Barium plaster.*

xylem The botanical name for wood, which supports a tree and brings the nutrients for growth.

xylem ray A wood *ray.*

xylol A *solvent* for synthetic *resins* and gums, distilled from coal tar.

Y

yacht varnish *Phenolic varnish.*

Yale lock A common *cylinder lock.*

Yankee gutter, arris g. A roof *gutter* made of *supported sheetmetal roofing* built directly on the *boarding* of a *pitched roof* near the *eaves*. The sheetmetal is held in place by an upright board near the edge of the roof.

Yankee screwdriver, pump s. A quick-action *screwdriver* worked by hand. It has an inner shaft with a steep thread, turned by pushing or pulling the handle. It works better screwing in than out and demands skill. See diagram, p. 67.

yard An *area.*

yard lumber (USA) Timber graded according to its size, length and intended use, stocked in a lumber yard. See dimension lumber.

Y-bend, Y-fitting A *wye.*

Y-branch (USA) A *wye.*

year ring An *annual ring.*

yellow pine Pitch pine, a European *whitewood.*

yelm, yelven A double handful of reeds or straw, laid on a roof as *thatch.*

yield (1) *Plastic* deformation of a material. See *C.*

(2) See volume yield.

yoke (1) (USA) A *wye.*

(2) A *column yoke.*

Yorky A *slate* with curved cleavage.

Z

Zalutite *Aluminium-zinc coated* steel sheet.

zax, slater's axe A tool with a straight flat blade like a butcher's chopper, with a point projecting from the back. It is used to cut *slates* or punch nail holes. See diagram, p. 332.

Z-bar, zee-bar (USA) A *Z-purlin*.

zeolite Minerals used in *water softeners* that function by *base exchange*.

zinc A hard silvery metal with good resistance to corrosion by unpolluted rain. It corrodes twenty-five times more slowly than bare steel and is used mainly as a protective coating to steel as *galvanizing*, *sherardizing*, electroplating, metal spraying, and in *zinc-rich paints*, alone or in *aluminium-zinc coatings*. Other uses include *zinc roofing*. Zinc is attacked by aluminium, copper, iron, or steel from *dissimilar metal contact*. See BRE Digest 305.

zinc chromate primer (1) A deep red *metal primer* for steel, usually a *two-part product*, containing phosphoric acid to etch the surface and react with resins in the *medium*. Its colour comes from *red oxide*.

(2) Bright yellow or green clear *metal primer* for aluminium, which gives a thin but fairly adherent *film*.

zinc oxide, z. white, Chinese w. A white *pigment* which can block *ultraviolet* and prevent *chalking* in paints.

zinc phosphate primer A general *metal primer* for steel that protects against water and oxygen by the *barrier effect*. The zinc phosphate is not an *inhibiting pigment*, so all rust must be removed before painting.

zinc-rich paint, cold galvanizing A *paint* that is almost entirely zinc metal, in the form of very fine dust, with only a small proportion of liquid synthetic *resin* to act as a *binder*. It is not as hard or firmly attached as *galvanizing*, but is easy to put on by brush or spray. It is used as a complete protective coating on steel, for *touching up* damage to galvanizing, or as a *metal primer*, e.g. under *tar epoxy paint*. See BS 4652.

zinc roofing *Supported sheetmetal roofing* of solid, plain zinc sheet that was popular in France and Europe until recently. Atmospheric pollution, against which efforts are being made, shortens its life in cities. In the country it may last fifty years or more. Zinc sheet for roofing is either titanium-zinc (Metizinc) for general purposes or lead-zinc for *flashings*. It should not be marked with *scribers* before folding and is best folded across its *grain*. It weathers to a dull lustrous grey and is usually left unpainted. See BS 6561.

zinc silicate primer A high-performance *metal primer* used on bare steel after wet *blast cleaning*.

zipper A stiff rubber strip pushed into a groove in the face of a *gasket glazing* so as to tighten its hold on the glass and make an airtight seal.

zoning (1) The reservation of land for certain uses, according to a master plan, and the enforcement of these uses by restrictions on building types by the town planning office of a *local authority*. In this way, some areas can be kept for light industry, others for heavy industry, dwellings, offices, shops, and so on, each area being called a zone. See **noise**.

(2) The grouping of the spaces in a building into separate areas, for convenience, to reduce noise, etc. Zoning includes the sub-division of *services* for a whole building into zones, which allows local control and creates savings in the length and size of pipes, ducts, and cables. Each service has its own zoning, which can cover part of a floor, a whole floor or several floors. See **services core**.

Z-purlin A steel purlin shaped like the letter Z, formed by folding galvanized sheet. It is used in factory and warehouse roofs.

PENGUIN ONLINE

READ MORE IN PENGUIN

In every corner of the world, on every subject under the sun, Penguin represents quality and variety – the very best in publishing today.

For complete information about books available from Penguin – including Puffins, Penguin Classics and Arkana – and how to order them, write to us at the appropriate address below. Please note that for copyright reasons the selection of books varies from country to country.

In the United Kingdom: Please write to *Dept. EP, Penguin Books Ltd, Bath Road, Harmondsworth, West Drayton, Middlesex UB7 0DA*

In the United States: Please write to *Consumer Sales, Penguin Putnam Inc., P.O. Box 12289 Dept. B, Newark, New Jersey 07101-5289.* VISA and MasterCard holders call 1-800-788-6262 to order Penguin titles

In Canada: Please write to *Penguin Books Canada Ltd, 10 Alcorn Avenue, Suite 300, Toronto, Ontario M4V 3B2*

In Australia: Please write to *Penguin Books Australia Ltd, P.O. Box 257, Ringwood, Victoria 3134*

In New Zealand: Please write to *Penguin Books (NZ) Ltd, Private Bag 102902, North Shore Mail Centre, Auckland 10*

In India: Please write to *Penguin Books India Pvt Ltd, 11 Community Centre, Panchsheel Park, New Delhi 110017*

In the Netherlands: Please write to *Penguin Books Netherlands bv, Postbus 3507, NL-1001 AH Amsterdam*

In Germany: Please write to *Penguin Books Deutschland GmbH, Metzlerstrasse 26, 60594 Frankfurt am Main*

In Spain: Please write to *Penguin Books S. A., Bravo Murillo 19, 1° B, 28015 Madrid*

In Italy: Please write to *Penguin Italia s.r.l., Via Benedetto Croce 2, 20094 Corsico, Milano*

In France: Please write to *Penguin France, Le Carré Wilson, 62 rue Benjamin Baillaud, 31500 Toulouse*

In Japan: Please write to *Penguin Books Japan Ltd, Kaneko Building, 2-3-25 Koraku, Bunkyo-Ku, Tokyo 112*

In South Africa: Please write to *Penguin Books South Africa (Pty) Ltd, Private Bag X14, Parkview, 2122 Johannesburg*

READ MORE IN PENGUIN

HISTORY

A History of Twentieth-Century Russia Robert Service

'A remarkable work of scholarship and synthesis ... [it] demands to be read' *Spectator*. 'A fine book ... It is a dizzying tale and Service tells it well; he has none of the ideological baggage that has so often bedevilled Western histories of Russia ... A balanced, dispassionate and painstaking account' *Sunday Times*

A Monarchy Transformed: Britain 1603–1714 Mark Kishlansky

'Kishlansky's century saw one king executed, another exiled, the House of Lords abolished, and the Church of England reconstructed along Presbyterian lines ... A masterly narrative, shot through with the shrewdness that comes from profound scholarship' *Spectator*

American Frontiers Gregory H. Nobles

'At last someone has written a narrative of America's frontier experience with sensitivity and insight. This is a book which will appeal to both the specialist and the novice' James M. McPherson, Princeton University

The Pleasures of the Past David Cannadine

'This is almost everything you ever wanted to know about the past but were too scared to ask ... A fascinating book and one to strike up arguments in the pub' *Daily Mail*. 'He is erudite and rigorous, yet always fun. I can imagine no better introduction to historical study than this collection' *Observer*

Prague in Black and Gold Peter Demetz

'A dramatic and compelling history of a city Demetz admits to loving and hating ... He embraces myth, economics, sociology, linguistics and cultural history ... His reflections on visiting Prague after almost a half-century are a moving elegy on a world lost through revolutions, velvet or otherwise' *Literary Review*

READ MORE IN PENGUIN

SCIENCE AND MATHEMATICS

Six Easy Pieces Richard P. Feynman

Drawn from his celebrated and landmark text *Lectures on Physics*, this collection of essays introduces the essentials of physics to the general reader. 'If one book was all that could be passed on to the next generation of scientists it would undoubtedly have to be *Six Easy Pieces*' John Gribbin, *New Scientist*

A Mathematician Reads the Newspapers John Allen Paulos

In this book, John Allen Paulos continues his liberating campaign against mathematical illiteracy. 'Mathematics is all around you. And it's a great defence against the sharks, cowboys and liars who want your vote, your money or your life' Ian Stewart

Dinosaur in a Haystack Stephen Jay Gould

'Today we have many outstanding science writers ... but, whether he is writing about pandas or Jurassic Park, none grabs you so powerfully and personally as Stephen Jay Gould ... he is not merely a pleasure but an education and a chronicler of the times' *Observer*

Does God Play Dice? Ian Stewart

As Ian Stewart shows in this stimulating and accessible account, the key to this unpredictable world can be found in the concept of chaos, one of the most exciting breakthroughs in recent decades. 'A fine introduction to a complex subject' *Daily Telegraph*

About Time Paul Davies

'With his usual clarity and flair, Davies argues that time in the twentieth century is Einstein's time and sets out on a fascinating discussion of why Einstein's can't be the last word on the subject' *Independent on Sunday*

READ MORE IN PENGUIN

SCIENCE AND MATHEMATICS

In Search of SUSY John Gribbin

Many physicists believe that we are on the verge of developing a complete 'theory of everything' which can reduce the four basic forces of nature – gravity, electromagnetism, the strong and weak nuclear forces – to a single superforce. At its heart is the principle of supersymmetry (SUSY).

Fermat's Last Theorem Amir D. Aczel

Here, weaving together history and science, Amir D. Aczel offers a thrilling, step-by-step account of the search for the mathematicians' Holy Grail. 'Mr Aczel has written a tale of buried treasure ... This is a captivating volume' *The New York Times*

Insanely Great Steven Levy

It was Apple's co-founder Steve Jobs who referred to the Mac as 'insanely great'. He was absolutely right: the machine that revolutionized the world of personal computing was and is great – yet the machinations behind its inception were nothing short of insane. 'A delightful and timely book' *The New York Times Book Review*

The Artful Universe John D. Barrow

This thought-provoking investigation illustrates some unexpected links between art and science. 'Full of good things ... In what is probably the most novel part of the book, Barrow analyses music from a mathematical perspective ... an excellent writer' *New Scientist*

The Jungles of Randomness Ivars Peterson

Taking us on a fascinating journey into the ambiguities and uncertainties of randomness, Ivars Peterson explores the complex interplay of order and disorder, giving us a new understanding of nature and human activity.

READ MORE IN PENGUIN

SCIENCE AND MATHEMATICS

The Fabric of Reality   David Deutsch

'Reading this book might just change your life ... The theory of everything, quantum mechanics, virtual reality, evolution, the significance of life, time travel, the end of our Universe: all these and much else find their place ... this is an awesome book' *New Scientist*

Mathematics: The New Golden Age   Keith Devlin

In the computerized world of today mathematics has an impact on almost every aspect of our lives, yet most people believe they cannot hope to understand or enjoy the subject. This comprehensive survey sets out to show just how mistaken they are and brilliantly captures the essential richness of mathematics' 'new golden age'.

Climbing Mount Improbable   Richard Dawkins

'Mount Improbable ... is Dawkins's metaphor for natural selection: its peaks standing for evolution's most complex achievements ... exhilarating – a perfect, elegant riposte to a great deal of fuzzy thinking about natural selection and evolution' *Observer*. 'Dazzling' David Attenborough

Brainchildren   Daniel C. Dennett

Thinking about thinking can be a baffling business. Investigations into the nature of the mind – how it works, why it works, its very existence – can seem convoluted to the point of fruitlessness. Daniel C. Dennett has provided an eloquent and often witty guide through some of the mental and moral mazes that surround these areas of thought.

Fluid Concepts and Creative Analogies   Douglas Hofstadter

'This exhilarating book is the most important on AI for the thoughtful general reader in years and ample proof that reports of the death of artificial intelligence have been greatly exaggerated' *The New York Times Book Review*

READ MORE IN PENGUIN

POPULAR SCIENCE

In Search of Nature Edward O. Wilson

A collection of essays of 'elegance, lucidity and breadth' *Independent*. 'A graceful, eloquent, playful and wise introduction to many of the subjects he has studied during his long and distinguished career in science' *The New York Times*

Clone Gina Kolata

'A thoughtful, engaging, interpretive and intelligent account ... I highly recommend it to all those with an interest in ... the new developments in cloning' *New Scientist*. 'Superb but unsettling' J. G. Ballard, *Sunday Times*

The Feminization of Nature Deborah Cadbury

Scientists around the world are uncovering alarming facts. There is strong evidence that sperm counts have fallen dramatically. Testicular and prostate cancer are on the increase. Different species are showing signs of 'feminization' or even 'changing sex'. 'Grips you from page one ... it reads like a Michael Crichton thriller' John Gribbin

Richard Feynman: A Life in Science John Gribbin and Mary Gribbin

'Richard Feynman (1918–88) was to the second half of the century what Einstein was to the first: the perfect example of scientific genius' *Independent*. 'One of the most influential and best-loved physicists of his generation ... This biography is both compelling and highly readable' *Mail on Sunday*

T. rex and the Crater of Doom Walter Alvarez

Walter Alvarez unfolds the quest for the answer to one of science's greatest mysteries – the cataclysmic impact on Earth which brought about the extinction of the dinosaurs. 'A scientific detective story par excellence, told with charm and candour' Niles Eldredge

READ MORE IN PENGUIN

POPULAR SCIENCE

How the Mind Works Steven Pinker

'Presented with extraordinary lucidity, cogency and panache ... Powerful and gripping ... To have read [the book] is to have consulted a first draft of the structural plan of the human psyche ... a glittering *tour de force*' *Spectator*. 'Witty, lucid and ultimately enthralling' *Observer*

At Home in the Universe Stuart Kauffman

Stuart Kauffman brilliantly weaves together the excitement of intellectual discovery and a fertile mix of insights to give the general reader a fascinating look at this new science – the science of complexity – and at the forces for order that lie at the edge of chaos. 'Kauffman shares his discovery with us, with lucidity, wit and cogent argument, and we see his vision ... He is a pioneer' Roger Lewin

Stephen Hawking: A Life in Science
Michael White and John Gribbin

'A gripping account of a physicist whose speculations could prove as revolutionary as those of Albert Einstein ... Its combination of erudition, warmth, robustness and wit is entirely appropriate to their subject' *New Statesman & Society*. 'Well-nigh unputdownable' *The Times Educational Supplement*

Voyage of the *Beagle* Charles Darwin

The five-year voyage of the *Beagle* set in motion the intellectual currents that culminated in the publication of *The Origin of Species*. His journal, reprinted here in a shortened version, is vivid and immediate, showing us a naturalist making patient observations, above all in geology. The editors have provided an excellent introduction and notes for this edition, which also contains maps and appendices.

READ MORE IN PENGUIN

ARCHAEOLOGY

The Penguin Dictionary of Archaeology
Warwick Bray and David Trump

The range of this dictionary is from the earliest prehistory to the civilizations before the rise of classical Greece and Rome. From the Abbevillian handaxe and the god Baal of the Canaanites to the Wisconsin and Würm glaciations of America and Europe, this dictionary concisely describes, in more than 1,600 entries, the sites, cultures, periods, techniques and terms of archaeology.

The Complete Dead Sea Scrolls in English Geza Vermes

The discovery of the Dead Sea Scrolls in the Judaean desert between 1947 and 1956 transformed our understanding of the Hebrew Bible, early Judaism and the origins of Christianity. 'No translation of the Scrolls is either more readable or more authoritative than that of Vermes' *The Times Higher Education Supplement*

Ancient Iraq Georges Roux

Newly revised and now in its third edition, *Ancient Iraq* covers the political, cultural and socio-economic history of Mesopotamia from the days of prehistory to the Christian era and somewhat beyond.

Breaking the Maya Code Michael D. Coe

Over twenty years ago, no one could read the hieroglyphic texts carved on the magnificent Maya temples and palaces; today we can understand almost all of them. The inscriptions reveal a culture obsessed with warfare, dynastic rivalries and ritual blood-letting. 'An entertaining, enlightening and even humorous history of the great searchers after the meaning that lies in the Maya inscriptions' *Observer*

READ MORE IN PENGUIN

ART AND ARCHITECTURE

Ways of Seeing John Berger

Seeing comes before words. The child looks before it can speak. Yet there is another sense in which seeing comes before words ... These seven provocative essays – some written, some visual – offer a key to exploring the multiplicity of ways of seeing.

The Penguin Dictionary of Architecture
John Fleming, Hugh Honour and Nikolaus Pevsner

This wide-ranging dictionary includes entries on architectural terms, ornamentation, building materials, styles and movements, with over a hundred clear and detailed drawings. 'Immensely useful, succinct and judicious ... this is a book rich in accurate fact and accumulated wisdom' *The Times Literary Supplement*

Style and Civilization

These eight beautifully illustrated volumes interpret the major styles in European art – from the Byzantine era and the Renaissance to Romanticism and Realism – in the broadest context of the civilization and thought of their times. 'One of the most admirable ventures in British scholarly publishing' *The Times*

Michelangelo: A Biography George Bull

'The final picture of Michelangelo the man is suitably three-dimensional and constructed entirely of evidence as strong as Tuscan marble' *Sunday Telegraph*. 'An impressive number of the observations we are treated to, both in matters of fact and in interpretation, are taking their first bows beyond the confines of the world of the learned journal' *The Times*

Values of Art Malcolm Budd

'Budd is a first-rate thinker ... He brings to aesthetics formidable gifts of precision, far-sightedness and argument, together with a wide philosophical knowledge and a sincere belief in the importance of art' *The Times*

READ MORE IN PENGUIN

REFERENCE

The Penguin Dictionary of the Third Reich
James Taylor and Warren Shaw

This dictionary provides a full background to the rise of Nazism and the role of Germany in the Second World War. Among the areas covered are the major figures from Nazi politics, arts and industry, the German Resistance, the politics of race and the Nuremberg trials.

The Penguin Biographical Dictionary of Women

This stimulating, informative and entirely new Penguin dictionary of women from all over the world, through the ages, contains over 1,600 clear and concise biographies on major figures from politicians, saints and scientists to poets, film stars and writers.

Roget's Thesaurus of English Words and Phrases
Edited by Betty Kirkpatrick

This new edition of Roget's classic work, now brought up to date for the nineties, will increase anyone's command of the English language. Fully cross-referenced, it includes synonyms of every kind (formal or colloquial, idiomatic and figurative) for almost 900 headings. It is a must for writers and utterly fascinating for any English speaker.

The Penguin Dictionary of International Relations
Graham Evans and Jeffrey Newnham

International relations have undergone a revolution since the end of the Cold War. This new world disorder is fully reflected in this new Penguin dictionary, which is extensively cross-referenced with a select bibliography to aid further study.

The Penguin Guide to Synonyms and Related Words
S. I. Hayakawa

'More helpful than a thesaurus, more humane than a dictionary, the *Guide to Synonyms and Related Words* maps linguistic boundaries with precision, sensitivity to nuance and, on occasion, dry wit' *The Times Literary Supplement*

READ MORE IN PENGUIN

REFERENCE

The Penguin Dictionary of Troublesome Words Bill Bryson

Why should you avoid discussing the *weather conditions*? Can a married woman be celibate? Why is it eccentric to talk about the aroma of a cowshed? A straightforward guide to the pitfalls and hotly disputed issues in standard written English.

Swearing Geoffrey Hughes

'A deliciously filthy trawl among taboo words across the ages and the globe' Valentine Cunningham, *Observer*, Books of the Year. 'Erudite and entertaining' Penelope Lively, *Daily Telegraph*, Books of the Year.

Medicines: A Guide for Everybody Peter Parish

Now in its seventh edition and completely revised and updated, this bestselling guide is written in ordinary language for the ordinary reader yet will prove indispensable to anyone involved in health care: nurses, pharmacists, opticians, social workers and doctors.

Media Law Geoffrey Robertson QC and Andrew Nichol

Crisp and authoritative surveys explain the up-to-date position on defamation, obscenity, official secrecy, copyright and confidentiality, contempt of court, the protection of privacy and much more.

The Penguin Careers Guide
Anna Alston and Anne Daniel; Consultant Editor: Ruth Miller

As the concept of a 'job for life' wanes, this guide encourages you to think broadly about occupational areas as well as describing day-to-day work and detailing the latest developments and qualifications such as NVQs. Special features include possibilities for working part-time and job-sharing, returning to work after a break and an assessment of the current position of women.

DICTIONARIES

Abbreviations
Ancient History
Archaeology
Architecture
Art and Artists
Astronomy
Biographical Dictionary of
  Women
Biology
Botany
Building
Business
Challenging Words
Chemistry
Civil Engineering
Classical Mythology
Computers
Contemporary American History
Curious and Interesting Geometry
Curious and Interesting Numbers
Curious and Interesting Words
Design and Designers
Economics
Eighteenth-Century History
Electronics
English and European History
English Idioms
Foreign Terms and Phrases
French
Geography
Geology
German
Historical Slang
Human Geography
Information Technology

International Finance
International Relations
Literary Terms and Literary
  Theory
Mathematics
Modern History 1789–1945
Modern Quotations
Music
Musical Performers
Nineteenth-Century World
  History
Philosophy
Physical Geography
Physics
Politics
Proverbs
Psychology
Quotations
Quotations from Shakespeare
Religions
Rhyming Dictionary
Russian
Saints
Science
Sociology
Spanish
Surnames
Symbols
Synonyms and Antonyms
Telecommunications
Theatre
The Third Reich
Third World Terms
Troublesome Words
Twentieth-Century History
Twentieth-Century Quotations